The Urban Book Series

The Urban Book Series is a resource for urban studies and geography research worldwide. It provides a unique and innovative resource for the latest developments in the field, nurturing a comprehensive and encompassing publication venue for urban studies, urban geography, planning and regional development.

The series publishes peer-reviewed volumes related to urbanization, sustainability, urban environments, sustainable urbanism, governance, globalization, urban and sustainable development, spatial and area studies, urban management, transport systems, urban infrastructure, urban dynamics, green cities and urban landscapes. It also invites research which documents urbanization processes and urban dynamics on a national, regional and local level, welcoming case studies, as well as comparative and applied research.

The series will appeal to urbanists, geographers, planners, engineers, architects, policy makers, and to all of those interested in a wide-ranging overview of contemporary urban studies and innovations in the field. It accepts monographs, edited volumes and textbooks.

Indexed by Scopus.

More information about this series at http://www.springer.com/series/14773

Poonam Sharma
Editor

Geospatial Technology and Smart Cities

ICT, Geoscience Modeling, GIS and Remote Sensing

Springer

Editor
Poonam Sharma
Department of Geography
Shaheed Bhagat Singh College, University of Delhi
New Delhi, Delhi, India

ISSN 2365-757X ISSN 2365-7588 (electronic)
The Urban Book Series
ISBN 978-3-030-71944-9 ISBN 978-3-030-71945-6 (eBook)
https://doi.org/10.1007/978-3-030-71945-6

This Springer imprint is published by the registered company Springer Nature Switzerland AG
The registered company address is: Gewerbestrasse 11, 6330 Cham, Switzerland

Preface

Urban regions and urban systems work in a wholesome pattern which includes multi-dimensional functional interconnections, mobility, growth, and high demand for resources. These places are pivotal to the high density of population, higher intensity of development, and complex spatial associations with immediate hinterland besides the global linkages and impact. The cities functions as nodal spatial places at various scales among different urban centers, as part of the urban region, group of a larger system, and urban systems themselves. The degree and scale of linkages within regional, national, or global levels can be explained in various perspectives of space, time, and also in the technological context related to growth and innovative methods of development. The interdependency, co-operation, competition among urban areas for innovations is important. The complex internal and external interconnectedness besides the continuing growth courses bring about the nature of the growth pattern of urban systems.

City functions as an anchor for radiating the diffusions of all kinds of developmental innovations and economic vibrancy to immediate and next order spatial levels. Urban areas also become locations of concentration for higher educational, infrastructure, health, other services, and thus multi-skilled human resource availability happens to be greater than the other surrounding regions. However, cities have their own set of challenges, viz. as places with high population growth and density due to migration of people besides the natural growth; the resource-consumption is on the higher side with dependence on immediate or next order regions. Cities are perpetually struggling to cope up with the rising demand for energy, water, and waste management. The mobility and transportations infrastructure crisis is posing a big issue that has a direct connection with the economic and industrial development of the city. The shortage of housing, insufficient medical and education services, congestion, and overcrowding are pressing concerns for large urban centers. The problems like pollution and other environmental issues are a great threat to human quality of life and well-being. The kind of ecological footprints the high resource consuming cities are creating is another major struggle for a future developmental path.

Smart city concept has created a vision to offer a kind of dream solution to a variety of challenges cities are facing worldwide. The technology-driven system is expected to handle the problems and make cities function more smoothly and efficiently. It seems to provide ways out through the bottlenecks that the city systems are facing today owing to their growth models. The geospatial technology, geographical information systems, remote sensing, global positioning systems, internet of things, artificial intelligence, sensors, big data, open data, and cloud computing, etc., are a variety of innovative technologies that have revolutionized urban planning and management of cities. The massive digitalization of all types of functioning of urban working has been very instrumental in streamlining the path toward smart development. Cities are not only expected to function smartly but also intelligent, efficient, and reasonable ways. The quality of life of citizens and their role as more enable and empowered to users of the infrastructure and services is considered important. Besides all these achievements, environmental sustainability is kept as one of the significant targets to achieve. It is also interesting to note that in a different part of the world, the smart city provisions have been implemented on a good scale and the impact has been seen remarkably fruitful. By now many cities of the world have been working toward achieving the smart city agendas.

The present book is an endeavor toward exploring the role of geospatial technology and smart city development. The papers included are covering various themes including energy, water, transportation, disaster, land transformation, environmental aspects, citizens, health, and well-being. The case studies are from different parts of the world viz US, Poland, Norway, Malaysia, and Iraq at the global level. The research papers discussing studies from different regions of India such as cities from northern, southern, eastern, and western parts of India have been included. The book would provide a wholesome mix of themes encompassing cities of different nature. I feel blessed and grateful to the Almighty to show the path for work. I am thankful to all authors for contributing the original piece of research work. I am also thankful to my family members and friends for motivation and inspiration to work toward the book.

New Delhi, India Poonam Sharma

About This Book

- Discusses wide coverage of the themes such as land transformation, hazard vulnerability, disaster management and preparedness; energy use and solutions; transportation studies, healthcare management, water, waste management, environmental and ecological sustainability.
- The case studies encompass different parts of the world, viz. the US, Poland, Norway, Malaysia, and Iraq at the global level and various cities from different parts of India.
- Academicians and experts from premier institutions like the University of California, Norwegian University, Poland, IITs, Town Planning Institutes, School of planning and architecture, and many other universities/institutes of national and international repute are contributors.

The book is an endeavor towards exploring the fast pace technological innovations in earth observation system, geographical information systems, geo-intelligence, and information communication technology have completely revolutionized the manner the solutions to city management are possible. The papers included are covering various themes including energy, water, transportation, disaster management, land transformation, environmental aspects, citizens, health, and well-being. The case studies that included are from different parts of the world, viz. the US, Poland, Norway, Malaysia, and Iraq at the global level. The book will provide a wholesome mix of themes that are encompassing cities of different natures and from different parts of the globe. It will be very useful for academicians, research scholar, professionals, planners, government officials, and practitioners. The book will provide resourceful content for the subject areas of geography, town planning, urban and regional planning, smart city, disaster management, infrastructure planners concerned with transportation, energy, water and health; environmental studies, policy design, and city managers.

Contents

Editor and Contributors

About the Editor

Dr. Poonam Sharma is an associate professor in Geography and Director, Centre for Disaster Management Studies, Shaheed Bhagat Singh College, University of Delhi. She has an experience of more than two decades of teaching and research. Her research interest includes urban geography, urban and regional planning, smart city, developmental studies, environmental studies, statistical methods in geography, remote sensing, and GIS. She has published work in national and international journals. She has two books to her credit which include one authored book titled Structure and Growth of Mega Cities: An Inter-Industry Analysis published by Concept Publishers and Sustainable Smart Cities: Challenges and Future perspectives as the first editor published by Springer. She has contributed as a content writer and recorded video lecture in e-PG pathshala program of University Grant Commission, Ministry of Human Resource Development, Government of India. She has also been involved as a subject expert for the preparation of geography dictionary under the Commission for Scientific and Technical Terminology, Ministry of Human Resource Development, Government of India. She has completed Research Program jointly with Institute for Studies in Industrial Development on "Urbanization and Human Capital Development in Assam" sponsored by the Indian Council of Social Science Research, Ministry of Human

Resource Development, Government of India, Co-Project Director. She has been invited to deliver lectures as a subject expert at various institutes and universities. She has organized national, international conferences, faculty a developmental programs as convener and co-convener. She has been awarded "Meritorious Teacher Award 2017–18" by the Department of Higher Education, Government of Delhi.

Contributors

Bharath Haridas Aithal Ranbir and Chitra Gupta School of Infrastructure Design and Management, Indian Institute of Technology Kharagpur, Kharagpur, West Bengal, India

Husam A. H. Al-Najjar Faculty of Engineering and IT, Centre for Advanced Modelling and Geospatial Information Systems (CAMGIS), University of Technology Sydney, Ultimo, NSW, Australia

Adrian Albert Lawrence Berkeley National Lab, Berkeley, CA, USA

Asif Department of Geography, Faculty of Natural Sciences, Jamia Millia Islamia, New Delhi, India

Leila Carvalho Department of Geography, University of California at Santa Barbara, Santa Barbara, CA, USA

Grazyna Chaberek Spatial Management Department, University of Gdansk, Gdansk, Poland

Richard L. Church Department of Geography, University of California at Santa Barbara, Santa Barbara, CA, USA

Debajit Datta Department of Geography, Jadavpur University, Kolkata, West Bengal, India

Yngve Frøyen Department of Architecture and Planning, Faculty of Architecture and Design, Norwegian University of Science and Technology, Trondheim, Norway

Mahak Garg Natural Resource Data Management System Division, Department of Science and Technology, Ministry of Science and Technology, New Delhi, India

Pooja Goel Shaheed Bhagat Singh College, University of Delhi, Delhi, India

Marta Gonzalez Lawrence Berkeley National Lab, Berkeley, CA, USA

Jayanta Goswami National Institute of Rural Development & Panchayati Raj-NERC, Guwahati, Assam, India

Nimish Gupta Ranbir and Chitra Gupta School of Infrastructure Design and Management, Indian Institute of Technology Kharagpur, Kharagpur, West Bengal, India

Piali Haldar School of Business, Sharda University, Greater Noida, India

Alfian Abdul Halin Faculty of Computer Science and Information Technology, Department of Multimedia, Universiti Putra Malaysia, Serdang, Selangor, Malaysia

Safraj Shahul Hameed CCDC, New Delhi, India

Hoang Thi Hang Department of Geography, Faculty of Natural Sciences, Jamia Millia Islamia, New Delhi, India

Arnab Jana Centre for Urban Science and Engineering, Indian Institute of Technology Bombay, Mumbai, India

Charles Jones Department of Geography, University of California at Santa Barbara, Santa Barbara, CA, USA

Bahareh Kalantar Goal-Oriented Technology Research Group, Disaster Resilience Science Team, RIKEN Center for Advanced Intelligence Project, Tokyo, Japan

Jasleen Kaur Philips Lighting Research USA, Cambridge, MA, USA

Deepak Kumar Amity Institute of Geoinformatics & Remote Sensing (AIGIRS), Amity University Uttar Pradesh, Noida, Uttar Pradesh, India

Dharmendra Kumar Indian Institute of Remote Sensing (IIRS), ISRO, Department of Space, Government of India, Dehradun, Uttarakhand, India

Vaibhav Kumar Centre for Urban Science and Engineering, Indian Institute of Technology Bombay, Mumbai, India

Virendra Kumar Remote Sensing Applications Centre-Uttar Pradesh, Lucknow, India

Babita Kumari Department of Geography, Faculty of Natural Sciences, Jamia Millia Islamia, New Delhi, India;
GIS Analyst at GeoSpectrum Technologies Pvt. Ltd., Bangalore, India

Paulose N. Kuriakose Department of Urban and Regional Planning, School of Planning and Architecture, Bhopal, India

Ammar S. Mahmoud Faculty of Engineering, Department of Civil Engineering, University of Babylon, Hillah city, Iraq

Ashish Mani Amity Institute of Geoinformatics & Remote Sensing (AIGIRS), Amity University Uttar Pradesh, Noida, Uttar Pradesh, India

Shattri Mansor Faculty of Engineering, Department of Civil Engineering, Universiti Putra Malaysia, Serdang, Selangor, Malaysia

Maged Marghany Department of Informatics, Faculty of Mathematics and Natural Sciences, Universitas Syiah Kuala, Banda Aceh, Aceh, Indonesia

Ajay Mishra Lucknow University, Lucknow, India

Prabuddh Kumar Mishra Department of Geography, Shivaji College, University of Delhi, Delhi, India

Shashank Shekhar Mishra Lucknow University, Lucknow, India

Suman Mitra Department of Geography, University of Calcutta, Kolkata, West Bengal, India

Tapas Mitra School of Planning and Architecture, Bhopal, India

Hossein Moayedi Department for Management of Science and Technology Development, Ton Duc Thang University, Ho Chi Minh City, Vietnam;
Faculty of Civil Engineering, Ton Duc Thang University, Ho Chi Minh City, Vietnam

Alan T. Murray Department of Geography, University of California at Santa Barbara, Santa Barbara, CA, USA

Chandan Mysore Chandrashekar Ranbir and Chitra Gupta School of Infrastructure Design and Management, Indian Institute of Technology Kharagpur, Kharagpur, West Bengal, India

Jyotika Nigam Urban Design Consultant, Opus Design Studio, Bhopal, Madhya Pradesh, India

Muslim Nouri Department of Geography, Payam Noor University, Tehran, Iran

Neha Pandey Department of Remote Sensing, Birla Institute of Scientific Research (B.I.S.R), Jaipur, Rajasthan, India

Rama U. Pandey School of Planning and Architecture, Bhopal, India

Shubha Pandey Natural Resource Data Management System Division, Department of Science and Technology, Ministry of Science and Technology, New Delhi, India

S. K. Patnaik Department of Geography, Rajiv Gandhi University, Doimukh, Itanagar, Arunachal Pardesh, India

M. P. Punia Department of Remote Sensing, Birla Institute of Scientific Research (B.I.S.R), Jaipur, Rajasthan, India

Atiqur Rahman Department of Geography, Faculty of Natural Sciences, Jamia Millia Islamia, New Delhi, India

Shyan Kirat Rai Centre for Urban Science and Engineering, Indian Institute of Technology Bombay, Mumbai, India

Suraj P. Rajeendran Department of Urban and Regional Planning, School of Planning and Architecture, Bhopal, India

Krithi Ramamritham Computer Science and Engineering, Centre for Urban Science and Engineering, Indian Institute of Technology Bombay, Mumbai, India

Dar Roberts Department of Geography, University of California at Santa Barbara, Santa Barbara, CA, USA

Asit Kumar Roy Department of Geography, Jadavpur University, Kolkata, West Bengal, India

Rostam Saberifar Department of Geography, Payam Noor University, Tehran, Iran

Kakoli Saha Department of Planning, School of Planning and Architecture Bhopal, Bhauri, Bhopal, India

Shahfahad Department of Geography, Faculty of Natural Sciences, Jamia Millia Islamia, New Delhi, India

Chilka Sharma School of Earth Sciences, Banasthali Vidyapith, Niwai, Rajasthan, India

Mukta Sharma IKG Punjab Technical University, Jalandhar, Punjab, India

Namita Sharma National Institute of Rural Development & Panchayati Raj-NERC, Guwahati, Assam, India

Poonam Sharma Department of Geography, Shaheed Bhagat Singh College, University of Delhi, New Delhi, India

Arjun Singh Remote Sensing Applications Centre-Uttar Pradesh, Lucknow, India

Ashok Kumar Singh Natural Resource Data Management System Division, Department of Science and Technology, Ministry of Science and Technology, New Delhi, India

Bhoop Singh Natural Resource Data Management System Division, Department of Science and Technology, Ministry of Science and Technology, New Delhi, India

Rashmi Singh Department of Geography, Delhi School of Economics, University of Delhi, New Delhi, Delhi, India

Reetanjali Singh Remote Sensing Applications Centre-Uttar Pradesh, Lucknow, India

Ankur Srivastava Department of Geography, Delhi School of Economics, University of Delhi, New Delhi, Delhi, India

Emanuele Strano MindEarth, Biel, Switzerland

Mohammad Tayyab Department of Geography, Faculty of Natural Sciences, Jamia Millia Islamia, New Delhi, India;
DDA, New Delhi, India

Kriti Trivedi School of Planning and Architecture, Bhopal, India

Bhavana N. Umrikar Department of Geology, Savitribai Phule Pune University, Pune, India

Deepank Verma Centre for Urban Science and Engineering, Indian Institute of Technology Bombay, Mumbai, India

Mrunmayi Wadwekar School of Planning and Architecture, Bhopal, India

Jing Xu Department of Geography, University of California at Santa Barbara, Santa Barbara, CA, USA

Abhishek Kumar Yadav Remote Sensing Applications Centre-Uttar Pradesh, Lucknow, India

Vinita Yadav Professor and Head, Department of Regional Planning, School of Planning and Architecture, New Delhi, India

Chapter 1
Analyzing the Role of Geospatial Technology in Smart City Development

Poonam Sharma, Rashmi Singh, and Ankur Srivastava

Abstract Geospatial technology helps in the creation, management, analysis, and visualization of spatial data. For Smart city management and functional applications; geospatial data and geospatial technology are instrumental. In this paper, geospatial technology and its role have been broadly discussed to assess its significance in smart city development. A smart city concept is considered to transform the quality of life in cities through the digitalization of different infrastructure sectors such as transportation, health, energy, education, and environment. Identifying and obtaining valuable information from large amounts of data that is generated in the growing urban areas. Smart city ideas have been implemented in many countries to seek solutions toward resource scarcities, congestion, and environmental issues. Concepts like open data, interconnected systems, internet of things, artificial intelligence, cloud computing, big data, and geospatial intelligence are innovative technologies that are expected to help in various fields of smart city development and give solutions to a variety of problems that the cities are facing.

Keywords Smart city · Internet of things · Open data · Artificial intelligence · Geomatics

Acronyms

AI	Artificial Intelligence
APIs	Application Programming Interfaces

P. Sharma (✉)
Department of Geography, Shaheed Bhagat Singh College, University of Delhi, New Delhi, India
e-mail: poonamsharma299@gmail.com

R. Singh · A. Srivastava
Department of Geography, Delhi School of Economics, University of Delhi, New Delhi, Delhi, India
e-mail: rashmisingh90@gmail.com

A. Srivastava
e-mail: ankurdse@gmail.com

P. Sharma (ed.), *Geospatial Technology and Smart Cities*, The Urban Book Series,
https://doi.org/10.1007/978-3-030-71945-6_1

BI	Business Intelligence
ELINT	Electronic Intelligence
FININT	Financial Intelligence
GEOINT	Geospatial Intelligence
GIS	Geographic Information Systems
GLCF	Global Land Cover Facility
GNSS	Global Navigation Satellite System
GPS	Global Positioning Systems
HUMINT	Human Intelligence
ICT	Information communication technologies
IMINT	Imagery Intelligence
IoT	Internet of Things
M2M	Machine-to-Machine
MASINT	Measurement and Signature Intelligence
ML	Machine Learning
OGC	Open Geospatial Consortium
OSINT	Open-source Intelligence
RFID	Radio Frequency Identification
SIGINT	Signals Intelligence
TECHINT	Technical Intelligence
UI	User Interface
UNCTAD	United Nations Conference on Trade and Development
UN-GGIM	Global Geospatial Information Management
USGS	United States Geological Survey

1.1 Smart City Concept

Urbanization is an ever-evolving pan-world dimension of population growth as observed from the expansion of megalopolis in different parts of the world (UN, World Urbanization Report 2008). As per UN projections by 2050, 68% population of the world will live in urban areas. The city systems have been in continuously evolving processes towards increasingly larger with complex linkages and interconnected systems. The UN's 11th Sustainable Development Goal (SDG) aims to make cities inclusive, safe, resilient, and sustainable; therefore, incorporating these targets should be an integral part of urban planning. The technology-driven smart city concept intends to provide the complete transformation from a traditional urban agglomeration to the high-tech twenty-first-century cities with complete automation of the system. Cities play a vital role as a unique social entity with a wider economic impact and have caused environmental problems too (Mori and Christodoulou 2012). Cities have emerged as centers of heavy resource consumptions and dependence on the outside regions. All these complex problems have forced cities to look for solutions to a variety of issues such as dynamic land use, stressed infrastructures, mobility, transport, environmental problems, and quality of life of its citizens (Ballas 2013). For developing ecologically viable urban systems, the smart city notion has been projected and advocated immensely. The technology at the base, works to provide faster ways to manage the challenges toward achieving the more livable and resilient societies (Innovation and Skills 2013). IBM is one of the notable corporates that has been associated with the smart city discussions and detailed research support.

A range of alternative adjectives is often used in of place smart cities such as an intelligent or digital city. (O'Grady and O'Hare 2012). Multilayer simultaneous working that includes the first layer as the technology base, specific applications at second, and user adoption is the third and most significant layer. A smart city is built upon the infrastructure of the digital City (UNDRR 2015). Information technology has brought a revolution in urban governance to develop and manage infrastructure and resource consumptions. The application of digital and electronic technologies has been used in the city and its communities to enhance the quality of life and the functional environments in the region (Husam Al Waer and Mark Deakin 2011). The new urban system works as a catalyst to economic growth with the core concept of better services to people along with sustainability. It also offers cities, a new investment and innovation in technology. It is vital to a smart city that provides the efficient and cost-effective infrastructure (Dimitri et al. 2012).

Numerous tech-based systems like user interface commutation network, machine learning, big data, internet of things, etc., are to name some. In fact, one and all the sections of urban system can be targeted to be managed on the concept of smart city initiative. The integration of information technology for creating and managing resources and utilities has been really significant (Klein and Kaefer 2008). For example, the smart parking meter enables digital payment as well as helps drivers to find available parking spaces at ease. Smart traffic management is used in the transportation arena and for optimizing traffic flow. The building and home intelligence systems have provided various solutions (Ghaffarian Hoseini et al. 2013). It is no longer limited to ICT alone but also the people and community quality of life are considered equally significant (Batty et al. 2012). Hancke and Silva (2013) discussed the significance of sensors in managing the infrastructure for city. A smart city seeks solutions with geospatial and human intelligence to improve the provisions of services to people (Giffinger and Gudrun 2010). However, some of the authors emphasized that new tech solutions will create an enabling environment for city managers to cope up with the growing scale of the demand for resources (Cugurullo 2013 and Vanolo 2014).

A smart city focuses on energy conservation and efficiency. Using smart streetlights, smart sensors and smart grid improvements in energy transmission from end to end point have been achieved. It can help in supply power on demand and monitor energy usages. Smart cities are expected to reduce carbon emissions and the efficiency of urban infrastructures through ICT. Environmental concern is a major initiative of smart city, aim to monitor and address issues like climate change, deforestation, air pollution, sanitation, and waste management with smart technology. Smart buildings are another important part of a smart city project. Connecting sensors to building infrastructure help in the identification and rectification of infrastructure maintenance

Table 1.1 Smart city concept implementations and impacts

Parameters	Levels achieved
Crime reduction	30–40%
Lower disease burden	8–15%
Minutes saved in daily commute	15–30 min
Water saved /person/day	25–30 L
Faster emergency response	20–35%
GHG emission reduction	10–15%
Water consumption reduction	20–30%
Un-recycled waste reduction	10–20%

Source Adapted from McKinsey Global Institute Report (2018a, b)

related to water supply pipelines, sewage, gas, and various others. For natural disaster preparedness and improved early warning systems for floods, storms, and droughts, these connected sensors have been instrumental. Technology will be instrumental for cities to sustain growth and better urban governance to provide a good quality of life to their citizens in the future. Environmental sustainability, EIA, public–private sector collocations aspect was studied by Satapathy et al. (2008), Mohan (2016), Sahely et al. (2005), and Wang TAO (2013).

Currently, a variety of ranking mechanisms and methods are used to explain the smartness of cities. Lombardi et al. (2012) developed a smart city ranking using 60 parameters. Lazaroiu and Roscia (2012) created a smart city ranking based on eighteen factors. Giffinger et al. (2007) prepared cities ranking on strength and weakness. The Natural Resources Defense Council, United States did the ranking based on environmental-related criteria (IDA 2015). The Japanese Institute for Urban Strategies has developed the Global Power City Index based on how a city behaves as a point of attraction. Joel Kotkin has worked to formulate a list of the global smart cities with Forbes. IBM and McKinsey Global Institute also keep bringing the comparative rankings of urban areas from time to time. McKinsey Global Institute Report has witnessed the positive impact of smart city initiatives as seen in Table 1.1.

1.1.1 Smart City Functioning

Spatial data, the geospatial ecosystem, and the multilevel working of different technologies are significant for developing and managing smart cities in terms of efficient solutions. The interconnected network of numerous devices creates the base for the aspired targets of a smart city.

For effective working, data follows four steps path and GIS tool searches, scrutinize, and filter data from various databases to organize it for better geo-data analysis.

For reaching the spatial patterns and associations with multiple data sets that happen in real-time, the GIS scenario includes various stages.

- Collection—Smart sensors generate data in real time.
- Analysis—At this stage insights and comprehension are developed.
- Communication—It goes to decision-makers for creating the usage and network.
- Application—The data is ready for institutions, infrastructure, and people to optimize operations and management.

1.1.2 Challenges and Concerns of Smart City

Smart city initiatives focus on the collaborative working of government and corporates. Since a humongous amount of data generation and usage is involved, data transparency, privacy, and security have become of vital significance to the people as well as for city governance. The presence of sensors, cameras, and massive data that is generated by users has also been argued against people's privacy. Another aspect of paramount significance is the digital divide that the city managers and governance are facing; when some section of the society is fully enabled to use all type of automation and on contrary other section is absolute non-users.

Moreover, the massive digitilization of everything connected with smart city targets, raises the challenge to meet up the environmental sustainability concern. Furthermore, public transit, traffic management, electricity, natural gas supply, various other infrastructure, and utilities would need some special provisions for maintenance as a system age with time. A continuously kept proper safeguarding and monitoring to ensure their proper functioning or supply is required.

1.2 Geospatial Technology

Geospatial technology gives inventive ways to everything around to empower the processes. It enables us to create, analyze, and visualize information as geo-information; locational analytics, business intelligence, data infrastructure, governance, people, physical infrastructure, utilities, and so on. Once the location is associated with data it becomes a powerful value addition for a variety of decision processes and the information is called geospatial data (UNCTAD 2016). An advantage of geospatial data is the ability to increase the accuracy of data collection and analysis. Multiple segments and areas provide solutions through geospatial technology and its expanded fields as geospatial industry, geospatial professionals, and geospatial industry's market. In Fig. 1.1 different stages involved in geospatial technology are illustrated.

It can create huge capital potential with a multi-trillion dollar global market (Gabrys 2014). The geospatial industry's market share was US$9 billion in 2014 which increased to US$22 billion in 2018 and is expected to double these numbers

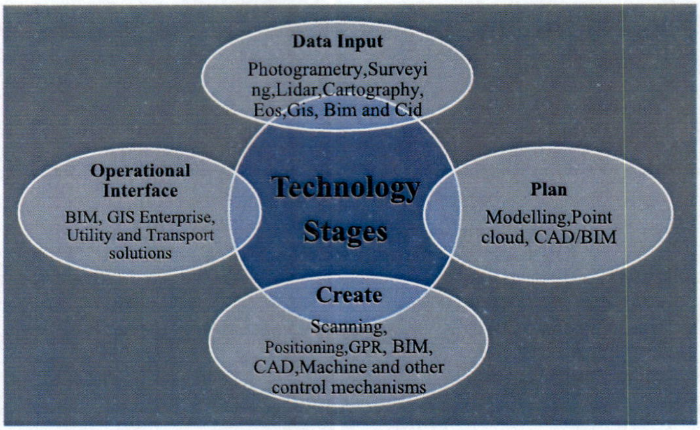

Fig. 1.1 Stages of smart city infrastructure

by 2022. There is an expectation of UD$1.6 trillion gain annually in investments in digitalization and enabling technologies (Buiz 2019 GEO-Report). The UN Global Geospatial Information Management (UN-GGIM) is global initiative in this direction. Crowdsourcing and collective intelligence is now seen as new ways of data availability (Good Child 2009). Geomatics is a multidisciplinary field and very significantly contributing to a variety of geospatial data sources like various survey methods, satellite systems, and geographical information systems (Pun Cheng 2001). The role of geomatics has further increased with the drive toward more integration of geospatial technologies with other related (Xue et al. 2002). Several studies (Hunter 2001; McDougall et al. 2006 and Aina 2009) have discussed to improve the learning experience in universities in this regard.

When the technological developments are discussed about, it is seen that the major turning points have brought tremendous and innovative transformations to the world. The current era of Industrial Revolution 0.4(IR 0.4) has seen innovations in multidisciplinary fields. India is the fourth country after the US, Japan, and China to join the league of IR 0.4 with the establishment of a center toward it. In the first Industrial Revolution, water and steam power were used to mechanize production. Tracing back to the Second Industrial Revolution electric power was used for mass production. The Third Industrial Revolution was mainly based on the development of computer technology. All previous innovations laid path for future revolutions. With IR 0.4, there has been a massive change in the global business, economy, and all other fields of development and infrastructure (Table 1.2).

Table 1.2 Navigating the industrial revolutions

Industrial revolution	Year	Information	Impact
First	1784	Steam, water, mechanical production equipment	Empowered Corporations
Second	1870	Division of labor, electrical mass production	
Third	1969	Electronics, IT, automated production	
Fourth	Current years	Cyber-physical systems	Empowered governance, corporations and people

Source Adapted from World Economic Forum

1.2.1 Types of Geospatial Technology

The technologies have been in ever-evolving mode since the first map was drawn. The satellites have provided to capture information on nature–man interaction along with the computer software that has contributed to the massive processing of spatial information. The technology advancements have created smooth interoperability, among diverse systems, platforms, data, processes, and services that work together efficiently. Interoperability is essential between heterogeneous machines and platforms for decision support services and spatial analytical functions. It is mainly dependent on standards, which are essential for advancing data access to various collaborations in governance, civic bodies, and departments. Remote sensing, geographical information system, global position system, radar, lidar, and a variety of other technologies have made it possible to manage the multi-platform and multidimension and real-time spatial data generation, management, analysis, and decision support system for governance and policymakers.

1.2.2 Open Data

Open data implies that data is available to anyone and everyone to be used freely with the provisions of further reuse and redistribution. Interoperability is an important feature that allows the open data system to happen. Through the usage of open data, governments and organizations may become more accessible and interactive (Pereira 2016). The importance of open data is well known as it ranges from improved proficiency of public administrations as well as economic growth. The smart city objectives of economic regeneration, social inclusion, improved governance, infrastructure, and utility management can be achieved with these ICT possibilities (Ojo et al. 2015). A McKinsey 2018, report states that open data like public information shared data from various private sources that can help in creating $3 trillion per annum of value. It is estimated that every day the world is creating huge data approximately 2.5 quintillion bytes. The integration of sensor web with spatial data and formulating by Open Geospatial Consortium (OGC) keep the entire platform open access and open source (Bhattacharya et al. 2017).

1.2.3 Internet of Things

IoT is about devices that are integrated and interconnected multi-tier machine communication. The multi-level and multi-platform information transmission is the key to efficiency (Gubbi et al. 2013). The IoT is a broadband network that uses standard communication protocols while its convergence point is the internet (Atzori et al. 2010). The working of this technology is primarily centered on massive data collected by the sensors and devices every second of the minute. The true potential of it depends on the data analysis, and on how information is leverage through Cloud-based applications. UNGGIM estimates see significant developments in the architecture of the Internet in the coming years. During 2014 alone, over UD $1.6 billion was invested into IoT companies by venture capitalists. Urban IoTs are designed for smart city development that works on the communication system for the administration of the city (Zanella et al. 2014). It is expected that IoT will be instrumental in providing more efficient, economic, and secure operations. Wireless sensing technology for the home environment and building intelligence have been showing results (Ghayvat et al. 2015). It has been estimated that significant efficiency may be achieved in power usage with IoT integration (Sikder et al. 2018). Smart systems of electricity consumption are becoming popular in different parts of the world (Neirotti et al. 2014 and Watteyne and Pister 2011).

Table 1.3 has discussed the twelve smart cities indices developed by different institutions and bodies to measure the development achievements of smart city initiatives. The top ten cities of those indices are included to see the pattern of their distribution (Fig. 1.2) and it reflects that these top ten cities are in Europe, the United States of America South East Asia, and Australia. Similarly, another statistic is prepared illustrating that how these cities are performing on the various indices (Table 1.4). It has been witnessed that Tokyo is among the top ten cities in eight of these indices. London, New York, and Singapore credits to be in the top ten positions on the seven out of ten indices. Cities like Paris, Helsinki, Seoul, Toronto, Los Angeles, Amsterdam, and Copenhagen have been able to manage their place in the top ten cities in four to five indices. In all these different types of indices, the importance of technology and geospatial technology has been clearly seen.

1.2.4 Artificial Intelligence (AI)

The International Data Corporation (IDC)'s Worldwide Internet of Things Forecast, 2015–2020, has predicted tremendous growth in the usage of this technology. The

Table 1.3 Top ten cities with smart city parameters

S.No.	High tech cities[A]	High GDP cities[B]	Smart transport cities[C]	Inclusive cities[D]	E-governance cities[E]	Livable cities[F]	Sustainable cities[G]	Intelligent cities[H]	Safe cities index[I]	Innovative cities [J]	Smartest cities[K]	Ubiquitous cities [L]
1	Tokyo	Tokyo	Singapore	Zurich	London	Vienna	Copenhagen	New York City	Tokyo	New York	Singapore	London
2	London	New York	Amsterdam	Vienna	Singapore	Melbourne	Amsterdam	London	Singapore	Tokyo	Zurich	New York
3	San Francisco-Silicon Valley	Los Angeles	London	Copenhagen	Seoul	Sydney	Stockholm	Paris	Osaka	London	Oslo	Paris
4	New York	Seoul	Shanghai	Luxembourg	New York	Osaka	Berlin	Tokyo	Amsterdam	Los Angles	Geneva	Tokyo
5	Los Angeles	London	New York	Helsinki	Helsinki	Calgary	Portland	Reykjavik	Sydney	Singapore	Copenhagen	Moscow
6	Singapore	Paris	Tokyo	Taipei	Montreal	Vancouver	San Francisco	Singapore	Toronto	Paris	Auckland	Dubai
7	Boston	Osaka	Helsinki	Oslo	Boston	Toronto	Cape Town	Seoul	Washington, DC	Chicago	Taipei	Singapore
8	Toronto	Chicago	Beijing	Ottawa	Melbourne	Tokyo	Helsinki	Toronto	Seoul	Boston	Helsinki	Barcelona
9	Paris	Moscow	Seoul	Kiel	Barcelona	Copenhagen	Vancouver	Hong Kong	Copenhagen	San Francisco - San Jose	Bilbao	Los Angeles
10	Sydney	Shanghai	Berlin	Geneva	Shanghai	Adelaide	Reykjavik	Amsterdam	Melbourne	Toronto	Dusseldorf	Rome

Source Compiled from various mentioned sources-: [A]www.businessinsider.in/slideshows/miscellaneous/the-50-most-high-tech-cities-in-the-world-in-2018/[B] World Economic Forum, 2020/ [C]www.businessinsider.com/15-cities-worlds-best-transport-solutions-2019/[D] The Prosperity & Inclusion City Seal and Awards (PISCA) Index (2019)/[E]www.smartcitygovt.com/ [F] The Economist Intelligence Unit, 2019/ [G] The Sustainable Cities Index, 2019/ [H]IESE Cities in Motion Index, 2019/ [I] The Safe Cities Index 2019/ [I]nnovation Cities Index 2019/ [K]IMD Smart City Index 2019/ [L]www.Bestcities.Org/Rankings/Worlds-Best-Cities

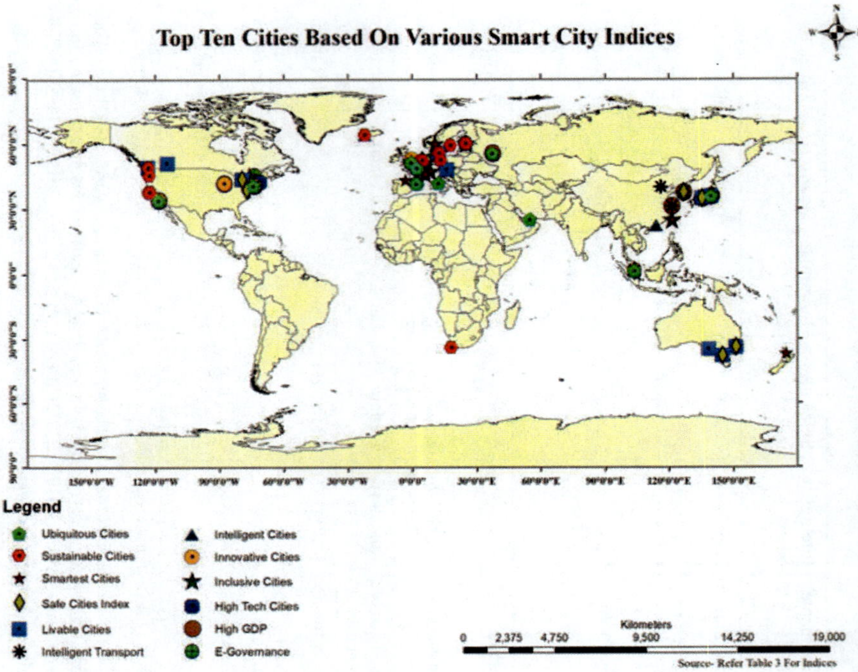

Fig. 1.2 Cities and smart city indices

IoT market size is forecast to grow from US$157 billion in 2016 to $661 billion by 2021. IoT is a network paradigm interconnected with multidimensional information flow (Xu et al. 2018). In the next two decades, it is estimated that AI will contribute immensely to the expansion of economic growth. Studies reveal that the countries like Sweden, Finland, the US, and Japan will gain the most from AI technologies (Geo Biz 2018 report). IOT will be instrumental in filtering the data from the mammoth volume and intricate flow of information (Chen et al. 2018). Variety of artificial intelligence methods have been successfully used in many fields like transportation, building management, medical, and many others (Abduljabbar et al. 2019). The cloud computing know-how will help better urban management potentials (Vashist et al. 2015) and transforming the health care sector (Turukalo et al. 2019). Intelligent video analysis for inquiry in various smart city applications is possible with AI-oriented large-scale video management (Lingyu et al. 2017). Sensors at different levels used for data collection are explained in Fig. 1.3.

The cognitive internet of things can be analyzed using cognitive computing to handle the flexibility issues. The interconnection between a large no of sensory devices and integration between the real-world situation and the virtual world is possible through IoT (Sheth 2016). An AI-based semantic IoT (AI-SIoT) hybrid service architecture is expected to support heterogeneous devices, and find applications in practical scenarios (Guo et al. 2018). Faster and optimal decision-making

Table 1.4 Cities on various indices of technology usage for smart city developments

Tokyo	High tech, High GDP, Smart transport, Livable, Intelligent, Safe, Innovative and ubiquitous
London	High Tech, High GDP, Smart Transport, E-governance, Intelligent, Innovative, and Ubiquitous
New York	High Tech, High GDP, Smart Transport, E-governance, Intelligent, Innovative, and Ubiquitous
Singapore	High Tech, Smart Transport, E-governance, Intelligent, safe, Innovative, and Ubiquitous
Paris	High Tech, High GDP, Intelligent, Innovative, and Ubiquitous
Helsinki	Smart Transport, Inclusive, E-Governance, Sustainable, and Smartest
Seoul	High GDP, Smart Transport, E-governance, Intelligent, and Safe
Toronto	High Tech, Livable, Intelligent, Safe, and Innovative
Los Angeles	High Tech, High GDP, Innovative, and Ubiquitous
Amsterdam	Smart Transport, Sustainable, Intelligent, and Safe
Copenhagen	Inclusive, Livable, Sustainable, Smartest, and Safe

Source: Compiled based on data from Table 1.3

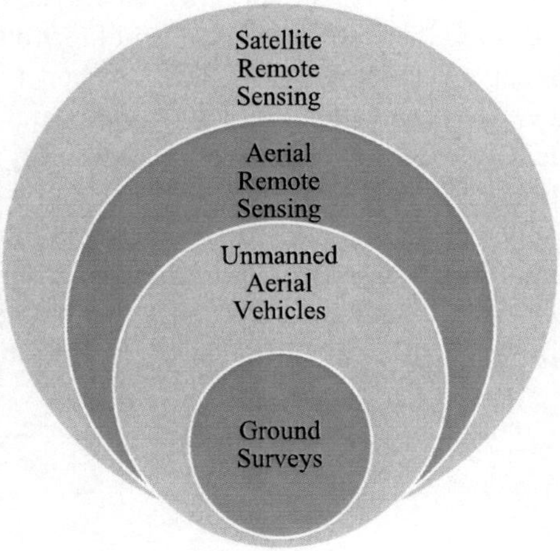

Fig. 1.3 Types of sensors used for geospatial data for smart cities

in real time has been seen in Google's AI application Alpha Go (Silver et al. 2016). The application of AI in the solutions of ecological and environmental problems will give way forward for urban science (Amditis and Lytrivis 2015).

1.2.5 Cloud Computing

When the world is expecting that around six billion people will be residing in cities (World Urbanization Prospects, 2014, UN) the volume of data generated would be massive. Cloud computing will create digital frames for storage and analysis for managing the cities. Cloud technology gives a platform to integrate the observation systems, all kinds' analytics, and decision support. An ID-based cryptosystem to provide grid safety and verifications technique was proposed by Li et al. (2009). Many of the studies relating government interfaces with citizens, "e-government", "24/7 government", local governance and service provision levels show the integrated observation systems (O'Looney 2000) and (Baud et al. 2009) Mobile Cloud Computing (MCC) and Heterogeneous Networks (HetNets) are viewed as infrastructures providing together with a key solution. Figure 1.4 explains the technological drivers and institutional drivers along with the geospatial technology creating the smart city environment city management attributes.

1.2.6 Wireless & Broadband

For an integrated real-time digital decision support system, three major components are important viz Internet GIS, Mobile GIS, and broadband wireless communication networks. The urban management functions, infrastructure provisions, and management will improve great deals. Abbott (2003), Perera et al. (2013), Gubbi et al. (2013). The challenges such as interoperability support among heterogeneous wireless networks, wireless solutions, mobility management, and high energy consumption are important to develop connectivity in smart cities (Yaqoob et al. 2017). Web applications are used for managing different arenas for wireless, digital, or ubiquitous city (Anthopoulos and Panos 2010). The concept of a small cell meets the emerging communication and networking requirements and user-centric perspective of forthcoming mobile technologies of future smart cities (Cimmino et al. 2013).

1.2.7 Big Data

The McKinsey Global Institute's report on big data in 2009 estimates the availability of huge amounts of location data with a growth of about 20–22 per cent annually. The

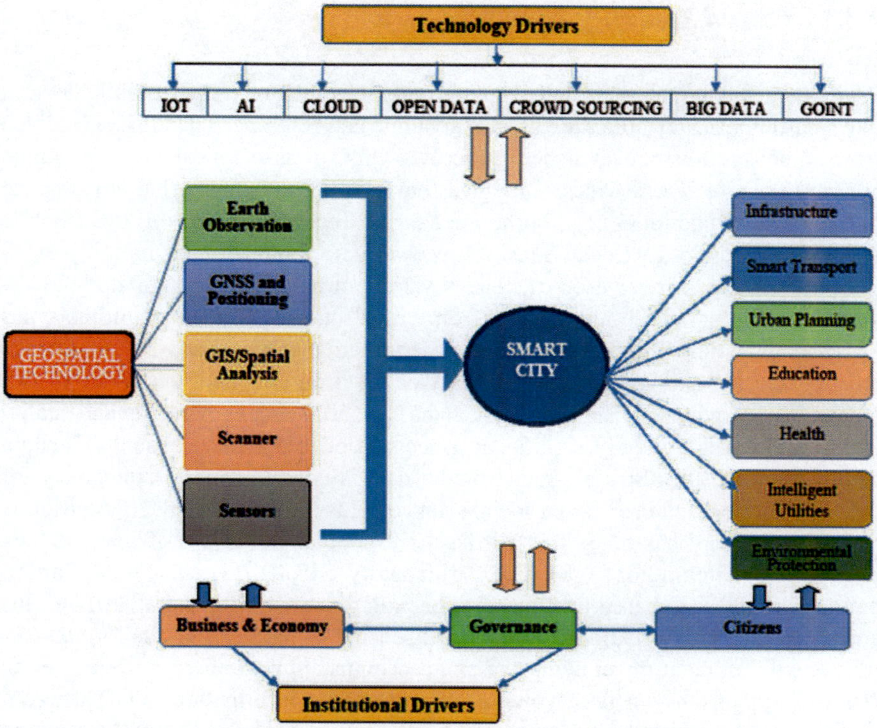

Fig. 1.4 Geospatial technology and smart city development. *Source* Authors

geospatial aspects of data are becoming important across all types of global industrial sectors. Big data is prospective for cities to process the massive data into usable information. The massive formless data generated through various platforms and its transfer requires a seamless and error-free mechanism (Borgia 2014). Besides, big data can be instrumental for policymakers. It is witnessed that much of geospatial data include photos, social media chats, and audio–video content. It adds innovation and sustainability. Big data has a multidimensional impact in catering to the answer-oriented capabilities to seek solutions to new urban economy (Batty et al. 2012). Cloud technology is also proving complementary to the big data technology framework (Hashem et al. 2015). One such example is the Hadoop framework that streamlines the process of the massive data set. Repositories and numerous types of visualization help in handling huge data (Kang et al. 2016). For decision-makers,' smart big data for smart grids will help to maintain optimum supply and future requirement projections (Al Nuaimi et al. 2015). In the transportation sector for end-users, environmental impact, and safety big data have a role to play (Ju et al. 2013).

1.2.8 Crowd Sourcing

While the users are on their daily routine mobility in the city, their mobile devices like smartphones and tablets are used to generate geo-referenced data. Data received from numerous users are later checked across with various technical measures which are vital for effective services. This gives new opportunities for urban governance to explore the functional neighborhood changes. Technology like mPASS (mobile Pervasive Accessibility Social Sensing) receives data about people's mobility experiences which are further used to create a variety of mappings (Prandi et al. 2014). Technically integrated usage of data from crowdsourcing, through multiple types of sensors and numerous other sources of geo-intelligent data, has become vital for creating specific and general maps for users. Crowdsourcing and sensors are instrumental in improving data density (Biancalana et al. 2013) but authenticity and quality of information are equally significant aspects to be taken into account (Flanagin and Mertzger 2008). Spatial -social media data is now days used extensively for research purpose and in different social science fields (Poorthuis et al. 2016). Mobile phone technology has made this all the more popular in the era of massive data phenomena (Kitchin 2014a, b). The Participatory GIS (PGIS) and Public Participation GIS (PPGIS)-based data has also been in great use (Bugs et al. 2010). This emerging paradigm is called mobile crowd-sensing or participatory sensing experiments with efficient travel plans and smart campus applications, etc. (Szabó et al. 2013). Another concept is of volunteered geographic information (VGI) where it is important to volunteer in data generation exercise (Goodchild 2007; Crampton et al. 2013). In such cases, people become important in dual capacity as data creator and user both (Simonofski et al. 2017). The challenges with crowdsourcing data for a smart city are to ensure security, user-friendliness, sustainability, and its smooth functioning (Degbelo et al. 2016). The role of social media in case of emergency or disaster situations becomes vital (Bird et al. 2012; Zook et al. 2010). Crowdsourcing was explained as effective and innovative for smart city development (Ballon et al. 2018). For intelligent management of urban systems, both parameters are required i.e. all types of spatial data and enabled citizens (Coe et al. 2001). The new perspectives of citizen participation along with governance and corporates as collective contributions have also been applied at places (Anttiroiko 2016).

1.2.9 Geospatial Intelligence (GEOINT)

Clapper gave the term GEOINT during his tenure at National Geospatial-Intelligence Agency. Geospatial intelligence (GEOINT) has been explained as the process of extracting geospatial information for phenomena and things on the earth in a space–time framework (Bacastow and Bellafiore 2009). The benefits of GEOINT were realized with the integration and analysis of three core capabilities as imagery, imagery intelligence, and geospatial information. It allows for in-depth crime pattern analysis through the use of descriptive and predictive models such as hot spots, etc. (Glasner and Leitner 2017). Geospatial hotspot mapping is a recognized methodology of determining crime patterns in space and time (Tompson and Townsley 2010). Satellite data is significant for various monitoring and mapping exercises for infrastructure, resources, disasters, and health (Suppasri et al. 2012; AlSaud 2010; Yang et al. 2011 and (Blaschke 2010). For various geophysical processes phenomenon studies were done by (Nittel 2009; Yoo et al. 2011). A common intelligence framework with data analytics, sensor fusion, augmented reality, cyber-security, 3D tracking, and predictive modeling can be developed to provide situational awareness on the many dimensions of megacities (Loper 2018).

The Internet of Things (IoT) , Artificial Intelligence, Cloud and Big Data play the major role in the process in a geospatial ecosystem to various platforms connect the outreach of geospatial technologies to end-users and other interconnected systems (Tiwari and Jain 2014). Sustainable transport, "predict and provide" and forecasted demand studies show the role of the technology-based solutions (Newman and Kenworthy 1999; Schiller et al. 2010; Blanco et al. 2009; Dimitriou 2006; Preston and Rajé 2007) and (World Bank 2002). Geoint has a vital role to play in technology connectivity, locational data, different types of intelligence and solutions to various aspects of smart city development (Fig. 1.5).

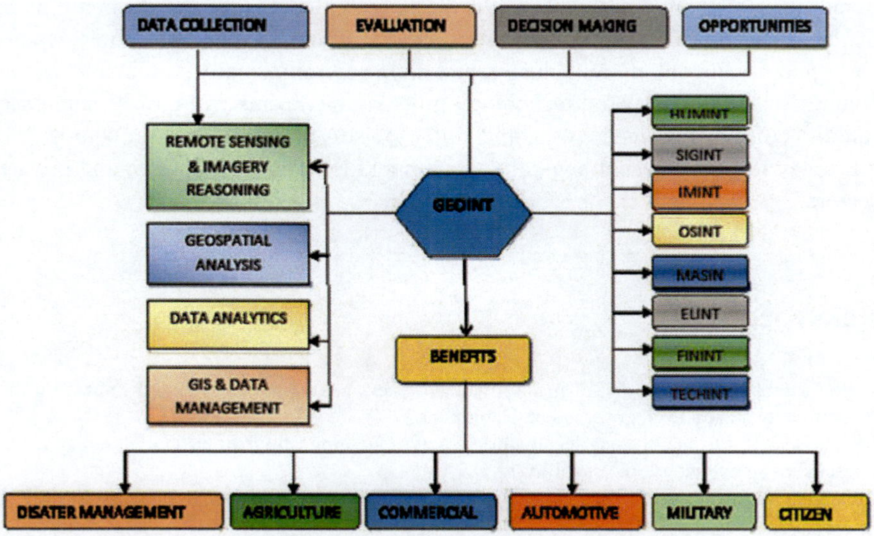

Fig. 1.5 Benefits of GEOINT. *Source* Authors

1.3 Conclusion

This chapter attempts to explore the multifaceted concept of the smart city. Smart cities seek technology-based solutions to the urbanization externalities related to infrastructure, economy, people, and environment. Smart cities are expected to create smarter choices to increase citizen's quality of life and a faster pace of urban socio-cultural and economic development. They are knowledge-based cities with the capability to raise the growth of their respective nation by planning, governance, management, and development systems where geospatial information is at the forefront of helping decision-makers. Geospatial technology plays a key role in the growth and management of infrastructures of smart cities. The main target is actually to provide efficient and cost-effective infrastructure; public–private collaborative creation, and management of city infrastructure management.

Smart cities are now moving towards a new phase with almost after a decade of experimentation where digital methods are becoming most powerful and cost-effective. The Mckinsey Global Institute, 2018 report analyzed dozens of current applications and showed that it improves quality-of-life indicators by 10–30%. Smart technologies will help cities cope up with the challenges to manage city governance. Cities are multifaceted intricate system which creates a massive pool of data. Moreover, making real-time information available to people and companies enables them to make better choices and work in the urban scenario. It is the insight in all kinds of data that contribute to making cities smarter, more livable, and more responsive, more intelligent, more inclusive, and more environmentally sustainable. In the coming years the geospatial technology in its various forms will play an immense role in adding convenience to people's life, tools to policymakers and companies; efficiency to governance along collaboration and investment by public and private sectors.

References

Abbott J (2003) The use of GIS in informal settlement upgrading: Its role and impact on the community and on local government. Habitat Int 27(4):575–593

Abduljabbar R, Dia H, Liyanage S, Bagloee SA (2019) Applications of artificial intelligence in transport: an overview. Sustainability 11:189

Aina YA (2009) Geomatics education in Saudi Arabia: status, challenges and prospects. Int Res Geogr Environ Educ 18(2):111–119. ISSN 1038-2046

Al Nuaimi E, Al Neyadi H, Mohamed N, Al-Jaroodi J (2015) Applications of big data to smart cities. J Internet Serv Appl 6(1):1–15

AlSaud M (2010) Assessment of flood hazard of Jeddah area 2009, Saudi Arabia. J Water Resour Prot 2(9):839–847. ISSN 1945-3108

Amditis A, Lytrivis P (2015). Towards automated transport systems: European initiatives, challenges and theway forward. In: Meyer G, Beiker S (eds) Road vehicle automation 2, Lecture notes in mobility. Springer:Cham, Switzerland

Anttiroiko AV (2016) City-as-a-platform: the rise of participatory innovation platforms in finnish cities. Sustainability 8:922

Atzori L, Iera A, Morabito G (2010) The internet of things: a survey. Comput Netw 54:2787–2805
Bacastow TS, Bellafiore DJ (2009) Redefining Geospatial Intelligence Am Intell J 27(1):38–40
Ballas D (2013) What makes a 'happy city'? Cities 32:S39-S50
Ballon P, Van Hoed M, Schuurman D (2018) The effectiveness of involving users in digital innovation: measuring the impact of living labs. Telematics Inform 35(5):1201–1214
Batty M et al (2012) From intelligent to smart cities. Intell Build Int 3:3
Baud ISA, Pfeffer K, Sridharan N, Nainan N (2009) Matching deprivation mapping to urban governance in three Indian mega-cities. Habitat Int 33(4):365–377
Biancalana C, Gasparetti F, Micarelli A, Sansonetti G (2013) An approach to social recommendation for context-aware mobile services. ACM Trans Intell Syst Technol 4(1)10:31
Bird D, Ling M, Haynes K (2012) Flooding facebook–the use of social media during the Queensland and Victorian floods. Aust J Emerg Manag 27(1):27
Blanco H, Alberti M, Forsyth A, Krizek KJ, Rodríguez DA, Talen E et al (2009) Hot, congested, crowded and diverse: emerging research agendas in planning. Prog Plan 71(4):153–205
Blaschke T (2010) Object-based image analysis for remote sensing. ISPRS J Photogramm Remote Sens 65(1):2–16
Borgia E (2014) The internet of things vision: key features, applications and open issues. Comput Commun 54:1–31
Bugs G, Granell C, Fonts O, Huerta J, Painho M (2010) An assessment of public participation GIS and web 2.0technologies in urban planning practice in Canela. Brazil Cities 27:172–181
Chen M, Herrera F, Hwang K (2018) Cognitive computing: architecture, technologies and intelligent applications. IEEE Access 6:19774–19783
Cimmino A et al (2013) The role of small cell technology in future smart city applications. Trans Emerg Telecommun Technol 2:1–10
Coe A, Paquet G, Roy J (2001) E-Governance and smart communities. Soc Sci Comput Rev 19:80–93
Crampton J, Graham M, Poorthuis A et al (2013) Beyond the geo-tag? deconstructing "'Big Data"' and leveraging the potential of the geoweb. Cartogr Geogr Inf Sci (CaGIS) 40(2):130–139
Cugurullo F (2013) How to build a sandcastle: an analysis of the genesis and development of Masdar city. J Urban Technol 20(1):23–37
Deakin M, Al Waer H (2011) From intelligent to smart cities. Intell Build Int 3(3):133–139
Degbelo A, Granell C, Trilles S, Bhattacharya D, Casteleyn S, Kray C (2016) Opening up smart cities: citizen-centric challenges and opportunities from GIScience. Int J Geo-Inf 5:16
Dimitriou HT (2006) Towards a generic sustainable urban transport strategy for middle-sized cities in Asia: lessons from Ningbo. Kanpur Solo Habitat Int 30(4):1082–1099
Dimitri S, Bastiaan B, De Marez L, Mechant P (2012) Smart ideas for smart cities: investigating crowdsourcing for generating and selecting ideas for ICT innovation in a city context. J Theor Appl Electron Commer Res 7:49–62
Flanagin J, Mertzger MJ (2008) The credibility of volunteered geographic information. Geo J 72(3–4):137–148
Gabrys J (2014) Programming environments: environmentality and citizen sensing in the smart city. Environ Plan: Soc Space 32(1):30–48
GEO-Buiz (2019) Report-Geospatial readiness and industrial outlook
Ghaffarian A, Hoseini U et al (2013) The essence of future smart houses: from embedding ict to adapting to sustainability principles. Renew Sustain Energy Rev 24:593–607
Ghayvat H, Mukhopadhyay S, Gui X, Suryadevara N (2015) WSN-and IOT-based smart homes and their extension to smart buildings. Sensors 15:10350–10379
Giffinger R, Gudrun H (2010) "Smart cities ranking: an effective instrument for the positioning of cities? City Environ 4(12):7–25
Giffinger R, Fertner C, Kramar H, Kalasek R, Pichler-Milanovic´, N. and Meijers, E. (2007) Smart cities: ranking of European medium-sized cities. Centre of Regional Science, Vienna

Glasner P, Leitner M (2017) Evaluating the impact the weekday has on near-repeat victimization: a spatiotemporal analysis of street robberies in the city of Vienna, Austria. ISPRS Int J GeoInformation 6(3):1–18.L

Goodchild MF (2009) Citizens as sensors: the world of volunteered geography. Geo J 69:211–221

Gubbi J, Buyya R, Marusic S, Palaniswami M (2013) Internet of Things (IoT): a vision, architectural elements, and future directions. Futur Gener Comput Syst 29(7):1645–1660

Guo K, Lu Y, Gao H, Cao R (2018) Artificial intelligence-based semantic internet of things in a user-centric smart city. Sensors 18(1341):2–22

Hancke GP, Silva BC (2013) The role of advanced sensing in smart cities. Sensors 13, 393–425

Hashem IAT, Yaqoob I, Anuar NB, Mokhtar S, Gani A, Khan SU (2015) The rise of "big data" on cloud computing: review and open research issues. Inf Syst 47:98–115

Hunter GJ (2001) Ensuring the survival of geomatics engineering at the University of Melbourne. Surv Land Inf Sci 61(4):255–262. ISSN 1538-1242

IDA Singapore, "iN2015 Masterplan" (2015).https://www.ida.gov.sg/~/media/Files/Infocomm%20Landscape/iN2015/Reports/realisingthevisionin2015.pdf

Innovation and Skill (2013) https://www.gov.uk/government/publications/bis-annual-report-and-accounts-2013-to-2014

Ju G, Cheng M, Xiao M, Xu J, Pan K, Wang X, Shi F (2013) Smart transportation between three phases through a stimulus-responsive functionally cooperating device. Adv Mater 25(21):2915–2919

Kang Y-S, Park I-H, Rhee J, Lee YH (2016) Mongo DB-based repository design for IoT-generated RFID/sensor big data. IEEE Sens J 16(2):485–497

Kitchin R (2014a) The data revolution: big data, open data, data infrastructures and their consequences. Sage, London, UK

Kitchin R (2014b) The real-time city? big data and smart urbanism. Geo J 79(1):1–14

Klein C, Kaefer G (2008) From smart homes to smart cities: opportunities and challenges from an industrial perspective, Next Generation Teletraff and Wired/Wireless Advanced Networking, Proceedings, Lecture Notes in Computer Science, 5174, 260

Bhattacharya et al (2017) Smart cities intelligence system (smacisys) integrating sensor web with spatial data. ISPRS Ann Photogramm, Remote Sens Spat Inf Sci, IV-4/W3, 21–28

Lingyu et al (2017) AI oriented large-scale video management for smart city: technologies, standards and beyond, 1–7. https://arxiv.org/pdf/1712.01432.pdf

Lazaroiu GC, Roscia M (2012) Definition methodology for the smart cities model. Energy 47(1):326–332

Li W, Batty M, Goodchild MF (2020) Real-time GIS for smart cities. Int J Geogr Inf Sci 34:311–324

Lombardi P, Giordano S, Farouh H, Yousef W (2012) Modelling the smart city performance. Innov Euro J Soc Sci Res 25(2):137–149

Loper ML (2018) Situational awareness in megacities. Springer, Technology and the Intelligence Community, pp 205–235

Mckinsey (2018a) Report

Mckinsey (2018b). Report: smart cities digital solution for more livable future

Mohan M (2016) Geospatial information from satellite imagery for geo-visualisation of smart cities in India. Remote Sens Spat Inf Sci XLI-B8, XXIII ISPRS Congress, 12–19 July 2016. Prague, Czech Republic

Mori K, Christodoulou A (2012) Review of sustainability indices and indicators: towards a new city sustainability index (CSI). Environ Impact Assess Rev 32(1):94–106

Neirotti P, De Marco A, Cagliano AC, Mangano G, Scorrano F (2014) Current trends in smart city initiatives: some stylised facts. Cities 38:25–36

Newman P, Kenworthy J (1999) Sustainability and cities: overcoming automobile dependence. Island Press Washington DC. ISBN 1-55963-660-2

Nittel S (2009) A survey of geosensor networks: advances in dynamic environmental monitoring. Sensors 9, 5664–5678. ISSN 1424-8220

O'Grady M, O'Hare G (2012) How Smart Is Your City? Science 335(3):1581–1582

O'Looney J (2000) Beyond maps: GIS and decision making in local government. ESRI Press
Ojo A, Curry E, Zeleti FA (2015) A tale of open data innovations in five smart cities. In: Proceedings of the 48th hawaii international conference on system sciences (HICSS). IEEE
Park E, Cho Y, Han J, Kwon SJ (2014) Comprehensive approaches to user acceptance of internet of things in a smart home environment. IEEE Internet Things J 4:2342–2350
Patil NB, Nalawade SS (2016) Applications of GIS to smart cities. Int J Recent Adv Eng Technol (IJRAET), 4(3):2347–2812. ISSN (Online)
Pereira GV, et al (2016) Delivering public value through open government data initiatives in a Smart City context. Inf Syst Front (Springer), 19, 213–229
Perera C et al (2013) Sensing as a service model for smart cities supported by internet of things. Trans Emerg Telecommun Technol 25(1), 02 Sept 2013. https://doi.org/10.1002/ett.2704
Poorthuis A, Zook M, Shelton T et al (2016) Using geotagged digital social data in geographic research. In: Clifford N, French S, Cope M et al (eds) Key methods in geography. Sage, London, pp 248–269
Prandi C, Salomoni P, Mirri S (2014) mPASS: integrating people sensing and crowdsourcing to map urban accessibility. In: Proceedings of the IEEE international conference on consumer communications and networking conference (CCNC '14). Las Vegas, Nevada, USA, 10–13
Preston J, Rajé F (2007) Accessibility, mobility and transport-related social exclusion. J Transp Geogr 15(3):151–160
Pun-Cheng LSC (2001) Knowing our customers: a quantitative analysis of geomatics students. Int Res Geogr Environ Educ 10(3):322–341. ISSN 1038-2046
Sahely HR, Kennedy CA, Adams BJ (2005) Developing sustainability criteria for urban infrastructure systems. Can J Civ Eng 32(1):72–85
Satapathy DR, Katpatal Y, Wate S (2008) Application of geospatial technologies for environmental impact assessment: an Indian scenario. Int J Remote Sens 29:355–386. https://doi.org/10.1080/01431160701269002
Schiller PL, Bruun EC, Kenworthy JR (2010) An introduction to sustainable transportation: policy. Planning and Implementation, Earthscan
Sheth A (2016) Internet of things to smart IoT through semantic, cognitive, and perceptual computing. IEEE Intell Syst 31(2):108–112
Sikder AK et al (2018) IoT-enabled smart lighting systems for smart cities. In: IEEE explore, IEEE 8th annual computing and communication workshop and conference (CCWC), pp 639–645
Silver D, Huang A, Maddison CJ (2016) Mastering the game of Go with deep neural networks and tree search. Nature 529:484–489
Simonofski A, Asensio ES, de Smedt J, Snoeck M (2017) Citizen participation in smart cities: evaluation framework proposal. In: Proceedings of the 2017 IEEE 19th conference on business informatics (CBI 2017). Thessaloniki, Greece, pp 227–236. 24–26 July 2017
Suppasri A, Koshimura S, Matsuoka M, Gokon H, Kamthonkiat D (2012) Application of remote sensing for tsunami disaster. In: Chemin Y (ed) Remote sensing of planet earth, pp 143–168. InTech, ISBN 978-953-307-919-6
Szabó R, et al (2013) Framework for smart city applications based on participatory sensing. In: IEEE 4th international conference on cognitive info communications, pp 295–300
Tiwari A, Jain J (2014) GIS steering smart future for smart Indian cities. Int J Sci Res Publ 4(8)
Tompson and M. Townsley, (2010) Looking Back to the Future: using space–time patterns to better predict the location of street crime. Int J Police Sci Manag 12(1):23–40
UN (2008) World urbanization prospects: the 2007 revision population database. https://esa.un.org/unup/
UNCTAD (2016) Report on smart cities and infrastructure
UNDRR (2015) Annual report
Vanolo A (2014) Smart mentality: the smart city as disciplinary strategy. Urban Stud 51(5):883–898
Vashist SK, Luppa PB, Yeo LY, Ozcan A, Luong JHT (2015) Emerging technologies for next-generation point-of-care testing. Trends Biotechnol 33:692–705

Wang TAO (2013) Interdisciplinary urban GIS for smart cities: advancements and opportunities. Geo-Spat Inf Sci 16(1):25–34. https://doi.org/10.1080/10095020.2013.774108

Watteyne T, Pister KS (2011) Smarter cities through standards-based wireless sensor networks. IBM J Res Dev 55(1.2):7–1

World Bank (2002) Report-building institutions for market

Xue Y, Cracknell AP, Guo HD (2002) Telegeoprocessing: the integration of remote sensing, geographic information system (GIS), global positioning system (GPS) and telecommunication. Int J Remote Sens 23(9):1851–1893. ISSN 0141-1161

Xu H, Yu W, Griffith D, Golmie N (2018) A survey on industrial internet of things: a cyber-physical systems perspective. IEEE Access 6:78238–78259

Yang X, Jiang G, Luo X, Zheng Z (2011) Preliminary mapping of high-resolution rural population distribution based on imagery from Google Earth: a case study in the Lake Tai Basin, Eastern China. Applied Geography 32, 221–227. ISSN 0143-6228

Yaqoob I, et al (2017) Enabling communication technologies for smart cities. IEEE Commun Mag 55, 1

Yoo J, Jazzar AB, Jeong KJ, Lee KW (2011) Development of real-time monitoring system integrated with GIS. In: Proceedings of the sixth national GIS symposium in Saudi Arabia (Online). Khobar, Saudi Arabia, 24–26 April 2011. www.saudigis.org/FCKFiles/File/6thSaudiGIS_Papers/T1_3.doc

Zanella NB et al (2014) Internet of things for smart cities. IEEE Internet Things J 1(1):22–32. https://doi.org/10.1109/JIOT.2014.2306328

Zook M, Graham M, Shelton T et al (2010) Volunteered geographic information and crowdsourcing disaster relief: a case study of the Haitian earthquake. World Med Health Policy 2(2):7–33

Part I
Urban Expansion and Infrastructure

Chapter 2
The Dark Side of the Earth: Benchmarking Lighting Access for All Cities on Earth and the *CityNet* dataset

Adrian Albert, Emanuele Strano, Jasleen Kaur, and Marta Gonzalez

Abstract In this paper, we present an analysis of urban form, defined as the spatial distribution of macroeconomic quantities that characterize a city such as population, built environment, and energy use. In particular, we develop a framework to study the question of "mismatch" between the spatial distribution of lighting levels observed in a city (which was previously shown to be a proxy for energy access and wealth levels) and that city's population density and built area distribution. This allows us to rank cities globally by their ability to, intuitively, "match people with lighting/energy access". For this, we develop and make available a derived dataset we call *CityNet* that is based on best-available open-source remote sensing data products for the world's largest 30,000 cities. We first describe how cities may be grouped into a few classes by the scale magnitude of these key quantities. Then we introduce simple quantities to measure their spatial distributions such as the *average radial profile*, the *discrepancy* between the radial profile of population and that of other quantities, and the *effort* of transforming the distribution of a given quantity to match that of the population density. To compare a given city against its "peers", we define a simple benchmark model of urban form. We use this model to rank cities by the relative magnitude of the lack of access to built and energy infrastructure. The key observation we make is that in many parts of the world, development (including built area density and energy access and use) does not follow the spatial population distribution.

Keywords Built environment · Energy use · Access to lighting · Urban footprint · Geographical regions

---

A. Albert · M. Gonzalez (✉)
Lawrence Berkeley National Lab, 1 Cyclotron Rd, Berkeley, CA 94720, USA
e-mail: martag@berkeley.edu

E. Strano
MindEarth, Biel, Switzerland

J. Kaur
Philips Lighting Research USA, 1 Charles Park, Cambridge, MA, USA
e-mail: jasleen.kaur@signify.com

P. Sharma (ed.), *Geospatial Technology and Smart Cities*, The Urban Book Series,
https://doi.org/10.1007/978-3-030-71945-6_2

Acronyms

VIIRS	Visible Infrared Imaging Radiometer Suite
SAR	Synthetic Aperture Radar
DLR	German Aerospace Center
DLR	German Aerospace
GIS	Geographical Information System

2.1 Introduction

The rapid global population growth with the consequent global urbanization and land use change are important drivers of several global societal and environmental issues such as economic development, climate change, and related resilience measures. In this context, an important aspect for understanding how urbanization drives economic development concerns the objective and uniform understanding of urban energy resource access. However, for most of the urban surface (particularly in the developing world), reliable statistics on how well energy resources are allocated are either non-existent, difficult to compile, unreliable, or inconsistent across cities and countries. For example, currently, there are no systematic attempts to rank, or benchmark, however, coarsely, *all* the world's major cities by their ability to "match" services such as access to lighting (or energy) to their existing population distribution. Such global benchmarking of urban areas is important on many dimensions, as the world is moving to cities: over half of the globe's population living in urban agglomerations as of 2008, producing over 80% of economic output globally, and consuming over 60% of the world's energy (Polly et al. 2016). Thus, better measurements and tools are needed to quantify the energy and environmental impact of the distribution and density of population and the built environment worldwide, which accounts for over 70% of global GHG emissions. Proposing a framework, dataset, and toolset for performing such a global-level analysis is one of the major contributions of this paper. In our task, we are aided by the emergence of two important trends in the last several years that promise to make this endeavor possible. First, there has been a dramatic increase in the availability of wide-area observational data on the Earth's urban areas acquired by remote sensing satellites, fueled primarily by the lowering costs of both sensor technology and of access to space. Second, the recent developments in machine learning and computing infrastructure have produced both powerful artificial intelligence algorithms for computer vision, and robust, high-performance, and open-source computational frameworks that lower the barriers for scientists for undertaking such an ambitious research task. While we believe that these more powerful computer vision and generative machine learning models can be of tremendous use in extracting insight and operationalizing the new remote sensing data on cities, our work presented here provides a necessary first-order, interpretable set of data for further analysis using more advanced algorithms. The first analysis by our team for the related problems of the complex spatial distribution of urban maps, and their relationship, using the remote sensing data and state-of-the-art generative machine learning models has been described in Albert et al. (2018).

Previous research has typically used small-scale, localized case studies of one or several cities. In one of the most comprehensive studies to date (Bertaud and Malpezzi 2003), Bertaud and Malpezzi analyze the average spatial profiles of population density for 48 cities worldwide from the perspective of the classic "exponential-decay" model of city density. They also offer compelling arguments on how the spatial distribution of population in certain developing regions has been heavily influenced by factors such as land use regulation and state intervention in the real estate markets. We build on this foundational work to extend the analysis to *all* of the over 2, 500 large metropolitan regions worldwide, and to show that, in many cases, the observed distribution of key macroeconomic variables describing a city—population, luminosity (a proxy for energy access and use), and building density—departs from the exponential decay model prevalent in the literature.

The macroeconomic quantities that we incorporate in our study: built environment, population density, and luminosity (as proxy for energy use or economic activity), are of key importance to a number of fields related to urbanization analysis, urban planning, and economic development studies.

The spatial distribution of the built environment has been the subject of much prior research as well. In Barrington-Leigh and Millard-Ball (2015), the authors analyze urbanization trends in the U.S. from the perspective of the road network (Barrington-Leigh and Millard-Ball 2015), and note that sprawl has peaked there around 1990. A recent study (Frolking et al. 2013) investigates the macro-scale changes in the urban structure worldwide, defined as the amount of horizontal change in cities (sprawl) compared to the amount of vertical change (indicating the concentration of urban build up areas). The authors note that cities have been developing both upwards and outwards, with different relative rates across different geographical regions. Nighttime luminosity is another major remote sensing data product that has been used in prior work to study economic activity and energy consumption and access. One of the first studies to make this link is Chen and Nordhaus (2011), where the authors use nighttime luminosity from NASA's VIIRS mission (NASA EOSDIS Land Processes DAAC, USGS Earth Resources Observation and Science (EROS) Center, Sioux Falls, South Dakota 2013) to recover several economic statistics such as economic output and its rate of change. More recently in Weidmann and Schutte (2017), luminosity is used to infer relative levels of wealth across different regions in several countries in Africa. Closer to the techniques proposed in this paper, a custom deep learning model was developed in Jean et al. (2016) to relate nightlights (as proxies of economic activity) to visible-band satellite imagery to infer poverty.

This paper showcases the first use of planetary-scale remote sensing data to analyze and benchmark "urban form" across all of the world's major cities. Here, we define *urban form* as the spatial distribution of the density of key macroeconomic variables describing the "vital signs" of a city: building infrastructure, population, and relative nighttime lighting levels (a proxy for access to and use of energy, economic activity, and wealth levels). This work provides a first step for the development of a new generation of urban economic analysis tools that can effectively leverage the vast, ever-increasing amount of observational data on the urban environment. Our contributions are as follows. First, we derive a global-scale dataset on the world's largest 30,000 metropolitan areas, by integrating *(i)* measurements of the spatial distribution of the built environment obtained from extensive analysis (Esch et al. 2013) of synthetic aperture radar (SAR) data obtained from the German Aerospace Center (DLR), *(ii)* a proxy for economic development and energy use from nighttime luminosity measurements obtained from NASA's Visible Infrared Imaging Radiometry Suite (VIIRS) mission, and *(iii)* population density maps obtained from the LandScan project (Oak Ridge National Laboratory 2014) developed at the Oak Ridge National Laboratory. Second, we introduce simple metrics that measure the *discrepancy* between spatial distribution of lighting and that of population density (as characterized by the average spatial profile), and the *effort* needed to match the two spatial distributions, that we use to rank and compare cities worldwide. Third, we use the spatial radial profiles of population density, built area density, and relative lighting levels to build a benchmarking model of cities' lighting profiles. We use this benchmarking model to compare cities against their "peers" in terms of patterns in population and built area density distributions and highlight the uneven geographical distribution worldwide of cities performance on the *discrepancy* and *effort* measures.

The paper is organized as follows. Section 2.2 present the data and the data preprocessing and the experimental setup, including the creation of the *CityNet* dataset. Section 2.3 formally introduces the problem of benchmarking urban form and lighting profiles, as well as the notation used throughout the paper. Section 2.4 presents our analysis on benchmarking city luminosity worldwide. Section 2.5 concludes the paper. All the codes and data are available at https://github.com/adrianalbert/urbanization-patterns.

2.2 The *CityNet* Dataset

2.2.1 Integrating Open-Source Remote Sensing Data Products on the Urban Environment

The "Global Urban Footprint" (GUF) Esch et al. (2017) is a novel dataset that maps the distribution of human settlements for the entire planet at an unprecedented resolution of $\sim$12 m/px. Here we show a first analysis aggregating this data at a coarser 750m/px. It has been obtained through extensive processing of several hundred TB of synthetic aperture radar (SAR) scenes obtained between 2011–2012 from the satellite missions TerraSAR-X and TanDEM-X operated by the German Aerospace Center (DLR). The main advantages of SAR (radar) sensing over visible-band data are that *(i)* it allows to detect vertical built structures and to distinguish them from roads or other pavement as opposed, and *(ii)* it works both during the day, as well as during the night, and is not affected by weather (such as clouds). The data consists of three classes: built areas, non-built land areas, and water areas.

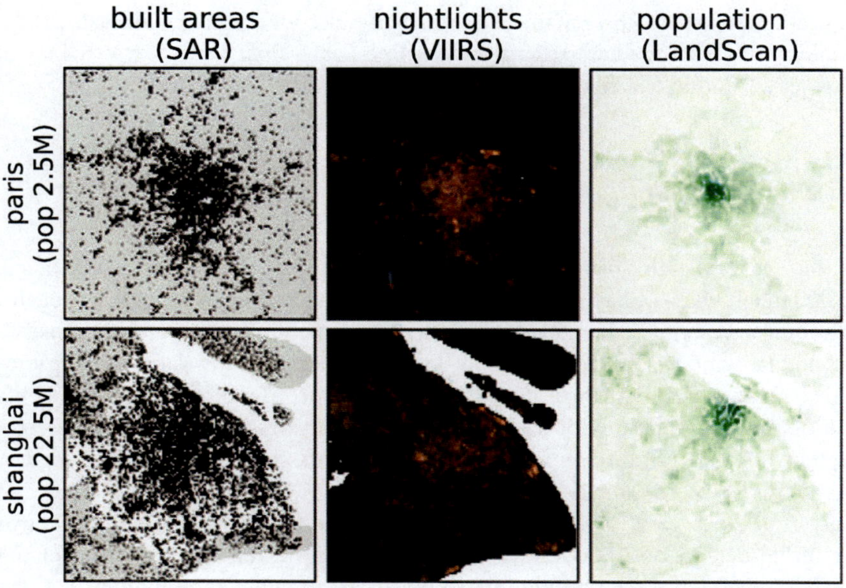

Fig. 2.1 Spatial maps (building, population, and luminosity density, columns) for two example cities in our analysis, Paris and Shanghai (rows). The luminosity and population density maps are shown on a logarithmic scale to improve contrast

Population density data. The *LandScan* data (Oak Ridge National Laboratory 2014) consists of population density estimates available worldwide at a $1km/px$ resolution. It has been produced yearly since 2000 at the Oak Ridge National Laboratory through a major modeling effort that used remote sensing imagery to disaggregate census surveys. For this analysis, we used the product for 2013. We preprocessed this data by setting obvious erroneous (negative or extremely large) values to 0 and transforming the data to log-scale. We also used the zero-values (prior to pre-processing) to derive a proxy mask of areas that cannot be developed, which we use later in the computation of the average radial profiles.

Nighttime luminosity data. The nightlights dataset has been obtained from NASA's Visible Infrared Imaging Radiometer Suite (VIIRS) (NASA EOSDIS Land Processes DAAC, USGS Earth Resources Observation and Science (EROS) Center, Sioux Falls, South Dakota 2013) satellite mission and consists of relative luminance values between 20:00 and 22:00 local time at a $750m/px$ resolution and on a scale from 0 (no lights) to 180. We used a yearly average (with further processing such as cloud removal) from 2013 for this analysis. As also this data has a highly-skewed distribution, we transform it to a log-scale.

Figure 2.1 presents several examples of our training set for the Shanghai and Paris metropolitan regions. For each city (column), we present the data sources used, from bottom to top: built areas, luminosity, and population density (the latter two on log-scale). Gray areas in the SAR images are water bodies.

In addition to these three main data sources, we also incorporate a GIS data product of the world's water bodies to extract masks m_i for each city $i = 1, \ldots, N$, to indicate whether a location can be developed or not, as described above in Sect. 2.3.

2.2.2 *Constructing an Analysis Dataset*

For the analyses in this paper, we chose to focus on the worldwide cities with at least 10,000 inhabitants—there are ~25,000 worldwide according to the UN. For each of these cities, we extracted samples of built areas, population density, and luminosity as rectangular windows of width $W = 200$ km around a city center (latitude, longitude) tuple.[1] Fixing a spatial scale of $W = 200$ km results in different image sizes for cities at different geographical locations, as this window size corresponds to different numbers of pixels at different latitudes on the Earth. For simplicity, we resize all images to 286×286 pixels. With these preparations, we create a final dataset for analysis of the largest 2,500 worldwide metropolitan areas, which we obtained from our initial dataset of 25,000 cities by first filtering out the cities with fewer than 100,000 inhabitants, then pruning the obtained list to remove those cities that were too close to each other (within a radius of 100 km). We obtained then a more spatially-uniform dataset of ~2,500 large metropolitan areas is shown on panel *c*) in Fig. 2.2.

2.3 Problem Formulation

We now formalize the problem of studying and benchmarking urban form and introduce the notation used in this paper.

Urban form. We observe spatial maps $\mathbf{x}^i$ for $i = 1, \ldots, N$ cities, each of size $W \times W \times S$, where S is the number of data sources available, as well as corresponding binary masks m_i, with $m_i(x, y) = 1$ if the land from city i at location (x, y) can be developed (here, we use water areas as our mask). In our analysis, we employ a dataset of the N ~30,000 largest cities globally to illustrate the patterns in spatial profiles characterizing the urban form. Based on the discussion in Sect. 2.2, here we have $W = 286px \times 286px$. S refers to the number of data sources, which are assumed distinct, if not independent, in the sense that each brings unique, interpretable information that is not contained in the others (as such, we expect significant correlations between data

[1] We chose this value as the square containing a circle of radius 100 km around the economic and administrative center of each city. This value for the radius has been chosen as the maximum distance most people would be willing to commute for work (a one-hour commute driving at 60 *mph*), see Dash Nelson and Rae (2016).

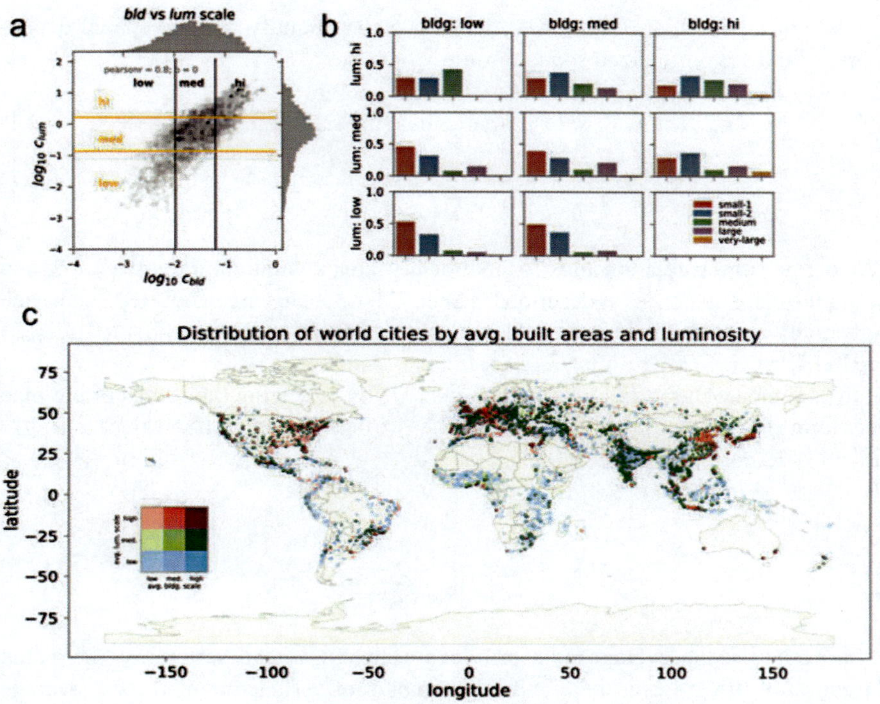

Fig. 2.2 The worldwide distribution of the city scale parameters **c** (average value) for relative luminosity and built area density: **a**—a segmentation of world cities by low/medium/high values of c_{lum} and c_{bld}, respectively. Note there are cities that have, for example, low luminosity levels and medium built area density levels; **b**—a further refinement of the 9 segments of cities by the city size (by population). Note the uneven distribution of cities in each of the classes that indicate the great variety of urban areas worldwide; **c**—the geographic distribution of the 9 segments of cities (color-coded) by their average luminosity and built area density. Note the specific clustering of cities clearly identifying major regions such as Sub-Saharan Africa, China, the U.S. East and West Coasts, Western and Eastern Europe, etc.

sources). Here, the number of channels is $S = 3$, corresponding to population density x_{pop}, nighttime luminosity x_{lum}, and building density x_{bld} (we drop the superscript i for simplicity when referring to a given city). With this, for a given city, we define the collection of spatial maps $\mathbf{x} \equiv [x_{pop},\ x_{lum},\ x_{bld}]$ as the *urban form* of the city.

The scale and spatial profile as "vital signals" of urban form. As the example cities in Fig. 2.1 illustrate, urban form can be tremendously complex, making them difficult to model parametrically. Past work has typically employed simplified, easily-interpretable representations of spatial distributions **x**. An often-used statistical representation of urban spatial maps is the *spatial profile* $x(d)$ with distance d from the city center (see, e.g., in Bertaud and Malpezzi (2003); Louf and Barthelemy (2016)). Here, we posit that urban form may be described by two key quantities, the *scale c* of a particular indicator (which is a scalar quantity with no spatial dependence), and the normalized spatial profile $\hat{x}_{\cdot}(d)$

$$x_{pop}(d) = c_{pop} \cdot \hat{x}_{pop}(d), \text{ where the scale parameter is} \tag{2.1}$$

$$c_{pop} = \frac{1}{\sum_{u,v} m(u,v)} \sum_{u,v} x(u,v) \cdot m(u,v) \tag{2.2}$$

We define similar relationships for luminosity `lum` and building density `bld`, and compute these quantities as described in Sect. 2.2. As such, we can now represent each given city in a simplified way by the quantity $(\mathbf{c}, \hat{\mathbf{x}}(d))$, with $\mathbf{c} \equiv (c_{bld}, c_{pop}, c_{lum})$ and $\hat{\mathbf{x}}(d) \equiv (\hat{x}_{bld}(d), \hat{x}_{pop}(d), \hat{x}_{lum}(d))$.

We compute the *radial average profile* $x(d)$ by averaging the values of the map x within rings of width Δd at at a distance d from the center, i.e., values $x(u, v)$: $(u, v) \in \mathcal{R}(d)$, with $\mathcal{R}(d) \equiv \{(u, v)|(u - u_0)^2 + (v - v_0)^2 > d^2$ and $(u - u_0)^2 + (v - v_0)^2 \leq (d + \Delta d)^2\}$

$$x(d) \equiv \frac{1}{|\mathcal{R}(d)|} \sum_{(u,v)\in\mathcal{R}(d)} x(u,v) \tag{2.3}$$

We note also that average radial profiles are but first-order descriptors of the spatial structure of urban form; there is a large amount of variance around these average spatial profiles that are not captured in $x(d)$ as it is clear from the varied spatial distributions observed for real cities as in Fig. 2.1. In this paper, we, however, do not address this aspect and focus on $x(d)$, as it is the typical quantity studied in the urban development literature.

Comparing spatial profiles: *discrepancy* and *effort*. To study the "mismatch" between a city's spatial distribution of population density and relative lighting levels, we introduce the two related measures of *discrepancy* and *effort*. We define discrepancy $\hat{\delta}$ as

$$\hat{\delta} \equiv \int_0^L (\hat{x}_{lum}(d) - \hat{x}_{pop}(d))\delta d, \tag{2.4}$$

i.e., the area under the graphs of $\hat{x}_{pop}(d)$ and $\hat{x}_{lum}(d)$, with $\Delta d = 0.75km$. This is illustrated in the right panel in Fig. 2.3. Positive values of t he discrepancy measure ($\hat{\delta} > 0$, shown as blue-shaded regions) indicate, intuitively, that there is an excess of lighting infrastructure and/or energy use relative to the population profile, while a negative value of discrepancy ($\hat{\delta} < 0$, shown as red-shaded regions) indicate a lack of access to lighting and energy infrastructure relative to the existing population distribution.

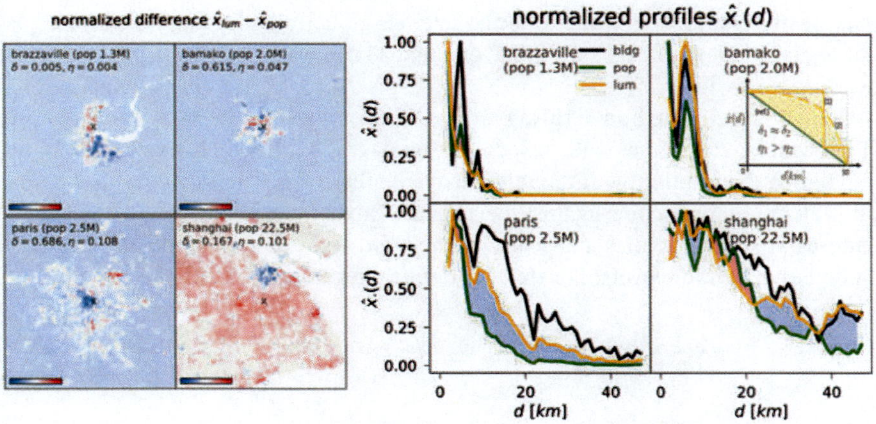

Fig. 2.3 Comparing the relative mismatch in the spatial distributions of city luminosity, population, and built areas density. *Left*: spatial maps of the normalized difference $\hat{x}_{lum} - \hat{x}_{pop}$ for four example cities: Brazzaville, Bamako, Paris, and Shanghai. The blue regions indicate locations with higher lighting levels relative to population density, whereas red regions indicate the opposite; *Right*: normalized spatial profiles for built areas density (black), relative luminosity (yellow), and population density (green) for the four example cities

As a related, but a distinct measure of mismatch between the lighting and population distributions for a given city, we define the *effort* $\hat{\eta}$ as

$$\hat{\eta} \equiv EMD(\hat{x}_{lum}(d), \hat{x}_{pop}(d)), \quad (2.5)$$

where $EMD(p, q)$ denotes the Earth Mover's Distance (Hastie et al. 2009). The Earth Mover's Distance (or Wasserstein-1 metric) is a measure of the distance between two distributions p and q, defined over the same domain D, that encodes the minimum cost of transporting mass ("earth") on D between the two distributions such that p matches q. It is a frequently-used distance measure in computer vision to compare histograms of pixel values in images, for example, image retrieval applications. Intuitively, η summarizes not just the average relative mismatch between the two spatial distributions as δ, but also captures how this mismatch is distributed in space with respect to the city center.

We note that a major drawback of these simple measures is that both implicitly make the much simplifying assumption that there is a *linear* relationship between the relative population density and relative luminosity levels for a given city. There are a host of reasons why that might be grossly inaccurate: the benefits in terms of economic activity from higher population density concentrations are super-linear and generally follow power laws (Gomez-Lievano et al. 2016), residential regions in a city with higher population densities may have lower relative illumination levels than commercial districts where people work during the day (Louf and Barthelemy 2016), etc. However, we believe that even these coarse measures are informative and useful in their simplicity because they allow a quick, consistent comparison and ranking of lighting infrastructure access in cities globally, and as Fig. 2.4 shows, give rise to clear geographical patterns that distinguish certain developed from developing regions worldwide.

Benchmarking urban lighting access. Since we are interested in comparing cities against each other with respect to the degree with which they "match" the population distribution with the relative availability of lighting, we propose a simple way to compare cities against their "peers", understood as cities having similar patterns of spatial profiles and scales of the build areas density, population density, and relative lighting levels. For this, we define a regression model of the form

$$\log \hat{x}_{lum}(d) \sim \log \hat{x}_{pop}(d) + \log \hat{x}_{bld}(d) + d + \text{region} + \text{subregion} + c_{bld} + c_{lum}. \tag{2.6}$$

In practice, we express this regression model using a random forest (Hastie et al. 2009) implementation available from the `Python` package `scikit-learn` (Pedregosa et al. 2011). For a given set of inputs (spatial profiles and scale parameters of built areas density and population, geographical region and subregion), the model then estimates the (log-scale) *benchmark* relative luminosity levels $\tilde{x}_{lum}(d)$ at a given distance d from city center. We can now compare the benchmark luminosity profile $\tilde{x}_{lum}(d)$ with the observed luminosity profile $\hat{x}_{lum}(d)$ and compute metrics of discrepancy and effort as above.

2.4 Results: Benchmarking City Lighting Profiles

2.4.1 Classes of Cities by Urban Form: The Scale and Spatial Profile Worldwide

As first-order descriptors of urban form, we consider the scale parameters of the three globally-monitored quantities we use in our analysis: the average relative lighting c_{lum}, the total population c_{pop}, and the average built area density c_{bld}. We present an analysis in Fig. 2.2. There, in panel c in the figure, we show the distribution of worldwide cities by (c_{lum}, c_{bld}), i.e., the dependence of average relative luminosity with the average built areas density. As expected, the more built-up a city is, the higher the lighting level, with a moderate degree of spread. We divided up the set of cities into quartiles of the respective distributions of c_{lum} and c_{bld} as marked on the figure with the low/medium/high, resulting in 9 city classes. Note that there are cities with low luminosity but medium levels of the built-up density—which offers a coarse indication about the "inequality" across the world's cities in terms of access to lighting (and, by extension, energy) infrastructure. That picture of unequal access and usage levels is reinforced in panel b in the figure, where we illustrate the distribution of cities in the 9 categories we consider according to their size in terms of total population (which we denote by c_{pop}). For example, most cities that have low levels of built-up area density but high level of luminosity are also mid-sized in terms of population, which may indicate cities in the developed world with both high access to resources (high luminosity) and low density levels (sprawled areas). The general trend to highlight in panel *b* is that the distributions of cities by population in each of the classes are generally uneven—which indicates a preference of cities of certain sizes towards certain levels of lighting and built area densities. This uneven distribution is even more apparent when mapped out geographically, as we show in panel *c* of Fig. 2.2. There, we color-coded the low/medium/high levels of luminosity by hues (blue/green/red), whereas the saturation encodes the low/medium/high levels of built area density for a given city. Clear geographical patterns emerge: sub-Saharan Africa is dominated by cities of low built-up density and low luminosity levels, whereas certain cities in South Africa, Northern Africa, and on the coasts are more developed under these criteria. The U.S. coasts are clusters of high-luminosity, high built density cities, whereas the heartland has more spread-out cities of high-luminosity but low built area density.

Figure 2.3 provides a closer look at a few example cities, and illustrates the spatial distribution of the relative population, built area, and luminosity density with distance from the official city center. We show four example cities: Brazzaville (Nigeria), Bamako (Mali), Paris (France), and Shanghai (China). The left set of panels in the figure shows the distribution of relative luminosity levels compared to the relative levels of population density, where red areas indicate locations where relative population levels are higher than those of luminosity (blue areas represent the inverse of that). On the right set of panels, we show the average radial profiles for each respective city for population density $\hat{x}_{pop}(d)$ (green), relative luminosity $\hat{x}_{lum}(d)$ (yellow), and built areas density $\hat{x}_{bld}(d)$ (green). We also indicate the values of the measures previously introduced of discrepancy δ and effort η in the left panels in the figure. This picture allows to contrast different regions within a city with respect to the usage of/access to lighting: in Brazzaville, there are two parts of the city, one more populous than luminous above the body of water, and another where this situation is inverse south of the river (a mainly industrial area and another mainly residential area); certain neighborhoods of Bamako are relatively better illuminated than others relative to the population density; in Paris, the core of the city is strongly illuminated relative to population density, whereas the periphery much less so; in Shanghai, the city center is again very populous relative to illumination levels, whereas the city stretches for miles of well-lit areas.

2.4.2 *Comparing Cities Worldwide by Relative Lighting Access Levels*

With these observations and tools in hand, we analyze the geographic distribution of levels of lighting relative to population density using δ and η. The analysis is summarized in Fig. 2.4 where we mapped and plotted the distribution of δ and η by macro regions, namely: Asia, Africa, Americas, Europe, and Oceania, and by urban

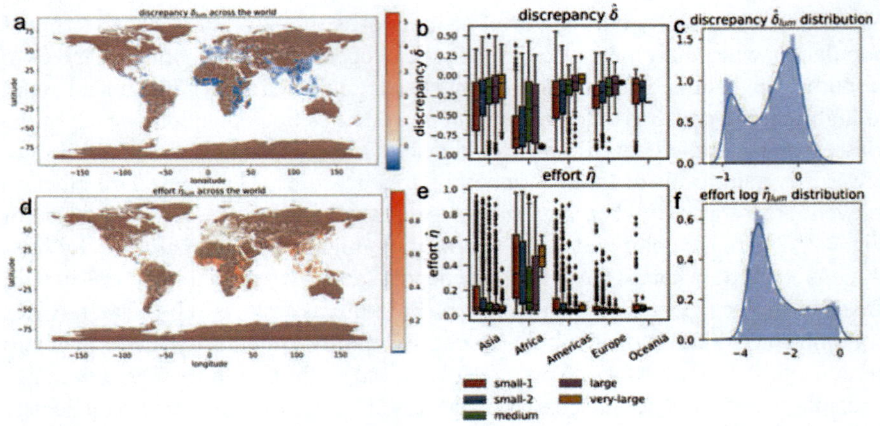

Fig. 2.4 The distribution across the world of the mismatch between the spatial distribution of lighting levels and population density as measured by the discrepancy $\hat{\delta}$ and effort $\hat{\eta}$ measures: **a**—the geographic distribution of $\hat{\delta}$ for world cities; **b**—the worldwide distribution of $\hat{\delta}$ by city size and major geographic region. Note the larger negative values for cities in sub-Saharan Africa, South and East Asia compared with those of more developed parts of the world also apparent in panel **a**; **c**—the overall distribution of $\hat{\delta}$. Panels **d–e** show similar plots for the effort $\hat{\eta}$

classes as previously defined by city size. It is possible to clearly observe an uneven distribution of δ and η with special regards to Africa that presents a considerable difference both on its overall distribution (Fig. 2.4b, e), and in its internal distribution by macro urban classes. In fact, for all macro regions a clear pattern emerges in which larger the city, bigger the discrepancy, and lower the effort, meaning that bigger town have better access to illumination and energy. While for Africa, this pattern is not valid, first the second moments are much wider than other macro areas, indicating internal inequalities of resource distributions, second the trend for larger cities seems to follow an opposite behavior in which larger the cities bigger the effort and lower the discrepancy. In a way, for African's larger cities, urbanization does not follow access to light and possibly economic development. Beyond the evident causality for such mismatch, due probably to the extremely rapid and informal urbanization process coupled with an evident lack of resources, it may be possible that African urbanization fosters dis-economy forces which breaking the agglomeration economy laws. Other observations can also be done for Europe, in which first and second moments are similar across urban classes indicating homogeneous and size-independent access to lights.

For a more direct comparison of the spatial distribution of relative lighting levels in cities worldwide, we employ the simple regression model in Sect. 2.3 to build benchmark city lighting models. As discussed in that section, we consider "peer" cities by patterns in the average levels and spatial distribution of population density, built area density, and geographic region to build benchmark profiles of relative lighting levels with distance from the city center, and compare with the actual spatial distribution (encoded in the average radial profile) of the lighting levels. We present

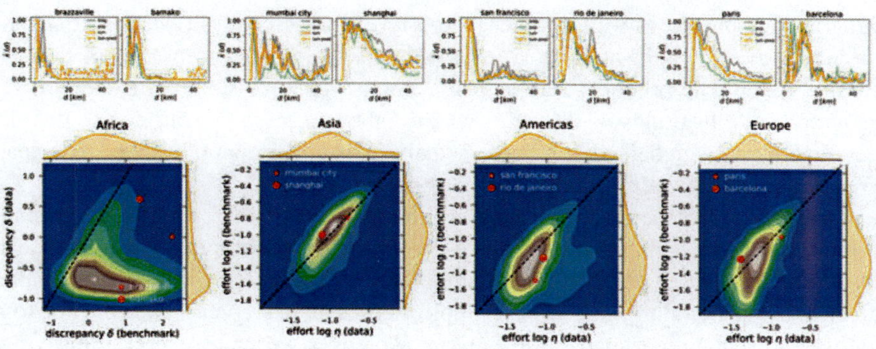

Fig. 2.5 Benchmarking the spatial distribution of city luminosity for the four major geographic regions across the world. We compare the discrepancy $\hat{\delta}$ between the model-predicted luminosity profile $\tilde{x}_{lum}(d)$ and the observed luminosity profile $\hat{x}_{lum}(d)$ for cities in Africa, Asia, the Americas, and Europe as a density contour plot. Note the specific patterns for the different regions in the density of cities in the $(\tilde{x}_{lum}(d), \hat{x}_{lum}(d))$ space. We plot two example cities for each region, and show on the top panels the luminosity profiles for each city, predicted (dashed line) and actual (solid line)

this analysis in Fig. 2.5 for each of the four major geographical regions of the world (Africa, Asia, the Americas, and Europe). There, the bottom row shows the density distribution of the (log-scale) effort measure η computed for the benchmark lighting profile (y-axis) against the value computed for the observed lighting profile (x-axis). We pick two major cities in each geographical region (indicated by red marks in each panel of the figure) and show the benchmark and actual lighting profiles in comparison with the population density profile in the top panel corresponding to each geographical region. At a high level, the density regions in the $(\tilde{\eta}, \eta)$ space indicates that cities vary greatly with respect to their "ability" to match their population density profile with the energy/lighting usage and access profiles. This provides a quick, intuitive visual benchmarking of where a given city is situated with respect to its "peers" in terms of the relative effort η to match its population distribution profile. For example, in Europe, Paris performs marginally worse than Barcelona with respect to the benchmark, whereas in the Americas San Francisco is much better than Rio de Janeiro.

2.5 Conclusion

This paper benchmarks urban lighting access at a planetary scale using remote sensing and geographic information systems (GIS) data, and simple machine learning techniques. We have focused our analysis using the classic radial profile model of urban form to show that cities globally show characteristic levels of mismatch between observed population distribution and relative lighting levels. These patterns vary by geographical region, as well as other macro-level characteristics (average luminosity levels, average built-up areas), observable via the global inventories of remote sensing-based data products that we use. Notably, the analysis highlights intuitive geographical clusterings of the cities where lighting access (by our coarse definition that allows a global, comprehensive analysis) is either high (Northern America, Western Europe, parts of Asia) or low (Sub-Saharan Africa).

In providing a planetary-level analysis based on globally available remote sensing data products, we had to make key trade-offs in the choice of measures and models to employ. We settled for the average radial model of urban form, widely used in the urban analysis literature, and defined simple metrics of *discrepancy* and *effort* on top of this representation of spatial distributions. Moreover, we made several simplifying assumptions: that "peer" cities are comparable by a few characteristics (average and spatial distribution of population density and built areas, geographical region); that the average radial profile is in some sense "sufficient" to describe complex spatial distributions; that our simple measures offer enough discriminatory power to reveal global patterns in access to lighting. Certainly, these much simplify the analysis and do not capture most of the complexity in the real data, but our focus was on providing a blueprint for comparing urban form and lighting access globally.

The analysis is made possible by a new remote sensing derived dataset on the urban form that we compiled on all major cities in the world that we call *CityNet*. We make this dataset available for other researchers to use[2] for further discussion on computer vision methods for studying cities.

References

Albert A, Strano E, Kaur J, Gonzalez M (2018) Modeling global urbanization patterns with generative adversarial networks. In: To appear in IEEE international geosciences and remote sensing symposium. Valencia, Spain

Barrington-Leigh C, Millard-Ball A (2015) A century of sprawl in the United States. Proc Natl Acad Sci 112(27):8244–8249. https://doi.org/10.1073/pnas.1504033112. https://www.pnas.org/content/112/27/8244.full.pdf

Bertaud A, Malpezzi S (2003) The spatial distribution of population in 48 world cities: implications for economies in transition. Technical report

Chen X, Nordhaus WD (2011) Using luminosity data as a proxy for economic statistics. Proc Natl Acad Sci 108(21:8589–8594. https://doi.org/10.1073/pnas.1017031108, arXiv:www.pnas.org/content/108/21/8589.full.pdf

Esch T, Heldens W, Hirner A, Keil M, Marconcini M, Roth A, Zeidler J, Dech S, Strano E (2017) Breaking new ground in mapping human settlements from space-the global urban footprint. ISPRS J Photogramm Remote Sens 134(2017):30–42

Esch T, Marconcini M, Felbier A, Roth A, Heldens W, Huber M, Schwinger M, Taubenböck H, Müller A, Dech S (2013) Urban footprint processor x2014; Fully automated processing chain generating settlement masks from global data of the TanDEM-X mission. IEEE Geosci Remote Sens Lett 10(6):1617–1621. 1545-598X https://doi.org/10.1109/LGRS.2013.2272953

[2] https://github.com/adrianalbert/urbanization-patterns/.

Frolking S, Milliman T, Seto KC, Friedl MA (2013) A global fingerprint of macro-scale changes in urban structure from 1999 to 2009. Environ Res Lett 8(2):024004. http://stacks.iop.org/1748-9326/8/i=2/a=024004

Gomez-Lievano A, Patterson-Lomba O, Hausmann R (2016) Explaining the prevalence, scaling and variance of urban phenomena. arXiv:physics.soc-ph/1604.07876

Hastie T, Tibshirani R, Friedman J (2009) The elements of statistical learning: data mining, inference, and prediction, 2nd edn. Springer New York. 9780387848587 2008941148 https://books.google.com/books?id=tVIjmNS3Ob8C

Jean N, Burke M, Xie M, Davis WM, Lobell DB, Ermon S (2016) Combining satellite imagery and machine learning to predict poverty. Science 353(6301):790–794

Louf R, Barthelemy M (2016) Patterns of residential segregation. PLOS ONE 11(6):1–20. https://doi.org/10.1371/journal.pone.0157476

NASA EOSDIS Land Processes DAAC, USGS Earth Resources Observation and Science (EROS) Center, Sioux Falls, South Dakota (2013) VIIRS. https://lpdaac.usgs.gov

Nelson GD, Rae A (2016) An economic geography of the United States: from commutes to megaregions. PLOS ONE 11(11):1–23. https://doi.org/10.1371/journal.pone.0166083

Oak Ridge National Laboratory (2014) LandScan global population dataset. Oak Ridge, Tennessee

Pedregosa F, Varoquaux G, Gramfort A, Michel V, Thirion B, Grisel O, Blondel M, Prettenhofer P, Weiss R, Dubourg V, Vanderplas J, Passos A, Cournapeau D, Brucher M, Perrot M, Duchesnay E (2011) Scikit-learn: machine learning in python. J Mach Learn Res 12(2011):2825–2830

Polly B, Kutscher C, Macumber D, Schott M, Pless S, Livingood B, Van Geet O (2016) From zero energy buildings to zero energy districts. ACEEEE Summer Study Energy Effic Build

Weidmann NB, Schutte S (2017) Using night light emissions for the prediction of local wealth. J Peace Res 54(2):125–140. https://doi.org/10.1177/0022343316630359, arXiv:10.1177/0022343316630359

Chapter 3
Object-Oriented Approach for Urbanization Growth by Using Remote Sensing and GIS Techniques: A Case Study in Hilla City, Babylon Governorate, Iraq

Ammar S. Mahmoud, Bahareh Kalantar, Husam A. H. Al-Najjar, Hossein Moayedi, Alfian Abdul Halin, and Shattri Mansor

Abstract High rate of urbanization coupled with population growth has led to unexpected land use and land cover changes in Hilla city, which is located in the Babylon governorate of Iraq. Understanding and quantifying the spatiotemporal dynamics of the urban land use and land cover changes, as well as the driving factors behind them, are therefore vital in order to design appropriate policies and monitoring mechanisms

A. S. Mahmoud
Faculty of Engineering, Department of Civil Engineering, University of Babylon, Hillah city 51002, Iraq

B. Kalantar (✉)
Goal-Oriented Technology Research Group, Disaster Resilience Science Team, RIKEN Center for Advanced Intelligence Project, Tokyo 103-0027, Japan
e-mail: Bahareh.kalantar@riekn.jp

H. A. H. Al-Najjar
Faculty of Engineering and IT, Centre for Advanced Modelling and Geospatial Information Systems (CAMGIS), University of Technology Sydney, Ultimo, NSW 2007, Australia
e-mail: Husam.AL-NAJJAR@student.uts.edu.au

H. Moayedi
Department for Management of Science and Technology Development, Ton Duc Thang University, Ho Chi Minh City, Vietnam

Faculty of Civil Engineering, Ton Duc Thang University, Ho Chi Minh City, Vietnam

H. Moayedi
e-mail: hossein.moayedi@tdtu.edu.vn

A. A. Halin
Faculty of Computer Science and Information Technology, Department of Multimedia, Universiti Putra Malaysia, Serdang 43400, Selangor, Malaysia
e-mail: alfian@ieee.org

S. Mansor
Faculty of Engineering, Department of Civil Engineering, Universiti Putra Malaysia, Serdang 43400, Selangor, Malaysia
e-mail: shattri@upm.edu.my

P. Sharma (ed.), *Geospatial Technology and Smart Cities*, The Urban Book Series,
https://doi.org/10.1007/978-3-030-71945-6_3

to govern urban growth. This study analyzes land use and land cover changes over Hilla city through remote sensing and GIS (Geographical Information System) techniques. IKONOS satellite imagery from years 2000, 2005, and 2011 was collected and pre-processed using ENVI and ArcGIS, which then goes through an object-based supervised image classification stage to generate land use and land cover maps. The classification is performed using the statistical machine learning algorithm, SVM (Support Vector Machine). The confusion matrix and kappa coefficients are used to evaluate the overall accuracy of the results. The statistical results obtained enable assessment of class changes from years 2000 to 2011 and also identify the gain and loss of the built-up areas in relation to other land cover classes. The results also allow assessment of the spatial trend of these built-up areas. Ultimately, forecasts can be made to predict expected future class changes in 2026 and 2036. Generally, the results of this study show increased expansions of built-up areas, i.e., from 8.14% in 2000 to 14.53% in 2005 and up to 18.36% in 2011. All this was at the expense of bare land areas. Simultaneously, there was an increased expansion of vegetation/agricultural land area, specifically from 36.14% in 2000 to 41.71% in 2005 and 45.13% in 2011. The spatial trend also shows that the growth of built-up areas is focused in the south-western part of Hilla city. In all, we foresee that the findings of this study can provide a good visual resource for decision-makers to perform more efficient urban planning.

Keywords Urbanization · Land use dynamics · Spatial trend · Built-up areas · Support vector machine

Acronyms

CSO	Central Statistics Office
QUAC	Quick Atmospheric Correction
GIS	Geographic Information System
HRSI	High-resolution satellite imagery
OBC	Object-oriented classification
RS	Remote Sensing
SVM	Support Vector Machine
UN	United Nations
UTM	Universal Transverse Mercator
WGS	World Geodetic System

3.1 Introduction

By 2017, Iraq was reported to have a population of over 38.27 million. Ranked as the fourth largest populated country in the Middle East, Iraq is not exempt from the phenomena of urban growth. Urbanization is occurring at a rapid pace throughout developing/third world countries during the last few years (Chauvin et al. 2017). According to the United Nations (UN), urban populations around the world are expected to increase to an astounding 6.3 billion over the next four decades, which is an increase from 3.6 billion in 2011 (UNFPA 2013). UN studies have also shown the proportions between the urban growth in Iraq by region of Asia and western Asia (Fig. 3.1). According to Rana (2011), rapid urban growth could occur due to two reasons, namely (i) higher birth rates compared to death rates and (ii) urban migrations. With regards to the latter, Barbero-Sierra et al. (2013) also mention pull (e.g., health care and education) and push (e.g., disaster and droughts) factors. Other factors such as physical and geological, economy, transportation, population, policy, and planning (Xi et al. 2010) also contribute to growth. The main downside is when urban growth is unplanned. This can lead to damage or overuse (or even abuse) of agricultural/vegetation lands. Since vegetation can be a scarce resource in many Middle Eastern nations, monitoring its use is crucial.

Change detection is one way for monitoring the loss of land, where vegetation/agricultural land change detection is one of the major focus areas in urbanization. Geographic Information System (GIS) and Remote Sensing (RS) offer critical tools and possible solutions for applying and analyzing change detection in this particular context. Obtaining valuable information from RS imagery can be done via several digital analysis techniques such as image classification, image transformation, and change detection (Jensen 2005). GIS data can also be integrated with RS in order to include biophysical and socioeconomic factors data, which can aid with the recognition of forces driving urban growth (Burgi et al. 2004). Spatial analysis

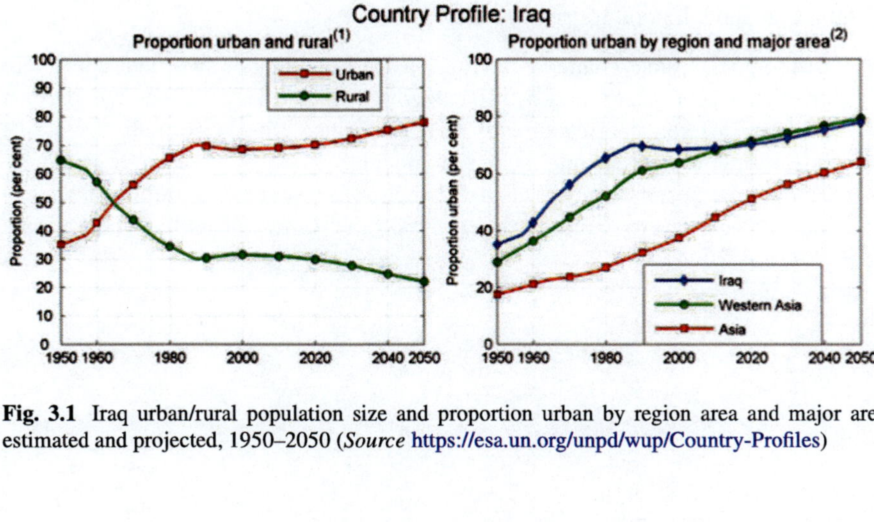

Fig. 3.1 Iraq urban/rural population size and proportion urban by region area and major area estimated and projected, 1950–2050 (*Source* https://esa.un.org/unpd/wup/Country-Profiles)

and modeling describe complex configurations of the objects or events using a mathematical equation that is utilized to discover the essential dynamics and complicate urban structures and evaluate diverse situations for future urban growth (Pijanowski et al. 2005). There are two main reasons for integrating RS and GIS, which are (i) GIS is based on the most currently sourced data (ii) GIS is more effective in data inventory and information management compared to spatial modeling.

3.1.1 Remote Sensing Sensors for Urban Growth

Various sensing equipment such as Landsat (TM & ETM+), ASTER, IKONOS, GeoEye, Quick bird, RapidEye, WorldView can be to obtain imagery data for RS. The main difference between these equipment are their spatial resolution, which can range from medium to high to very high-resolution data. The type of imagery is selected based on the user's need, the scale and characteristics of the study area, and the analyst's knowledge in using the selected imagery. Urban growth analysis, for example, requires scale classification at a local level. Therefore, high spatial resolution data such as IKONOS and SPOT 5 HRG data are preferred. At a regional scale, medium spatial resolution data such as Landsat TM/ETM+, and Terra ASTER would be more suitable. At a continental or global scale, coarse spatial resolution data such as AVHRR, MODIS, and SPOT Vegetation are preferable (Lu and Weng 2007).

3.1.2 Methods and Approaches for Change Detection

Change detection algorithms can be used to detect urban growth and to analyze the results of the changes. These algorithms can be classified into two categories: bi-temporal and temporal trajectory analysis.

i. Bi-temporal image change detection basically compares two datasets. There are three approaches under this category, which are (i) direct comparison between different datasets directly, (ii) comparing derived information, and (iii) integrating all datasets into a uniform model.
ii. Temporal trajectory analysis change detection focuses on determining the orientation of change by building the 'curves' or 'profiles' of multi-temporal data (Lu et al. 2004).

3.1.3 *Impact of Urbanization Expansion on Agricultural/Vegetation Area*

Urban growth has a great effect on the change of land use and land cover in many areas around the world, especially in developing/third world countries. Unprecedented population growth combined with unplanned development activities (residential, commercial, etc.) can cause damage to agricultural/vegetation lands. Other possible concerns include threats to natural and environmental resources (Raddad et al. 2010; Rana 2011). Moreover, since urbanization can involve natural and agricultural lands being converted to urban land, this can lead to environmental and ecological problems in both the urban and rural environments (Raddad et al. 2010).

Within the last four decades, Iraq has been embroiled in continuous conflict. Events such as the Iran–Iraq War (1980–1988), the Gulf War (1990–1991), and economic sanctions (1990–2003) have inversely affected urban growth in various governorates across Iraq. This study will focus on one of the governorates, namely Babylon. Among the main motivations for this study are

i. The local government lacks updated maps and information regarding the distribution of land use/cover (LULC)
ii. There is no objective assessment for urban growth in Babylon from years 2000 to 2011, and
iii. To develop the geospatial information system for Babylon by analyzing the effect of urban growth on the vegetation such as agriculture sector in the future.

Note that the scope of this study only involves built-up areas. The research questions being asked are

i. Were there any major changes in built-up areas within the study periods?
ii. How did each land cover class contribute towards the growth of built-up areas?
iii. What will be the trends of urban growth be after 20 years?

Specifically, the aim of this chapter is to give a good understanding of urbanization growth by Using Remote Sensing and GIS Techniques at Hilla City, Babylon Governorate, Iraq. This objective is to be achieved through analyzing collected data as follows: (i) to extract the thematic maps for the study area. (ii) To identify the urbanization expansion area. (iii) To assess the impact of the urbanization growth on the vegetation/agricultural lands in order to predict how the urban growth will be expanded at the expense of agricultural lands up to the year 2026 and 2036.

3.2 Study Area and Dataset

The study area chosen is Hilla city, which is the capital of the Babylon governorate. It is located in the central part of Iraq on the Euphrates River, adjacent to the ancient city of Babylon. The geographical location is 32° 40′ 00″ N and 44° 35′ 00″ E,

Table 3.1 Total population of Hilla city 1987–2014

Year	Urban	Rural	Total
1987	217,902	50,932	268,834
1997	259,499	90,221	349,720
2010	393,919	114,418	508,337
2014	424,065	124,531	548,596

Source Central Statistics Office (CSO) 2015

which is also surrounded by the governorates of Baghdad, Wassit, Qadissiya, Najaf, and Karbala. Hilla city is situated in a predominantly agricultural region. It is well irrigated by the Hilla River, which is one of the branches of the Euphrates River. This region is one of the most important agricultural areas of Iraq and has very fertile soil. The organic matter content of the topsoil is normally low. The climate is semi-arid with mean annual rainfall ranging from less than 50 mm to roughly 200 mm. The temperature in summer can reach a high of 50 °C during the daytime. Geologically, the study area is characterized by Holocene sediments, which mainly consist of clay, silt, and sand as typical sediments of flood plain systems (Manii 2014). The first national population and housing census were conducted in 1987 where the reported population of Hilla city was 268,834. Ten years later in 1997, the second census reported a total population increase to 349,720. The total population has since ballooned to 508,337 in 2010 and to 548,596 in 2014 (Table 3.1).

In this study, IKONOS imagery was acquired in the same season and the same level of resolution for the periods of 2000, 2005, and 2011 (as illustrated in Table 3.2). The images were obtained from the local government of Babylon (Fig. 3.2). The spatially referenced in the Universal Transverse Mercator (UTM) projection with datum World Geodetic System (WGS) 1984 UTM zone 38N are used for these images. The properties of each IKONOS source are summarized in Table 3.2.

Table 3.2 The characteristics of IKONOS satellite data used in this study

Sensor	Acquisition time	Spatial resolution (m)	Producer
IKONOS	25/05/2000	1	Babylon Governorate
IKONOS	12/06/2005	1	Babylon Governorate
IKONOS	27/06/2011	1	Babylon Governorate

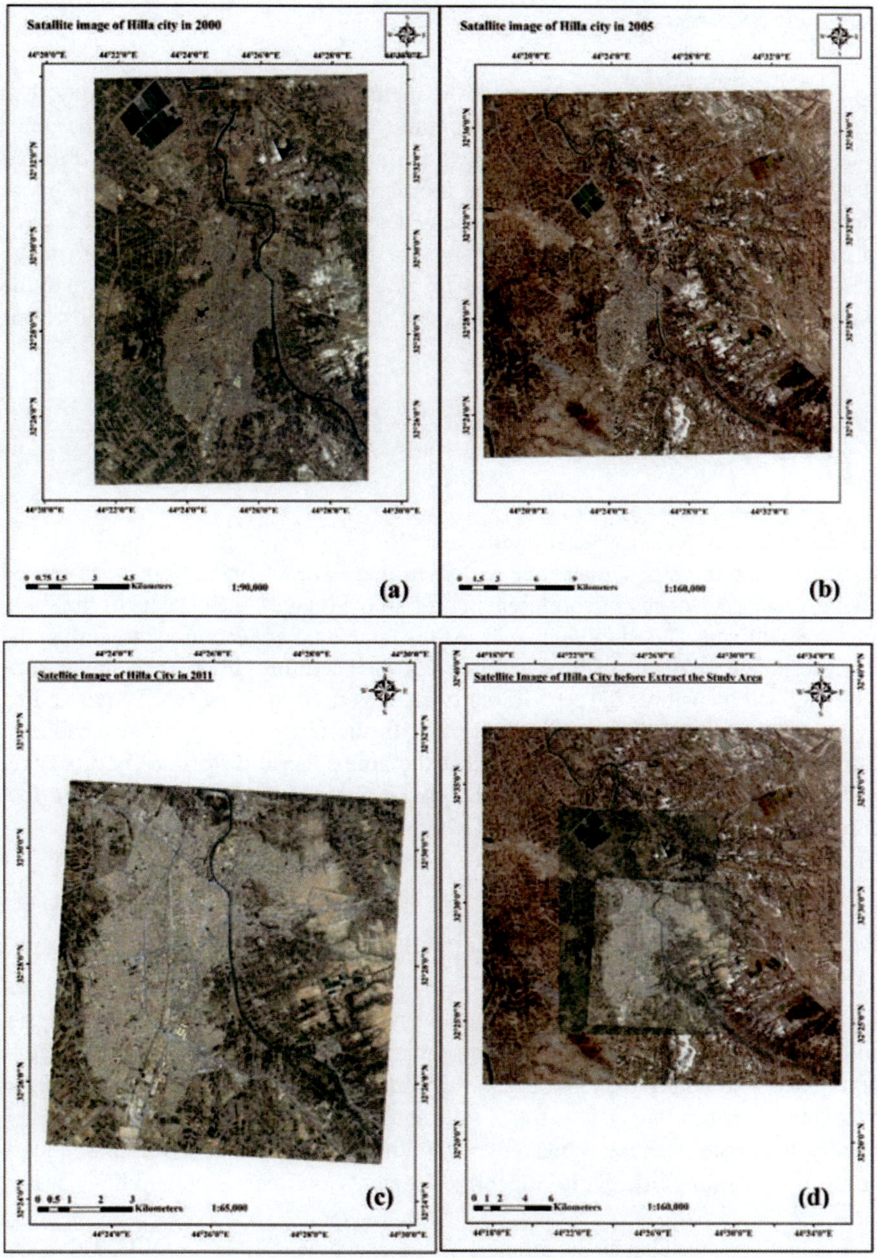

Fig. 3.2 The original IKONOS images of Hilla city in **a** 2000, **b** 2005, **c** 2011, **d** 2000, 2005 and 2011

3.3 Methodology

This section is mainly concerned with the methodology applied in this study. The satellite images are pre-processed through three steps: (i) geo-referencing, (ii) atmospheric correction, and (iii) boundary extraction of the study area. Object-oriented classification is then performed on the processed images. In this work, we adopt the Support Vector Machine (SVM) machine learning algorithm for all three images in order to obtain accurate measurements. Lastly, the change detection of image classification was obtained by using statistical results as well as the prediction of the changes in classes for the years 2026 and 2036. The stepwise methodology flowchart is shown in Fig. 3.3.

3.3.1 Pre-processing

In this study, the IKONOS satellite image in 2011 was geometrically corrected. It is selected as the reference image and the other images (2000 and 2005) as the sensed images. After performing georeferencing for all the images, the borders of the study area are obtained. The satellite images in 2000 and 2005 are captured larger than the satellite image in 2011 as illustrated in Fig. 3.2. Therefore, in order to standardize all image's dimensions, the ArcGIS software was used to extract the required area by using the feature "Extract by Polygon." In the next step, Quick Atmospheric Correction (QUAC) is applied to correct the atmospheric distortion by retrieving surface reflectance and engage topographic correction as well as adjacency effect correction (Bernstein et al. 2012).

3.3.2 Object-Oriented Classification

Object-oriented classification (OBC) involves image segmentation, feature analysis, and the classification task itself. Image segmentation is the part where the image is segmented into contiguous pixels sharing similar spatial properties (such as color or texture). Feature analysis extracts relevant features from each of the segments. Finally, the classification task makes sense of the features, either via rule-based logic or machine learning in order to categorize segments as either being buildings, water, trees, etc.

In this paper, we choose the Support Vector Machine (SVM) for OBC. The SVM is as a supervised statistical machine learning technique, which makes no assumption about the underlying data distribution (Mountrakis et al. 2011). During training, an SVM will try to discover an optimal hyperplane in feature space that is able to classify (or separate) the features into the two classes ω_1 and ω_2. This separating hyperplane is optimal if there is a maximum margin to the nearest training examples of each

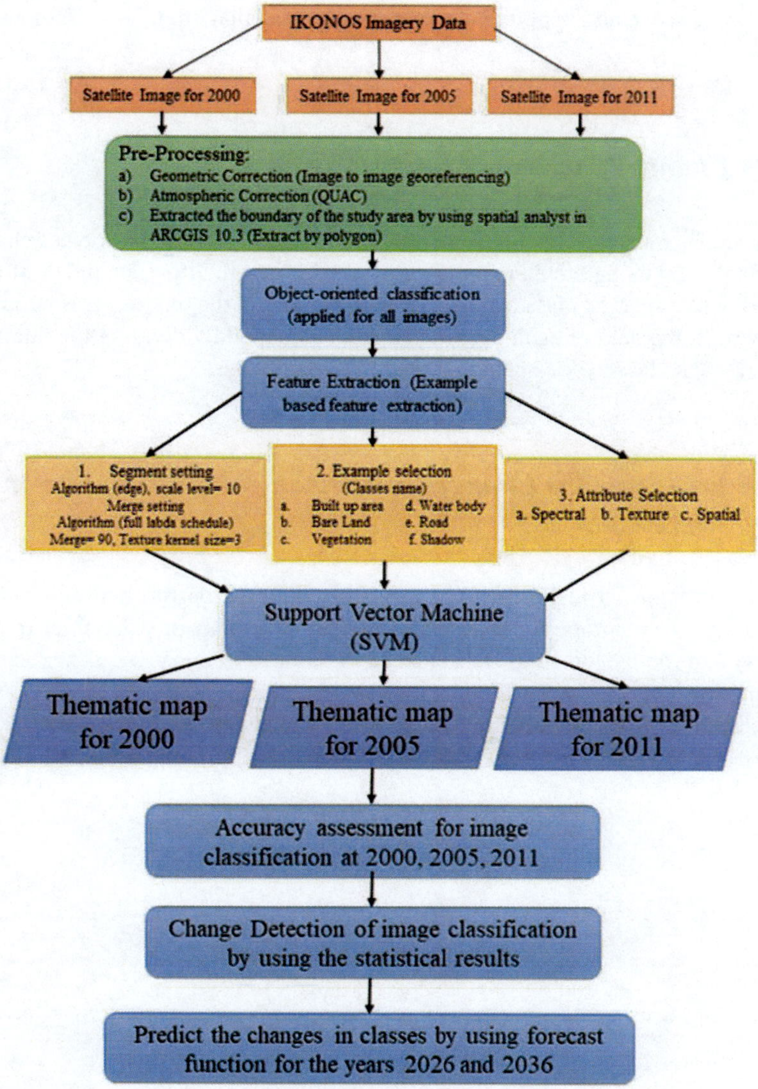

Fig. 3.3 Flow chart of the methodology

class. This can be written as

$$\omega_i(w \cdot x_i + b) \geq 1, \forall i \tag{3.1}$$

where w is the weights and ba bias/threshold value. SVMs are originally intended for binary (two-class) linear classification problems. However, they can perform non-linear and multi-class classification by transforming the data into a higher dimensional space and applying the kernel trick (Kalantar et al. 2017; Ashour et al. 2015).

3.3.3 Change Detection of Image Classification

In this study, statistical results from each image class are used to detect changes. Specifically, by using the thematic map properties in ArcGIS, the number of pixels belonging to a specific class can be calculated based on the unique class number. In other words, the area of each classification is calculated by dividing the number of pixels of each class by the spatial resolution of the image.

3.3.4 Predicting the Changes in Thematic Map Using Linear Regression

Linear regression basically involves modeling the relationship between a dependent variable (e.g., property price) and one or more explanatory variables (e.g., the property's of rooms, location, area, etc.). In the context of this work, the explanatory variable considered is the year, which is used to predict the area changes (percentage) for each class. Specifically, we look at changes occurring in years 2026 and 2036. The forecast function will calculate a new y-value using the simple straight-line equation:

$$y = a + bx \tag{3.2}$$

$$a = \bar{y} - b\bar{x} \tag{3.3}$$

$$b = \frac{\sum (x - \bar{x})(y - \bar{y})}{\sum (x - \bar{x})^2} \tag{3.4}$$

where the value of x and y are the sample means (the year and average of the area changes for each class).

3.3.5 Accuracy Assessment

Classification accuracy is evaluated using the error/confusion matrix. This matrix is visualized as a square array or table where the rows represent the predicted values (class) and the columns are the actual class labels (note that a vice versa case is also acceptable). The confusion matrix makes it possible to see if a classifier mistakenly confuses between two classes (e.g., misclassifying a building as a tree). For remote sensing, it is very useful to summarize two types of thematic errors namely omission and commission. Omission refers to pixels in the reference map identified as something else. Commission on the other hand are pixels that are incorrectly classified as a class in a row (Senseman et al. 1995; Maingi et al. 2002). In this work, we refer to the correctly allocated cases as a percentage. Based on this, accuracy is ultimately the probability that a given pixel can be found in the ground as it is in the classified image, whereas producer's accuracy refers to the percentage of a given class that is correctly identified on the map (Yesserie 2009).

3.4 Results and Discussion

The LULC maps in Fig. 3.4 are generated through the object-based supervised SVM classifier. The study area is classified into six classes: built-up area, bare land area, vegetation/agricultural, water body, road, and shadow. The results in Fig. 3.5 clearly show an increase of built-up areas with respective values of 8.14% of the study area in 2000 to 14.53% in 2005 and 18.36% in 2011. However, there has been a decrease of bare land areas from 38.39% in 2000 to 23.16% in 2005 and 16.3% in 2011. Because of the successive decrease of bare land areas, built-up areas, as well as the vegetation/agricultural areas, have drastically increased in the study periods. This could be attributed to increased population associated with high demand for land and urban supplies. Nonetheless, it can be considered as a positive indicator of proper government planning. It is also clear from Fig. 3.4 and Table 3.3 that water bodies have shown an increase from 3.27% of the study area in 2000 to 4.5% in 2005 and a slight decrease of about 3.94% of the study area in 2011. Road areas increased from 10.01% in 2000 to 12.91% in 2005 and 15.24% in 2011. Not only did urban areas increase, but vegetation/agricultural areas also increased. Vegetation/agricultural areas increased from 36.14% in 2000 to 41.71% in 2005 and this increment continued to reach 45.13% in 2011.

IKONOS satellite images can be considered as high-resolution satellite imagery (HRSI). It offers great possibilities for urban mapping. Unfortunately, shadows cast by buildings in high-density urban environments obscure much of the detail in the image leading that corrupted classification results causing misinterpretation (Dare 2005). The percentage of shadow area was 4.05%, 3.19%, and 1.03%, respectively, in 2000, 2005, and 2011 causing misclassifications.

3.4.1 Accuracy Assessment of the Land Cover Maps

The classification accuracy is assessed using the confusion matrix and Kappa coefficient. The accuracy assessment in this study was made on the basis of comparing

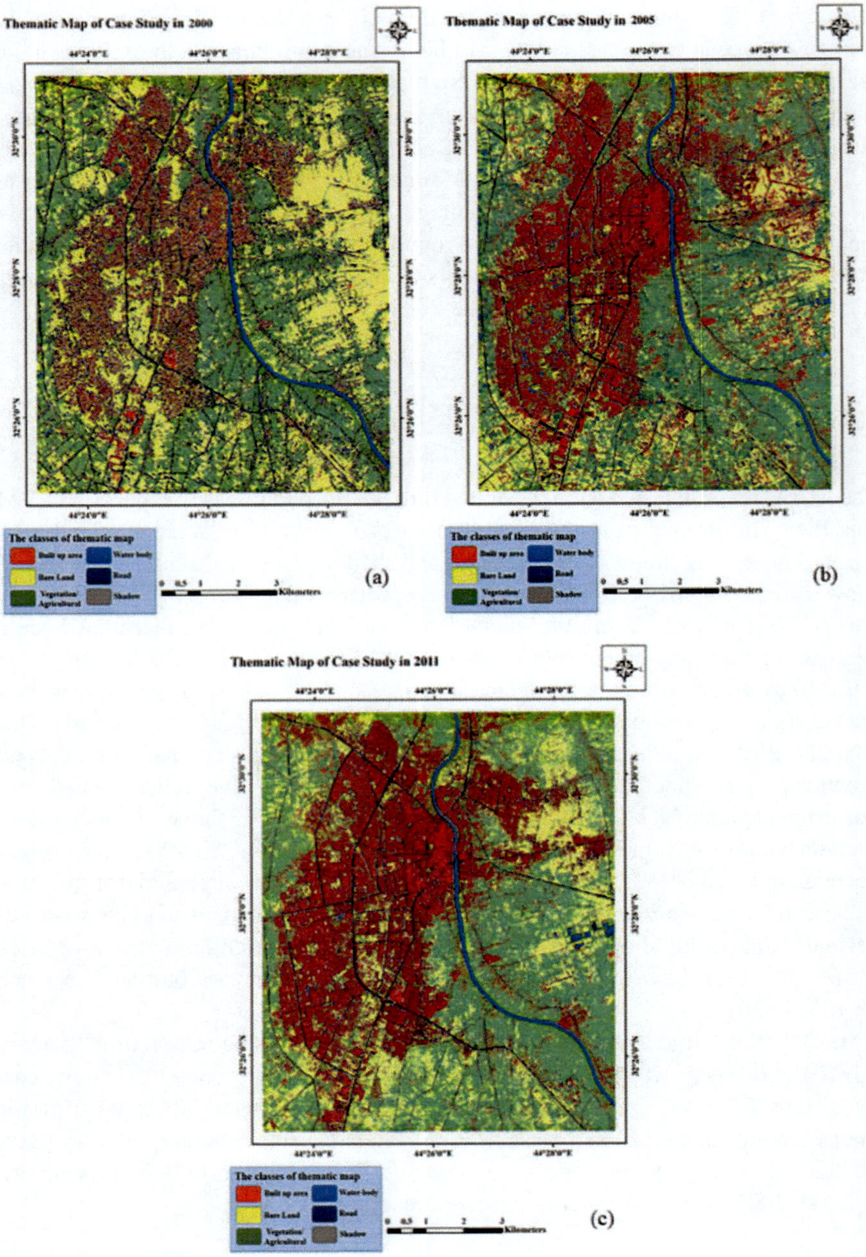

Fig. 3.4 The thematic map of study area in **a** 2000, **b** 2005, **c** 2011

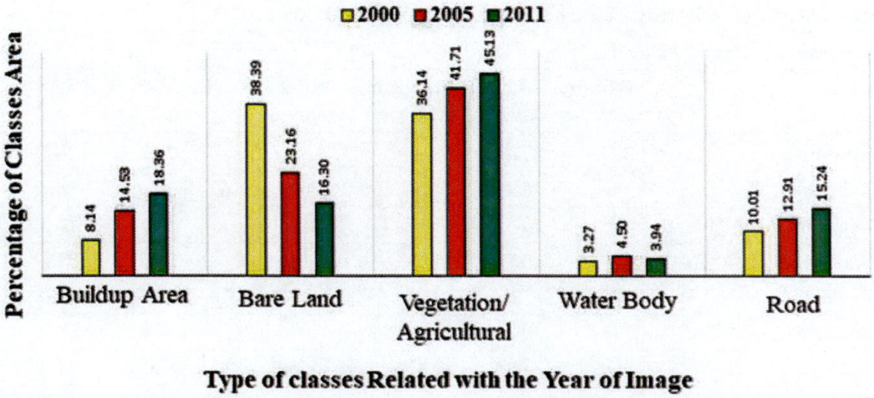

Fig. 3.5 The changes in area percentage of classes in 2000, 2005 and 2011

Table 3.3 Area statistics of the land use land cover units from 2000 to 2011

Land cover classes	2000		2005		2011	
	Area (ha)	%	Area (ha)	%	Area (ha)	%
Built-up areas	823.12	8.14	1469.27	14.53	1856.56	18.36
Bare lands	3882.00	38.39	2341.94	23.16	1648.26	16.30
Vegetation/Agricultural	3654.48	36.14	4217.72	41.71	4563.55	45.13
Water bodies	330.66	3.27	455.04	4.50	398.41	3.94
Roads	1012.21	10.01	1305.46	12.91	1541.07	15.24
Shadows	409.54	4.05	322.57	3.19	104.15	1.03
Total	10112.00	100	10112.00	100	10112.00	100

with the original thematic map images from the years 2000, 2005, and 2011, based on ROIs (Region of Interest). As shown in Table 3.4, the overall accuracy of the image in year 2000 was 82.96% with a Kappa coefficient of 0.78. The image in year 2005 had an accuracy of 84.52% with a Kappa coefficient of 0.79. The accuracy for the image from year 2011 was 87.34% and the kappa coefficient was 0.84. From the results, the best accuracy is obtained for year 2011. We believe that this is due to the image being clearer compared to the others and also due to less shadow.

3.4.2 Land Use Changes of Built-up Areas

The results for in this part are shown in Table 3.3. It is clear that the built-up area has increased from 8.14% in 2000 to 18.36% in 2011. We believe that this occurred mainly due to:

Table 3.4 Confusion matrix for land cover map in 2000, 2005, and 2011

Class	Built-up area	Bare land	Vegetation	Water body	Road	Shadow	Sum
Confusion matrix for land cover map of 2000							
Built-up area	2073	720	0	0	73	11	2877
Bare land	933	6631	14	2	1263	39	8882
Vegetation/Agricultural	0	0	10,181	50	7	140	10,378
Water body	0	0	162	3531	0	76	3769
Road	195	1256	70	34	3018	15	4588
Shadow	198	0	9	44	36	599	886
Sum	3399	8607	10,436	3661	4397	880	31,380
Producer	60.99	77.04	97.56	96.45	68.64	68.07	
User	72.05	74.66	98.10	93.69	65.78	67.61	
Overall accuracy %	**82.96**						
Kappa	**0.78**						
Confusion matrix for land cover map of 2005							
Built-up area	1706	100	74	0	365	7	2252
Bare land	98	5947	681	0	198	0	6924
Vegetation/Agricultural	41	498	7506	26	4	115	8190
Water body	7	0	104	1773	0	400	2284
Road	0	654	20	0	1624	0	2298
Shadow	0	0	61	0	0	294	355
Sum	1852	7199	8446	1799	2191	816	22,303
Producer	92.12	82.61	88.87	98.55	74.12	36.03	
User	75.75	85.89	91.65	77.63	70.67	82.82	
Overall accuracy %	**84.52**						
Kappa	**0.79**						
Confusion matrix for land cover map of 2011							
Built-up area	2829	340	0	0	100	22	3291
Bare land	0	6003	0	0	36	0	6039
Vegetation/Agricultural	9	211	6569	1152	13	8	7962
Water body	0	0	2	3419	5	38	3464
Road	456	1001	0	0	4574	0	6031
Shadow	37	0	0	0	2	291	330
Sum	3331	7555	6571	4571	4730	359	27,117
Producer	84.93	79.46	99.97	74.80	96.70	81.06	
User	85.96	99.40	82.50	98.70	75.84	88.18	
Overall accuracy %	**87.34**						
Kappa	**0.84**						

(1) Dramatic population growth in Hilla city, which seems to correlate with the increase in the number of built-up areas. Figure 3.6 shows the amount of population growth in the city from 1997 to 2014.
(2) Increase in annual income of Iraqi citizens correlates with the increasing rate of urbanization after 2005; Fig. 3.7 shows the rate of increase in the annual income of the Iraqi citizens from 1999 to 2013.
(3) The abundance of construction materials that are locally manufactured (sand, gravel, brick, cement, and rebar) and their moderate prices helped constructing new buildings or rehabilitating old ones, this, in turn, led to an increase in a built-up area.

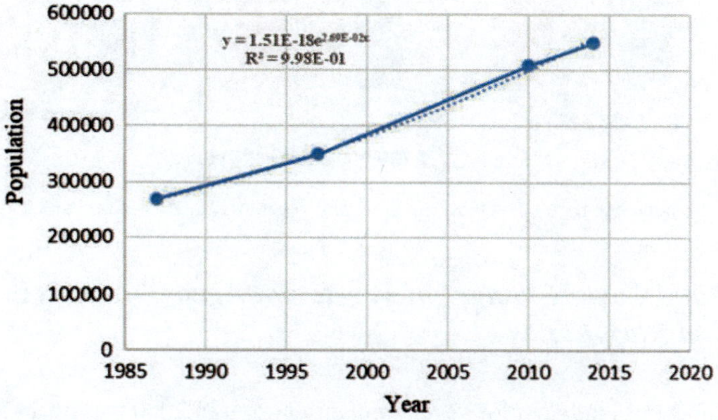

Fig. 3.6 The population growth of Hilla city from 1987 to 2014

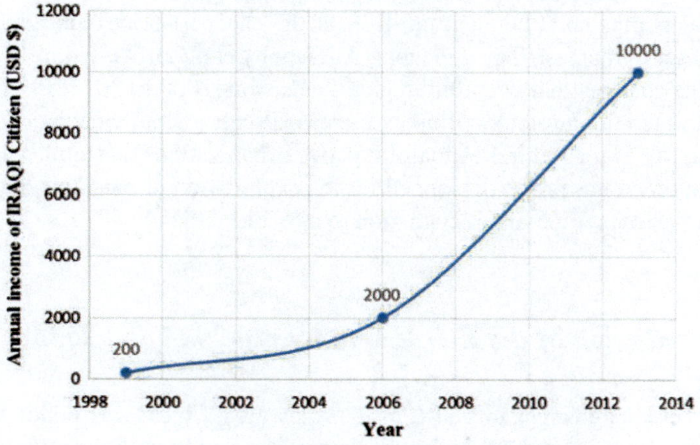

Fig. 3.7 Annual income of Iraqi citizen in 1999, 2006, and 2013 (*Source* https://www.economist.com/)

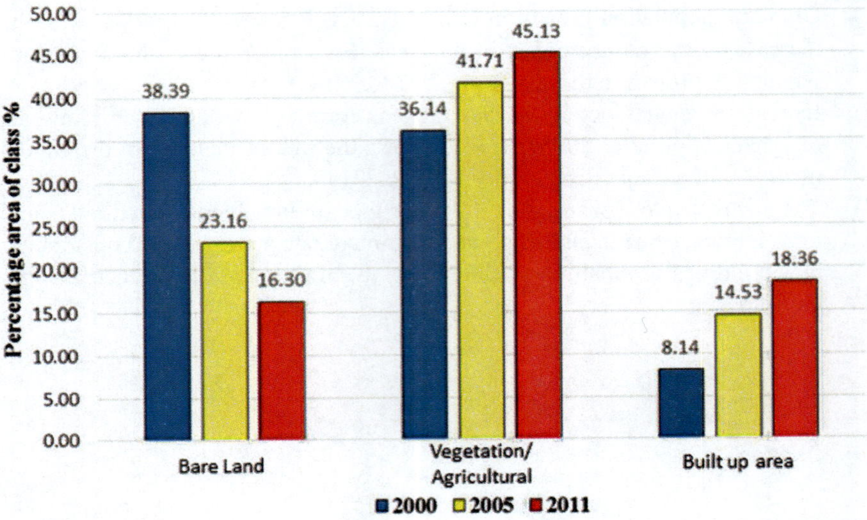

Fig. 3.8 The changes in bare lands, built-up area, and vegetation in 2000, 2005 and 2011

3.4.3 *Land Cover Changes of Vegetation/Agricultural Areas and Bare Lands*

The percentage of land area for vegetation/agricultural in 2000 was 36.14%. It rose to 41.71% in 2005 and continued rising to 45.13% in 2011. The previous and current governments have supported and developed the agricultural sector by providing irrigation canals to the agricultural lands and supporting the farmers by giving them mechanical pumps and chemical fertilizer at the cheapest prices and giving them loans in order to increase their products (Al-Haboby et al. 2014). Figure 3.8 clearly explains the changes that occurred in bare lands from 2000 to 2011. Both built-up areas and vegetation/agricultural areas increased with a simultaneous decrease in bare lands; this is a good indication of positive urbanization. It also means that the local government has properly planned for the exploitation of bare land appropriate for both urban expansion and reclamation to agricultural land.

3.4.4 *Prediction of the LULC Changes in 2026 and 2036*

In order to predict changes in LULC of the study area, Linear Regression was used based on data from 2000 to 2011. The six classes involved are built-up area, bare land, vegetation/agricultural, road, water body, and shadow. The classification task distinguishes land areas to be from one of the classes. These six variables are presented in three assumptions:

(1) The shadow area changes are fixed because there are related to the satellite images; the maximum area percentage of shadow class is considered as the same as it was in 2011.
(2) There are no changes in water body areas.
(3) The pattern of urban growth will continue in the same tendency that happened in the period from 2000 to 2011. This means that the increase in classes will be at the expense of bare lands only. Therefore, the variables will be only the built-up area, road, and bare land whereas the vegetation/agricultural is a by-product of all of these three variables. The results, shown in Fig. 3.9 and Table 3.5 are obtained with the use of the forecast function in Excel.

From Table 3.5, there is the possibility that there will no longer be bare land by 2020. This can result in urban expansion into vegetation/agricultural areas. If this happens, Hilla city will lose 8.93% and 13.76% from its vegetation/agricultural areas in 2026 and 2036, respectively.

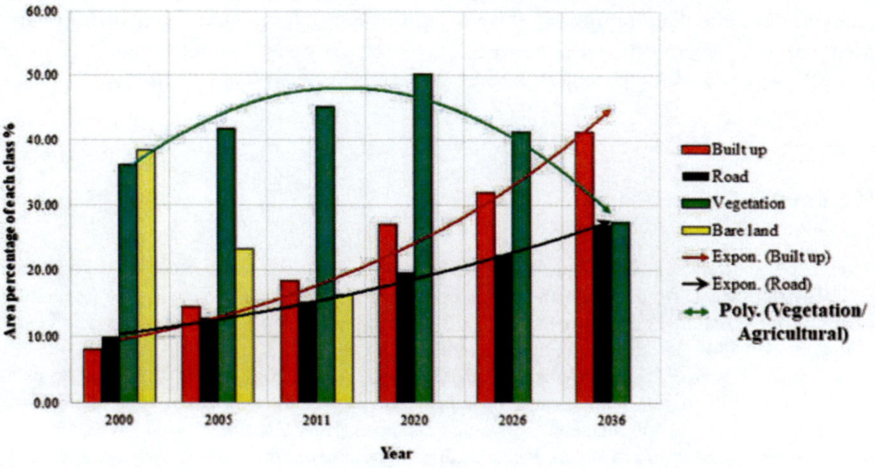

Fig. 3.9 The prediction of the classes changes in 2026 and 2036

Table 3.5 The expected area percentage of each class in 2026 and 2036

Year	Built-up area	Bare land	Road	Vegetation/Agricultural	Water body	Shadow	Total
2000	8.14	38.39	10.01	36.14	3.27	4.05	100
2005	14.53	23.16	12.91	41.71	4.5	3.19	100
2011	18.36	16.30	15.24	45.13	3.94	1.03	100
2020	26.16	0.00	19.65	50.13	3.94	1.03	100
2026	31.96	0.00	22.27	41.2	3.94	1.03	100
2036	41.33	0.00	27.05	27.44	3.94	1.03	100

3.5 Conclusion

Urbanization and its relation with the agricultural lands have a wide range of consequences on all spatial and temporal scales. Because of these consequences, it has become one of the major problems for environmental change as well as natural resource management. The overall accuracy of land use and land cover maps generated in this study have got an acceptable value of above the minimum threshold (80% for the accuracy of every class mapped and assessed). From the remote sensing of image classification result, the study showed that the proportion of built-up areas were increased. There was a rapidly change in built-up areas from 8.14% in 2000 to 14.53% in 2005 and 18.36% in 2011. Bare land areas were a good resource for built-up areas. Bare land areas showed a continuous decrease from 38.39% in 2000 to 23.16% in 2005 and finally had a value of 16.30% in 2011. The conversion of bare land to built-up areas could be related to the increment of population and faster economic development in Hilla city. On the other hand, the vegetation/agricultural areas were increased from 36.14% in 2000 to 41.71% in 2005 and 45.13% in 2011, respectively. The development in the agriculture sector happened because of government support. Accuracy assessments of classified images show better results with an overall accuracy of 82.96% in 2000, 84.52% in 2005, and 87.34% in 2011.

References

Al-Haboby A, Breisinger C, Debowicz D, El-Hakim AH, Ferguson J, van Rheenen T, Telleria R (2014) Agriculture for development in Iraq estimating the impacts of achieving the agricultural targets of the national development plan 2013–2017 on economic growth, incomes, and gender equality, vol 1349. International Food Policy Research Institute

Ashour MW, Khalid F, Halin AA, Abdullah LN (2015) Machining process classification using PCA reduced histogram features and the support vector machine. In: 2015 IEEE international conference on signal and image processing applications (ICSIPA). IEEE, pp 414–418

Barbero-Sierra C, Marques MJ, Ruíz-Pérez M (2013) The case of urban sprawl in Spain as an active and irreversible driving force for desertification. J Arid Environ 90:95–102

Bernstein LS, Jin X, Gregor B, Adler-Golden SM (2012) Quick atmospheric correction code: algorithm description and recent upgrades. Opt Eng 51(11):111719

Burgi M, Hersperger AM, Schneeberger N (2004) Driving forces of landscape change—current and new directions. Landscape Ecol 19(8):857–868

Chauvin JP, Glaeser E, Ma Y, Tobio K (2017) What is different about urbanization in rich and poor countries? Cities in Brazil, China, India and the United States. J Urban Econ 98:17–49

Dare PM (2005) Shadow analysis in high-resolution satellite imagery of urban areas. Photogr Eng Remote Sens 71(2):169–177

Jensen JR (2005) Introductory digital image processing: a remote sensing perspective, 3rd edn. Prentice Hall, Upper Saddle River, NJ

Kalantar B, Mansor SB, Sameen MI, Pradhan B, Shafri HZ (2017) Drone-based land-cover mapping using a fuzzy unordered rule induction algorithm integrated into object-based image analysis. Int J Remote Sens 38(8–10):2535–2556

Lu D, Weng Q (2007) A survey of image classification methods and techniques for improving classification performance. Int J Remote Sens 28(5):823–870

Lu D, Mausel P, Brondizio E, Moran E (2004) Change detection techniques. Int J Remote Sens 25(12):2365–2401

Maingi JK, Kepner SE, Edmonds WG (2002) Accuracy assessment of 1992 Landsat-MSS derived land cover for the Upper San Pedro Watershed (US/Mexico). Sponsored by Environmental Protection Agency, Las Vegas, NV. National Exposure Research Lab

Manii JK (2014) Using GIS to study the probability pollution of surface soil in Babylon province, Iraq. J Appl Geol Geophys 2:14–18

Mountrakis G, Im J, Ogole C (2011) Support vector machines in remote sensing: a review. ISPRS J Photogramm Remote Sens 66(3):247–259

Pijanowski BC, Pithadia S, Shellito BA, Alexandridis K (2005) Calibrating a neural network-based urban change model for two metropolitan areas of the Upper Midwest of the United States. Int J Geogr Inf Sci 19(2):197–215

Raddad S, Salleh AG, Samat N (2010) Determinants of agriculture land use change in Palestinian urban environment: urban planners at local government's perspective. Am Eurasian J Sustain Agric 4(1):30–38

Rana MMP (2011) Urbanization and sustainability: challenges and strategies for sustainable urban development in Bangladesh. Environ Dev Sustain 13(1):237–256

Senseman GM, Bagley CF, Tweddale SA (1995) Accuracy assessment of the discrete classification of remotely-sensed digital data for land cover mapping (No. CERL-TR-EN-95/04). Construction Engineering Research Lab (Army) Champaign Il

UNFPA (2013). https://www.unfpa.org/pds/urbanization.htm. Accessed Dec 2013

Xi FM, He HS et al (2010) Spatiotemporal pattern of urban growth and its driving forces in urban agglomeration of central Liaoning Province, China. Chin J Appl Ecol 21(3):707–713

Yesserie AG (2009) Spatio-temporal land use/land cover changes analysis and monitoring in the Valencia municipality, Spain

Chapter 4
Designing Streets for Smart Cities

Grazyna Chaberek

Abstract Cities today are an attractive living environment for people because of the services they provide. But cities, primarily, are their streets. Streets shape the structure of cities are responsible for their appearance and allow mobility and transport, which are necessary for the implementation of urban services. Streets are not only routes, but also areas along which most of the technical infrastructure of the entire city extends. The latest digital technology solutions implemented in cities also focus their technical devices (cameras, sensors, light-boards, etc.) within streets. However, nowadays, the streets of cities that strive to be smart are not only spaces equipped with the latest technology devices, but also very high-quality public spaces. The streets of smart cities must combine the infrastructure of various means of transport and communication, but at the same time be pedestrian-friendly, provide security, complement urban greenery, as well as create conditions to stimulate the business. The purpose of the chapter was a comparative study of postulates regarding the appearance, construction, and technical infrastructure of streets according to the latest urban approach in the spirit of smart cities with the real structure and appearance of streets in sample cities. The chapter contains the results of field studies conducted in cities of Europe, India, and the United States in the years 2017–2018.

Keywords Smart city · Smart streets · Smart traffic solutions · Quality urban space · Infrastructure development

4.1 Introduction

The problems in cities can be different in the world, depending on the level of economic and social development, or the threat of violent meteorological phenomena. But cities around the world also have similar problems regardless of geographic location and size. The problem common to all cities is the deteriorating condition of the residents' living environment caused by poor air quality, noise, and lack of sense of security. Solving these problems is the basis for further development. Cities must

G. Chaberek (✉)
Spatial Management Department, University of Gdansk, Gdansk, Poland

P. Sharma (ed.), *Geospatial Technology and Smart Cities*, The Urban Book Series,
https://doi.org/10.1007/978-3-030-71945-6_4

develop economically. The population and businesses inflow shouldn't be limited. However, cities must develop wisely. In the USA such development has been called the Urban Smart Growth (Durand et al. 2011; Wlodarczak 2012; Molavi and Roshan 2018).

First of all, it is needed to understand what "smart cities" are. The most common definitions bring the idea of smart city to focus on urban management, using the latest technical and digital advances, which in fact allow for improving the life conditions (Stawasz 2015). Unfortunately, this concept can be understood and imported in practice not much more than to gadgets and the tendency to unconventional conduct of conventional matters, that is, electronically, digitally, and virtually. The digitization is inseparable from the smart concept. The concept of smart cities adds new services such as monitoring, traffic control, and parking information. The concept of being smart also involves transferring of many traditional urban services into the Internet environment. It is primarily an e-city hall, but also many public and private apps, created by the users themselves, and aimed at improving city traffic, such as tracking urban transport in real time with timetables, the ability to purchase tickets for public transport, or payment for the car parking, etc. (Stawasz 2015). Such an understanding, though important, is trivial. The idea of smart city is about more than using computers for what was previously done using print media, radio, and television (Barber 2013).

Smart city is all together: smart economy, smart mobility, smart environment, smart people, smart governance, and smart living (Stawasz and Sikora-Fernandez 2015; Singh and Singh 2019). Cities should show high productivity as a gigantic network of connections, they should combine all their resources with high speed using both traditional transport and digital communication while rational use of existing infrastructure. A smart city should carry out activities that reduce the emission of pollutants into the environment, and the management of resources based on the principle of sustainable development. The initiators of changes in cities should be their residents, who with the appropriate technical support are able to prevent excessive energy consumption, environmental pollution, and strive to improve the quality of life. Smart growth "seeks to revitalize the already-built environment and, to the extent necessary, to foster efficient development at the edges of the regions, in the process creating more livable communities" (Scott 2007). Smart growth requires the creation of an appropriate city management system, elaboration of procedures requiring the cooperation of local authorities and other city users, and the use of modern technologies in the functioning of the city. At the same time, it is necessary to provide a friendly living environment, in particular to ensure wide access to public services, technical and social infrastructure, as well as a high level of security and having an appropriate cultural and entertainment offer as well as care for the environment and green spaces.

The problem of communication accessibility is common for all cities (especially time and quality of service). Transport accessibility is the basis of economic and social development, hence smart development of cities must guarantee fast and convenient communication accessibility. The high share of cars in satisfying transport needs is also common for all cities, which to a large extent causes congestion, noise, pollution, health, and life risks for urban users (Behan et al. 2008; Wlodarczak 2012; McCormack et al. 2014). There are a lot of cars due to the fact that it is the fastest and most flexible way of commuting, especially in sprawl cities. The phenomenon of urban sprawl for many years justifies the dominance of the car in meeting transport needs in urban areas. It is the fastest way of commuting when there is no good alternative and/or because cities allow to move fast by car. However, this would not be possible without the favorable conditions that planners and urbanists themselves create by designing and building streets dedicated only for cars.

4.2 Thesis and the Concept

Smart cities are an intelligent community—creative, enterprising, frugal human capital. Smart cities are intelligent people using the latest technologies. However, urban space can support or limit the possibilities of creating human capital. Social capital is created where people meet. Where they have space and amenities to be with each other personally and to cooperate (Brdulak and Chaberek-Karwacka 2018). Speck (2012) argues that creating social ties is favored by spaces where people meet often face-to-face, and not where people spend most of their time in private cars. Smart city first of all is a city for people (Gehl 2010). Cities are actually starting to discover information technology to change the future of functioning. But, as it turns out, this technology has nothing to do with, for example, fantastic new vehicles using to commute. The key role is played by new ways of thinking, sharing information, and modifying the methods of using the resources that already exist. In terms of satisfying the needs of transport and communication, the "smart" manifests itself through the use of means of transport, used for many years (Montgomery 2013). Today, "to be smart" in the city means for example to use electric scooters and skateboards which allow to move faster than on foot, but it easy to take them anywhere, if necessary, and overcome a longer distance by public transport.

City for people means safe, useful (to satisfy needs), life-friendly (green), and quiet city. Safe means zero accidents, zero deaths, and harms. Useful means city which allows to satisfy all needs, also needs like connecting from one place to another, meeting people, transport goods, etc. Transportation needs have to be satisfy associated with the lowest cost. Cost is not only money needs to pay but also time and effort/comfort.

Why the smart city should be primarily for people and not for cars? Of course, the car is an important means of flexible and fast communication and it is not about cars completely disappear from city streets. The idea of a "city for people" assumes only a change of priorities. Reorganization of the urban environment in such a way that it would be primarily friendly to pedestrians and users of soft, non-emission forms of communication, such as a bicycle or scooter. However, in order for it to be possible, users must have an urban space conducive to being on the streets and moving in a comfortable and safe way.

The considerations in this chapter are based on the thesis that infrastructure determines behavior. Forester, and later his successors, conducted many years of research into the influence of deterministic infrastructure on behavior (Forrester 2009). "Causes are usually found, not in prior events, but in the structure and policies of the system" (Forrester 1969). Cars dominate in cities, because the streets are dedicated primarily to cars, facilitate fast driving, and even "encourage" to ignore pedestrians. Therefore, if cities want to develop in an smart, or sustainable way, must create the appropriate infrastructure, including the structure of the streets on which the main part of urban life takes place. That is why smart cities require proper infrastructure, human infrastructure (Rashmi 2019), in other words smart cities require smart streets.

The purpose of the chapter was a comparative study of postulates regarding the appearance, construction, and technical infrastructure of streets according to the latest urban approach in the spirit of smart cities with the real structure and appearance of streets in sample cities. The implementation of the goal required the creation of a smart streets model. The chapter contains the streets' construction and appearance model for smart cities based on current United Nations (UN General Assembly Distr. 2016) postulates and based on the literature.

The nature of the adopted assumptions required qualitative and comparative research. In qualitative research, great importance is attached to the context and specific cases as factors explaining the examined issue. Lots of qualitative research are case studies or their compilations, and often specific cases (their history and complexity) are an important context for understanding the studied area (Banks 2009). Therefore, in order to show differences between streets that are more or less "smart", the study was used the evidences of streets from selected cities were presented and analysis of their adjustment to the model. For this purpose, were used results of field studies conducted in cities of Europe, India, and the United States in the years 2017–2018. The presented comparative analysis takes the form of a photo-essay.

4.3 The Model of Smart Streets

Digitization introduces changes in the appearance of urban streets, because the technical infrastructure necessary to collect and transmit a data is placed within streets. These are primarily cameras monitoring both the image from the streets and, for example, the speed of vehicles. Solutions related to traffic control require information light-boards and additional signaling devices. However, smart streets are not just additional digital devices. According to the smart city's assumptions, a completely new approach to street design is required.

New approach to street design meets the demand of today and the challenges of tomorrow based on the principle that streets are public spaces for people as well as arteries for traffic and transportation (National Association of City Transportation Officials 2013). "Street design must meet the needs of people walking, driving, cycling, and taking transit, all in constrained space. The best street design also adds to the value of businesses, offices, and schools located along the roadway" (National Association of City Transportation Officials 2013).

People want to be smarter, more aware, and adapted. Smart cities are primarily made up of smart people. Innovations are created in a collaborative environment. Various studies prove that people are more creative and collaborate better when they do it face-to-face than using electronic media. The ability to easily meet other people and unhindered interaction creates a creative and innovative environment. What really creates a smart city? Whether it is not creative people? Face-to-face collaboration is, of course, possible in any setting. But it is easier in a walkable city (Speck 2012).

The Urban Streets Design Guide (National Association of City Transportation Officials 2013) reveals the six most important aspects to keep in mind when designing streets smartly: (1) streets are public spaces, (2) streets are places of business location, (3) traffic organization can be flexible and variable depending on needs, (4) streets must be designed so as to ensure safety for all users, (5) it should be remembered that streets are ecosystems, where "man-made systems interface with natural systems", (6) projects of transforming streets do not have to be time-consuming and cost-consuming—simple solutions from available resources should be used. *The Urban Streets Design Guide* developed in the USA by National Association of City Transportation Officials (2013) contains exact technical guidelines on how streets should be designed. Many practical tips and specific solutions for designing streets for people give, among others Gehl (2010), Speck (2012), United Nations (UN General Assembly Distr. 2016), Sadik-Khan and Solomonow (2016), Brdulak and Brdulak (2017).

All these postulates in the field of street design can be collected in a smart streets design model. Smart streets are those that meet the following criteria:

1. Multi-functionality.
2. Reduction of the road width in favor of bicycle paths, pedestrian routes—force drivers to speed reduction.
3. Easy access to public transport.
4. Removal of barriers in pedestrian routes.
5. Adding greenery.
6. Creating places for being on streets (benches, squares, etc.).

What does it mean to redesign of the streets according to the technical guidelines outlined above? Multifunctional streets force drivers to low speed and extreme caution. On such streets, the driver knows that he has to take into account the presence of a pedestrian or cyclist and is more attentive and, above all, reduces speed. Multifunctional streets provide a lot of good quality public spaces, or places to be within. Such streets have spaces for greenery, a place to sit, meet, spend time. The possibility of meeting and staying in the public urban space contributes to the creation of local social capital (more at this topic in Brdulak and Chaberek-Karwacka 2018). New assumptions in the streets design regulate the way of parking cars, which contributes to the aestheticization of urban space. In addition, in many solutions parked cars become a natural protective barrier for pedestrians and cyclists.

Of course, it should be remembered that not all streets in the city should be and will be designed with the way of presented model. This will be primarily related to the function of a given street and its location in the urban tissue. Some of the streets, especially those on the peripheries of cities, whose dominant function is the communication and transport function, also the link between local and out-of-urban traffic, will be designed in such a way as to best fulfill this task and facilitate traffic and accessibility. Intelligent changes in the street area therefore firstly require street classification by function. An example of such an approach is London (Roads Task Force 2013). Before the changes in terms of the appearance and functionality of the streets were made there, a thorough analysis was carried out for which purpose the given street is used, their location and dedicated functions. As a result of this analysis, streets designated for the complete closure of car traffic were selected, as well as those streets dedicated only to public transport, as well as multifunctional or purely transit streets.

4.4 Evidences of More and Less Smart Streets

Below are examples of streets from different cities. The first part presents examples of solutions that do not meet the requirements of smart streets.

In the literature of the subject, it can be noted that many research works in the USA are devoted to this topic. Indeed, the US cities have been built entirely for cars since the 1950s. It consisted in such adaptation of the streets that they would be mainly very capacious for cars—means wide, with many parking spaces from which all the shops and services were available. The space necessary to build roads and parking lots disperses and distances everything, making space very less dense and forcing even more intensive use of cars (Jacobs 1961).

The heritage of many decades of such consistent urban planning and construction not only streets, but also entire cities are very wide streets—impossible to pass, no sidewalks, poorly developed, or no public transport. An additional consequence is the city turned backside to the streets. It is easier to get to commercial points or public services from an underground park lot than to enter from the street. Figure 4.1 shows an example of street in Long Beach, California.

Additionally in Fig. 4.1 it is worth noting that there is a pedestrian crossing over Seaside Way. This passage is sometimes used, because it is not only use to cross the street, but it is also a communication route to the hotel and entertainment center. Overground pedestrian crossings are another element of street construction that gives priority to cars. Some planners try to justify the construction of overground crossings, arguing that is safer for pedestrians. However, this is not true. The overground pedestrian crossing is a very big barrier to pedestrian traffic. For all those who are in a hurry, not to mention children or older people, for whom climbing up the stairs is often not only very difficult, but often not feasible. Pedestrians naturally striving to reduce the

Fig. 4.1 Pine Ave and Seaside Way, Long Beach, California, USA, February 2018. *Source* Author's own archive

cost of their movement will always be willing to choose ways which are faster and require the least effort. In this way, pedestrian crossings pose an even greater threat because pedestrians try to get to the other side of the street in places where drivers do not expect them. Figure 4.2 shows an example of overground crossing in Mumbai. During the observations carried out, no one used this overground passage, while a lot of people were crossing the street, despite the fact that a concrete wall separating the roadways in both directions needed to be crossed.

Thinking about implementing IT technology on city streets cannot assume that it will solve all urban problems. An example of these dilemmas is the issue of introducing systems of automatic toll system for entering by private car to the some areas of the city, usually to the zone of the strict center. Mangalore is an example of the Indian city, which in its latest development strategy has set itself the main goal to become smart city as soon as possible. The city puts a lot of emphasis on planning, financing, and current activities on digitization, remote traffic management, construction of multi-level car parks, etc. (Mangaluru Smart City), but also assumes the introduction of the model of many other cities—the automatic toll system for entry to the center (information gained in the Mangaluru City Corporation 2019). The intention of the city authorities is to reduce the car traffic in the center and reduce the nuisance associated with it. Certainly, it will work in some way, however, it is important to emphasize that the idea of introducing fees for entering particular

Fig. 4.2 P D'mello Road, Mumbai, India, June 2018. *Source* Author's own archive

city zones is a fairly controversial idea. Such a method of limiting the number of cars in the city raises some ethical doubts. Whether public property, such as urban streets, should be reserved in the first place for those who can afford an additional fee (Matricadi 2006). Part of the dilemma is to invest the funds obtained from fees for the development of public transport. However, such demand management projects can do little to improve the balance of security and access to urban streets (Montgomery 2013). Therefore, it is possible to discuss whether digitization solutions are sufficient for streets and the city to be smart. It's worth taking a look at how streets are constructed in the very center of Mangalore. Figure 4.3 shows one of the streets. Barriers in the form of fences separating pedestrians and cars are a common solution in the city. They are to protect pedestrians and organize traffic, in fact, enable drivers to develop high speeds, while pedestrians make it very difficult or completely impossible to cross the road. Walking around the city is so difficult that it is much easier to travel even a few meters by rickshaw or taxi than on foot.

A similar approach can be found in Gdansk. It is a polish city of similar size as Mangalore, which aspires to be the smart city. At the end of 2015, Gdansk launched the Integrated Traffic Management System TRISTAR. The system includes specially designed traffic lights, sensors in the roadway, computers in public transport vehicles, transceivers, cameras, speed cameras, and photo-recorders, as well as variable message light-boards. The whole is coordinated by a team working in the center of traffic management. The introduction of the system allowed for the management of

Fig. 4.3 The street in Mangaluru center, India, June 2018. *Source* Author's own archive

urban traffic using the latest IT technologies and increasing the capacity of roads by 20–30% (Gdansk.pl 2015). Thus, it encourages even more residents to use cars. Despite this advanced digitalization, the streets of Gdansk are mostly highways, dominated by two or three lanes in one direction and fences making it impossible for pedestrians to cross. Pedestrian traffic is directed to overground crossings or worse to the underground crossings (Fig. 4.4). Both solutions are not only a serious barrier to pedestrians, they also disintegrate the city tissue and make the pedestrians do not stay on the streets but only sneak between their destinations.

All the above examples of streets give priority to cars and motorized users. They are designed to encourage drivers to high speed. On such streets pedestrians must overcome the barriers, as well as are very vulnerable to loss of health or life, which is rather discouraging walking. These streets are characterized by low aesthetics and lack of space for a good quality public space. Most often, all squares and roadsides within these streets are surrounded by parked cars.

Fortunately, the awareness of the importance of street construction in the fight against excessive motorization has been steadily growing for several years and there are already many positive changes in many cities around the world. An example can certainly be the streets of Copenhagen (Denmark). On the streets of Copenhagen, there is a clear division and separation of pedestrian, car, and tram routes, with the privileged availability for bicycles. There is marked domination of space dedicated to pedestrians at the reduction of the space for cars (Fig. 4.5). Narrowing of the

Fig. 4.4 Waly Jagielonskie street, Gdansk city center, Poland, March 2017. *Source* Author's own archive

Fig. 4.5 Vesterbrogade street, Copenhagen, Denmark, September 2017. *Source* Author's own archive

Fig. 4.6 Vester Voldgade street, Copenhagen, Denmark, September 2017. *Source* Author's own archive

carriageway for cars by separating bicycle paths and the parking lane (Fig. 4.6), or as in Fig. 4.7 reducing the area available for cars by replacing lanes with a greenery and park space.

The Downtown example in Houston (USA) (Fig. 4.8) shows how can be change a wide, uncrossable for pedestrians street into a friendly multifunctional communication route by adding a tramway and narrowing the lanes, leaving more space for pedestrians.

As already mentioned, flexible solutions, which use existing resources in various ways, are particularly smart solutions. The smart solution is streets flexibly adapting to the needs of users changing throughout the day. It can be movable (depending on the hour during the day) parking spaces, or loading bays or the roadway being at the same time a car road and track of city tram. Figure 4.9 shows an example of the street in the center of The Hague (the Netherlands). For most of the time it is a pedestrian promenade with allowed free bicycle traffic. It has a tram infrastructure, so when the tram arrives it is a tramway. Steel barriers that prevent cars from entering most of the time, when a resident, business owner, or cleaning services of the city appears, these columns automatically hide in the ground and allow the car to enter.

Figure 4.10 illustrates another way to improve pedestrian traffic, used in Pasadena, California, USA. The solution is simple and cheap. It required only a bit of paint and resetting the traffic lights cycle. Traditionally, by crossing, pedestrians can move as

Fig. 4.7 Sankt Anne Place, Copenhagen, Denmark, September 2017. *Source* Author's own archive

shown in Fig. 4.10a. In such an arrangement, the diagonal passing usually requires dividing this movement into two steps of passage first through one road and then through second one. It is connected with waiting for two cycles of light on two passes. The solution, which was used in Pasadena (Figs. 4.10b and 4.11), allows the pedestrian to cross the road in all directions on one traffic lights signal. When traffic lights turn green for pedestrians, cars traffic from all four directions is stopped.

The examples of streets more or less smart could be given much more, however, due to the small volume of the chapter, only the above ones were selected. The author realizes that this approach does not completely cover the topic, if only because every city manager and every urban planner can look for various solutions, depending on the existing situation, the character and significance of the street, or available financial resources. The most important thing is to create streets so that they are a tool for creating smart people and smart cities.

4.5 Conclusion

The digitization of cities is very expensive. It requires a lot of financial resources and infrastructure development. It requires software and utility applications, and for this purpose also sharing and managing of public big data. In order the technology

Fig. 4.8 Main Street, Houston Downtown, Texas, USA, February 2018. *Source* Author's own archive

to be used in the right way, it requires adequate equipment of streets and residents in devices that will be able to receive, process, and share data with the infrastructure and other users. And this, of course, is the direction that needs to be developed, as IT technologies improve and facilitate life in the city, help manage the city's resources or crisis situations. However, it should not be forgotten that digitization will not solve all problems and will not replace wisely managed urban space. There are a number of cheap and fast infrastructural solutions, which with a little effort on the part of the city authorities can immediately improve the situation on the streets, especially in those parts of congested, central cities and all those where people must and want to be. These solutions can easily contribute to increasing the quality of life of residents and reducing the negative impact of the city on the environment.

Fig. 4.9 Spui Street The Hague, the Netherlands, July 2018. *Source* Author's own archive

Fig. 4.10 **a** Crossroad Colorado Blvd i Fair Oaks Ave, Pasadena, California, USA, February 2018; **b** Crossroad Colorado Blvd i Raymont Ave, Pasadena, California, USA. *Source* Google maps (25.06.2019)

Fig. 4.11 Crossroad Colorado Blvd i Raymont Ave, Pasadena, California, USA, February 2018. *Source* Author's own archive

References

Banks M (2009) Materiały wizualne w badaniach jakościowych. PWN, Warszawa

Barber BR (2013) If mayors ruled the world: Dysfunctional nations. Yale University Press, Rising Cities

Behan K, Maoh H, Kanaroglou P (2008) Smart growth strategies, transportation and urban sprawl: simulated futures for Hamilton, Ontario. Can Geogr 52(3):291–308

Brdulak A, Chaberek-Karwacka G (2018) Kultura organizacyjna start-upów jako narzędzie rozwoju zrównoważonego miast. In: Bujak A, Topolska K, Kolasińska-Morawska K (eds) Logistyka w naukach o zrządzaniu, część 1. Przedsiębiorczość i zarządzanie XIX (11) part 3, SAN, Warszawa-Lodz

Brdulak A, Brdulak H (eds) (2017) Happy city—how to plan and create best livable area for the people. Springer

Durand CP, Andalib M, Dunton GF et al (2011) A systematic review of built environment factors related to physical activity and obesity risk: implications for smart growth urban planning. Obes Rev 12(5):173–182

Gdansk.pl (2015) Wdrożenie Zintegrowanego Systemu Zarządzania Ruchem TRISTAR w Gdańsku, Gdyni i Sopocie. https://www.gdansk.pl/inwestycje-miejskie/tristar-inteligentny-system-sterowania-ruchem,a,17755. Accessed 26 June 2019

Gehl J (2010) Cities for people. Island Press, Washington

Google maps. https://www.google.pl/maps/@34.1457995,-118.1494155,106m/data=!3m1!1e3?hl=pl&authuser=0. Accessed 03 June 2019

Jacobs J (1961) The death and life of great American cities. Random House

Mangaluru Smart City (2019) Command Ctrl Centre. https://mangalurusmartcity.net/smart-street. Accessed 30 June 2019
Matricadi AB (2006) Los Angeles missed the bus: solutions to issues in transportation equity. J Transp Law Logist Policy 73(4):410–449
McCormack E, Goodchild A, Bassok A (2014) Smart growth and urban goods movement. TR News 295:34–38
Molavi M, Roshan A (2018) Spatial-physical analysis of the urban smart growth indicators (Case Study: Districts of Rasht). Urbanism Architecture Constr 9(4):311–326
Montgomery C (2013) Happy city: transforming our lives through urban design. Farrar, Straus and Giroux
National Association of City Transportation Officials (2013) Urban street design guide. Island Press, New Orleans, Los Angeles
Rashmi (2019) The status of research on smart cities: a review. In: Sharma VR, Chandrakanta (eds) Making cities resilient. The urban book series. Springer.
Roads Task Force (2013) The vision and direction for London's streets and roads. Executive summary. www.tfl.gov.uk/roadtaskforce. Accessed 27 Dec 2016
Sadik-Khan J, Solomonow S (2016) Streetfight: handbook for an urban revolution. Penguin Books
Scott JW (2007) Smart growth as urban reform: a pragmatic 'Recoding' of the new regionalism. Urban Stud 44(1):15–35
Singh P, Singh P (2019) Smart cities: milestone of new era. In: Sharma VR, Chandrakanta (eds) Making cities resilient. The urban book series. Springer
Speck J (2012) Walkable city. How downtown can save America, one step at a time. North Point Press, New York
Stawasz D (2015) Problemy współczesnych miast i możliwości ich rozwiązania zgodnie z koncepcją *smart city*. In: Stawasz D, Sikora-Fernandez D (eds) Zarządzanie w polskich miastach zgodnie z koncepcją *smart city*. Placet, Warszawa
Stawasz D, Sikora-Fernandez D (2015) Koncepcja *smart city* w teorii i praktyce zarządzania rozwojem miast. In: Stawasz D, Sikora-Fernandez D (eds) Zarządzanie w polskich miastach zgodnie z koncepcją *smart city*. Placet, Warszawa
United Nations (2016) General Assembly Distr. Draft outcome document of the UN habitat III-conference on housing and sustainable urban development. A/CONF.226/4. United Nations. https://habitat3.org/wp-content/uploads/Draft-Outcome-Document-of-Habitat-III-E.pdf. Accessed 25 June 2019
Wlodarczak D (2012) Smart growth and urban economic development: connecting economic development and land-use planning using the example of high-tech firms. Environ Plan A 44(5):1255–1269

Chapter 5
An Automated Approach to Facilitate Rooftop Solar PV Installation in Smart Cities: A Comparative Study Between Bhopal, India and Trondheim, Norway

Kakoli Saha and Yngve Frøyen

Abstract The Indian Smart city concept thrives on providing assured electricity while maintaining the livability of a city. Since rooftop solar PV can contribute to 24×7 supply of electricity to cities while keeping the air pollution at bay, it plays an important role in Smart City planning. The main advantage of rooftop solar PV is that it can be installed on roof areas of buildings thus minimizing the distribution loss. This chapter proposes an automated method to extract available roof area for rooftop solar PV panel installation. The available roof area was then used to estimate the solar energy potential of a city. Saha and Mandal (Curr Urban Stud 4:356–375, 2016) has already devised an automated method to extract urban roof areas of Bhopal city using object-oriented classification approach. For the purpose of the study, a neighborhood, namely, Minaal residency was selected, roof area of which was extracted using the above-mentioned method. The same was applied to extract residential roofs of Trondheim city of Norway which is also a potential Smart City. Several adjustments were needed since the urban roofs of Trondheim city are different from the roofs of Bhopal city. To assess the accuracy of the automated method, a visual comparison was performed between the automatically and manually extracted roof areas. The visual comparison was followed by a statistical comparison for further confirmation. The extracted roof area was then used to estimate the solar energy potential of Trondheim city through rooftop solar PV.

Keywords Digital Surface Model · Rooftop PV · Smart city · Object-oriented classification

K. Saha (✉)
Department of Planning, School of Planning and Architecture Bhopal, Neelbad Road, Bhauri, Bhopal 462030, India
e-mail: kakolisaha@spabhopal.ac.in

Y. Frøyen
Department of Architecture and Planning, Faculty of Architecture and Design, Norwegian University of Science and Technology, Alfred Getz v. 3, N-7491 Trondheim, Norway
e-mail: yngve.froyen@ntnu.no

P. Sharma (ed.), *Geospatial Technology and Smart Cities*, The Urban Book Series,
https://doi.org/10.1007/978-3-030-71945-6_5

5.1 Introduction

Because there is no universally accepted definition of a Smart City, the conceptualization of Smart City varies. The variation depends on several factors such as the level of development, willingness to change and reform, resources and aspirations of the city residents. In India, the objective of Smart City Mission (2015) is to promote cities that provide core infrastructure and give a decent quality of life to its citizens (Smart City Mission Statement and Guidelines 2015). The core infrastructure included assured electricity supply through renewable sources of energy. As per the guideline, every Smart City proposal in India should have the provision of at least 10% of the Smart City's energy requirement coming from solar. Among the shortlisted potential Smart Cities of India, Bhopal City of Madhya Pradesh has the highest potential of harnessing solar energy due to its geographical location. Like other cities in India, Bhopal also lacks the open space for setting up large solar PV panel which can provide solar electricity to the city. To solve the problem, rooftop solar PV can be used as it does not require any extra space.

The process of Solar PV energy generation starts with converting solar light into direct current (DC) electricity using the photovoltaic (PV) effect of photovoltaic material within solar cells in solar modules (Kumar and Sudhakar 2015). Rooftops of residential, commercial or industrial premises can be used for installing rooftop PV or RTPV systems. The electricity generated from RTPV systems can be used for self-consumption and can also be supplied into the grid. Rooftop PV provides benefits at different levels. For example, it reduces dependency on grid power, minimizes the losses during transmission distribution and conversion of electricity. It may also help to generate local employment (Saha and Mandal 2016).

To estimate the solar energy potential, Saha andMandal (2016) have devised an automated method to extract roof areas of Bhopal city, Madhya Pradesh. The automated method used reflectance pattern of pixels and contextual information of image objects. Slope and aspect values were derived from digital elevation model (DEM) and analyzed using eCognition Developer (v.8.7.2). The rule-set developed for automated extraction consists of multiresolution segmentation which divides the entire image into image objects. The image objects were further classified into given class which is "Final Rooftop" in this case. Polygons that represent the roof area were then extracted using automated method. Accuracy assessment was performed using visual recognition and statistical tools. The results of accuracy assessment confirmed the reliability of the method (Saha and Mandal 2016).

In this research, the focus is to test the flexibility and robustness of the automated method. For that purpose, Trondheim City, Norway was selected because it is also a potential Smart City with different roof orientation and typology compare to Bhopal City.

Across Europe, over 240 cities with population over 100,000 are now pursuing smart policy initiatives (Euractiv 2017). Amsterdam Smart City Plan (The Netherlands) aims to reduce emissions of CO_2 and improve the environmental record of the city following the partnership strategy among different sectors like administrative, academia, citizens of Amsterdam (Angelidou 2017). Barcelona Smart City (Spain) focuses on "international promotion," "international collaboration" and "local projects" which is more than 100. The strategy is structured in same way like Netherland mentioned above (Barcelona Smart City official website 2014). Smart London Plan (United Kingdom) revolves around seven key themes in the domains of services for citizens, development of businesses, citizen engagement, networking among stakeholders and smart infrastructure (Greater London Authority 2013). PlanIT Valley (Portugal) is a private, planned Smart City to be developed in Portugal. It showcases the "Urban Operating System" which was developed by the software company Living PlanIT. The system is comprised of sensors which are placed within the city and receive various information. The information then processed and fed to devices that will help to monitor activities within the city (Living PlanIT SA official website 2013). Stockholm Smart City (Sweden) uses environmental and information technologies to monitor and control city's infrastructure to improve the ecosystem comprised of residents, industry and the public sector (Stockholm Smart City official website 2014). While the above-mentioned potential Smart Cities of Europe have detailed out the strategies, the Smart City Plan of Trondheim, Norway is in inception stage. The Smart Cities group is formed to pursue high value integrated design to support cities, citizens and society. The Smart Cities group works in close collaboration with Norwegian University of Science and Technology (NTNU) Sustainability and NTNU Energy. The Smart Sustainable Cities project of NTNU is a cross-disciplinary project in which architects, planners, designers and artists work together to develop a sustainable low carbon future (NTNU Smart Sustainable Cities 2018). NTNU energy is committed to develop technology for producing new renewable energy, especially solar energy, bio-energy, wave power and offshore wind power (NTNU Energy 2018).

Generating electricity using solar PV is still not very common in Norway. According to scholars like Andresen (2008), Halvorsen et al. (2011), and Nord et al. (2016), Norway receives enough solar energy to meet countries energy demand. In spite of the fact, generation of electricity from solar PV is still low in Norway (Merlet and Thorud 2015). But the trend is changing as there is a significant increase in the number of solar PV installations due to growing investment in solar technology during the past decade. Solar PV panels were installed both in private and public buildings. The automated method mentioned in this research can aid in this process by extracting available roof area for solar panel installation.

5.2 Background Work

In recent years, scholars and professionals have been publishing large number of literature on various aspects of Smart Cities. While some (Albino et al. 2015; Hollands 2008; Mora and Bolici 2017) worked on basic concepts and definition of Smart

City, others (Angelidou 2014, 2016; Bolici and Mora 2015) concentrated on policies and strategies. Angelidou (2017), Van Winden and Van den Buuse (2017) did a detailed study on characteristics, dimensions and conditions of a Smart City. While researchers like Stratigea et al. (2015), Streitz (2011), Tsarchopoulos et al. (2017) broadly discussed about tools and technologies for planning and development of Smart City, Elmaghraby and Losavio (2014) particularly emphasized on cybersecurity challenges of Smart City. Social aspects of Smart Cities were also studied by Carvalho (2015) and McNerney and Zhang (2011). A few of scholars (Madreiter and Haunold 2012; Mora and Bolici 2017; Viitanen and Kingston 2014; De Falco et al. 2018) dealt with issues of sustainability and resilience in the context of Smart Cities. A couple of scholars (Bakici et al. 2013; Mora and Bolici 2017; March and Ribera-Fumaz 2016) studied existing Smart Cities as case examples. A few researchers (Anthopoulos 2017; Wiig 2015; Datta 2015; Townsend 2013; Shelton et al. 2015) questioned the present concept of Smart City and compared it with the concept of "Urban Utopia". Researchers like Kitchin (2015) addressed present shortcomings of Smart Cities. Despite this growing body of literature on Smart Cities, research involving the application of remote sensing to estimate rooftop solar potential of a Smart City is limited. In this research, an automated method developed in object-oriented environment has been tested to extract available urban roof area for solar PV installation.

Application of object-based method is growing in the context of delineating urban features from remotely sensed images (Shackelford and Davis 2003). The main advantage of object-based method is that contextual, spatial and textural information can be obtained along with spectral information as the features are extracted as objects rather than pixels. Besides, it lacks "salt-and-pepper" effect which is the common drawback of pixel-based classification as features are extracted as objects (Saha et al. 2011). Another advantage would be information like asymmetry, size and orientation can easily be acquired when urban features are extracted as image objects. Researchers like Aldred and Wang (2011) applied the object-oriented classification to quantify urban roof area of Canada. Saha and Mandal (2016) used object-oriented method to extract urban roof area of Bhopal, Madhya Pradesh, India and later used the data to estimate rooftop solar PV potential. In this paper, the method applied to the Bhopal rooftops was applied to extract urban roof area of Trondheim City of Norway. The quantified roof area was further used to estimate the potential of rooftop solar PV in Trondheim City.

5.3 The Study Areas: Bhopal and Trondheim City

For the purpose of this research, residential areas within Bhopal Municipal Corporation (BMC) have been taken into consideration. Within BMC, a medium sized gated residential neighborhood, namely, "Minaal Residency" was chosen as a test area (Fig. 5.1A). The area covered by the gated neighborhood is 0.89 km^2 and it has a population of 7344. The site is comprised of residential buildings where the

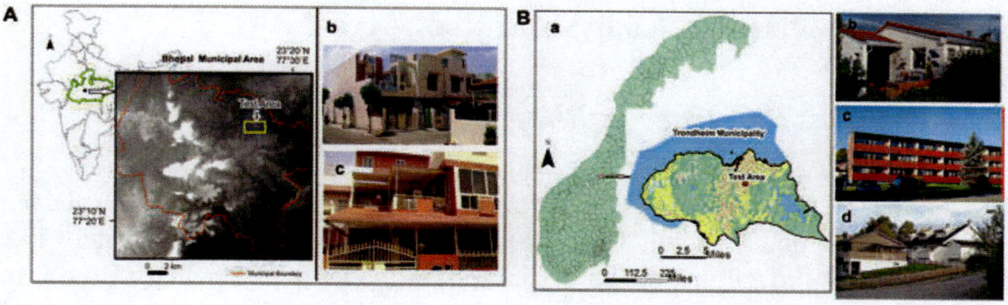

Fig. 5.1 **A** Location of residential neighborhood of Bhopal, **B** location of residential neighborhood of Trondheim City

building typology is row housing. The plot sizes of the dwelling units are variable and there are five types of plots. The site does not have any tall trees or high rise buildings. The dwelling units are one or two-storied high (Fig. 1A, C). As a result, residential buildings get ample sunlight and have high potential for harvesting solar energy through rooftop PV.

Trondheim city is located in central Norway (Fig. 5.1). To match with test site of Bhopal city, a residential neighborhood, namely, "Steindal" was chosen within Trondheim Municipality. The roof areas of "Steindal" are different from Bhopal in the context of morphology (Fig. 5.1B–D).

5.4 Data Used

Remotely sensed images in this research include World View 2 (WV2) stereo pairs and multispectral images. WV2 satellite is operated by Digital Globe Inc. the stereo images are used to extract elevation data which further helped to isolate building roofs from roads (refer Sect. 5.5.1.1. Image Pre-Processing).

The Norwegian Mapping Authority (www.kartverket.no) provided the high resolution (4 cm spatial resolution) orthophoto for performing analysis on Steindal. 25 cm spatial resolution digital surface model (DSM) and digital terrain model (DTM) were also obtained from Trondheim municipality (www.trondheim.kommune.no).

In this research, temperature data of Bhopal and Trondheim is not used as inputs as it is already established that micro-climatic parameters including ambient temperature are very suitable for PV module in Bhopal (Shukla et al. 2016) and also in Norway (Adaramola et al. 2015).

5.5 Object-Oriented Analysis of Roof Area

5.5.1 *Minaal Residency in Bhopal City*

The automated method devised to extract roof areas of Bhopal city is explained in detail in Saha and Mandal (2016). The process of extraction included steps like deriving digital surface model and digital terrain model of Bhopal city from 2 m spatial resolution digital elevation model, followed by digitization on remotely sensed image to prepare a reference map. The final step involved segmenting the input image into image objects followed by classification.

5.5.1.1 Deriving Normalized Digital Surface Model

Satellite sensors obtain information about the rooftops and roads same way because they are composed of similar material. Elevation data is used to isolate rooftops from roads. To get Elevation data, 2 m digital surface model or DSM (Fig. 5.2A) is derived from WV2 stereo pairs with 0.5 m spatial resolution. A DSM contains the elevation of every natural and/or artificial object like building, vegetation, etc. on the earth which includes earth's own elevation also. A digital terrain model or DTM is further derived from DSM using the filtering option. A DTM (Fig. 5.2B) contains elevation information of bare earth. The normalized digital surface model or nDSM (Fig. 5.2C) is derived using both DTM and DSM. The method employed was subtraction. nDSM contains information of any object standing over earth's surface

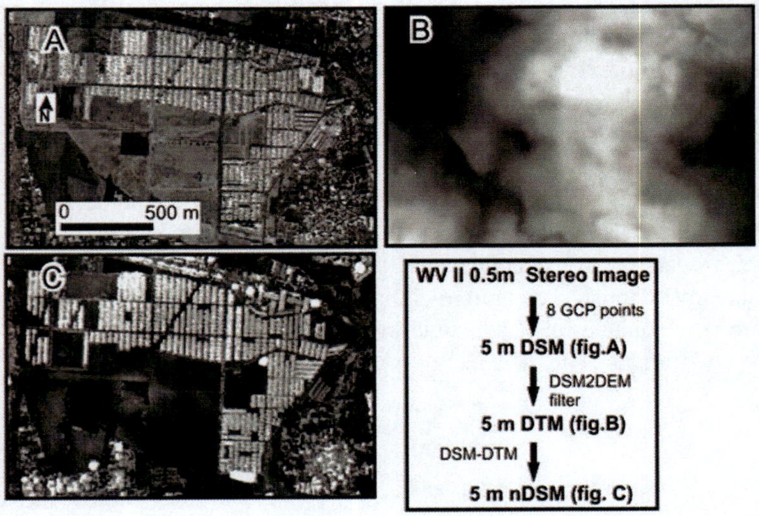

Fig. 5.2 Flow chart and intermediate outputs showing process of deriving normalized digital surface model.

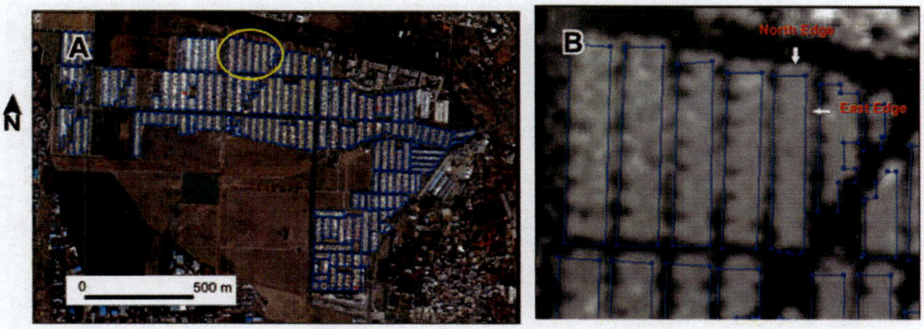

Fig. 5.3 Creating reference roof polygons: **A** digitized building polygons on multispectral image; **B** zoomed in version showing digitized block in single polygon

without including the earth's elevation. A reference map or a base map was prepared through manual digitization of building polygons of the test area within eCognition Developer software.

5.5.1.2 Production of Reference Map

The high resolution (2 m spatial resolution) World View 2 multispectral image is used to prepare a reference map or base map. The site has the housing typology of row housing. Each row contains 8 to 36 dwelling units and separated from other rows by circulation paths. In the image (Fig. 5.3A), it can be observed that individual dwelling units are not visible. Each row was digitized as an individual polygon and is referred as building polygon in subsequent analysis. The digitized polygons were put on the top of nDSM layer and it was found that areas outside north and east boundaries of buildings are considered as part of the buildings (Fig. 5.3B). They are termed as north and east edges (Fig. 5.3B).

5.5.1.3 Extracting Final Roof Area as Classified Polygons

Relief shaded image and near infra red (NIR) band are used to identify north and east edges. Sobel edge detection algorithm available through Rolta Geomatica is used for that purpose. The resultant image shows edges in a brighter tone (Fig. 5.4A, B). Segmentation is followed by classification using information regarding spectral reflectance and orientation has been used (Fig. 5.4C). Two classes were created, namely, east edge and north edge. The north edge and east edge classes were then exported to vector files in. shp file format. The shape file is further used to extract polygons representing final roof area for each row. The building polygons were then classified using membership function in building height and contextual information. The context was defined in terms of relative border to both Edges (Fig. 5.4D).

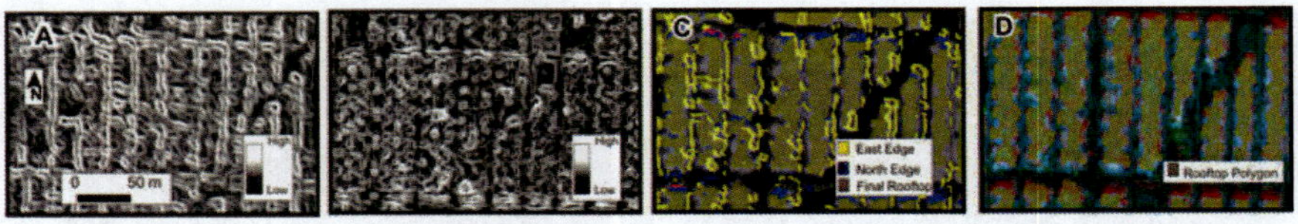

Fig. 5.4 Extracting edges from relief shaded image and NIR band **A** east edge, **B** north edge, **C** north edge and east edge classes, **D** classified building polygons

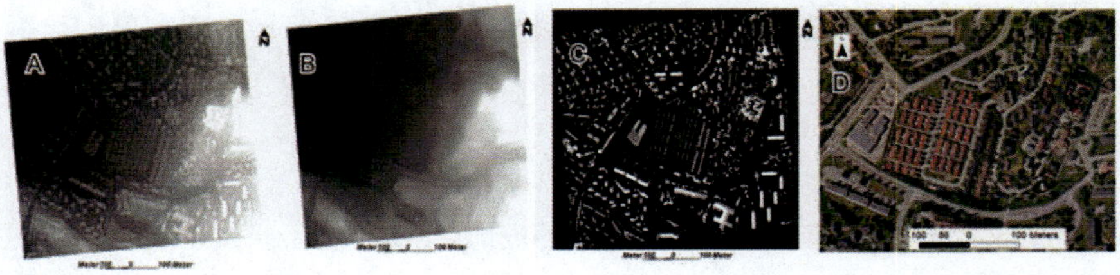

Fig. 5.5 Image pre-processing, **A** DSM, **B** DTM, **C** nDSM, **D** orthophoto of Steindal

5.5.2 Steindal in Trondheim City

Same process like Bhopal is also followed to extract urban roof area of Trondheim city.

5.5.2.1 Deriving Normalized Digital Surface Model for Steindal

The Trondheim Municipality has light detection and ranging data (LIDAR) for the entire city of Trondheim. They have generated both DSM (Fig. 5.5A) and DTM (Fig. 5.5B) from that LIDAR data. As a part of the image pre-processing, nDSM (Fig. 5C) is generated by subtracting DTM from DSM. Since the orthophoto used to extract roof areas of Steindal is of very high resolution, the individual roof areas are visible (Fig. 5D). Because of this reason, the method for automated extraction focused on extracting individual roof area instead of rows of building.

5.5.2.2 Extracting Individual Roof Area for Steindal

Because of climate, residential units at Steindal have chimneys on their roof. To extract roof edge and chimneys, the reflectance in the red band has been used. To increase the contrast between roof edges from the surrounding the Sobel edge detection filter was used on red band. Chimney and roof boundary were represented in brighter tone in the filtered red band (Fig. 5.6A). Since roofs have different slope compare to trees, slope information (Fig. 5.6B) is used to segregate trees from the roof.

Fig. 5.6 **A** Filtered red band, **B** slope layer to differentiate between trees and roofs, **C** merged image

Then the all layers of normalized digital surface model, slope layer, filtered red band and orthoimage were stacked to create a merged image (Fig. 5.6C).

The merged image was then segmented using Chessboard Segmentation technique in eCognition developer software. The resultant image contains squared shape image objects of given size (Fig. 5.7A). Using information from filtered red band and slope, the chimneys are separated (Fig. 5.7B). After this, the image was further segmented using multiresolution (MS) segmentation technique where values of normalized digital surface model were used (Fig. 5.7C). MS segmentation merges the pixels in the image into bigger objects using homogeneity criteria such as color (spectral information) and compactness (representing shape). These criteria can be set in combination or individually. The user can also set the scale of the output using the scale parameter. If the scale value is set to high number, the resultant image objects will be larger and vice versa (eCognition Developer 2014). To utilize information from normalized digital surface model, values for color were set high (refer flow chart of Fig. 5.7). After image objects for roof area were extracted, they were classified under "Roof" class (Fig. 5.7D). Classified image objects were then exported as shape file and used in subsequent accuracy assessment.

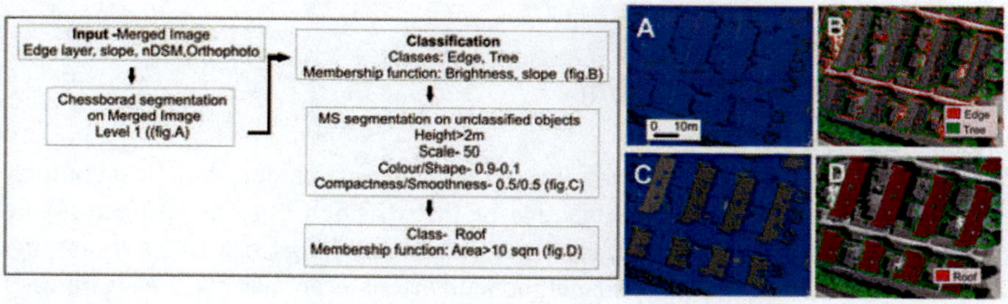

Fig. 5.7 Automated extraction of urban roofs of Steindal

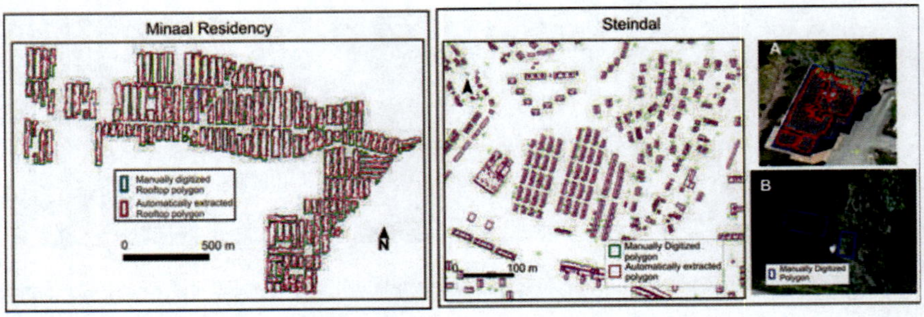

Fig. 5.8 Visual comparison between automated and manually extracted roof areas

5.6 Accuracy Assessment

5.6.1 Visual Matching

The accuracy assessment of the proposed method starts with visual comparison. For that purpose, roof polygons of Minaal Residency, Bhopal City and Steindal, Trondheim City that are automatically extracted were overlaid on the top of manually digitized polygons (Fig. 5.8). For Minaal Residency, the reference polygons were obtained by performing digitization in eCognition Developer software itself, whereas for Steindal the reference roof polygons were acquired from the Trondheim Municipality. The results of matching show that there is a 100% overlap between automatically and manually extracted roof areas. According to Fig. 5.8, some of the manually delineated roof polygons were not identified by the automated method. The reason might be by the time the aerial photographs were taken, the roofs were covered by vegetation (Fig. 5.8A, B). It proves the applicability of the automated method to map real-time data.

5.6.2 Statistical Comparison

After getting a satisfactory result in visual comparison, detailed statistical comparison was performed. Since the objective of this research is to quantify available rooftop areas for PV installation, accuracy assessment is performed by comparing the area extracted by the automated method with that of reference maps. Area for each building polygons extracted through two different methods was calculated in the software itself (Table 5.1). Then basic statistics were calculated for Minaal Residency, Bhopal City, India and Steindal, Trondheim City, Norway.

According to Table 5.1, values for building polygons of Minaal Residency are higher than Steindal. This is because each row of Minaal Residency is digitized as a single polygon. Each row contains a number of dwelling units ranging from 8 to 36. Individual units were not visible in satellite images due to its spatial

Table 5.1 Means of the area calculated for building polygons of two sites

Basic statistics				
	Mean	Standard deviation	Maximum	Minimum
Minaal residency				
Area (sqm)				
Manual	2219.49	1030.00	4676.00	132.00
Auto	1367.05	1149.68	4896.00	12.00
Steindal				
Area (sqm)				
Manual	132.96	250.65	3271.59	10.08
Auto	139.52	195.59	1894.87	10.06

Table 5.2 Comparison of mean values

	Z score for U score	*P* value when a = 0.05	*MWW* decision
Minaal residency			
Area (sqm)	−7.03	0.000	Different
Steindal			
Area (sqm)	−0.586	0.56	Same

resolution. Table 5.1 shows that mean values for area extracted in two different methods are different. To check whether the difference between means values is statistically significant or not, non-parametric test for comparing means, namely, Mann–Whitney–Wilcoxon (MWW) test (Table 5.2) was performed.

Table 5.2 shows that the area measured using the two different methods is statistically similar for Steindal but statistically different for Minaal Residency.

To investigate the matter more closely, values for area extracted in the two methods are plotted for both Minaal Residency and Steindal (Fig. 5.9). Vertical bars of histogram represent manual data and the superimposed markers show automated data.

In histogram also, automated method identifies more building polygons with small area compared to the manual method in Minaal Residency. Detailed investigation was performed to find out the reason behind this difference. It was found that the manual method tends to digitize boundaries of building polygons in smooth straight lines which are not necessarily a true representation of building outline. On the other hand, the automated method extracts polygon boundaries following the input information regarding elevation and reflectance value. As a result, building polygons have irregular sides (Fig. 5.10).

On the other hand, histogram of Steindal shows that the automated method routinely identified both larger and smaller values than the manual method.

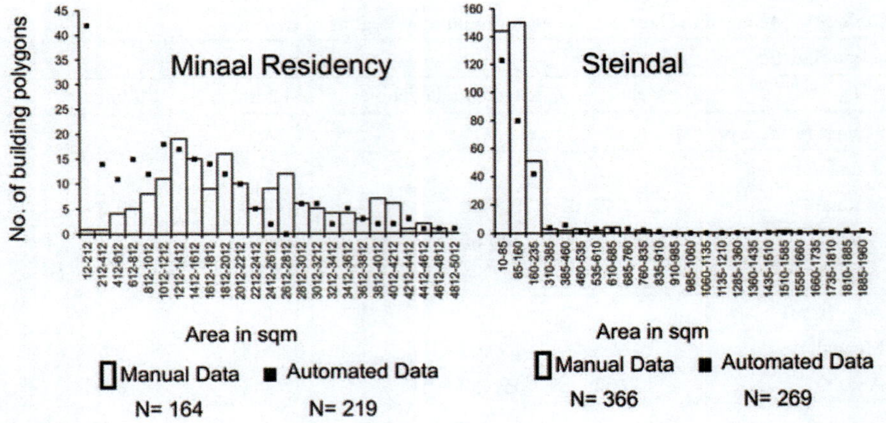

Fig. 5.9 Histogram comparing areas of the automatically and manually delineated building polygons

Fig. 5.10 Jagged and irregular boundary of automatically extracted polygon

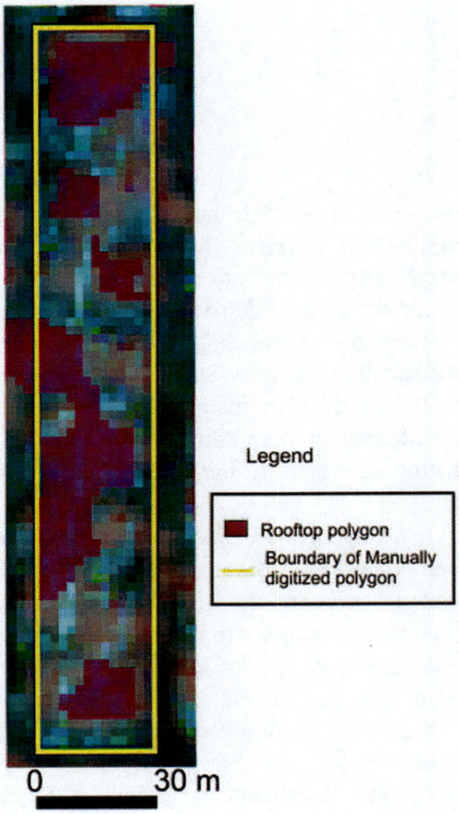

5.7 Estimating Solar Energy Potential from Rooftop PV

5.7.1 Minaal Residency, Bhopal, Madhya Pradesh

After obtaining the total roof area of a region, the next step was to estimate the area that could be available for solar photovoltaic applications. The potential for solar power output depends on area available for installing rooftop solar PV. Factors like shading, orientation, installation process can affect the area available for Rooftop PV. For example, shades from neighboring building or trees may obstruct solar energy to reach the rooftop PV. Same way, orientation of roofs such as whether pitched or flat can have a significant impact on amount of solar radiation it receives. Besides these, roof spaces are used for other purposes such as installing ventilation, heating/air conditioning, and dormers or chimneys. In India, roofs are also used by its residents for hanging out, taking rest, drying clothes, etc. (Saha 2017). As a result, entire roof cannot be taken into consideration while estimating area available for rooftop panel installation. Performing simulation or statistical analysis for obtaining reduction factor was not within the scope of this research. Several literatures by Scartezzini et al. (2002), Lehmann and Peter (2003), Ghosh and Vale (2006), Pillai and Banerjee (2007), Izquierdo (2008), Wiginton et al. (2010) were studied and reduction factor of 0.30–30% was taken from the literature review (Pillai and Banerjee 2007). It accounted for the factors mentioned above.

The equation for calculation available roof area is given below:

$$APV = 0.3 * ARoof \tag{5.1}$$

$$When\, APV = Roof\ Area\ Available,\ A\,Roof = Total\ roof\ area,\ 0.3 = Chosen\,fraction$$

Since the total roof area automatically extracted for the Test Area is 0.29 km^2

$$APV = (0.3) * 0.29\,\text{km}^2 = 0.087\,\text{km}^2$$

The derived value is then used to calculate solar energy output/day via equation:

$$E = Imd * e * APV \tag{5.2}$$

$$When,\ Imd = Mean\ Daily\ Global\ Insolation\ on\ a\ horizontal\ plane,\ e = module\ efficiency$$

Mean Daily Global Insolation is calculated as of 5.85 kWh/m^2/day (MNRE) and the module efficiency for polycrystalline silicon panels, calculated as an average of 15% (TATA Solar 2018). Table 5.3 summarizes the daily energy demand on a per

Table 5.3 Comparison between solar energy output and energy consumption (Minaal Residency)

Total energy output potential (kWh/day)	Energy output potential Per capita (kWh/day)	Per capita energy consumption (kWh/day)
76,342.50	10.39	3.91
The tentative total population of Minaal Residency is 7344. (76,342.50/7344) = 10.39		

capita basis at household level. Electricity department of Bhopal divides the entire municipal area into different zones. These zone boundaries do not overlap with administrative boundaries of municipality. Single feeder is installed at each zone and energy consumption data is recorded from zone-wise feeder. It may be possible that a single feeder can supply energy to multiple administrative units. Because of this reason, a sample survey was conducted in the test area to obtain the energy consumption data. Random sampling was done keeping house types in mind. For each type, monthly electricity bills for one calendar year were obtained from ten households and daily average was further calculated (Table 5.3).

As may be noted from Table 5.3, 10.40 kWh/capita energy can be produced daily from rooftop solar PV with efficient polycrystalline-silicon panels within a rooftop PV module. The table also indicates that per capita daily energy consumption in the test area is 3.91 kWh. From this calculation, we can safely assume that the rooftop solar PV in the test area has the potential of providing 265% of daily energy requirement. If the area adopts net metering system, it can supply the extra energy to the grid after full filling its daily energy requirement.

5.7.2 Steindal, Trondheim, Norway

Since we already excluded the chimney, vegetation while extracting the roof area for Steindal dwelling units, no reduction factor is applied to estimate available roof area of Steindal houses.

$$APV = ARoof \tag{5.3}$$

$$APV = 0.037\,\text{km}^2$$

The next step was to obtain solar energy potential for Steindal using Eq. 5.2.

The mean daily global insolation on a horizontal plane for Trondheim, Norway is 2.53 kWh/m^2/day; 15% is considered for "e" as the efficiency of a PV panel does not depend on its location, rather it depends on technology and materials of the PV panel. Table 5.4 shows potential energy output for Steindal.

As may be noted from Table 5.4, 15.00 kWh/capita energy can be produced daily from rooftop solar PV with efficient polycrystalline- silicon solar PV panels. The table also indicates that per capita daily energy consumption in Steindal is 25.82

Table 5.4 Potential energy output and consumption pattern of Steindal

Potential solar energy output (kWh/day)	Potential Per capita energy output (kWh/day)	Per capita energy consumption (kWh/day
14,257.96	15.00	25.82
Total population of Steindal which is 950 (Statistics Norway 2016). (14,257.96/950) = 15.00		

kWh (Statistics Norway 2015). The high consumption rate is due to the heating requirement during the long winter months in Trondheim City. According to Table 5.4, solar energy can full fill 58.09% of per day energy demand.

5.8 Conclusion and Wider Implications

The above explanatory analysis points to the flexibility and robustness of the automated method to extract urban rooftop. The automated method was developed in eCognition Developer using object-oriented approach by Saha and Mandal (2016) using eCognition Developer and based on a set of rules. A small residential neighborhood namely Minaal Residency located within Bhopal municipal boundary (BMC) was selected as a test area to apply the proposed method. After successful extraction of total roof area of Minaal Residency, an accuracy assessment was performed using manually digitized building polygons as reference. The accuracy assessment shows that morphometry calculated from automatically extracted polygon is close but not identical to morphometry calculated from manually digitized building polygons. Detailed investigation reveals that automated method produces results close to actual measurements compare to manual digitization. Manual digitization has its own biases and errors which may influence the shape of the building polygons. Next, roof area available for solar PV installation was calculated by multiplying the total roof area by a reduction factor taken from the literature. The potential of solar energy production from this available roof area was estimated for specific solar PV modules. It is found that the rooftop solar PV in the test area has the potential of providing 265% of daily energy requirement.

The automated method was then tested on urban roofs of Trondheim, Norway from a high resolution orthophoto. The test area within Trondheim was Steindal which is identical to Minaal Residency in terms of building use. The rule-set developed for extraction of roofs of Minaal Residency was partly modified because of the difference in roof typology. While the roofs of Minaal residency are flat, roofs of Steindal are the combination of both flat and pitched. Because of this, shadows of buildings were considered as the main factor for extracting rooftops of Minaal residency. On the other hand, slope, elevation and reflectance pattern in red band played a major role to extract roofs of Steindal. After successful extraction of total roof area in Steindal, an accuracy assessment was performed with reference to manually digitized total roof area and it was concluded that automated method is both reliable and flexible as the results are consistent for different roof types.

Detailed analysis also shows that the automated method performed better in orthophoto of Steindal than satellite images of Minaal Residency. For example, automated method captured details like vegetation cover, chimneys on individual roofs of Steindal. As a result, the total roof area extracted was considered as the available roof area for solar PV panel installation and application of reduction factor was not necessary. It proves the fact that the efficiency of the automated method increases with increase of spatial resolution of the image. The roof area data were further used to estimate the solar energy potential of Steindal and it was found that rooftop solar PV can contribute significantly to cater energy demand of that area.

Effective, efficient and resilient urban infrastructure is one of the major constituents of any Smart City strategy. Though energy sector is the main pillar of urban infrastructure, it is also one of the major reasons for urban pollution as most of it generated through non-renewable sources. Rooftop solar PV can be the solution to this problem. That is the reason why most of the Smart City plans across the world have energy generation through rooftop solar PV as a major component. The automated method can enhance planning and implementation of rooftop solar PV installation projects. The research presented in this paper shows that the method is flexible and thus can be adopted by any city to estimate city's potential of solar energy through rooftop solar PV. Thus the method can play a vital role to make the energy infrastructure of a Smart City efficient and resilient.

Acknowledgements The authors of this paper would like to thank and acknowledge the Department of Science and Technology, Government of India for funding this research via Project No. SR/FTP/ES-24/2012 and Trondheim Kommune for providing the geospatial data of Steindal, Trondheim, Norway.

References

Adaramola MS, Vågnes EE (2015) Preliminary assessment of a small-scale rooftop PV-grid tied in Norwegian climatic conditions. Energy Convers Manage 90:458–465

Albino V, Berardi U, Dangelico RM (2015) Smart cities: definitions, dimensions, performance and initiatives. J Urban Technol 22(1):3–21

Aldred DA, Wang J (2011) A method for obtaining and applying classification parameters in object-based urban rooftop extraction from VHR multispectral images. Int J Remote Sens 32:2811–2823. https://doi.org/10.1080/01431161003745590

Andresen I (2008) Planning of solar heating for low-energy and passive houses—an introduction. Sintef Building Infrastructure

Angelidou M (2014) Smart city policies: a spatial approach. Cities 41(Supplement 1):S3–S11

Angelidou M (2016) Four European Smart City Strategies. Int J of Soc Sci Stud 4(4):18–30

Angelidou M (2017) The role of smart city characteristics in the plans of fifteen cities. J Urban Technol 24(4):3–28

Anthopoulos L (2017) Smart utopia vs. smart reality: learning by experience from 10 smart city cases. Cities 63:128–148

Barcelona Smart City official website, "BCN Smart City" (2014). https://ajuntament.barcelona.cat/estrategiadigital/en. Accessed 20 April 2017

Bakici T, Almirall E, Wareham J (2013) A smart city initiative: the case of Barcelona. J Knowl Econ 4(2):135–148

Bolici R, Mora L (2015) Urban regeneration in the digital era: how to develop smart city strategies in large European cities. TECHNE-J Technol Archit Environ 10:110–119

Carvalho L (2015) Smart Cities from Scratch? A socio-technical perspective Cambridge. J Regions, Econ Soc 8(1):43–60

Datta A (2015) New urban utopias of postcolonial India: 'Entrepreneurial urbanization' in Dholera smart city. Gujarat. Dialogues Hum Geogr 5(1):3–22

de Falco S, Angelidou M, Addie JPD (2018) From the "smart city" to the "smart metropolis"? Building resilience in the urban periphery. Eur Urban Reg Stud 0969776418783813

eCognition Developer (2014) Userguide Version 9 (270 p). Trimble Germany GmbH Munich

Elmaghraby AS, Losavio MM (2014) Cyber security challenges in smart cities: safety, security and privacy. J Adv Res 5(4):491–497

Euractiv (2017) How many smart cities are there in Europe? https://www.euractiv.com/section/digital/infographic/how-many-smart-cities-arethere-in-europe/. Accessed 30 November 2017

Ghosh S, Vale R (2006) Domestic energy sustainability of different urban residential patterns: a New Zealand approach. Int J Sustain Dev 9(1):16–37

Greater London Authority, Smart London Plan (UK: Greater London Authority, 2013). https://www.london.gov.uk/sites/default/files/smart_london_plan.pdf. Accessed 21 April 2017

Halvorsen UM, Bernhard P, Fritjof S, Lars B, Inger A, Simonsen I (2011) Mulighetsstudie—Solenergi i Norge. Sintef Building and Infrastructure, KanEnergi

Hollands RG (2008) Will the real smart city please stand up? City 12(3):303–320

Izquierdo S, Rodrigues M, Fueyo N (2008) A method for estimating the geographical distribution of the available roof surface area for large-scale photovoltaic energy-potential evaluations. Sol Energy 82:929–939

Kitchin R (2015) Making sense of smart cities: addressing present shortcomings. Camb J Reg Econ Soc 8(1):131–136

Kumar BS, Sudhakar K (2015) Performance evaluation of 10 MW grid connected solar photovoltaic power plant in India. Energy Rep 1:184–192

Lehmann H, Peter S (2003) Assessment of roof and façade potentials for solar use in Europe. Institute for Sustainable Solutions and Innovations (ISUSI), Aachen, Germany. https://sustainable-soli.com/downloads/roofs.pdf.

Living PlanIT SA Official Website (2013). https://www.living-planit.com. Accessed 20 April 2017

Madreiter T, Haunold V (2012) Smart city Vienna: sustainable success. In: Widmann H (ed) Smart city: Viennese expertise based on science and research. Schmid Verlag, Vienna, pp 35–40

March H, Ribera-Fumaz R (2016) Smart contradictions: the politics of making Barcelona a self-sufficient city. Eur Urban Reg Stud 23(4):816–830

McNerney PJ, Zhang N (2011) Smarter cities: making societies smarter. ACM XRDS: crossroads 18(2):48–48

Merlet S, Thorud B (2015) Solenergi i Norge: Status og framtidsutsikter, in: NorskKlimastiftelse. In: Klimastiftelse N (ed) Hva Betyr Solenergirevolusjonen? Norwegian Climate Foundation, Bergen, pp 30–34

Mora L, Bolici R (2017) How to become a smart city: Learning from Amsterdam. In: Bisello A, Vettorato D, Stephens R, Elisei P (eds) Smart and sustainable planning for cities and regions: results of SSPCR 2015. Springer, Cham, pp 251–266

MNRE. India Solar Resource MAPs (2018). https://mnre.gov.in/sec/solar-assmnt.htm. Last Accessed 06 March 2018

Nord N, Qvistgaard LH, Cao G (2016) Identifying key design parameters of the integrated energy system for a residential Zero Emission Building in Norway. Renew Energy 87:1076–1087. https://doi.org/10.1016/j.renene.2015.08.022

NTNU Energy (2018). https://www.ntnu.edu/energy/research-area. Accessed 31 October 2018

NTNU Smart Sustainable Cities (2018). http:// www.ntnu.edu/smartcities. Accessed 31 October 2018

Pillai IR, Banerjee R (2007) Methodology for estimation of potential for solar water heating in a target area. Sol Energy 81(2):162–172

Saha K, Wells NA, Munro-Stasiuk M (2011) An object-oriented approach to automated landform mapping: a case study of drumlins. Comput Geosci 37(9):1324–1336

Saha K, Mandal NR (2016) Estimating solar PV potential for sustainable energy planning in Tier-II cities of India: case of Bhopal City. Curr Urban Stud 4:356–375. https://doi.org/10.4236/cus.2016.43024

Saha K (2017) A remote sensing approach to smart city development in India: case of Bhopal City, Madhya Pradesh. In: Baguma R, De R, Janowski T, Nielsen MM (eds) Proceedings of the special collection on egovernment innovations in India (ICEGOV '17). ACM, New York, NY, USA, pp 70–75. https://doi.org/10.1145/3055219.3055232

Scartezzini J-L, Montavon M, Compagnon R (2002) Computer evaluation of the solar energy potential in an urban environment. In: Paper presented at EuroSun2002. Bologna, Italy

Shackelford A, Davis CH (2003) A combined fuzzy pixel-based and object-based approach for classification of high-resolution multispectral data over urban areas. IEEE Trans Geosci Remote Sens 41:2354–2363. https://doi.org/10.1109/TGRS.2003.815972

Shelton T, Zook M, Wiig A (2015) The 'actually existing smart city.' Camb J Reg Econ Soc 8(1):13–25

Shukla AK, Sudhakar K, Baredar P (2016) Design, simulation and economic analysis of standalone roof top solar PV system in India. Sol Energy 136:437–449

Smart city mission statement and guidelines, Government of India, Ministry of Urban Development, June 2015

Statistics Norway (2015) Electricity balance. https://www.ssb.no/en/energi-ogindustri/statistikker/elektrisitet/aar/2015-12-22#content. Retrieved 5 May 2016

Statistics Norway (2016) Population projections, 2016–2100. https://www.ssb.no/en/befolkning. Retrieved 30 June 2016

Stockholm smart city official website (2014) The smart city. https://international.stockholm.se/city-development/the-smart-city/. Accessed 20 April 2017

Stratigea A, Papadopoulou CA, Panagiotopoulou M (2015) Tools and technologies for planning the development of smart cities. J Urban Technol 22(2):43–62

Streitz NA (2011) Smart cities, ambient intelligence and universal access. In: International conference on universal access in human-computer interaction (pp 425–432). Springer, Berlin, Heidelberg

TATA Solar (2018). https://www.greenice.in/download/Datasheet%20-%20TP250.pdf. Last Accessed March 2018

Townsend AM (2013) Smart Cities: big data, civic hackers, and the quest for a new utopia. WW Norton & Company, New York

Tsarchopoulos P, Komninos N, Kakderi C (2017) Accelerating the uptake of Smart City applications through cloud computing. Int J Soc, Behav, Educ, Econ, Bus Ind Eng 11(1):129–138

Van Winden W, Van den Buuse D (2017) Smart city pilot projects: exploring the dimensions and conditions of scaling up. J Urban Technol 24(4):51–72

Viitanen J, Kingston R (2014) Smart cities and green growth: outsourcing democratic and environmental resilience to the global technology sector. Environ Plan A 46(4):803–819

Wiig A (2015) IBM's smart city as techno-utopian policy mobility. City 19(2–3):258–273

Wiginton LK, Nguyen HT, Pearce JM (2010) Quantifying rooftop solar photovoltaic potential for regional renewable energy policy. Comput Environ Urban Syst 34:345–357

www.kartverket.no. Last Accessed September 2016

www.trondheim.kommune.no. Last Accessed September 2016

Chapter 6
Analyzing and Predicting Urban Expansion and Its Effects on Surface Temperature for Two Indian Megacities: Bengaluru and Chennai

Chandan Mysore Chandrashekar, Nimish Gupta, and Bharath Haridas Aithal

Abstract Anthropogenic activities have significantly transformed landscape over the past century, such that it has affected the natural processes in atmosphere. Understanding temporal land use change pattern helps in defining and managing environmental cycles, changes in climate, surface temperature alterations, human health, biodiversity, etc. These effects are mostly linked with cities sprawling beyond their administrative boundaries and jurisdiction limits to form an extortionate place for a residential and commercial purpose. Urban expansion is a dynamic process involving expansion of urban imprint as a result of multifaceted interaction of hominids with their surroundings, in response to the increased demographic pressure. Urban population has been substantially increasing and unplanned urbanization is causing urban sprawl, urban heat island effect, increased population, negative environmental and geographical influences, increased pollution levels, climate change, and increased surface temperature. This rise in surface temperature is due to the increased solar influx and trapped heat between adjacent building causing alterations in pattern of rainfall that affect the overall heat budget of the earth's surface leading to thermal discomfort. Researchers across the globe are trying to combat these perilous challenges considering sustainability as a top priority. Policymakers, planners, local administrative authorities, and government bodies of Indian cities are excessively concentrating on understanding and visualizing spatio-temporal land use changes, pattern of urbanization, effects of landscape alterations on surface temperature and climate change, and prediction of future urban landscape for better planning of future cities. This research considers two Indian megacities—Bengaluru and Chennai, primarily to define urban growth, its temporal urban footprint and its relation to thermal comfort of the residents through land surface temperature analysis. The outcome of this analysis would aid in improving urban decision-making in terms of support systems through visualization.

C. Mysore Chandrashekar · N. Gupta · B. H. Aithal (✉)
Ranbir and Chitra Gupta School of Infrastructure Design and Management, Indian Institute of Technology Kharagpur, Kharagpur 721302, West Bengal, India
e-mail: bharath@infra.iitkgp.ac.in

P. Sharma (ed.), *Geospatial Technology and Smart Cities*, The Urban Book Series,
https://doi.org/10.1007/978-3-030-71945-6_6

Keywords Urban · Land use analysis · Land surface temperature · Modeling land use · Cellular automata

6.1 Introduction

6.1.1 Urbanization and Urban Growth, Global and Indian Perspective

The term "urban" is derived from Roman word "Urbanus". Definition varies based on context, country, demography, physical, economic, social factors, etc. Urban area in a generic sense refers to a continuous stretch of paved surfaces, forming towns to cities, cities to megacities, and megacities to metropolis (Scott 2001; Basten 2010; Lang and Dhavale 2005). Urban areas are characterized by a core area with mostly concrete and paved structures. Public buildings, such as administrative offices, courts, commercial centers, schools, and hospitals, stand out prominently in an urban landscape. Urbanization can be termed as a process that may cause significant impact on economic and social aspects and is termed as irreversible in definition. The dynamic nature of the urban growth would directly impact the functions of the landscape, its structure (Dietzel et al. 2005; Sun et al. 2014). Due to a global rise in population and issues relating environment, it necessities researchers to work in the domain of urbanization with a multidisciplinary approach comprising of Sociology, Statistics, Geography, Economics, Political Science, etc. with respect to different countries and states.

6.1.2 Patterns of Urban Growth in India

Significance of urbanization process is to be viewed as a global phenomenon because more than half of the earth's population is already a part of urban areas. Urban areas, consistently increasing with urban population, have given impetus to understand urban morphology in depth and its implications on surrounding nature. Rapid urban growth in Indian scenario can be attributed to rural to urban migration, re-classification of cities, improved health facilities, etc. in urban areas. With the commencement of various urban development policies and schemes such as Smart Cities, AMRUT (Atal Mission for Rejuvenation and Urban Transformation), JNNURM (Jawaharlal Nehru National Urban Renewal Mission), NERUDP (North Eastern Region Urban Development Programme), and Capacity Building for Urban Local Bodies have promoted migration to core, transition or hinterland areas (Shaw and Das 2017). Figure 6.1 shows population trends of major Indian cities for the year 1991–2001–2011. The change in population is described in Fig. 6.2. Growth of pre-existing metropolitan cities had become slow while newer UA's, for instance, Bengaluru, Ahmedabad, Hyderabad, and Pune grew faster. India already has three

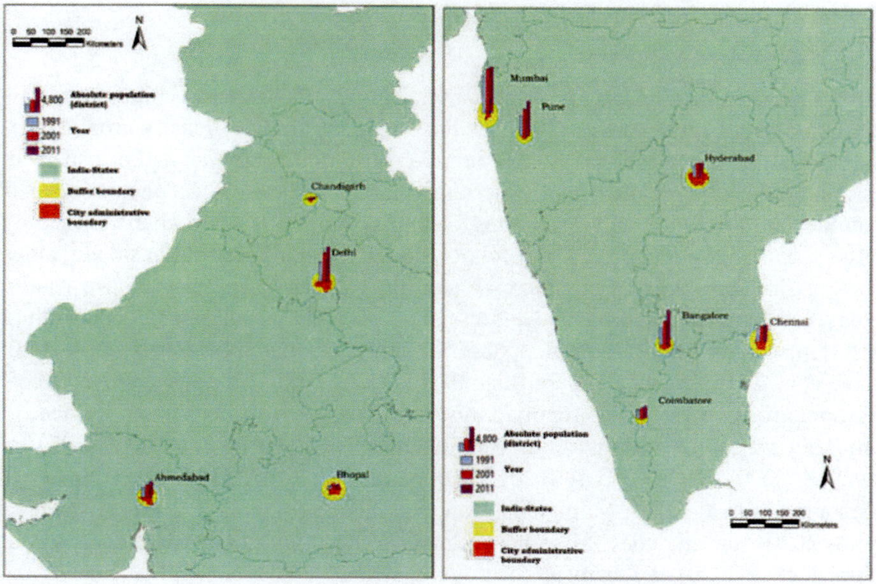

Fig. 6.1 North (left) and South (right) Indian cities and populations 1991–2001–2011 (*Source* Data obtained from Census of India. Map created by author)

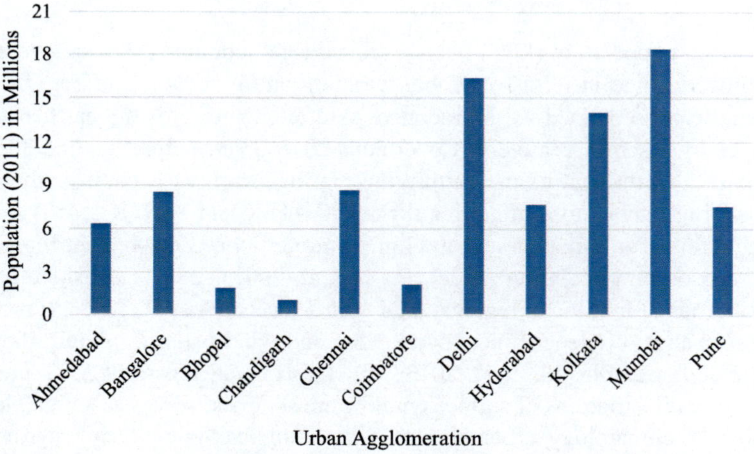

Fig. 6.2 Population of major cities in India (2011)

megacities (Mumbai, Delhi, and Kolkata urban agglomeration) of population greater than 10 million and 42 Conurbation areas or million plus cities (Census of India 2011). By 2025, Asia single-handedly will have at least 28 megacities. The rate of urbanization in India shows moderate progression (24% in 1981, 26% in 1991, 28% in 2001 and crossing 31.16% as per 2011 census).

6.1.3 Effects of Unplanned Urbanization

Urban sprawl is the direct outcome of unplanned urbanization. Urban sprawl is defined by Taubenbock et al. (2009) as the dominance of low-density urban settlements and the change of previously mono-centric compacted city into dispersed, discontinuous, fragmented, and polycentric urban patterns. This phenomenon is common in developed and developing countries; sprawl is often characterized by largely scattered population along countryside with poor access to basic amenities and missing key hubs of commercial activities. Urban investigators have used a combined geospatial approach to address urban sprawl all-round the globe (Sudhira and Ramachandra 2007; Jat et al. 2008; Batisani and Yarnal 2009; Bhatta et al. 2010; Jiang et al. 2016; Yue et al. 2016; Bharath et al. 2017). The wide spectrum of effects of rapid urbanization is also reported in studies ranging from ecosystem services to climate, agriculture, and environment (Taubenbock et al. 2009; Bharath et al. 2013a, b, 2018, 2020; Pandey and Seto 2015; Bounoua et al. 2015; Mehmood et al. 2016; Kantakumar et al. 2016; Muthamilselvan et al. 2016; Brian et al. 2017; Aishwarya et al. 2020) that includes urban heat island (Weng et al. 2004; Chen et al. 2006; Nimish et al. 2020) and many applications as such.

6.1.4 Land Surface Temperature

Land surface temperature (LST) plays a crucial role in defining the energy balance and climate change in terms of global warming and greenhouse effect (Jia et al. 2007) and can be defined as the radiative skin temperature of the earth's surface as viewed by a sensor remotely (Copernicus 2018; Ese Sentinel Online 2018). It is the basic determinant of the earthly thermal behavior, as it controls the effective blistering temperature of earth's surface (AATSR and SLSTR, 2018) and is a vital input for various models such as climate variability model, weather forecasting model, and ocean circulation model (Dash et al. 2002). LST is considered as an important indicator at local, regional, and global levels (Li et al. 2013). It modifies energy exchange, biogeochemical cycle, crop and wind pattern, rainfall, biodiversity, and ecology (Bharath et al. 2013a; b). Information regarding LST provides spatio-temporal variations of surface equilibrium state and serves as a vital element in subjects of climatology, urban climate, vegetation health monitoring, agrometeorology, estimation of greenhouse gases, hydrological modeling, urban heat islands, etc. (Schmugge and Becker 1991; Li and Becker 1993; Anderson et al. 2008; Li et al. 2013; Khandelwal et al. 2018). It provides information about spatio-temporal disparities of surface equilibrium and is a significant factor in various fields such as climate change, urban climate and environmental studies, vegetation monitoring, agrometeorology, assessment of hydrology, and evapotranspiration rates (Li et al. 2013). Estimation of LST depends upon albedo, moisture content in soil and atmosphere, health and density of vegetation, surface characteristics, season-of-year, and time-of-day

(Sun and Kafatos 2007; Copernicus 2018). Climatic changes and increased thermal discomfort due to increasing demographic pressure, various activities concentrated in urban areas such as construction, transportation, power generation at non-point sources to satisfy the needs of hominids is escalating the concentration of pollutants in the form of particulate matter (PM10, PM2.5, PM1, and Suspended PM) and Greenhouse Gas such as CO_2, CO, NO_X, SO_X, O_3, and CH_4,. These act as a source for an increase in LST and lead to the emergence of urban heat islands. To conceptualize urban climate, interactions between LULC, atmospheric parameters, and LST have to be understood (Santra and Mitra 2017) and with the introduction of sensors in thermal infrared band the possibility to measure LST across the globe spatially rather than ground point-based values. LST requires sensors which can acquire data from 8 to 15 μm in electromagnetic radiation (EMR) spectrum. Some of the satellites that provide data in this range are Landsat series (Landsat 4, 5, 7, and 8), NOAA, GOES, MODIS, ASTER, AATSR, SEVIRI (Tomlinson et al. 2011). Efforts have been carried out in this field and a number of algorithms have been developed for estimating LST. Single window, Split-window, Radiative transfer equation, etc. are few common algorithms used. Researchers have been working to improvise these algorithms for getting better and accurate temperature values.

6.1.5 Urbanization and Sustainable Development

Last decade has seen extended debates, research, and huge literature studies about growing cities and their local to global scale impact. Sustainability-related questions raised in WCED (1987), followed by UNCED famously known as Rio summit (1992), WSSD (2002), UNCSD (2012), and the latest UNSDS (2015) marked historic milestone with title "Transforming our world: the 2030 Agenda for Sustainable Development" in thinking of global community in adopting the Agenda for Sustainable Development (SD) through vision 2030. Agenda included a widely visualized framework to cater to the development and sustenance of natural resources through seventeen development goals called sustainable development goals (SDGs) with 169 sub-targets. In this context, recently Indian Ministries, for instance, Ministry of Commerce and Industry (MoCI) and Ministry of Urban Development (MoUD) are two major nodal ministries responsible for building resilient infrastructure, sustainable industrialization, and liveable indicators for cities (VNRR, GOI 2017). Keeping above-mentioned facts intact, it is necessary to develop smart, green, sustainable, and self-sufficient cities for future generations. It is possible only when one can understand the present scenario of urbanization, its quantification, its linkage with climatic changes, and its future aspects.

6.1.6 Capturing Real-Time Urbanization and Simulation Using Geospatial Technology

Satellite remote sensing has been credited for its quickest and lesser cost available method of mapping large areas. The availability of high-resolution and multi-spectral satellite imagery provides the best accuracy needed for land use mapping correspondingly and this has been widely recognized by various countries and regional bodies across the world (Chen et al. 2000; Ji et al. 2001). Satellite technology can help urban mapping at finer scales and thereby making it easier for policymakers and planners to understand the growth dynamics and sprawl with great detail (NASA 2001).

Urban modeling can be defined as a developing a systematic process of using a mathematical model that was originated from a scientific theory to visualize the process of Urban growth (Batty 2009) using geospatial techniques (U. S., EPA 2000; Sante et al. 2010; Verburg et al. 2002). SLEUTH is one mathematical model based on theory of cellular automata (CA). SLEUTH is based on five specific landscape modeling features that influence the model that includes slope, land use, excluded, urban, transportation, and hillshade (Dietzel and Clarke 2004; Rafiee et al. 2009). This study utilizes this open source model to understand and visualizes the future urban transition.

6.2 Method

The study involved an integrated approach to analyze land use, land cover, land surface temperature and predict future urban growth. Figure 6.3 depicts a stepwise working procedure to achieve the aforementioned objectives of the study. Resultant

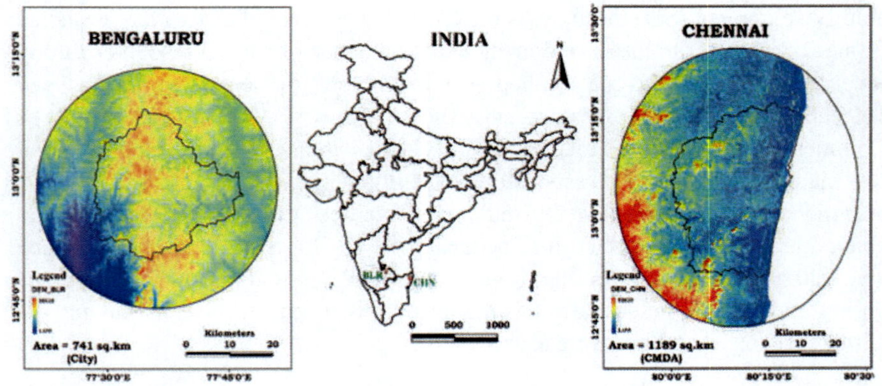

Fig. 6.3 Site location of Bengaluru and Chennai. Background image is DEM from ASTER data, showing mean sea level elevation

layers of each subprocess are highlighted, taken as input for LST and SLEUTH model as outlined in the following sections.

6.2.1 Study Area

Urban growth pattern identification, linkage with LST and simulation of urban expansion have been analyzed for two major metropolitan cities of India: Bengaluru and Chennai. Reason for selecting these cities is that both cities have accounted for drastic change in population in the last two decades and has housed various commercial sectors such as IT industry, raw material manufacturing, large-scale factories, and automobile industry. Ministry of Commerce and Industry, GoI has also been focusing on mega infrastructure projects throughout the country of which Chennai–Bengaluru Industrial Corridor Project is also listed with top priority. Primary aim of this project is to improve commercial activities between South India and East Asia by connecting several industrial hubs starting from Chennai, Sriperumbudur, Chittoor, Kolar, Hoskote, and Bengaluru. Figure 6.3 shows the geographical location of the cities along with detailed description of administrative boundaries, buffer boundary, and digital elevation model (DEM) as background.

Bengaluru is the capital of Karnataka state. Administrative boundary of greater Bengaluru includes 198 wards encompassing a total of 741 km^2 to become India's 5th largest metropolitan city. The topography of the region can be visualized in Fig. 6.3 and it is undulating with a variable altitude ranging from 740 to 960 m above mean sea level (MSL) . The city has recorded a population of 8.64 million (2011), however, post-2000 Bengaluru has witnessed a huge population increase due to the establishment of numerous information technology companies and hence gets the nickname "Silicon Valley of India". The city also houses important public sector companies in large number.

Chennai is the capital city of Tamilnadu state, India. Geographically the city is located sandwiched between two major rivers, i.e., Coovum and Adayar with less varied altitude levels from 6.7 and 60 m above MSL. Chennai is also called "Detroit of India" as it houses major automobile industry. Chennai metropolitan area (CMA) encompasses about 1189 km^2 comprising Chennai city district and partially extending to two districts Kancheepuram and Tiruvallur. Population of Chennai was 4.34 million as per census 2001 and has been increasing rapidly and is about 4.68 million as per census 2011.

6.2.2 Datasets

The primary data (Remote Sensing) used in the study includes Landsat 5, 7, and 8. Landsat data and ASTER data obtained from public domain: USGS Earth Explorer. The spatial resolution of these data is 30 m. Landsat data was pre-processed to rectify geometric errors, used to assess and analyze land use temporally (years chosen depends on the city and open data availability). Ground truth collection was done using global positioning system (GPS). Datasets included polygons of industries, educational institutes, healthcare units, ecologically sensitive areas, coastal zones, polylines of road and railway network, etc. for modeling purpose. Satellite data was cropped pertaining to the study area by considering administrative and buffer boundaries. Topographic maps downloaded from the survey of India were used to obtaining base layers of the administrative boundary.

6.2.3 Land Use Analysis to Understand the Urban Land Use Pattern

Land use map generation is based on image classification, involving various steps were performed as per Ramachandra et al. (2014). The classifier proves best among others because it takes into account the probability density function to evaluate the land use class of every pixel under consideration (Duda et al. 2000). Four land use classes were chosen for the study based on literature, they are urban, vegetation, water, and others as shown in Table 6.1 (Anderson et al. 1976; Nimish et al. 2017). GMLC is mathematically represented in Eq. 6.1. Pixel X belongs to class j only when the probability is maximum of those four classes.

$$X \in C_j \; if \; p\left(C_j/X\right) = \max[p(C_1/X), p(C_2/X), \ldots, (C_m/X)] \quad (6.1)$$

where $p(C_i/X)$ denotes the conditional probability of pixel X being a member of class.

$\max[p(C_1/X), \; p(C_2/X), \; \ldots, \; p(C_m/X)]$ is a function that returns the largest probability among four categories.

Accuracy assessment was performed by comparing the classified map with the validation map. Error/confusion matrix generated helps to estimate overall accuracy and kappa coefficient.

Table 6.1 Classes categorized under each land use

Class	Description of the land use included
Urban	Residential area, industrial area, paved surfaces, all built-up areas
Vegetation	Forest, cropland, nurseries
Water bodies	Tanks, lakes, rivers, reservoirs
Others	Rocks, quarry pits, open ground at building sites, un-metalled road

6.2.4 Estimation of LST

The study incorporates the most common but efficient method for measurement of surface temperature—Single window algorithm based on single thermal band ranging from 10.4 to 12.5 μm into account. The steps involved in the estimation of LST are as shown.

1. **Calculation of at-satellite brightness temperature**

Digital number was converted into top-of-atmosphere spectral radiance using Eq. 6.2.

$$L_\lambda = (\text{Gain} * DN) + \text{Offest} \tag{6.2}$$

here, L_λ is spectral radiance; Gain is band specific multiplicative factor; Offset is band specific additive factor (Gain and offset can be obtained from metadata).

Spectral radiance was then converted into at-satellite brightness temperature as shown in Eq. 6.3.

$$T_B = \frac{K_2}{\ln\left(\frac{K_1}{L_\lambda} + 1\right)} \tag{6.3}$$

T_B is at-satellite brightness temperature; K_1 and K_2 are thermal conversion constants and their values are as shown in Table 6.2

2. **Estimation of emissivity**

Emissivity is an important parameter for quantification of LST and NDVI threshold method was used to quantify it. The method distinguishes emissivity values for a pure class of water, soil, and vegetation, and takes into account mixed pixel composed of soil and vegetation. Emissivity values for soil, water, and vegetation were taken directly from literature (CMAP 2018) and mixed pixel was estimated as shown in Eq. 6.4 (Sobrino and Raissouni 2000).

$$\varepsilon_{SV} = \varepsilon_V P_V + \varepsilon_S(1 - P_V) + C \tag{6.4}$$

here, ε_{SV} is emissivity of soil + vegetation; ε_V is emissivity of vegetation; ε_S is emissivity of soil; P_V is the proportion of vegetation and can be calculated using

Table 6.2 Thermal conversion constants

Satellite	K_1	K_2
Landsat 5	607.76	1260.56
Landsat 7	666.09	1282.71
Landsat 8 (Band 10)	774.88	1321.08

Eq. 6.5; C is a constant that defines surface characteristics and can be estimated using Eq. 6.7.

$$P_V = \left(\frac{\text{NDVI} - \text{NDVI}_S}{\text{NDVI}_V - \text{NDVI}_S}\right)^2 \tag{6.5}$$

NDVI is normalized difference vegetation index that can be defined as the amount of vegetation present in the region and can be calculated using Eq. 6.6. NDVI_S, NDVI_V denotes NDVI thresholds for soil and vegetation, respectively.

$$\text{NDVI} = \frac{\text{spectral reflactance (NIR)} - \text{spectral reflactance (Red)}}{\text{spectral reflactance (NIR)} + \text{spectral reflactance (Red)}} \tag{6.6}$$

$$C = (1 - \varepsilon_S)\varepsilon_V F(1 - P_V) \tag{6.7}$$

F is geometrical factor that depends on surface geometry (usually considered 0.55).

3. **Quantification of land surface temperature**

Emissivity values for each land use class were estimated using Eq. 6.8.

$$\text{LST} = \frac{T_B}{1 + \left(\frac{\lambda T_B}{\rho} X \ln(\varepsilon)\right)} \tag{6.8}$$

λ denotes the wavelength at which maximum relative response is observed; $\rho = \frac{hc}{\sigma} = (1.438 * 10^{-2})$ mK (h is Plank's constant, c is speed the of light and σ is Stefan Boltzmann constant).

Coefficient of variation

$$\text{COV} = \frac{\sigma}{\mu} * 100 \tag{6.9}$$

COV = Coefficient of Variation (%), σ = Standard deviation, μ = mean.

6.2.5 *SLEUTH Urban Growth Model*

Input layers for the model were acquired from various data sources. Figure 6.4 depicts the general workflow of SLEUTH model with the following steps.

- Input dataset preparation: Layers were together taken into a single working directory with identical rows, columns, similar coordinate reference system, same map extent for all input layers, and a standard resolution of 30 m (Dietzel and Clarke

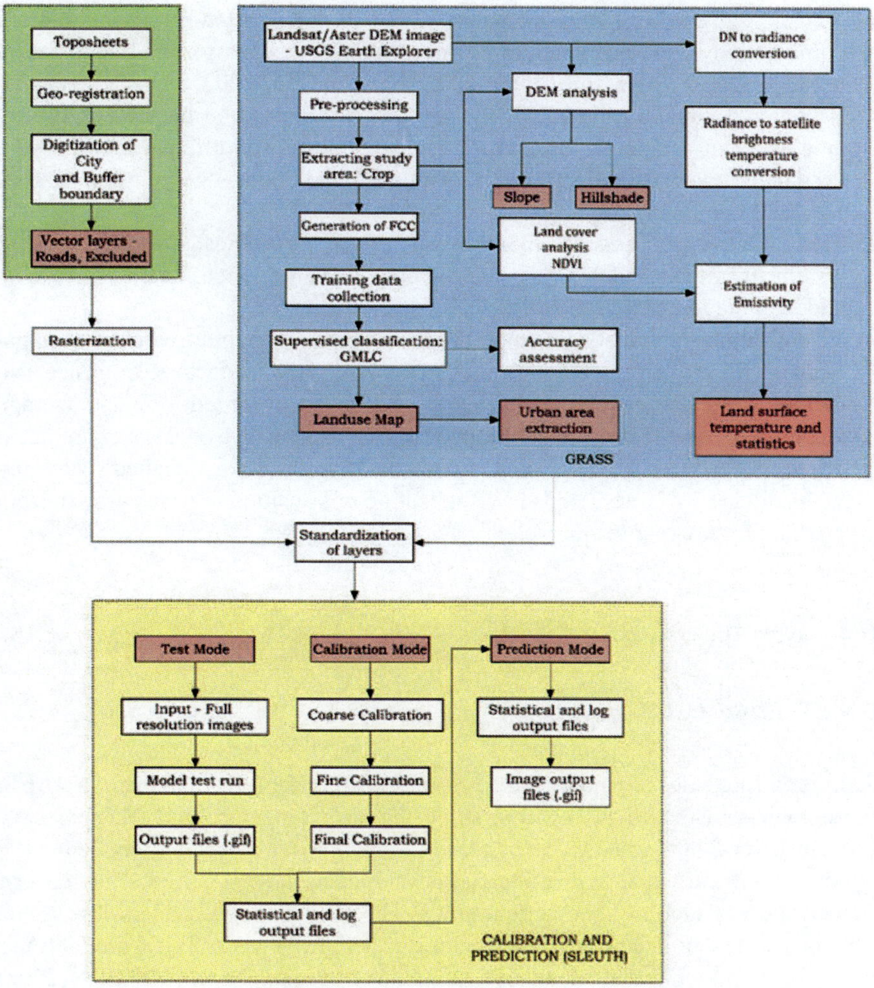

Fig. 6.4 Method adopted to analyze patterns of urban growth and LST

2004). These layers need to be compiled and resampled to resolutions of 120, 60, and 30 m for three phase calibration as coarse, fine, and full calibration.

- Test phase: Model test phase helps to verify the dataset compatibility. Random coefficient values were set with four Monte Carlo iterations. Test output provides statistic log files as well as image files along with animated gif for specified years.
- Calibration phase: To achieve the best range of values of coefficient that is necessary to predict urban growth several iterations were performed. Various indicators based on literature were used such as Lee-salee metric, a goodness of fit measure. These were arranged in decreasing order to find range values for further fine calibration and final calibration phase.

- Selection of goodness of fit statistic: Various researchers have used the selected combination of these metrics to estimate optimum values for predicting future urban scenario (Silva and Clarke 2002; Yang and Lo 2003; Jantz et al. 2004; Sakieh and Salmanmahiny 2016). These metrics depend on the data and study area under investigation. Based on various literature, six different metrics were used in the analysis such as product, compare, population, edges, clusters, cluster size, and Lee-Salee.
- Prediction phase: Lee-salee metric is one of the best indicator and is robust to test the fit statistic and this was used to obtain five unique coefficient values to be added in the final calibration phase.
- Visualization and statistical output: Prediction phase outputs are essentially statistical log file containing the best coefficient values of five factors. Log files also include input image validation, growth details with recorded time and average values of the five factors. In addition to log files, output also consists of predicted urban areas at user-specified time step, cumulative urban for specified year along with animated gif image, aids in visualization of transition of urban growth from prediction start year to stop year.

6.3 Results and Discussion

6.3.1 Land Use Dynamics

Results and statistics of land use analysis are shown in Table 6.2 and Fig. 6.5. Urban footprint has increased in both the study regions. Bengaluru and Chennai have accounted for 348% and 1188% increase in urban cover. Especially in Bengaluru region the infrastructure and residential development have occurred at the cost of vegetated areas such as large agricultural fields, plantations, secondary forest, etc. This phenomenon of land use conversion can be clearly visualized from land use maps (Fig. 6.6), while statistics show the decline of vegetation from 17.01% (1992) to 5.79% (2017). Waterbody has also shown a steep decrease from 2.42% (1992) to mere 0.7% (2017). The number of lakes has also considerably reduced, either converting to other forms of land use or dried over the period of time due to poor maintenance and encroachment. Considering Chennai's scenario, other land use categories have faced the maximum loss. Land features such as barren land, wetland, catchment area of Puzal and Chembarambakkam lake, Sholinganallur wetland area have rapidly transformed to urban structures, especially post-2000 due to administrative, economic policy changes and industry oriented planning. Alterations in land use cause microclimate change and therefore becomes a vital factor affecting the land surface temperature. Table 6.3 shows overall accuracy and kappa statistics for the classified images. High level of accuracy was achieved for both regions ranging from 86 to 94% while kappa range was between 0.78 and 0.92, proven to be satisfactory.

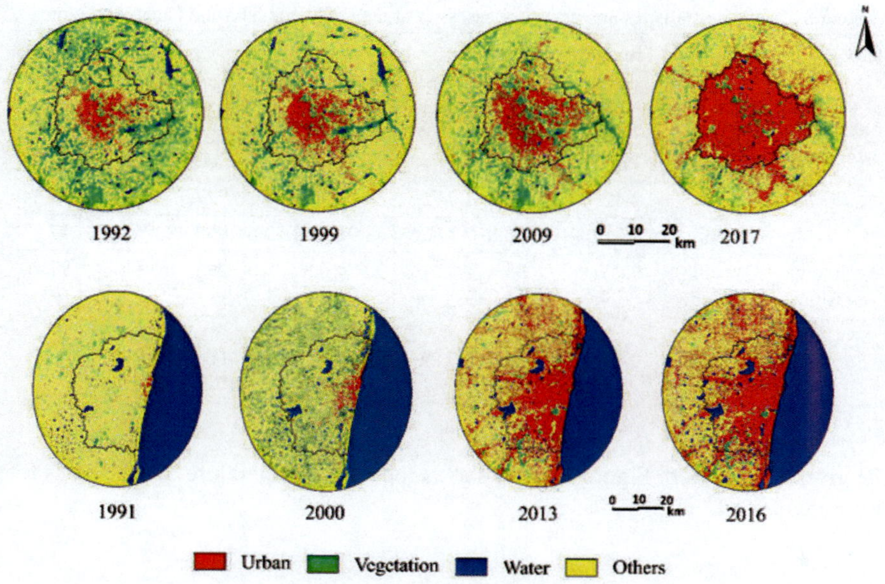

Fig. 6.5 Land use dynamics of Bengaluru and Chennai region, 1991–2017

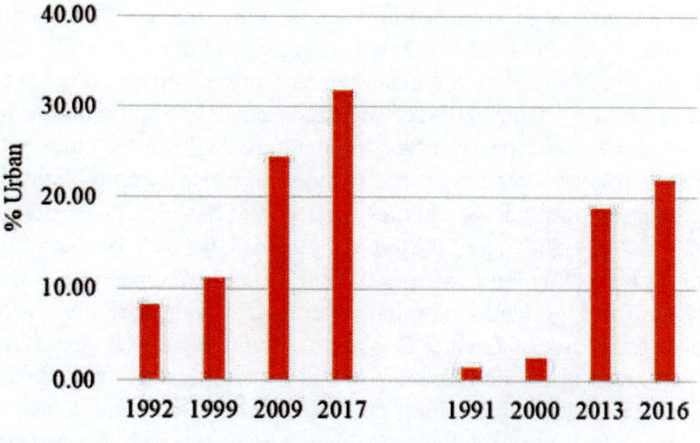

Fig. 6.6 Urban growth of Bengaluru and Chennai region, 1991–2017

6.3.2 Land Surface Temperature

Greater Bengaluru (Bruhat Bengaluru Mahanagar Palike) and Chennai metropolitan area boundary with buffer region were considered for observing spatio-temporal land surface temperature pattern for three decades as illustrated in Fig. 6.7. Bright reddish tone in map represents the region with high temperature and as the tone shifts to blue, the temperature reduces. The changes in LST can be inferred to LULC alterations in

Table 6.3 Land use statistics and accuracy assessment data of Bengaluru and Chennai region

City		Bengaluru				Chennai			
Year		1992	1999	2009	2017	1991	2000	2013	2016
Land use categories	Urban	5.47	8	13.48	24.53	1.46	2.52	18.81	22
	Vegetation	17.01	9.26	11.48	5.79	1.38	0.8	2.76	1.83
	Water	2.42	1.94	0.78	0.7	27.64	27.7	27.92	28.34
	Others	75.1	80.8	74.26	68.98	69.52	68.98	50.51	47.83
Accuracy assessment	Overall accuracy (%)	94	94	92	90	92	91	86	87
	Kappa statistic	0.87	0.83	0.82	0.78	0.92	0.9	0.78	0.81

the given time period. Figure 6.7 also shows the mean temperature of each class for the study period.

Bengaluru

The average surface temperature of the city with 10 km buffer for the summer season (Mar–May) has increased from 33 °C (1992) to 41 °C (2017). Mean surface temperature with respect to each class was estimated and statistical analysis was performed. Surface temperature range, mean surface temperature, and coefficient of variation were estimated as shown in Table 6.3. Due to a significant rise in built-up area, i.e., 5.47% (1992)–24.53% (2017), the mean temperature for urban class has increased from 33.08 to 41.14 °C. Conversion of vegetation cover and water bodies into urban pockets has led to rise in mean surface temperature for both the class by 7.68 and 9.13 °C. Rise in water bodies can be inferred to solid waste dumping into lakes along with oil discharges from garages in the vicinity. The barren area in Bengaluru has experienced a rise of 7.56 °C. Surface temperature for 2017 was considered to identify places with high and low temperatures. Jakkur airfield, Kempagowda international airport, HAL airport, goods loading terminal in Kadugodi, agricultural fields near Shankanipura (outskirts of Greater Bengaluru boundary), dried Hesaraghatta lake and few open areas in the city shows the highest temperature ranging from 44 to 51 °C. Green areas in Bengaluru golf course, Freedom Park, and dense vegetation patches in Indian Institute of Science, Cubbon park shows reduced surface temperatures (33–35 °C). Minimum surface temperature was observed for water bodies (Sankey tank, Ulsoor lake, Bellandur lake, etc.) in the region that ranges from 30 to 33 °C. Table 6.4 shows the range, mean, standard deviation, and coefficient of temperature for Greater Bengaluru. Coefficient of variation was estimated as it shows the intraclass variability that is highest for water body and the reason can be inferred as presence of sand particles and pollutants and also the variation in depth of water bodies. Vegetation, others, and urban category show moderate values that conclude a non-significant difference in density and type of vegetated areas, open areas, and construction material.

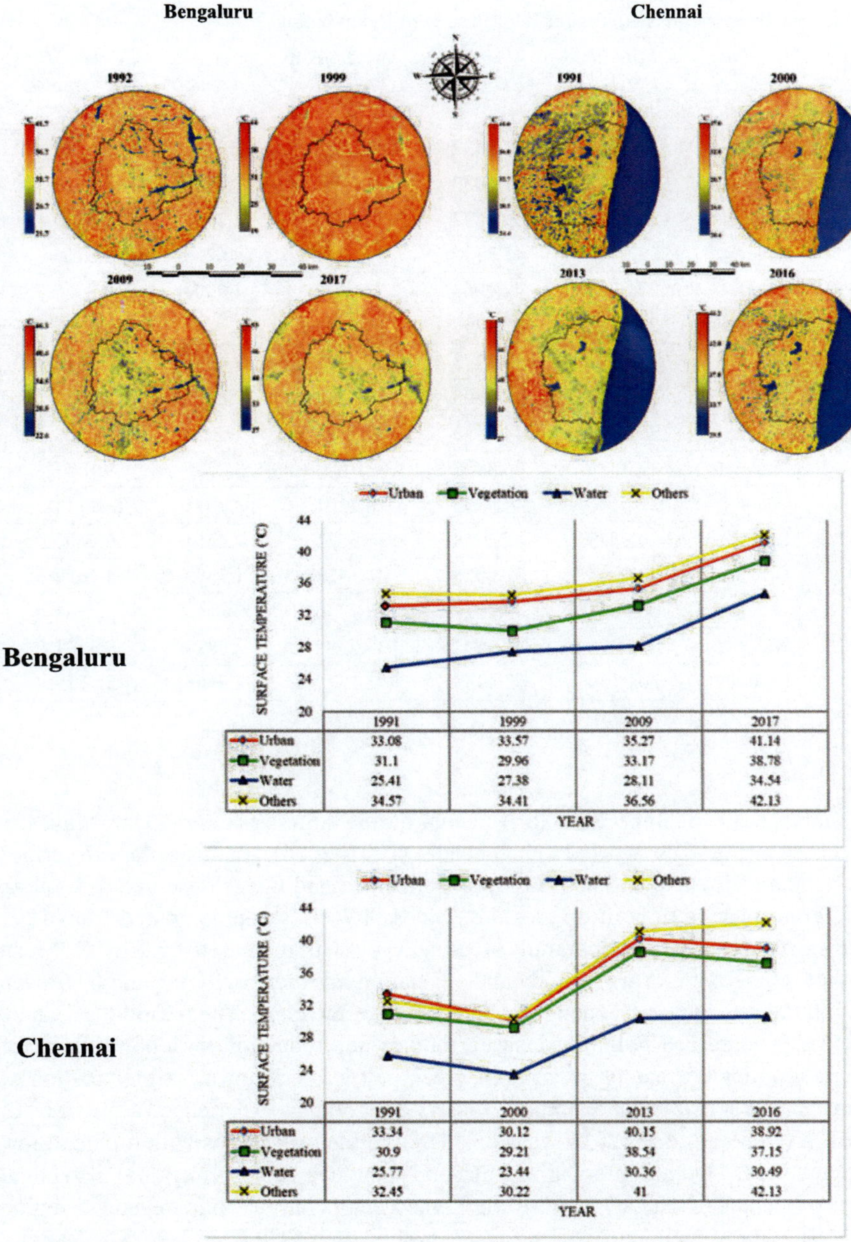

Fig. 6.7 Temporal land surface temperature graphs for Bengaluru and Chennai

Table 6.4 Temperature statistics for Bengaluru with 10 km buffer

Year	Land use class	Min. temp. (°C)	Max. temp. (°C)	Mean temp. (°C)	Standard deviation	Coefficient of variation (%)
1992	Urban	22.9969	41.7432	33.0766	1.8295	5.5312
	Vegetation	22.9475	41.7432	31.0980	2.2738	7.3117
	Water	21.7371	39.7837	25.4078	2.7012	10.6315
	Others	22.1766	41.3534	34.5724	2.0213	5.8466
1999	Urban	22.5723	42.9064	33.5656	1.6301	4.8566
	Vegetation	21.6813	40.5707	29.9555	2.0248	6.7593
	Water	22.6146	40.1778	27.3762	3.4279	12.5215
	Others	18.5875	44.0605	34.4116	2.0322	5.9056
2009	Urban	25.2107	44.4433	35.2693	1.7037	4.8306
	Vegetation	23.3849	46.3425	33.1653	3.0295	9.1347
	Water	22.6146	44.8250	28.1101	3.7025	13.1714
	Others	23.3849	45.2058	36.5625	2.4351	6.6600
2017	Urban	27.0065	52.5054	41.1427	1.9592	4.7619
	Vegetation	30.7761	49.7217	38.7756	2.2243	5.7364
	Water	30.0896	47.1504	34.5389	2.8877	8.3607
	Others	30.7159	52.4933	42.1277	2.5165	5.9735

Chennai

The mean surface temperature for Chennai during summer season (Mar–May) has increased by 5.8 °C in two and a half decades (1991–2016). Analysis of mean surface temperature for each land use class was performed and it was observed that due to prodigious change in built-up area, i.e., by 4684.76% (when compared from 1991 levels), mean surface temperature of urban class has increased by 5.6 °C. Mean surface temperature of water body and vegetation has increased by 5 and 6 °C even though there is not much change in percentage of the class. The reason that can be inferred is increased pollution in water bodies and reduction in depths, while for vegetation, approximately 50 acres of green belt was cut down for the creation of Thermal power plant near Kattupalli. Others category has experienced a rise of 6 °C. Surface temperature for 2016 was considered to identify places with high and low temperatures. Chennai international airport, open area near Pallikaranai marshland and city center (T Nagar) showed the highest temperature while vegetated region such as Guindy national park and water bodies such as Puzhal lake, Cholavaram tank, Chembarambakkam lake showed moderate and lower temperature respectively. Table 6.5 signifies the statistical parameter estimated for Chennai. Coefficient of variation shows moderate values for all the classes and it can be inferred that there are not many changes in vegetation density and type, depth of water bodies, construction material, and open areas.

Table 6.5 Temperature statistics for Chennai with 10 km buffer

Year	Land use class	Min. temperature (°C)	Max. temperature (°C)	Mean temperature (°C)	Standard deviation	Coefficient of variation (%)
1991	Urban	25.1195	40.1778	33.3419	2.6865	8.057429
	Vegetation	25.5495	38.9925	30.9034	1.8069	5.846929
	Water	24.3513	39.7837	25.7725	1.2085	4.689107
	Others	24.6881	40.9625	32.4525	2.8649	8.827979
2000	Urban	22.6146	34.9686	30.1226	1.7436	5.788345
	Vegetation	23.3849	36.9949	29.2062	1.8516	6.33975
	Water	20.4090	35.7826	23.4406	1.7285	7.373958
	Others	20.8533	36.5920	30.2200	2.0419	6.756784
2013	Urban	29.7964	52.2788	40.1524	2.1660	5.394447
	Vegetation	31.9702	49.4913	38.5364	1.8741	4.863194
	Water	27.0915	50.9636	30.3609	2.0078	6.613111
	Others	27.4180	52.4813	40.9990	2.8697	6.999439
2016	Urban	29.9431	46.1502	38.9241	1.7308	4.446602
	Vegetation	32.6322	42.8474	37.1460	1.2519	3.370215
	Water	29.5108	44.6877	30.4933	1.3801	4.525912
	Others	30.2116	45.6491	38.6702	2.0346	5.261416

6.3.3 Relationship Between LST and LULC

Temperature profile graphs were created to understand the variation of surface temperature with respect to each land use class as depicted in Fig. 6.8. One transact for each city was considered and the profile graphs were created. In case of Bengaluru, transacting from A to B, initially the temperature is high due to the presence of others classes (agricultural field), after that, a dip is observed due to the presence of Yelahanka kere. Moving further the temperature is high as the transact passes through the urban area. Small dips in middle indicate the presence of vegetated lands and a sharp dip is observed in the end due to the presence of Hullimavu kere. In case of Chennai, moving on transact C to D, the temperature profile initially shows lower values due to the presence of sown agricultural fields. Post that a rise is witnessed due to the presence of open area. Sharp dip after that indicates the presence of Cholavaram tank, Puzhal lake, and Korattur tank. Presence of an open area between these water bodies is the reason for increased temperature in the profile. Post that the profile shows high temperature due to the presence of urban area. A small dip at the end of transact is due to the presence of Guindy National park and sea.

It was observed that the region with water bodies or vegetated areas exhibited lower surface temperatures than others and urban category. The reason for this can be inferred as the amount of sensible heat energy absorbed by open areas, barren fields, urban structures is more when compared to water bodies and vegetated fields

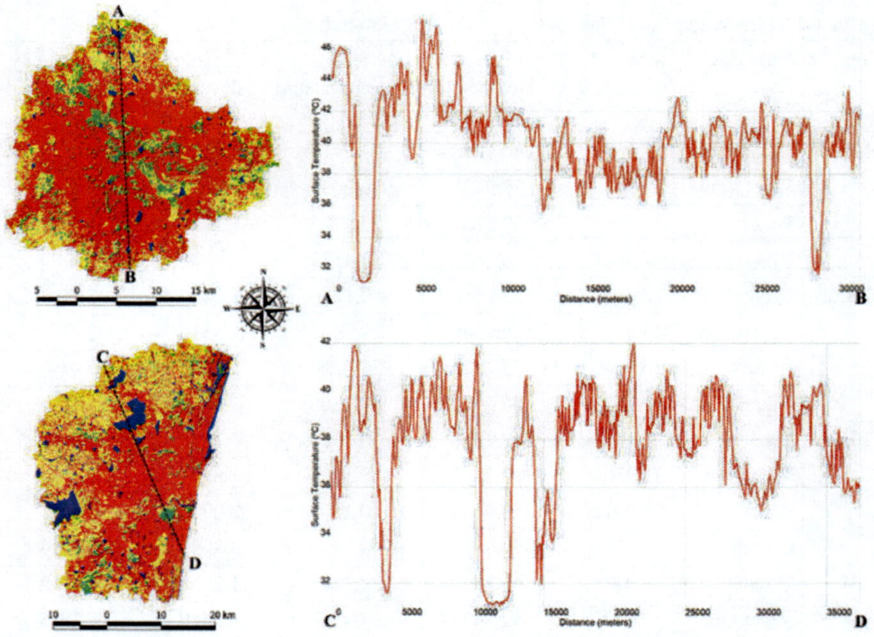

Fig. 6.8 Temperature profile graph for Bengaluru and Chennai region with transect

that converts this sensible heat into latent heat by the process of evaporation and evapotranspiration. Both study areas were blessed with a significant amount of water bodies and green spaces but unplanned urbanization has led to the conversion of these spaces into urban pockets and it can be inferred as one of the major and important factors for the overall mean surface temperature rise.

6.3.4 Geo-visualization of Urban Growth Using SLEUTH

Test mode was successfully conducted for all standardized datasets with full resolution of 30 m. Figure 6.10 represents SLEUTH model urban growth predicted for the year 2025 along with major road networks. It also depicts the annual probability of urban growth between the range 50–100%, for instance, brown color code indicates chance of 95–100% urbanization. Based on Lee-Salee metric, the final coefficient values obtained for Bengaluru showed least value for diffusion (1) and breed (4), low value for slope (20). Diffusion and breed coefficient controls outward dispersive urban growth. It is evident in Bengaluru that growth is infilling type, rapid and monotonous within the administrative boundary compared to growth outside the boundary (buffer region). Slope resistance can be observed in fewer parts (as visualized in Figs. 6.3 and 6.10) such as areas lying between SH9 and NH7 due north of Bengaluru city (Fig. 6.9).

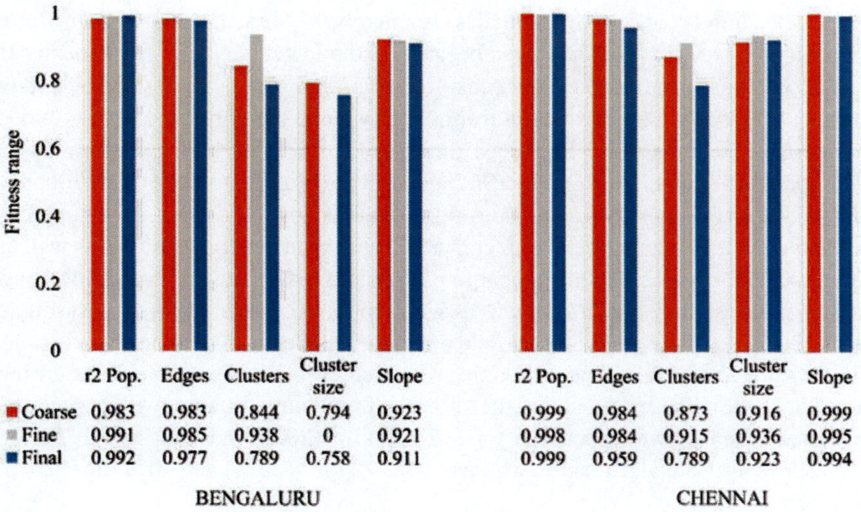

Fig. 6.9 Statistical fit values during calibration phase

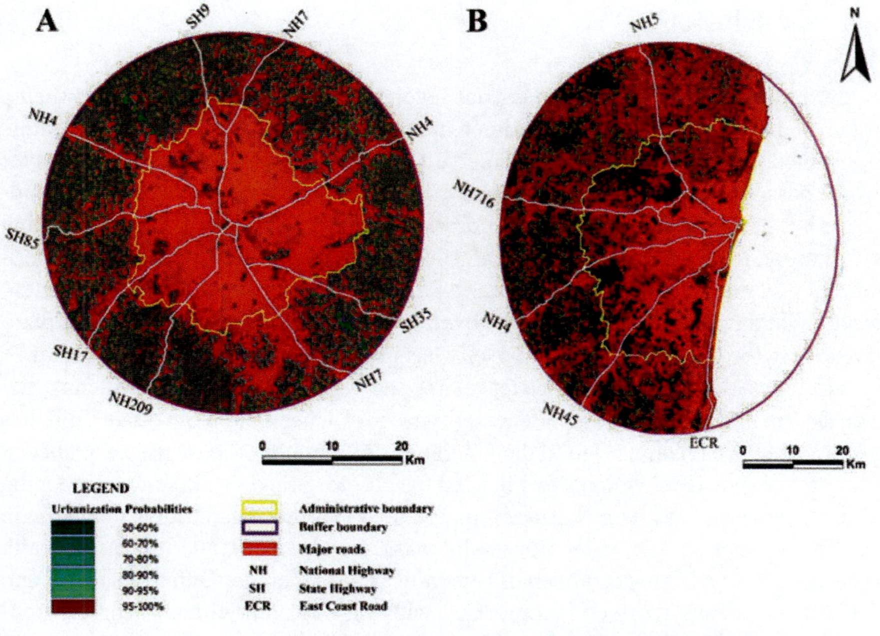

Fig. 6.10 Modeled output along with major road network of **a** Bengaluru and **b** Chennai

Modeled trend observed in Chennai is completely different than that of Bengaluru. Least value observed for slope (1) is because of the lower elevation from mean sea level as Chennai is situated on the coastline. Minimum diffusion (7) indicates rare chances of outward depressiveness from the city since the core is near surrounded by ocean in Northeast and Southeast directions. Extremely high value for breed (93), spread (98), and road gravity (96) suggests the likelihood of new settlements, detached from existing ones, spreading along major roads is highly predominant. It is to be noted that Chennai is very well connected with road network, railway as well as port, enhancing the movement and cargo between inland Indian states and Southeast Asian countries. Statistical fitness was assessed for both cities in terms of least square regressed values of r^2 population, urban edges, urban cluster, cluster size, and slope. Highly satisfactory values nearing unity were achieved in most of the cases during three calibration phases indicating the success of modeling urban growth procedure. Predicted urban growth accounted for 1323.30 km^2 (2025) in contrast with 727.88 km^2 (2017) for Bengaluru region whereas 2375.73 km^2 (2025) and 1077 km^2 (2016) for Chennai region.

6.4 Conclusion

In the past few decades, post-industrial revolution including new industries being set up and development of information technology, India has experienced a dynamic unplanned urbanization. India has emerged as one of the fastest growing countries which has experienced an impressive growth in urban population with changing landscape but has not fulfilled any basic criteria of providing basic amenities including clean water, fresh air, safe environment, and proper housing facility. With the experience of other countries and self-evaluation, India has understood that it is difficult to sustain without having a sustainable governance of resources and appropriate planning. As in the case of Bengaluru, the city has been sprawling since the outburst if IT sector. It has already lost many water bodies and vegetation on the cost of construction activities for building and connectivity (roads). Chennai, on the other hand, has experienced urbanization due to the emergence of number of industries and special economic zones in its vicinity that has led the city to sprawl. These cities are facing dire problems such as rising surface temperature (Thermal discomfort), depletion in quality of water and air, indecorous solid waste management, and increased health issues and are unable to get other basic amenities due to their unplanned development. The finding of this research incorporated with various other climatic and environmental variables will be useful in development of mitigation measures to combat the changing climate patterns due to unplanned urbanization for Indian cities thus, accomplishing the agenda of Goal 11 by United Nations Sustainable Development.

6.5 Scope of Further Research

Single window method utilized for quantification of land surface temperature has improved with specific emissivity obtained, but better methods that include various other atmospheric parameters can be used. Existing urban modeling techniques such as CA-Markov, constrained-CA, multi-criteria evaluation-CA, and logistic-CA have advantages, for instance, generation of richer forms of cells at individual class levels (Batty 2005) but they fail to calibrate and optimize factors responsible for cell state changes and lack expansion with intermixing of cells. Reformed CA approach such as SLEUTH, ABM-CA, and Genetic Algorithm (Clarke 2008; Spencer 2009; Torrens and Benenson 2005) has their greatest extensibility and adaptability for datasets because of optimization and calibration processes improving the output generated by enhanced self-modified cell transition rules (Clarke et al. 2007). It is therefore viable to adopt CA and reinforced with Monte Carlo iteration SLEUTH model, along with genetic algorithm to optimize parameter selection and thereby reducing computation time with reliable results. SLEUTH and genetic algorithm, a special and unique case of CA models, requires more detailed and intriguing research.

Acknowledgements We are grateful to SERB, India, Ministry of Science and Technology, Government of India, Ranbir and Chitra Gupta School of Infrastructure Design and Management, Sponsored research in Consultancy cell, Indian Institute of Technology Kharagpur and West Bengal Department of Higher Education for the financial and infrastructure support. We thank (i) United States Geological Survey and (ii) National Remote Sensing Centre (NRSC Hyderabad) for providing temporal remote sensing data.

References

Aishwarya N, Sutapa D, Vinay S, Bharath HA (2020) Adaptive capacity assessment for a flood vulnerable region through land use modelling and socio economic and physical indicators. J Environ Inf Lett 3(2):107–119. https://doi.org/10.3808/jeil.202000031

Anderson JR, Hardy EE, Roach JT, Witmer RE (1976) A land use and land cover classification system for use with remote sensor data. Development 964(671):1–28

Anderson MC, Norman JM, Kustas WP, Houborg R, Starks PJ, Agam N (2008) A thermal-based remote sensing technique for routine mapping of land-surface carbon, water and energy fluxes from field to regional scales. Remote Sens Environ 112:4227–4241

Basten L (2010) The polycentric metropolis: learning from mega-city regions in Europe—written and edited by Peter Hall and Kathy Pain. Econ Geogr 86(3):323–324

Batisani N, Yarnal B (2009) Urban expansion in Centre County, Pennsylvania: spatial dynamics and landscape transformations. Appl Geogr 29(2):235–249

Batty M (2005) Agents, cells, and cities: new representational models for simulating multiscale urban dynamics. Environ Plan A 37(8):1373–1394

Batty M (2009) Urban modeling. In: International encyclopedia of human geography. Elsevier, Oxford, UK, pp 51–58

Bharath HA, Vinay S, Durgappa S, Ramachandra TV (2013) Modeling and simulation of urbanisation in greater Bangalore, India. Paper presented at National Spatial Data Infrastructure 2013 conference, November 29–30, Mumbai, India. https://wgbis.ces.iisc.ernet.in/

Bharath S, Rajan KS, Ramachandra TV (2013) Land surface temperature responses to land use land cover dynamics. Geoinf Geostat Overview 1(4):1–10

Bharath HA, Chandan MC, Vinay S, Ramachandra TV (2017) Intra and inter spatio-temporal patterns of urbanisation in Indian megacities. Int J Imag Robot 17(2):28–39

Bharath HA, Vinay S, Chandan MC, Gouri BA, Ramachandra TV (2018) Green to gray: silicon valley of India. J Environ Manage 206:1287–1295

Bharath HA, Soumya KD, Prakash PS (2020) Urban 3D structure reconstruction through a generative adversarial network model. Arab J Sci Eng. https://doi.org/10.1007/s13369-020-04850-7

Bhatta B, Saraswati S, Bandyopadhyay D (2010) Urban sprawl measurement from remote sensing data. Appl Geogr 30(4):731–740

Bounoua L, Zhang P, Mostovoy G, Thome K, Masek J, ... Toure AM (2015) Impact of urbanization on US surface climate. Environ Res Lett 10(8)

Brian FT, James SF, Felix WL, David NW, Noah PM, Donald FA (2017) GRACE groundwater drought index: evaluation of California central valley groundwater drought. Remote Sens Environ 198:384–392

Census of India (2011) 2011 census data. https://censusindia.gov.in/. Accessed 29 May 2018

Chen S, Zheng S, Xie C (2000) Remote sensing and GIS for urban growth in China. Photogramm Eng Remote Sens 66(10):593–598

Chen XL, Zhao HM, Li PX, Yin ZY (2006) Remote sensing image-based analysis of the relationship between urban heat island and land use/cover changes. Remote Sens Environ 104(2):133–146

Clarke KC, Gazulis N, Dietzel C, Goldstein NC (2007) A decade of SLEUTHing: lessons learned from applications of a cellular automaton land use change model. Classics IJGIS Twenty Years Int J Geograph Inf Sci Syst (1st edn):413–425

Clarke KC (2008) A decade of cellular urban modeling with SLEUTH: unresolved issues and problems. In: Brail RK (ed) Planning support systems for cities and regions. Lincoln Institute of Land Policy, Cambridge, MA, pp 47–60

CMAP (2018) Automated generation of land surface temperature estimates. https://datahub.cmap.illinois.gov/dataset. Accessed 10 Oct 2018

Copernicus (2018) Land surface temperature. https://land.copernicus.eu/global/products/lst. Accessed 29 May 2018

Dash P, Gottsche FM, Olesen FS, Fischer H (2002) Land surface temperature and emissivity estimation from passive sensor data: theory and practice-current trends. Int J Remote Sens 23(13):2563–2594

Dietzel C, Clarke KC (2004) Replication of spatio-temporal land use patterns at three levels of aggregation by an urban cellular automata (523–532). Paper presented at 6th international conference on cellular automata for research and industry, Amsterdam, Netherlands, October 25–28. https://link.springer.com/chapter/10.1007/978-3-540-30479-1_54

Dietzel C, Herold M, Hemphill JJ, Clarke KC (2005) Spatio-temporal dynamics in California's central valley: empirical links to urban theory. Int J Geogr Inf Sci 19(2):175–195

Duda RO, Hart PE, Stork DG (2000) Pattern classification, 2nd edn. Wiley-Interscience Publication, New York

Ese Sentinel Online (2018) Land surface temperature. https://sentinel.esa.int/web/sentinel/user-guides/sentinel-3-slstr/overview/geophysical-measurements/land-surface-temperature. Accessed 10 March 2018

Jantz CA, Goetz SJ, Shelley MK (2004) Using the SLEUTH urban growth model to simulate the impacts of future policy scenarios on urban land use in the Baltimore-Washington metropolitan area. Environ Plann B Plann Des 31(2):251–271

Jat MK, Garg PK, Khare D (2008) Monitoring and modelling of urban sprawl using remote sensing and GIS techniques. Int J Appl Earth Obs Geoinf 10(1):26–43

Ji C, Liu Q, Sun D, Wang S, Lin P, Li X (2001) Monitoring urban expansion with remote sensing in China. Int J Remote Sens 22(8):1441–1455

Jia YY, Tang B, Zhang X, Li ZL (2007) Estimation of land surface temperature and emissivity from AMSR-E data. Paper presented at the conference of geoscience and remote sensing symposium (IGARSS), July 23–27. Barcelona, Spain. https://ieeexplore.ieee.org

Jiang G, Ma W, Qu Y, Zhang R, Zhou D (2016) How does sprawl differ across urban built-up land types in China? A spatial-temporal analysis of the Beijing metropolitan area using granted land parcel data. Cities 58:1–9

Kantakumar LN, Kumar S, Schneider K (2016) Spatiotemporal urban expansion in Pune metropolis, India using remote sensing. Habitat Int 51:11–22

Khandelwal S, Goyal R, Kaul N, Mathew A (2018) Assessment of land surface temperature variation due to change in elevation of area surrounding Jaipur, India. Egypt J Remote Sens Space Sci 21(1):87–94

Lang RE, Dhavale D (2005) Beyond Megalopolis: exploring America's new "Megapolitan" geography. Metropolitan Institute Census Report Series\r, vol 5. https://digitalscholarship.unlv.edu/brookings_pubs/38

Li ZL, Becker F (1993) Feasibility of land surface temperature and emissivity determination from AVHRR data. Remote Sens Environ 43(1):67–85

Li Z-L, Tang B-H, Wu H et al (2013) Satellite-derived land surface temperature: current status and perspectives. Remote Sens Environ 131:14–37

Mehmood R, Mehmood SA, Butt MA, Younas I, Adrees M (2016) Spatiotemporal Analysis of urban sprawl and its contributions to climate and environment of Peshawar using remote sensing and GIS techniques. J Geogr Inf Syst 8:137–148

Muthamilselvan A, Srimadhi K, Ramalingam N, Pavithra P (2016) Urbanization and its related environmental problem in Srirangam Island, Tiruchirappalli district of Tamil Nadu, India-thermal remote sensing and GIS approach. Environ Earth Sci 75(9):765

NASA (2001) Satellite maps provide better urban sprawl insight. Science Daily. https://www.sciencedaily.com/releases/2001/06/010605072451.htm

Nimish G, Setturu B, Ramachandra TV (2017). Modelling landscape dynamics in protected areas, Karnataka. Paper presented at Indian Biodiversity Congress, vol 4, March 10–12, Puducherry, India

Nimish G, Bharath HA, Lalitha A (2020) Exploring temperature indices by deriving relationship between land surface temperature and urban landscape. Remote Sens Appl Soc Environ 100299

Pandey B, Seto KC (2015) Urbanization and agricultural land loss in India: comparing satellite estimates with census data. J Environ Manage 148:53–66

Rafiee R, Mahiny AS, Khorasani N, Darvishsefat AA, Danekar A (2009) Simulating urban growth in Mashad City, Iran through the SLEUTH model (UGM). Cities 26(1):19–26

Ramachandra TV, Bharath HA, Beas B (2014) Urbanisation pattern of incipient mega region in India. Tema. J Land Use Mob Environ 7(1):83–100

Sakieh Y, Salmanmahiny A (2016) Treating a cancerous landscape: implications from medical sciences for urban and landscape planning in a developing region. Habitat Int 55:180–191

Sante I, Garcia AM, Miranda D, Crecente R (2010) Cellular automata models for the simulation of real-world urban processes: a review and analysis. Landsc Urban Plan 96(2):108–122

Santra A, Mitra SS (2017) Remote sensing techniques and GIS application in earth and environmental studies. IGI Global, USA

Schmugge TJ, Becker F (1991) Remote sensing observations for the monitoring of land-surface fluxes and water budgets. In: Schmugge TJ, André JC (eds) Land surface evaporation. Springer, New York, USA, pp 337–347

Scott AJ (2001) Globalization and the rise of city-regions. Eur Plan Stud 9(7):813–826

Shaw R, Das A (2017) Identifying peri-urban growth in small and medium towns using GIS and remote sensing technique: a case study of English Bazar Urban Agglomeration, West Bengal, India. Egypt J Remote Sens Space Sci 21(2):159–172

Silva EA, Clarke KC (2002) Calibration of the SLEUTH urban growth model for Lisbon and Porto, Portugal. Comput Environ Urban Syst 26(6):525–552

Sobrino JA, Raissouni N (2000) Toward remote sensing methods for land cover dynamic monitoring: application to Morocco. Int J Remote Sens 21(2):353–366
Spencer D (2009) Cities and complexity: understanding cities with cellular automata, agent-based models, and fractals. J Archit 14(3):446–450
Sudhira HS, Ramachandra TV (2007) Characterising urban sprawl from remote sensing data and using landscape metrics. Paper presented at 10th international conference on computers in urban planning and urban management, Parana, Brazil
Sun D, Kafatos M (2007) Note on the NDVI-LST relationship and the use of temperature-related drought indices over North America. Geophys Res Lett 34(24)
Sun S, Parker DC, Huang Q, Filatova T, Robinson DT et al (2014) Market impacts on land-use change: an agent-based experiment. Ann Assoc Am Geogr 104(3):460–484
Taubenbock H, Wegmann M, Roth A, Mehl H, Dech S (2009) Analysis of urban sprawl at mega city Cairo, Egypt using multisensoral remote sensing data, landscape metrics and gradient analysis. Paper presented at ISRSE conference, Stresa, Italy
Tomlinson CJ, Chapman L, Thornes JE, Baker C (2011) Remote sensing land surface temperature for meteorology and climatology: a review. Meteorol Appl 18(3):296–306
Torrens PM, Benenson I (2005) Geographic automata systems. Int J Geogr Inf Sci 19(4):385–412
U.S. EPA (2000) Projecting land-use change: a summary of models for assessing the effects of community growth and change on land-use patterns. Environmental Protection. https://nepis.epa.gov/
UNCED (1992) Earth Summit. Agenda 21: The United Nations Programme of Action from Rio. Agenda. UNDPI, New York. https://sustainabledevelopment.un.org/content/
UNCSD (2012) The Future We Want, Report of the UNCSD. https://sustainabledevelopment.un.org/. Accessed 7 April 2018
UNSDS (2015) New York summit: transforming our world: the 2030 agenda for sustainable development. https://www.un.org/sustainabledevelopment/summit/. Accessed 7 April 2018
Verburg PH, Soepboer W, Veldkamp A, Limpiada R, Espaldon V, Mastura SSA (2002) Modeling the spatial dynamics of regional land use: the CLUE-S model. Environ Manage 30(3):391–405
VNRR (2017) Report on the implementation of sustainable development goals. Presented to the high-level political forum on sustainable development, New York. https://niti.gov.in/writeread data/. Accessed 8 Nov 2017
WCED (1987) Our common future. Oxford University Press, United Kingdom
Weng Q, Lu D, Schubring J (2004) Estimation of land surface temperature-vegetation abundance relationship for urban heat island studies. Remote Sens Environ 89(4):467–483
WSSD (2002) Johannesburg summit. https://sustainabledevelopment.un.org/milesstones/wssd. Accessed 7 Mar 2018
Yang X, Lo CP (2003) Modelling urban growth and landscape changes in the Atlanta metropolitan area. Int J Geogr Inf Sci 17(5):463–488
Yue W, Zhang L, Liu Y (2016) Measuring sprawl in large Chinese cities along the Yangtze River via combined single and multidimensional metrics. Habitat Int 57:43–52

Chapter 7
Analyzing New Frontiers in Urban Preference and Perception Research

Deepank Verma, Arnab Jana, and Krithi Ramamritham

Abstract The assessment of people's perception of surroundings has been given less regard in top-down urban planning approaches. The earlier body of research aimed toward understanding preferences has been quintessential in highlighting modes and methods to appraise urban environments. Related studies have been focused on small-scale experiments based on evaluation of mainly visual realm due to unavailability of technological means and workforce to conduct research. We explore and review the possibilities of using recent advancements in technology for the collection of different sensory datasets required to gauge people's perception and preference toward urban spaces on a large scale. This paper provides fresh perspectives on the integration of traditional research practices with the latest approaches in sensory data collection and analysis.

Keywords Urban perception · Deep learning · Image classification · Sound classification · Urban planning and design

Acronyms

ANN	Artificial Neural Network
CNN	Convolutional Neural Networks
FAST	Frontier Areas of Science and Technology
GSV	Google Street View
HOG	Histogram of Oriented Gradients

D. Verma · A. Jana (✉) · K. Ramamritham
Centre for Urban Science and Engineering, Indian Institute of Technology Bombay, Mumbai, India
e-mail: arnab.jana@iitb.ac.in

D. Verma
e-mail: deepank@iitb.ac.in

K. Ramamritham
e-mail: krithi@cse.iitb.ac.in

P. Sharma (ed.), *Geospatial Technology and Smart Cities*, The Urban Book Series,
https://doi.org/10.1007/978-3-030-71945-6_7

IRCC	Industrial Research and Consultancy Centre
LSTM	Long-Short Term Memory
MHRD	Ministry of Human Resource Development
RNN	Recurrent Neural Networks
SVM	Support Vector Machines
SIFT	Scale-Invariant Feature Transform

7.1 Introduction

The assessment of individual's perception regarding urban spaces forms a critical part of the research in urban planning and design. Advancements in technology, mainly in the form of data analytics, have extended the viability and enhanced the generalizability of traditional research methodology. Fresh perspectives are required to integrate large-scale data capturing, analysis, and computation, along with well-established research practices. The inclusion of subfields of computer science and technology such as Machine Learning may unlock the new possibilities in understanding rapid urban change and its effect on the population. This research discusses the involvement of visual and auditory senses in the perception of urban surroundings. It further overviews the current state-of-the-art methods in auditory and visual data analysis, which can be integrated with traditional approaches in studying individuals' perception and preferences.

Studies in environment perception as a research domain begun as a quest to understand human psychology and behavior toward the landscapes. The domain of such studies overlapped with the subjects such as Environmental psychology, Behavioral science, Architecture, Urban planning, and design. The empirical studies in the visual assessment of landscapes started becoming apparent in the early '70s. The majority of these were focused on the identification and establishment of different variables and environmental descriptors (Kasmar 1970). Over the years, a significant effort has been put into defining, testing, and bridging the gap between the theory and the practice.

Berlyne (1951) and later, Wohlwill (1976) studied the factors which determined the level of arousal in a person while being in a particular environment such as (a) Complexity (Presence of diverse set of elements in the scene), (b) Incongruity (inconsistency between the elements), (c) Novelty (presence of unique elements), and (d) Surprisingness (presence of unexpected elements) (Chang 2009; Kaymaz 2012). Berlyne's theory of Arousal discussed environmental perception as a process of "explorative behavior" and "transmission of information" (Chang 2009). Similarly, Appleton (1996) proposed that people prefer those places which offer possibilities to hide and supervise another location from vantage points. The feelings of safety and pleasure are derived from the environments that provide views and sense of enclosure (Dosen and Ostwald 2016). He based these observations along the lines of human evolution, known as "prospect-refuge" theory (Appleton 1996). Information processing theory proposed by Kaplan et al. (1989) discussed that the preference of landscape depends upon the two cognitive processes such as the need to understand the surroundings and desire to explore the same (Pazhouhanfar and Kamal 2014). These studies assisted in assimilating variables to evaluate distinctive attributes and properties of the environment. The variables for the assessment of arousal provided by Berlyne; Environmental assessment indicators such as prospect and refuge by Appleton (1996); coherence, complexity, legibility, and mystery by Kaplan (1979) became the baseline for the environmental assessment for future studies. Collectively, these variables were grouped by Nasar (1989) into five different "classes of variables" (Fig. 7.1) such as (a) Collative variables, which measure the "comparison of scenes with degree and nature of similarity or difference" (Berlyne 1971), (b) Organizing variables, which provide structure to the scene and decrease uncertainties, (c) Spatial variables include variables such as prospect and refuge (Appleton 1996), (d) Ecological variables help in the content assessment of the scenes (Berlyne 1971), and (e) Psychophysical variables deal with the visual measurement of the scenes.

A majority of the empirical studies in environment assessment can be classified into two sub-domains: Evaluation of (a) Urban-natural spaces, and (b) Urban spaces (Table 7.1). Studies in urban-natural spaces primarily focused on the assessment of green areas such as urban forests, parks, and playgrounds in an urban setting, while studies on urban spaces concentrated mainly on the evaluation of architectural aesthetics of built structures (Oostendorp and Berlyne 1978) and variability in urban scenes (Herzog et al. 1982). These studies identified the dissimilarities between the preferences of people toward natural and urban scenes. One of the earliest research on quantifying preferences was conducted by Kaplan et al. (1972) in which the urban and natural scenes were compared with the help of two high-level variables: complexity and preference. Studies which involved assessment of urban landscapes mainly included (a) evaluation of streetscapes featuring facades of buildings (Gjerde 2010); (b) Scenes comprising of built structures in the city as stimuli to gather ratings on familiarity, complexity, and preference (Herzog et al. 1976); and (c) Perceived

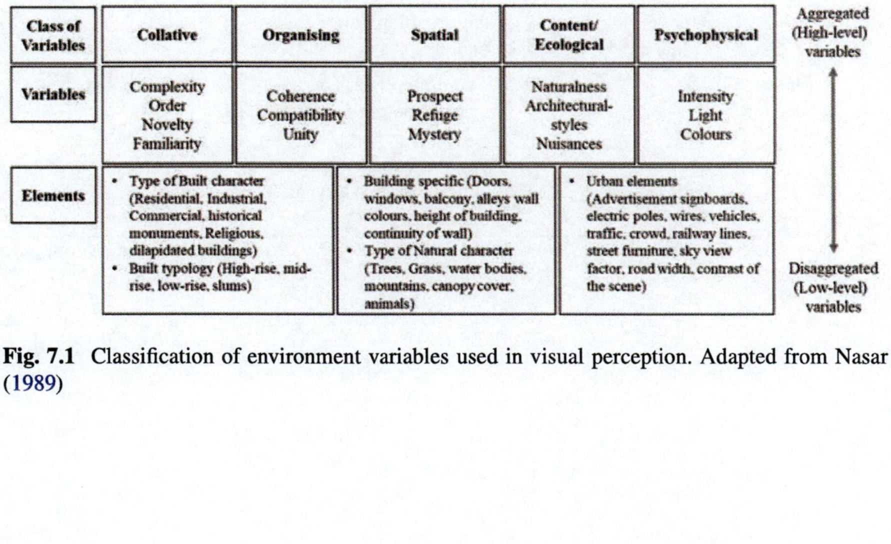

Fig. 7.1 Classification of environment variables used in visual perception. Adapted from Nasar (1989)

Table 7.1 Example of some of the empirical studies in environment perception

S.no	Study	Year	Main objectives	Stimulus
1	Kaplan et al. (1972)	1972	The relationship between complexity and preference of the environment is studied in which natural scenes are favored to urban scenes	Urban-natural scenes
2	Herzog et al. (1976)	1976	Assessment is done to study the effect of familiarity of the environmental scenes in the judgement ratings of complexity and preferences	Urban scenes
3	Oostendorp and Berlyne (1978)	1978	Buildings of various architectural styles are used as stimulus to judge the ratings on similarity-dissimilarity, collative and affective perceptions	Architectural buildings
4	Herzog et al. (1982)	1982	The preference of unfamiliar environments is studied with the help of variables such as complexity, coherence, identifiability, and mystery	Urban-natural scenes
5	Nasar (1984)	1984	Scenes from urban streets of USA and Japan are rated by participants from both the countries. The measurements were taken according to the characteristics of the street views such as visibility to nature, dilapidation, and vehicles	Urban scenes
6	Herzog (1989)	1989	The preferences for urban scenes containing natural elements are studied with variables such as refuge, coherence, mystery, and complexity	Urban-natural scenes

(continued)

Table 7.1 (continued)

S.no	Study	Year	Main objectives	Stimulus
7	Herzog (1992)	1992	The preferences for urban spaces regarding the various scene categories are studied such as open, spacious-structured, enclosed and blocked and corridors	Urban scenes
8	Loewen et al. (1993)	1993	The perception of safety is studied in urban environment based on prospect and refuge theory. The responses are taken from scenes having comprising of various attributes such as light, open space, and accessibility	Urban scenes
9	Laumann et al. (2001)	2001	Comparison between ratings of natural and urban scenes is done to identify restorative components of environments	Urban-natural scenes
10	Herzog and Stark (2004)	2004	The preferences to the positively valued, parks, and negatively valued urban spaces are studied	Urban-natural scenes
11	Gjerde (2010)	2010	The preferences of urban streetscapes with varying characteristics is studied with the help of variables such as visual interest, order, complexity, and maintenance	Urban scenes
12	Kardan et al. (2015)	2015	Ratings on perceived naturalness and esthetic preferences are taken to study the effect of low-level image features in individuals' perception	Urban-natural scenes
13	Shrivastava et al. (2017)	2017	Automated analysis of low-level features in natural scenes to judge degree of naturalness, openness, expansion, and color, etc	Natural scenes

(continued)

Table 7.1 (continued)

S.no	Study	Year	Main objectives	Stimulus
14	Wang and Zhao (2017)	2017	Study of differences in visual preference among various demographic groups	Natural scenes

safety assessment from different urban environments (Loewen et al. 1993). Herzog (1989) investigated the preference for urban environments with natural elements like trees, shrubs, flowers, weed, and grass. Kardan et al. (2015) studied the importance of low-level visual features (a combination of spatial and color properties) in the perception of natural versus human-made scenes. Similarly, Shrivastava et al. (2017) created a model to extract low-level features from natural and human-made images, which can emulate human perception in scene classification tasks.

Herzog (1992) investigated the feasibility of utilizing the environment variables primarily used in the perception of natural spaces in urban settings. Herzog and Stark (2004) studied the variance of preferences for positively valued natural and negatively valued urban environments. Laumann et al. (2001) developed a rating scale to judge the variables of "Information Processing" theory in different natural and urban scenes. The comparison between the preferences of the urban environment is studied from participants of two different cultural backgrounds (Nasar 1984). In a similar way, Wang and Zhao (2017) studied the relationship between demography and the low-level landscape descriptors such as colors, vertical structure, accessibility of the vegetation, buildings, the shape of the water body.

These studies presented an exhaustive work in exploration of a variety of environmental indicators and methods used in quantification of qualities of the environment; providing a concrete literary foundation for subsequent research to follow. However, such studies have slightly been biased toward the selection of specific variables, choice of study areas and participants, due to which no universal consensus on the results and conclusions can be made. Most of the studies have based their investigations under the ambit of the visual realm and rarely included auditory and olfactory cues in determining perception and preferences. Environment perception studies are closely linked to the individual's experience of the surroundings. Therefore, it is difficult to frame conclusions based only on the individual's visual understanding (Herzog et al. 1976). The environmental variables are high-level constructs relating to human behavior and psychology, the definitions of which may not be consistent across different studies. Further, it is difficult to ascertain the availability of similar background knowledge among the participants to understand such definitions.

The studies in recent years (Porzi et al. 2015; Dubey et al. 2016; Rossetti et al. 2019) have been relatively useful in establishing the relationship between the presence of low-level (disaggregated) variables (Fig. 7.1) and people's perceptions such as safety and liveliness. Such studies have been able to determine the feelings associated with the presence of particular environmental attributes. With evolving newer technologies, the gap between data availability and accessibility to state-of-the-art algorithms has been widely reduced (Verma et al. 2019). We look into the significance of visual and auditory realms in environment perception and the techniques to capture and analyze these sensory datasets in subsequent sections.

7.2 Sensory Realms

The visual environment constitutes the entities which are distinguishable and can be expressed with detailed clarity. Describing landscape scenes is, therefore, straightforward and exact in environmental assessment. The presence of natural elements such as trees, grass, mountains, and water bodies provide relaxation effect while generally, the opposite is true for most of the urban elements (Kaplan 2001). However, this notion might not hold if the multimodal/multisensory analysis is performed in the evaluation of such a place. In this section, we discuss the emerging tools and techniques to quantify the entities present in these realms.

a. ***Vision***

Human eyes are capable of involuntarily identifying distinct regions present inside the scene, which are then extracted and analyzed by the brain (Borji et al. 2014). The preference of the landscape largely depends upon the presence of physical elements and their attributes in the view. The "Ecological" variables (Fig. 7.1) such as "naturalness," "architectural style," and "nuisances" have been widely discussed in visual perception of a place. (a) The naturalness is the presence of natural features such as trees, grass, playgrounds, and parks; (b) The architectural style refers to the outlook of the built typology such as temporary structures, high and mid-rise buildings; (c) Nuisances degrade the environment quality and consists of elements of dilapidation, littering, garbage, advertisement signboards, etc. The urban landscapes are filled with the blend of these elements. The nuisance-inducing components may not cover the entire view of the landscape from observer's point of view, but even its presence in smaller quantities persuades the viewer to prefer or ignore the scene (Winkel et al. 1970; Nasar 1984). It has been established that the presence of specific elements within the scenes is more favored than others. The near accurate prediction of people's preferences is possible if the features of such elements are automatically determined within the scene. Scaling such preference-based studies to a regional or a city level, therefore, would require the assistance of computer-based algorithms and techniques.

Computer vision techniques such as Scale-Invariant Feature Transform (SIFT), Histogram of Oriented Gradients (HOG), filtering, and thresholding have already been utilized in small-scale analyses to extract and label elements in a photographic dataset used in studying preferences (Hagerhall et al. 2004; Kardan et al. 2015). However, such algorithms require manual tuning of parameters for specific tasks; utilizing which to assess large and complex datasets such as urban scenes is difficult (Suleiman et al. 2017). Recent advancements in machine learning models such as

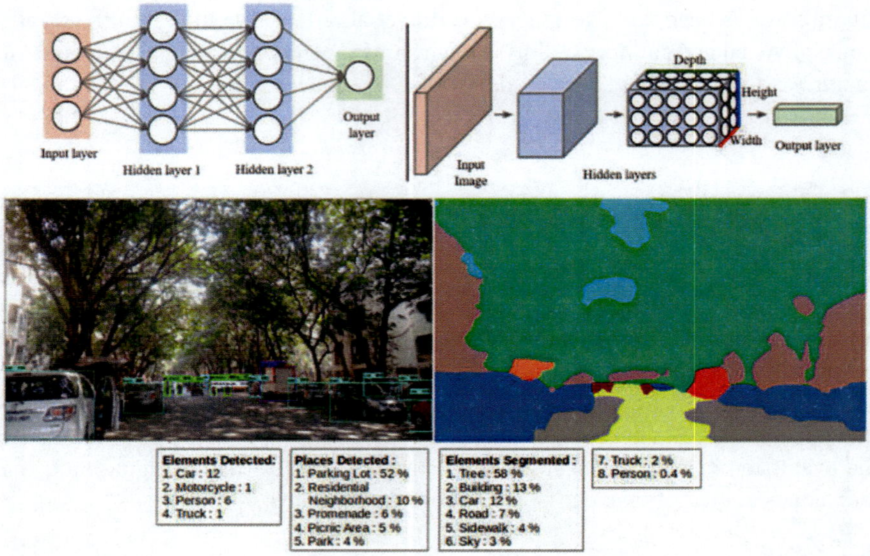

Fig. 7.2 [Top] Architecture of simple ANN and CNN (*Source* https://cs231n.github.io/convolutional-networks/. [Bottom] Output example of visual scene understanding task (*Source* Authors)

Artificial Neural Network (ANN) have shown exceptional efficiency and accuracy in feature extraction, classification and prediction tasks in large datasets (He et al. 2017). ANNs are the computational models which are biologically inspired and are designed to simulate the process of human neurons in responding to stimuli. These models perform by gathering knowledge based on the inherent relationships present in the datasets and through the training experience. These models develop a correlation between inputs and outputs on their own without the need for any explicit programming (Beresford and Agatonovic-Kustrin 2000).

The networks are shaped by the input layer, a hidden layer, and an output layer (Fig. 7.2). The components (or nodes) in the layers are interconnected with each other. Every such association has some weight, which is learned by the model over its training time. The model tries to match its outputs with the labels provided and persistently acquires a set of better weights to arrive at a certain level of accuracy. After which, the model is said to be trained. The trained model is used to provide inference to the similar new datasets with some likelihood (for more details refer Goodfellow et al. (2016)).

Convolutional Neural Networks (CNN) are one such variant of this simple architecture which has several applications in image processing. In CNN, the pixels of the images are provided as input to the model. The image is learned by the model with the help of filters which extract the edges, geometry, and other such image attributes and utilize it to understand its constituents. Different variants of CNN are used in various computer vision tasks such as object detection, localization, and image classification. These models fall into the specific category of machine learning called Deep Learning (DL) (Lecun et al. 2015).

The computer vision tasks of scene classification, object detection, scene segmentation can be collectively called "visual scene understanding" tasks. Scene classification refers to the task of assigning a provided image a label from a fixed set of categories. The object detector model detects the objects in the image on which it has been trained (Fig. 7.2). The outputs produced by the model are the bounding boxes or the masks around the detected object. Scene segmentation divides the images into different labels, which help in understanding the constituents of the image in depth. It provides the mask boundaries over the identified objects.

DL tasks generally require large-scale annotated datasets and massive computing power to produce results. Fortunately, the datasets such as ImageNet, Places, PASCAL VOC, Microsoft COCO, ADE20K, and Mapillary (Neuhold et al. 2017) have been made openly available to the researchers which aid in a variety of experiments and creation of products. Many of such datasets contain classes commonly present in outdoor scenes and therefore are most relevant in understanding urbanscapes. These urban-based datasets are vastly researched and documented, mostly due to their relevance toward research and development in self-driving cars (Yang et al. 2018). These datasets, after sufficient training process, predict the elements in the images with near human-level accuracy (Russakovsky et al. 2015).

By creating a pipeline to input a large array of such street-level views into the model, a detailed summary of the study area can be generated (Fig. 7.2). Inferences can be obtained regarding the presence of (a) types of urban and natural elements (cars, people, trees, buildings, etc.), (b) percentage of visual area covered by elements, (c) percentage of natural features at specific areas, and (d) classification of a place (such as park, promenade, and street). These outputs can be georeferenced and plotted in maps for better understanding (Verma et al. 2018). Utilizing this approach will provide a low-level (Fig. 7.1) element-wise knowledge of the urban surroundings, which will help in the quantification of the visual realm. Different images from the same location can be analyzed to study temporal changes in urban surroundings.

b. ***Sound***

Sound environments created by the sounds of the crowd, open markets, moving vehicles, chirping birds, etc. influence how people perceive urban spaces. However, compared to the assessment of visual scenes, the perception-based research in auditory environments is relatively recent. The emphasis on the evaluation of auditory landscape was given by Murray Schafer, where he coined the term "Soundscape" (Schafer 1993), which refers to the "acoustic environment as perceived or experienced and/or understood by a person or people" (ISO 2014). In other words, it is the comprehensive evaluation of sound as characteristics of sound waves, space as a functional entity, people as a social dimension and physical environment (Zhang and Kang 2007; Axelsson et al. 2010). According to Pijanowski et al. (2011), the different

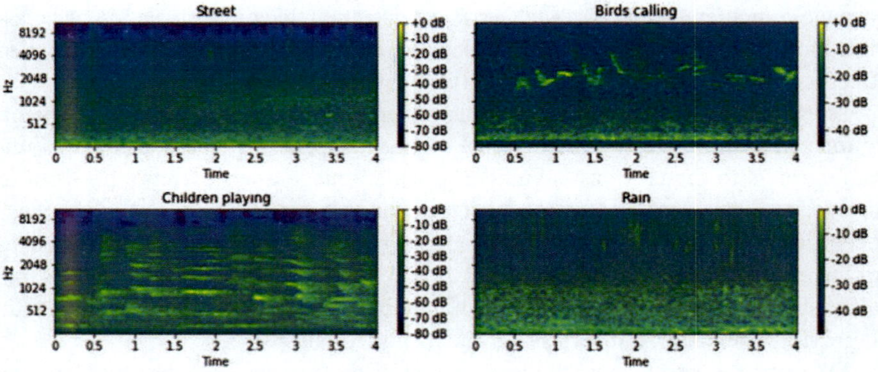

Fig. 7.3 Plots showing Mel spectrogram representation of recorded sounds of the live street, birds calling, children playing, and rain

sources of sounds can be differentiated as biological (birds and animals), geophysical (rain, thunder, and wind), and anthropogenic (human-made) (Fig. 7.3). Similar to the contribution of the visual realm in the interpretation of scenes, sounds play a significant role in individuals' perception (Axelsson et al. 2010). However, much of the research in urban sounds is conveniently linked to noise and its ill effects (Shepherd et al. 2013; Chew and Wu 2016; Park 2017). Urban management has been ineffective in handling sounds in urban areas, mostly due to its inefficiency in understanding and differentiating sounds from noises (Raimbault and Dubois 2005). Research on soundscapes in cities commenced due to the growth in vehicles and other mechanical equipment used daily in every household. The main policy interventions revolved around reducing noise and finding out its sources (Commission 2002). The loudness component of the sound measured as decibel remained the sole criterion of tagging a location with noise-related issues (Pijanowski et al. 2011). While it has been established that pleasant sounds have a positive impact on people (Zhang and Kang 2007; Hong and Jeon 2017), the management of sounds and their sources has not got sufficient interest from the authorities.

Studies have utilized the live sound monitoring approach (Park 2017) where a microphone along with the relay device is installed permanently at specific locations. Data collected by which is then used by the authorities to measure noise levels. These devices have been essential in creating noise maps at a temporal scale. Similarly, such devices can be installed to record sounds, which can be interpreted later by specialists to understand soundscapes. However, these devices have to be cheap, efficient, and precise enough to cover a substantial area and to provide accurate results (Mydlarz et al. 2017). The underlying infrastructure support provided by the local authority is also crucial for such a project. Due to the absence of such platforms and devices to manage audio data collection and analysis, the researchers have conducted sound walks to capture information regarding soundscapes while walking in the streets (Yong Jeon et al. 2013; Kang et al. 2018). Researchers have collected (a) survey responses to the preferences for different sounds such as traffic, human, and nature at specific intervals and locations in the city, and (b) binaural sound recording with the help of handheld recorders and microphones. These recordings have been further used to artificially simulate real-life scenes to understand the sonic environment.

Automated sound characterization and extraction of various sources have been tried and tested in bits and pieces due to the inherent complexity in sound analysis. Sounds have three perceptual layers (Schafer 1993) such as (a) Background sound, which is a constant hum present in the environment, (b) Intermediate sound is the collection of different sounds changing continuously at a distance, and (c) the Unique sounds are particular to any given location. The constant hum in the background is absent altogether in entirely natural areas but present almost elsewhere in urban areas. The sound recorded in general unstructured environment as in urban areas is polyphonic, in which different layers of sound overlap. The monophonic sounds are not present in surroundings unless otherwise created or recorded in artificial environments.

Earlier studies in sound analysis and acoustics were more focused on monophonic sound classification with the help of the state-of-the-art classification algorithms such as Support Vector Machines (SVM). Recently, DL-based architectures such as CNN and RNN (Recurrent Neural Networks) have shown near-human accuracy in classifying these sounds. Classification of monophonic sounds, however, does not have much relevance in real-life urban scenes. Therefore, modified DL architectures have been utilized to classify polyphonic sounds. The task of sound classification aimed at identifying specific events such as vehicle passing and human conversation is called Sound Event Detection (SED) task. Models such as CRNN (combination of CNN and RNN model), Long-Short Term Memory, a variant of RNN) have been able to provide almost twice the accuracy in SED tasks than the previous state-of-the-art Algorithms such as Hidden Markov Model and Gaussian Markov Models (HMM-GMM) (Cakir et al. 2017).

The interest in Sound classification and analysis among the research community has significantly grown after the introduction of real-world annotated datasets such as TUT Sound events,[1] Urban-sound dataset,[2] and AudioSet.[3] TUT Sound events are focused toward SED tasks, while Urban-sound and AudioSet data are primarily utilized in sound classification tasks. TUT dataset consists of binaural audio of total duration of about 1000 min comprising of recordings of 10 scenes such as Basketball stadium, hallway, office, restaurant, and shop. On the other hand, the Urban-sound dataset is composed of 27 h of audio related to the ten everyday sound events in urban areas such as the jackhammer, children playing, and car horn. The AudioSet (Gemmeke et al. 2017) is the most extensive open-source collection of everyday sounds. The dataset consists of 632 manually labeled audio classes which are prepared with clipped YouTube videos, each of 10 s duration. From such a large collection of different sound classes, the subset of everyday urban sounds can be made and trained using discussed algorithms. The introduction of Competitions such as

[1] https://zenodo.org/record/400516.

[2] https://urbansounddataset.weebly.com/.

[3] https://research.google.com/audioset/.

DCASE[4] has provided a common platform for the research community to test and evaluate sound classification and event detection tasks. Several new techniques have been experimented to get the varying level of accuracy. SED tasks are beneficial in understanding polyphonic sounds in urban areas and have the potential to classify sounds at a large scale with minimum human intervention.

Currently, the critical limitation in SED is the availability of accurately labeled datasets than the performance of algorithms. The categories used in the datasets are often quite ambiguous, which are subject to uncertainty. Each annotation task requires annotators to mark the onset and the end of the particular event (e.g., a passing vehicle, children playing) in the recorded audio clip. For which, the accuracy is dependent on the precision with which annotations are labeled. Further, commonly available SED datasets do not follow common ontology while labeling sound events, e.g., the audio event "car passing by" can be labeled as "car," "engine sound," "vehicle," etc. Utilizing such annotated data is often difficult to train and prone to misclassification in many cases.

7.3 Discussion

Urban surroundings are diverse; apart from the built infrastructure, cross-cultural perspective, traditions, and demography play a crucial role in the perception of the landscapes. The large databases such as Google and Tencent Street view images have become the de facto source for assessment and comparison of visual characteristics of cities and neighborhoods in recent studies (De Nadai et al. 2016; Cheng et al. 2017). Place pulse (Salesses et al. 2013) is one such example which used Street view scenes in the creation of a crowdsourced dataset. It provides a perception-based comparison of Street view scenes with the help of perceptual attributes such as liveliness, boredom, wealthiness, and safety. This dataset comprises Google street view scenes of different cities for which ratings are provided by the participants from all over the world. These studies have further utilized DL methods and ranking algorithms to compare various scenes and to predict the perception of the people in new scenes.

However, the limitations of using Google Street View (GSV) images are worth noting. The perceived qualities of the place such as liveliness and safety are highly dependent upon the duration of the day, the presence of people and vehicles on the streets (Thomas and Bromley 2000; Mehta 2007). GSV images are fixed data points with no temporal dimension, hence are unsuitable for detailed analysis at the neighborhood level (Goel et al. 2018). Similarly, crowdsourcing perception-based responses, such as in place pulse, have several drawbacks. The judgments on the environmental qualities by the participants may depend upon multiple factors such as nationality, culture, and traditions. The degree of familiarity (observer being a

[4]https://www.cs.tut.fi/sgn/arg/dcase2017/dcase.

resident or a visitor) of the place might affect the judgment of the observer toward the same.

Compared to well-maintained visual datasets in the form of street view photographs to experiment with, the availability of open sound datasets is somewhat limited. Researchers have relied on smaller datasets and soundwalk methods to collect environmental sounds pertinent to their studies. Such studies have been successful in creating soundscape maps and predicting of perception of people in the auditory realm through the use of different algorithms. The collected sound clips are used to extract psychoacoustic parameters and identification of sound sources with the help of participants (Hong and Jeon 2017). The studies involving automated identification of sound sources with the help of annotated machine learning datasets have not been conducted at the neighborhood or city level. This calls for an opportunity for researchers to experiment with the DL algorithms, which can provide consistent results in sound events and sources' detection. However, the continuous collection of sound clips and the choice of annotation classes may determine the accuracy and viability of such research.

Tagging geographic information to the array of sensory datasets may help in classifying places with respect to the morphology of the surroundings and perceived affective qualities. It will provide an opportunity to conduct longitudinal spatial assessments and create large-scale spatial models. The sensuous character of places fluctuates with time. Therefore, synchronized data collection from sensory realms will help in a comprehensive documentation of the area, which can be further used to answer different urban management and planning questions.

7.4 Conclusion

Urban landscapes represent the cultural and social identity of the city. These landscapes are continuously changing to house infrastructures and people at the cost of balancing the ecological, social, and cultural integrity. With the help of advanced technological means, urban perception studies can be scaled to include cities and regions. The use of different sensory realms in understanding the maps may help authorities and city managers to realize the state of the urban surroundings. The significant outcome of such studies could be the creation of a tool by which the surroundings can be ranked by their visual aesthetics or soundscapes. It will inform the authorities about the places which require makeover and special attention to restore and rebuild the pleasing, healthy and sustainable environment. Particular revenue streams can be generated by authorities in exchange for providing such provisions in the neighborhoods. Cities often invest vastly in the tourism sector where the aesthetics, novelty, and variety of the environment are the factors to invite visitors to the city. Apart from representing the maps with buildings and streets, perceptual layers can be included to depict cities from a different perspective.

Acknowledgements The authors would like to thank the Ministry of Human Resource Development (MHRD), India and Industrial Research and Consultancy Centre (IRCC), IIT Bombay for funding this study under the grant titled Frontier Areas of Science and Technology (FAST), Centre of Excellence in Urban Science and Engineering (grant number 14MHRD005).

References

Appleton J (1996) The experience of landscape. Wiley

Axelsson Ö, Nilsson ME, Berglund B (2010) A principal components model of soundscape perception. J Acoust Soc Am 128:2836–2846. https://doi.org/10.1121/1.3493436

Beresford R, Agatonovic-Kustrin S (2000) Basic concepts of artificial neural network (ANN) modeling and its application in pharmaceutical research. J Pharm Biomed Anal 22:717–727

Berlyne DE (1951) Attention, perception and behavior theory. Psychol Rev 58:137–146. https://doi.org/10.1037/h0058364

Berlyne DE (1971) Aesthetics and psychobiology. Appleton-Century-Crofts

Borji A, Cheng M-M, Hou Q et al (2014) Salient object detection: a survey 1–20. https://doi.org/10.1109/TIP.2015.2487833

Cakir E, Parascandolo G, Heittola T et al (2017) Convolutional recurrent neural networks for polyphonic sound event detection. IEEE/ACM Trans Audio Speech Lang Process 25:1291–1303. https://doi.org/10.1109/TASLP.2017.2690575

Chang H (2009) Mapping the web of landscape aesthetics: a critical study of theoretical perspectives in light of environmental sustainability

Cheng L, Chu S, Zong W et al (2017) Use of tencent street view imagery for visual perception of streets. ISPRS Int J Geo-Inf 6:265. https://doi.org/10.3390/ijgi6090265

Chew YR, Wu BS (2016) A soundscape approach to analyze traffic noise in the city of Taipei, Taiwan. Comput Environ Urban Syst 59:78–85. https://doi.org/10.1016/j.compenvurbsys.2016.05.002

Commission E (2002) Environmental noise directive. https://ec.europa.eu/environment/noise/directive_en.htm. Accessed 26 May 2019

De Nadai M, Vieriu RL, Zen G et al (2016) Are safer looking neighborhoods more lively? A multimodal investigation into urban life. In: Proceedings of 2016 ACM multimedia conference—MM '16, pp 1127–1135

Dosen AS, Ostwald MJ (2016) Evidence for prospect-refuge theory: a meta-analysis of the findings of environmental preference research. City Territ Archit 3:4. https://doi.org/10.1186/s40410-016-0033-1

Dubey A, Naik N, Parikh D et al (2016) Deep learning the city: quantifying urban perception at a global scale. In: Lecture notes in computer science (including subseries Lecture notes in artificial intelligence and lecture notes in bioinformatics), pp 196–212

Gemmeke JF, Ellis DPW, Freedman D et al (2017) Audio set: an ontology and human-labeled dataset for audio events. In: 2017 IEEE international conference on acoustics, speech and signal processing (ICASSP). IEEE, pp 776–780

Gjerde M (2010) Visual aesthetic perception and judgement of urban streetscapes. In: W101-special track 18th CIB World Build Congress 11

Goel R, Garcia LMT, Goodman A et al (2018) Estimating city-level travel patterns using street imagery: a case study of using Google Street View in Britain. PLoS ONE 13:e0196521. https://doi.org/10.1371/journal.pone.0196521

Goodfellow I, Bengio Y, Courville A (2016) Deep learning. MIT Press

Hagerhall CM, Purcell T, Taylor R (2004) Fractal dimension of landscape silhouette outlines as a predictor of landscape preference. J Environ Psychol 24:247–255. https://doi.org/10.1016/j.jenvp.2003.12.004

He Y, Mudur S, Poullis C (2017) Multi-label pixelwise classification for reconstruction of large-scale urban areas 6:1–9

Herzog TR (1989) A cognitive analysis of preference for urban nature. J Environ Psychol 27–43

Herzog TR (1992) A cognitive analysis of preference for urban spaces. Environ Psychol 1:237–248. https://doi.org/10.1146/annurev-psych-010213-115048

Herzog TR, Kaplan S, Kaplan R (1982) The prediction of preference for unfamiliar urban places. Popul Environ 5:43–59. https://doi.org/10.1007/BF01359051

Herzog TR, Kaplan S, Kaplan R (1976) The prediction of preference for familiar urban places. Environ Behav 8:627–645. https://doi.org/10.1177/001391657684008

Herzog TR, Stark JL (2004) Typicality and preference for positively and negatively valued environmental settings. J Environ Psychol 24:85–92. https://doi.org/10.1016/S0272-4944(03)00038-0

Hong JY, Jeon JY (2017) Exploring spatial relationships among soundscape variables in urban areas: a spatial statistical modelling approach. Landsc Urban Plan 157:352–364. https://doi.org/10.1016/j.landurbplan.2016.08.006

ISO (2014) Soundscape. https://www.iso.org/standard/52161.html. Accessed 26 May 2019

Kang J, Aletta F, Margaritis E, Yang M (2018) A model for implementing soundscape maps in smart cities. Noise Mapp 5:46–59. https://doi.org/10.1515/noise-2018-0004

Kaplan R (2001) The nature of the view from home: psychological benefits. Environ Behav 33:507–542. https://doi.org/10.1177/00139160121973115

Kaplan R, Rachel Kaplan SK, Kaplan S (1989) The experience of nature: a psychological perspective. Cambridge University Press

Kaplan S (1979) Perception and landscape: conceptions and misconceptions, pp 241–248

Kaplan S, Kaplan R, Wendt JS (1972) Rated preference and complexity for natural and urban visual material *. Percept Psychophys 12:354–356

Kardan O, Demiralp E, Hout MC et al (2015) Is the preference of natural versus man-made scenes driven by bottom–up processing of the visual features of nature? Front Psychol 6:1–13. https://doi.org/10.3389/fpsyg.2015.00471

Kasmar J (1970) The development of a usable lexicon of environmental descriptors. Environ Behav 2:153–169. https://doi.org/10.1177/001391657000200202

Kaymaz I (2012) Landscape perception. In: Landscape planning. Elsevier, pp 252–276

Laumann K, Garling T, Stormark KM (2001) Rating scale measures of restorative components of environments. J Environ Psychol 21:31–44. https://doi.org/10.1006/jevp.2000.0179

Lecun Y, Bengio Y, Hinton G (2015) Deep learning. Nature 521:436–444. https://doi.org/10.1038/nature14539

Loewen LJ, Steel GD, Suedfeld P (1993) Perceived safety from crime in the urban environment. J Environ Psychol 13:323–331. https://doi.org/10.1016/S0272-4944(05)80254-3

Mehta V (2007) Lively streets. J Plan Educ Res 27:165–187. https://doi.org/10.1177/0739456X07307947

Mydlarz C, Salamon J, Bello JP (2017) The implementation of low-cost urban acoustic monitoring devices. Appl Acoust 117:207–218. https://doi.org/10.1016/j.apacoust.2016.06.010

Nasar JL (1984) Visual preferences in urban street scenes. J Cross Cult Psychol 15:79–93. https://doi.org/10.1177/0022002184015001005

Nasar JL (1989) Perception, cognition, and evaluation of urban places. In: Altman I, Zube EH (eds) Public places and spaces. Springer, US, Boston, MA, pp 31–56

Neuhold G, Ollmann T, Bul SR, Kontschieder P (2017) The mapillary vistas dataset for semantic understanding of street scenes. Iccv, pp 4990–4999

Oostendorp A, Berlyne DE (1978) Dimensions in the perception of architecture: III Multidimensional preference scaling. Scand J Psychol 19:145–150. https://doi.org/10.1111/j.1467-9450.1978.tb00314.x

Park TH (2017) Mapping urban soundscapes via citygram. In: Thakuriah P, Tilahun N, Zellner M (eds) Springer International Publishing, Cham, pp 491–513

Pazhouhanfar M, Kamal M (2014) Effect of predictors of visual preference as characteristics of urban natural landscapes in increasing perceived restorative potential. Urban For Urban Green 13:145–151. https://doi.org/10.1016/j.ufug.2013.08.005

Pijanowski BC, Villanueva-Rivera LJ, Dumyahn SL et al (2011) Soundscape ecology: the science of sound in the landscape. Bioscience 61:203–216. https://doi.org/10.1525/bio.2011.61.3.6

Porzi L, Rota Bulò S, Lepri B, Ricci E (2015) Predicting and understanding urban perception with convolutional neural networks. In: Proceedings of the 23rd ACM international conference on multimedia—MM '15. ACM Press, New York, New York, USA, pp 139–148

Raimbault M, Dubois D (2005) Urban soundscapes: experiences and knowledge. Cities 22:339–350. https://doi.org/10.1016/j.cities.2005.05.003

Rossetti T, Lobel H, Rocco V, Hurtubia R (2019) Explaining subjective perceptions of public spaces as a function of the built environment: a massive data approach. Landsc Urban Plan 181:169–178. https://doi.org/10.1016/j.landurbplan.2018.09.020

Russakovsky O, Deng J, Su H et al (2015) ImageNet large scale visual recognition challenge. Int J Comput Vis 115:211–252. https://doi.org/10.1007/s11263-015-0816-y

Salesses P, Schechtner K, Hidalgo CA (2013) The collaborative image of the city: mapping the inequality of urban perception. PLoS ONE 8:e68400. https://doi.org/10.1371/journal.pone.0068400

Schafer RM (1993) The soundscape: our sonic environment and the tuning of the world. Simon and Schuster

Shepherd D, Welch D, Dirks KN, McBride D (2013) Do quiet areas afford greater health-related quality of life than noisy areas? Int J Environ Res Public Health 10:1284–1303. https://doi.org/10.3390/ijerph10041284

Shrivastava P, Bhoyar KK, Zadgaonkar AS (2017) Bridging the semantic gap with human perception based features for scene categorization. Int J Intell Comput Cybern 10:387–406. https://doi.org/10.1108/IJICC-09-2016-0035

Suleiman A, Chen YH, Emer J, Sze V (2017) Towards closing the energy gap between HOG and CNN features for embedded vision (Invited paper). Proc—IEEE Int Symp Circuits Syst. https://doi.org/10.1109/ISCAS.2017.8050341

Thomas CJ, Bromley RDF (2000) City-centre revitalisation: problems of fragmentation and fear in the evening and night-time city. Urban Stud 37:1403–1429. https://doi.org/10.1080/00420980020080181

Verma D, Jana A, Ramamritham K (2019) Intelligent human systems integration. In: Karwowski W, Ahram T (eds) Intelligent human systems integration. Springer International Publishing, Cham, pp 852–857

Verma D, Jana A, Ramamritham K (2018) Quantifying urban surroundings using deep learning techniques: a new proposal. Urban Sci 2:78. https://doi.org/10.3390/urbansci2030078

Wang R, Zhao J (2017) Demographic groups' differences in visual preference for vegetated landscapes in urban green space. Sustain Cities Soc 28:350–357. https://doi.org/10.1016/j.scs.2016.10.010

Winkel G, Malek R, Thiel P (1970) A study of human response to selected roadside environments. In: Proceedings of 1st EDRA conference, pp 224–240

Wohlwill JF (1976) Environmental aesthetics: the environment as a source of affect. In: Altman I, Wohlwill JF (eds) Human behavior and environment: advances in theory and research, vol 1. Springer. US, Boston, MA, pp 37–86

Yang Z, Zhang Y, Yu J et al (2018) End-to-end multi-modal multi-task vehicle control for self-driving cars with visual perceptions. In: Proceedings—international conference on pattern recognition 2018, August, pp 2289–2294. https://doi.org/10.1109/ICPR.2018.8546189

Yong Jeon J, Young Hong J, Jik Lee P (2013) Soundwalk approach to identify urban soundscapes individually. J Acoust Soc Am 134:803–812. https://doi.org/10.1121/1.4807801
Zhang M, Kang J (2007) Towards the evaluation, description, and creation of soundscapes in urban open spaces. Environ Plan B Plan Des 34:68–86. https://doi.org/10.1068/b31162

Chapter 8
Land Transformation and Future Projections of Land Consumption Using High-Resolution Remote Sensing Data for Allahabad, India

Virendra Kumar, Abhishek Kumar Yadav, and Arjun Singh

Abstract In almost developing countries of the world, the per capita per person ratio of land resource is going to be decreased because of increasing population and the country like India is losing their agricultural dominant identity due to transformation of rural area into urban centers. This paper focuses the results on urban sprawl, landuse change, land transformation and land consumption of Allahabad cit based on Survey of India topographical maps of 1973, IKONOS satellite data of 2009 and 2014. The results show that in 1973, the urban built-up area of city based on SOI topographical map was 4760.09 ha, based on IKONOS satellite imageries of 2009 and 2014, the urban built-up area of the city is 7803.16 ha and 8138.39 ha, respectively. This has increased as 3043.07 ha at an interval of 36 years from 1973–2009 to 2009–2014, growth in urban area observed as 335.23 ha. Agricultural land, orchard/plantation, open space, water bodies, and nala converted into urban land. The trend change between 2009 and 2014 has been observed as remarkable decreased to −22.63% in agricultural land, −30.55% in open space, waterbodies −3.91%, and nala-0.51%, respectively. The Land consumption rate and absorption coefficient between the period from 1973 to 2009 and 2014 is 0.006 ha and 0.007 ha, respectively. During 2009–2014, land consumption rate is observed 0.006 ha, land absorption coefficient is 0.001 ha and land consumption for 2014–2021, estimated as 0.005 ha. The digital database generated for Allahabad city in GIS would be very useful for better landuse planning/sustainable development of cities.

Keywords Urbanization · Land transformation · IKONOS satellite data · GIS techniques

Acronyms

ADA Allahabad Development Authority
GIS Geographical Information System

V. Kumar (✉) · A. K. Yadav · A. Singh
Remote Sensing Applications Centre-Uttar Pradesh, Lucknow, India

P. Sharma (ed.), *Geospatial Technology and Smart Cities*, The Urban Book Series,
https://doi.org/10.1007/978-3-030-71945-6_8

GPS	Global Positioning System
LAC	Land Absorption Coefficient
L.A	Land Absorption
LCR	Land Consumption Ratio
L.C	Land Consumption

8.1 Introduction

Low and lower middle economy group countries in the world are facing the problem of low level of urbanization. In these countries, the ever-increasing population and its need for shelter and other economic activities are engulfing our land resources in urban fringe areas. In recent year launching of high-resolution remote sensing data of Indian and foreign satellites coupled with GIS and GPS techniques are more helpful to know the present and part status of urban sprawl, landuse/landcover change, land transformation, urban environment and future perspective of land consumption and absorption coefficient for people. High-resolution satellite imageries having Pan and multispectral merged data with less than 1 m spatial resolution provide lucid and effective means of information of our earth surface because of its synoptic view. The continuous acquisition of satellite imageries at regular interval is very necessary for monitoring the urban sprawl and land transformation. Batty (1996), Subudhi et al. (1998), Bhatta (2009), Xu (2008), Griffiths et al. (2010), Anderson et al. (1976) suggested that an up to date landuse map prepared based on high-resolution satellite data in GIS is useful to monitor the changes in infrastructure, built-up urban/rural area environment. Gopalan (2009), Patkar (2003) have suggested that due to fine spatial resolution and spectral reflectance of high-resolution remote sensing data each natural or manmade object can easily be identified and mapped to generate the digital database for any developmental planning, management activities.

India is close to 7933 urban local bodies as per Census of India-2011. Uttar Pradesh in India is the fourth largest state in geographical area (240,928 km^2), and in population, this is the largest state in the country with 22 crores of people. The State has around 724 cities including some million-plus population cities as per Census of India, 2011, i.e., Nagar Panchayat/Nagarpalika Parishad/municipal corporations/industrial township and cantonments boards. Population growth rate in all such cities are high and numbers of colonies have been developed by housing and urban planning department and those colonies are being taken up by urban local bodies, in order to assess and monitor the urban spatial growth, landuse/cover trends of change, land transformation, land consumption rate, ratio and future population projections and per capita/person land consumption, absorption, coefficient for study area, an endeavor has been made to generate the digital database for Allahabad city using IKONOS satellite 1 m high-resolution data in GIS environment, which would be useful to Urban Development Authority for better planning & management of land resources for cities.

Conceptual Background of The Term Per Capita

The word "**Per capita**" is a Latin term that explains into "by head." Per capita means the average per person and is often used in place/land of per person in statistical observances. The phrase is used with economic data or reporting but is also applied to almost any other occurrence of population description.

Land consumption per capita

Per capita land consumption is measured as the difference between two or more time period data in habitat or developed area per **person**. Negative numbers indicate **per capita land consumption** declined over the decade.

Land Absorption Coefficient

Higher/lower values of Land Absorption Coefficient (LAC) and Land Consumption Ratio (LCR) values mean in explaining urban growth, sprawl, and sustainability in terms of population distribution and density.

8.2 Literature Review

Urban planners required the information on the rate and pattern on urban expansion for proper urban landuse planning and management policy directions. To evaluate the dynamics and spatial pattern of any city, multi-date satellite imageries are used to analyze the changes in landuse/cover. Ayele Almaw Fenta et al. (2017), analyzed the urban sprawl growth, per capita land consumption rate (ha per person) of Mekelle city, Northern Ethiopia. Robert I. McDonald,[1,*] Richard T. T. Forman,[2] and Peter Kareiva, (1990–2000), has published a research paper titled, "Open Space Loss and Land Inequality in United States' Cities," in which the per capita land consumption (m^2/person) of most cities indicates the per capita loss of open space. Theobald (2001, 2005), McDonald (2008), McDonald et al. (2008) have analyzed the urban landuse/cover, changes and per capita land consumption and absorption coefficient in cities. The majority of developed area in cities is in low-density neighborhoods housing a small proportion of urban residents, with Gini coefficients that quantify this developed land inequality averaging and results suggest conservation funding and reform-minded zoning decrease per capita open space loss.

8.3 Objectives

The broad objective of the present study is to assess the urban sprawl, area estimation under landuse/landcover, trend change of landuse, land transformation, land consumption rate, ratio and land absorption coefficient for Allahabad city and its environs using high-resolution remote sensing data of IKONOS satellite and Geographical Information System (GIS) techniques.

8.4 Study Area

The Allahabad city is situated on the bank of Ganga River. The city is located at 25°26′ North latitude and 81°50′ East longitude (Fig. 8.1). Allahabad city in Uttar Pradesh is a district headquarter and commissionary. Study area is located at the confluence of Ganges, Yamuna and Saraswati, popularly known as Triveni Sangam, which is famous for its rich culture heritage and religious importance. Allahabad is well connected with railway line and road network with capital town of the state and from other cities of different states of India and national capital New Delhi.

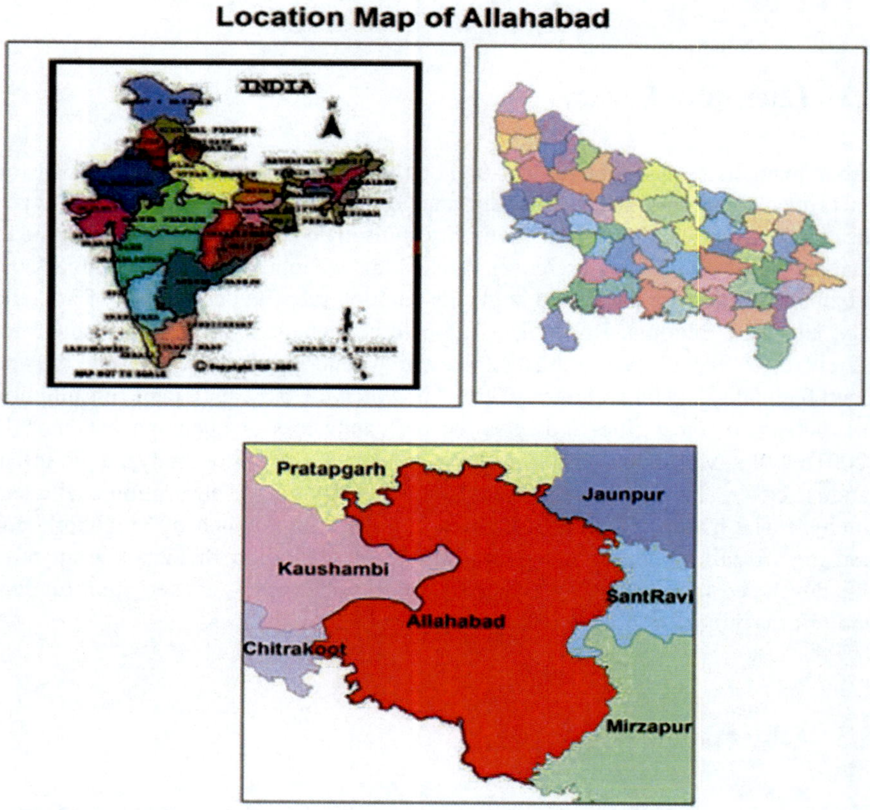

Fig. 8.1 Location map of study area

8.5 Materials and Data Used

The following datasets have been used to meet out the set objectives.

- Survey of India topographical map sheet No. 63G/10, 63G/11, 63G/12, 63G/14, 63G/15, 63G/16, 63H/13, 63K/2, 63K/4, 63K/6, 63K/7, and 63L/1 of Allahabad city and its surroundings on 1:25,000/1:50,000 scales surveyed in 1973.
- Allahabad city-guide map on 1:20,000 scale.
- IKONOS Satellite 1 m spatial resolution data of 2009 and 2014 downloaded from Google Earth.
- Ground truth/field data collected using Global Positioning System (GPS).
- Census of India population data—1971, 2001, and 2011.

Software and Equipment

- Arc-GIS software 10.1 V
- GPS Mobile Mapper 10.0 V

8.6 Methodology

To meet the set objective of the study area, at first road/transport network map was prepared using Survey of India topographical map corresponding to study area. Primary road-state/national highway and major roads and canals have been prepared as polygon layer, whereas minor roads of the city were delineated as linear feature. IKONOS satellite's 1 m spatial resolution data of 2009 and 2014 has been used for study area to update the road/transport network. Urban sprawl map was demarcated based on S.O.I. topographical map sheets surveyed in 1973 & IKONOS satellite data 2009 and 2014 (Fig. 8.2 and Table 8.1), and status of growth for different period has been calculated in Arc-GIS software. Landuse/landcover and land transformation and change map for the study area has been prepared using S.O.I. topographical map sheet No. 63G/10, 63G/11, 63G/12, 63G/14, 63G/15, 63G/16, 63H/13, 63K/2, 63K/4, 63K/6, 63K/7, 63L/1, and IKONOS Satellite data of 2009 and 2014. Trends of landuse change and land transformation, land consumption rate, ratio and future population projection and land consumption ratio and land absorption coefficient have also been generated using statistical method. Each and every urban spatial growth/layers was superimposed to know the status of spatial growth in different years from 1973 to 2014. Landuse/landcover and land transformation area estimation has been calculated for each layer. Each and every layer has been given attributes coding, land consumption rate, ratio and the future land consumption that has been predicted based on future population projection of the city. Ground truth/field survey was carried out based on printed maps and Global Positioning System (GPS) to know the locational extent of different landuse/landcover categories and field information has been incorporated before finalization of maps.

Fig. 8.2 Urban sprawl map of study area-1973–2014

Table 8.1 Urban sprawl growth rate and percentage per annum

Year	Data	Area (ha)	Sprawl growth (ha)	Growth %	Growth per annum %	Growth per annum (ha)
1973	SOI toposheet	4760.09	–	–	–	–
2009	IKONOS data	7803.16	3043.07	63.92	1.77	84.52
2014	IKONOS data	8138.39	335.23	4.29	0.85	67.04

8.6.1 Statistical Method Applied

The comparative study on landuse/landcover assists in identifying the trend and percentage of changes between 1973 to 2009 and 2009 to 2014. In achieving this, the first task was developed for preparation of tables showing the area in hectare and percentage of change between 1973 and 2009 and 2009 to 2014 measured in given each and every landuse/landcover category. The change of percentage was calculated to determine the trend of changes that has been can be calculated by dividing observed changes by the sum of changes. Land consumption rate was calculated by dividing the area by population where land absorption coefficient has been calculated by subtracting the areal extent of early year and later year and population of early year and later year and dividing of the same, respectively. Various tables for different times were prepared to know the percentage and trend of changes.

The future expansion for year 2021, land absorption coefficient was forecasted by correlating the 2014, i.e., current population (2011) of Allahabad city with help of growth per annum.

8.7 Results and Discussion

The results and discussions related to urban sprawl, landuse/landcover change and land transformation, land consumption rate, ratio and land absorption coefficient in the present study have been discussed under the following subheads.

8.7.1 Urban Sprawl, Rate, and Direction

Cities in India are getting overcrowded and expanding uncontrollably due to sustained migrations. In the process, these urban centers are affected by both natural and human activities in the absence of any planning policy. The physical extension of the cities is engulfing the productive agricultural lands for urban areas. This type of haphazard growth of urban built-up land over a period could be explained in terms of waves of urban sprawl (both physical and human). The quantum and direction of waves depend upon various centripetal and centrifugal forces working from city and adjoining area. It has been observed that the outward expansion of urban areas poses a threat to the landuse pattern. The rapid pace of urbanization combined with the explosive population growth has made urban and its surrounding areas dynamic. As the limited land of the city gets used, the ever-increasing demand creates pressure on surrounding fertile vulnerable lands in and around the city causing faster rate of land conversion from non-urban to urban use. This results in uncontrolled expansion of city as well as problem of providing basic public services and facilities. Monitoring the rate and direction of urban sprawl is necessary for urban planning. Estimation as well as updating the data of any region through conventional methods has severe limitations. Satellite remote sensing data with their repetitive and synoptic coverage makes a reliable source of information. This type of data on a particular area helps in understanding the physical processes and changes in the landuse and landcover in space and time. Mapping urban sprawl provides a "picture" of where this type of growth is occurring, and helps to identify the environmental and natural resources threatened by such sprawls, and suggests the likely future directions and patterns of sprawling growth. Analyzing the sprawl over a period of time will help in understanding the nature and growth of this phenomenon. Urban local bodies/municipal corporation has the power to manage the sprawl issues in terms of will and ability.

The growth of urban area over a period was determined by computing the area of all the settlements from topographical map sheet of 1973, and this area has been comparing it with the area obtained from the interpreted satellite imagery for the built-up urban area. Since the sprawl is characterized by an increase in the built-up area along the urban and rural fringe, this attribute gives considerable information for understanding the behavior of such sprawls. This is also influenced by parameters such as population density and population growth rate, etc.

In 1973, the city grew in concentric manner, after 1973, IKONOS satellite imageries of 2009 and 2014 have also been used to create vector layer to monitor the growth of city through overlaying of these vector layers of three-time period data, the area for 1973, 2009 and 2014 has been calculated in GIS, which is 4760.09 ha in 1973, 7803.16 ha in 2009 and 8138.39 ha in 2014, respectively. Sprawl growth, per annum growth, and percentage of growth have also been calculated, which is 3043.07 ha (84.52 ha) per annum between the years of 1973 and 2009 at an interval of 36 years. Similarly, Spatial growth, per annum growth, and percentage of growth have also been calculated between the year 2009 and 2014 which is 335.23 ha (Table 8.1). It has been observed from Fig. 8.2 and Table 8.1 that the city has grown mainly toward west direction and marginally toward north-east, with most of the prime agricultural land and vegetative area getting constructed into built-up land and between the periods of 1973–2009, and percentage growth and per annum growth for the period is 63.92% (1.77% per annum) in 36 years. Similarly, spatial growth, per annum growth, percentage growth, and per annum percentage growth have been calculated for the period of 2009–2014 which is 335.23 ha (4.29%) and 67.04 ha and 0.85% per annum, respectively. The urban growth in present study is defined as consisting of all urban landuses, i.e., Built-up-residential, industrial, institutional/utilities, commercial, etc. The present study considers only the physical factor influencing under sprawl growth, i.e., road/transport network distance from city core. The demographic setup and socio-economic factors have not been considered.

8.8 Landuse/Landcover

Landuse/landcover statistics of the study area has been calculated for different period data in GIS environment is discussed as under.

8.8.1 Landuse/Landcover (1973)

The urban landuse/Landcover map of Allahabad city for year 1973 using SOI topographical map sheet no 63G/10, 63G/11, 63G/12, 63G/14, 63G/15, 63G/16, 63H/13, 63K/2, 63K/4, 63K/6, 63K/7 and 63L/1 has been prepared, and eight landuse/landcover categories identified and mapped using Arc-GIS software (Fig. 8.3 and Table 8.2). They are built-up, orchard plantation, waterbody/pond, are occupying the area of 4262.66 ha (89.55%), 45.09 ha (0.94%), 19.20 ha (0.22%), agricultural land 116.18 ha (2.44%), open space 186.47 ha (3.92%), nala 1.71 ha (0.04%),

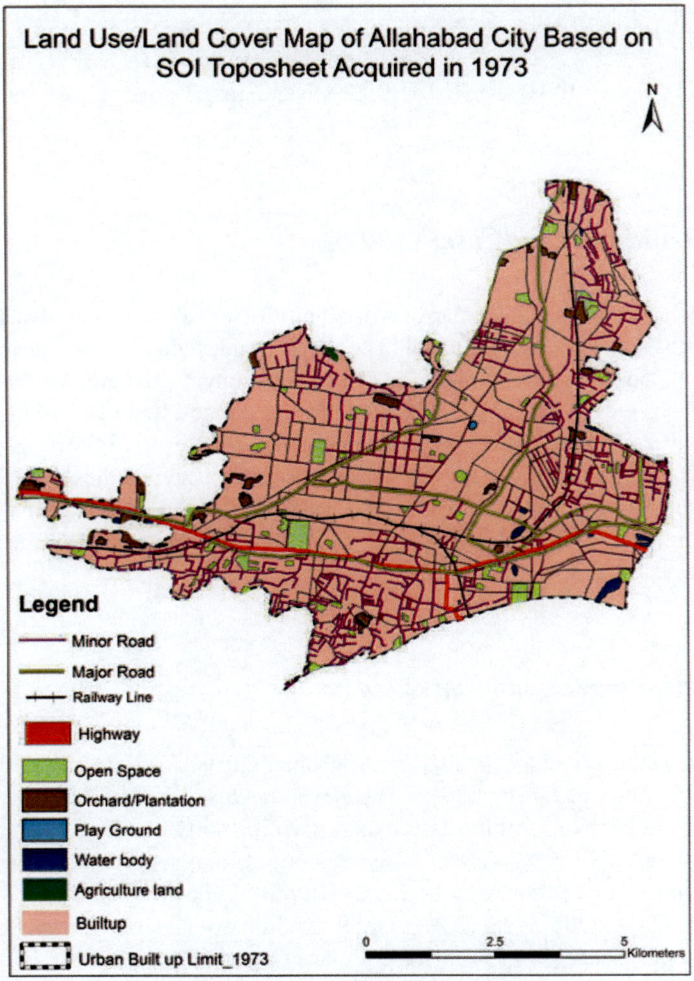

Fig. 8.3 Landuse/Landcover map-1973

Table 8.2 Landuse/Landcover statistics of study area (1973)

S. no.	Class name	Area (ha)	%
1	Agriculture land	116.18	2.44
2	Built-up	4262.66	89.55
3	Nala	1.71	0.04
4	Open space	186.47	3.92
5	Orchard/Plantation	45.09	0.94
6	Waterbody	19.20	0.41
7	NH	10.50	0.22
8	Major road	118.29	2.48
	Total	4760.09	100

National Highway (NH) 10.50 ha (0.22%), and major road 11829 ha (2.48%), respectively.

8.8.2 Landuse/LandCover (2009)

The urban landuse/Landcover map of Allahabad city for year 2009 based on IKONOS Satellite imagery has been prepared. The eleven landuse/landcover categories area were estimated in Arc-GIS software. They are occupying the area under different landuse categories in total geographical area as agriculture Land 162.18 ha (2.08%), built-up 6302.56 ha (80.77%), orchard/plantation 240.49 ha (3.08%), playground 111.89 ha (1.45%), open space 647.21 ha (8.29%), water bodies/pond 54.59 ha (0.69%), airport 97.74 ha (1.25%), canal 1.86 ha (0.03%), nala 5.05 ha (0.06%), national highway 15.53 ha (0.19%), and major road 164.06 ha (2.11%), respectively (Fig. 8.4 and Table 8.3).

8.8.3 Landuse/Landcover (2014)

The urban landuse/Landcover map of Allahabad city has been prepared based on IKONOS Satellite imagery of 2014. The eleven landuse/landcover categories identified and mapped in satellite data using Arc-GIS software (Fig. 8.5 and Table 8.4). These different categories in study area out of total geographical area occupy the area under agricultural land 86.29 ha (1.06%), built-up 6805.09 ha (83.62%), orchard/plantation 235.25 ha (2.89%), park 35.34 ha (0.44), playground 122.81 ha (1.51%), open space 544.79 ha (6.69%), water body 41.46 ha (0.51%), airport 102.94 ha (1.26%), canal 2.83 ha (0.04%), nala 4.54 ha (0.05%), national highway 26.57 ha (0.33%), and major road 165.81 ha (2.04%), respectively.

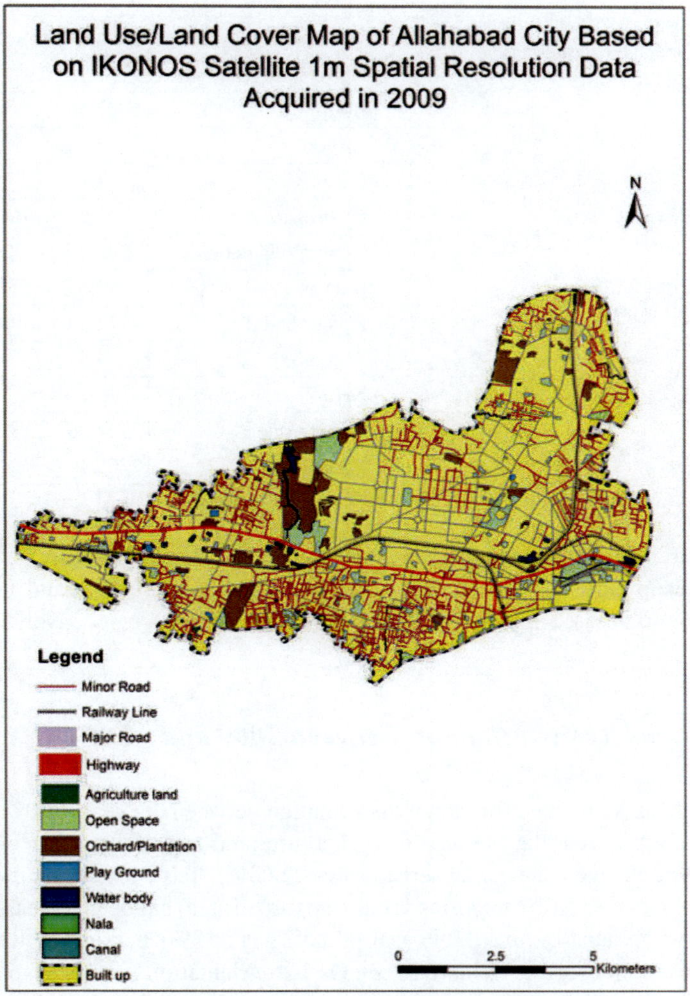

Fig. 8.4 Landuse/Landcover map-2009

8.8.4 Land Transformation Between 1973 and 2009

Results obtained for land transformation between years 193 and 2009 are summarized in Table 8.5, It has been observed that trend change in built-up (urban area) of the city has been increased to 2039.84 ha (67.03%)., and trend change in open spaces 460.74 ha (15.14%), orchard/plantation 195.40 ha (6.42%), agricultural land 46.00 ha (1.51%), major road 45.77 ha (1.51%), waterbodies 35.39 ha (1.16%), nala 3.34 ha (0.11%), national highway 5.03 ha (0.16%) has been observed, respectively. Table 8.5 shows the trend of changes in various landuse categories in between two time period data and four new additional landuse categories have been identified and geographical

Table 8.3 Landuse/Landcover statistics of study area (2009)

Sl. no.	Class name	Area (ha)	%
1	Agriculture land	162.18	2.08
2	Built-up	6302.56	80.77
3	Airport	97.74	1.25
4	Canal	1.86	0.03
5	Open space	647.21	8.29
6	Orchard/Plantation	240.49	3.08
7	Nala	5.05	0.06
8	Playground	111.89	1.45
9	Waterbody	54.59	0.69
10	NH	15.53	0.19
11	Major road	164.06	2.11
	Total	7803.16	100

area and trend change in these categories have been calculated as airport/restricted area are occupying 97.74 ha (3.21%), and trend change of canal, parks and playground are 1.86 ha (0.06%), 33.34 ha (1.095%), 111.89 ha (3.68%), respectively.

8.8.5 *Land Transformation Between 2009 and 2014*

From the results obtained for land transformation between the years 2009 and 2014, it has been observed that the area of agriculture land has decreased to −75.89 ha and trend of change has been observed as (−22.63%). Table 8.6 shows the trend of changes of other landuse categories as built-up (urban area) has been increased 502.53 ha (149.90%), canal 0.97 ha (0.29%), airport 5.2 ha (1.55%), national highway 11.04 ha (3.29%), major road 1.75 ha (0.52%), Orchard/Plantation decreased to −5.24 ha (−1.56%), similarly, open space has been decreased as −102.42 ha (−30.55%), water bodies −13.13 ha (−3.91%), and nala −0.51 ha (−0.15%), respectively.

Since 1973–2014, in span of 41 years various residential colonies at Allahabad city has been developed due to launching of different residential/commercial-market area schemes by Allahabad Development Authority (ADA), U.P. Housing Board & many cooperative housing societies/private builders and the area occupied by these colonies has been transformed into built-up (urban area) permanently.

8.8.6 *Land Consumption Rate and Future Projection*

It have been observed from Table 8.7 that the rate of Land Consumption (L.C) and Land Absorption (L.A) per capita/person between the periods of 1973 and 2009 is

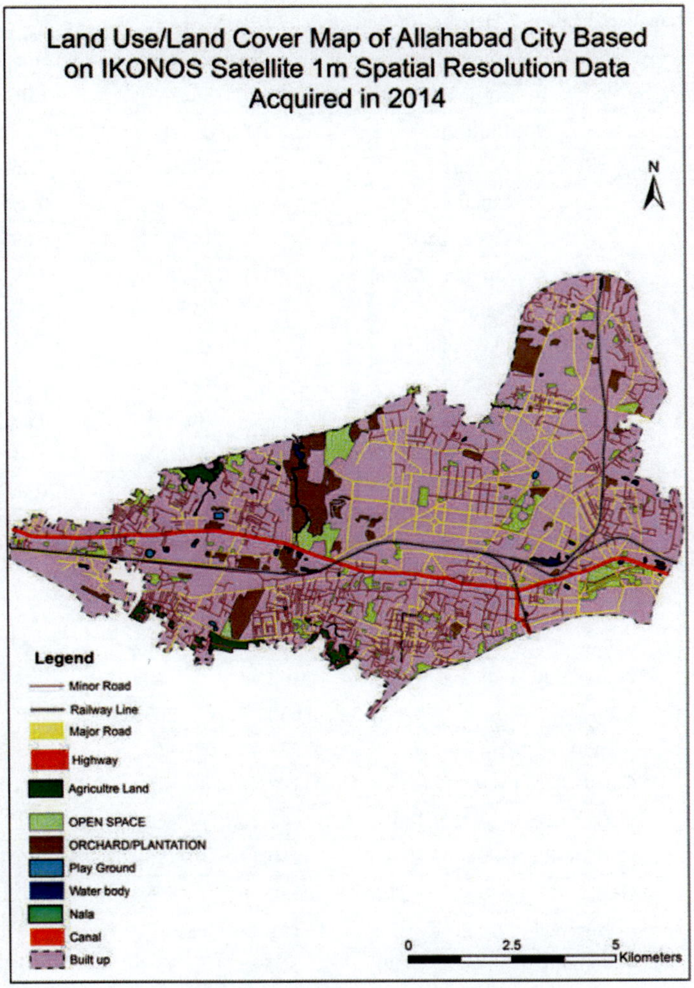

Fig. 8.5 Landuse/Landcover map-2014

0.006 ha and 0.007 ha, respectively. Similarly, L.C and L.A between the periods of 2009 and 2014 is 0.006 ha and 0.001 ha, respectively. L.C for the year 2014 is 0.005 ha, and L.A between the periods of 2014 and 2021 has been estimated as 0.0005 ha. Table 8.7 explains the results about per capita per person Land Consumption Rate and Land Absorption Coefficient from 1973 to 2014 and future projections of Land Consumption up to 2021 for the study area.

Table 8.8 shows the population of study area of 1991, 2001 and 2011, and future population projection for 2021.

Table 8.4 Landuse/Landcover statistics of study area (2014)

S. no.	Class name	Area (ha)	%
1	Agriculture land	86.29	1.06
2	Built-up	6805.09	83.62
3	Airport	102.94	1.26
4	Canal	2.83	0.04
5	Open space	544.79	6.69
6	Orchard/Plantation	235.25	2.89
7	Nala	4.54	0.05
8	Playground	122.81	1.51
9	Waterbody	41.46	0.51
10	NH	26.57	0.33
11	Major road	165.81	0.04
	Total	8138.39	100

Table 8.5 Land transformation statistics of study area between 1973 and 2009

S. no.	Class name	Area (ha) (1973)	Area (ha) (2009)	Change area (ha)	Trend change %
1	Agriculture land	116.18	162.18	46.00	1.51
2	Built-up	4262.66	6302.56	2039.84	67.03
3	Airport	–	97.74	–	3.21
4	Canal	–	1.86	–	0.06
5	Open space	186.47	647.21	460.74	15.14
6	Orchard/Plantation	45.09	240.49	195.40	6.42
7	Nala	1.71	5.05	3.34	0.11
8	Playground	–	111.89	–	3.68
9	Waterbody	19.20	54.59	35.39	1.16
10	NH	10.50	15.53	5.03	0.165
11	Major road	118.29	164.06	45.77	1.51
	Total	4760.09	7803.16	3043.07	–

8.9 Conclusion

The study demonstrates the use of high-resolution remote sensing data and GIS technique for mapping, monitoring of urban sprawl growth and land transformation. The measurement of landuse/landcover change is very useful for future realistic planning at local and global level. Although the urban growth cannot be stopped through proper planning and management, it can be restricted and directed in a desirable and sustainable manner to protect arable land, water and biological resources. Further,

Table 8.6 Land transformation statistics of study area between 2009 and 2014)

S. no.	Class name	Area (ha) 2009	Area (ha) 2014	Change area (ha)	Trend change %
1	Agriculture land	162.18	86.29	−75.89[a]	−22.63
2	Built-up	6302.56	6805.09	502.53	149.90
3	Airport	97.74	102.94	5.2	1.55
4	Canal	1.86	2.83	0.97	0.29
5	Open Space	647.21	544.79	−102.42[a]	−30.55
6	Orchard/Plantation	240.49	235.25	−5.24[a]	−1.56
7	Nala	5.05	4.54	−0.51[a]	−0.15
8	Playground	111.89	122.81	10.92	3.25
9	Waterbody	54.59	41.46	−13.13[a]	−3.91
10	NH	15.53	26.57	11.04	3.29
11	Major road	164.06	165.81	1.75	0.52
	Total	7803.16	8138.39	335.23	100

[a]Area shown (−) in this table indicates decreased in geographical area of individual categories

Table 8.7 Land consumption rate and absorption coefficient

Year	Land consumption rate (ha per person)	Year	Land absorption coefficient (ha per person)
1973	0.006	1973–2009	0.007
2009	0.006	2009–2014	0.001
2014	0.005	2014–2021[a]	0.0005[a]
2021[a]	0.0005[a]		

[a]Estimated

Table 8.8 Population of Allahabad city 1991, 2001, 2011, and 2021

Year	Population	Source
1991	7,92,858	Census of India
2001	12,06,785	Census of India
2011	14,72,873	Census of India
2021[a]	20,00,000[a]	

[a]Estimated

some recent high-resolution satellite data of sub-meter accuracy of KOMPSAT, Tripplesat, Worldview-2, Worldview-3, Deimos-1 and 2 can be used to know the actual status of our land resources in urban periphery. It should be planned by the government that outgrowth of city should be as per the laws and standards decided by housing and urban planning/development authorities.

Acknowledgements The authors are grateful to Director, Remote Sensing Applications Centre-Uttar Pradesh, Lucknow for providing the support to publish this research paper. The authors are also thankful to Deptt. of Planning, Govt. Of Uttar Pradesh for providing the funds to conduct this study.

References

Anderson JR, Hardy EE, Roach JT, Witmer RE (1976) A land use and land cover classification system for use with remote sensor data. United States Geological Survey Professional Paper

Batty M (1996) Urban change. Environ Plan B 23:513–514

Bhatta B (2009) Analysis of urban growth pattern using remote sensing and GIS: a case study of Kolkata, India. Int J Remote Sens 30:4733–4746

Census of India, population data-1971, 2001, 2011

Gopalan AKS (2009) High resolution imagery for developmental planning with spatial references to development. GIS@development.net https://www.GISdevelopmentmet.net/technology/rs/techrsr0014pf.htm.

Griffiths P, Hostert P, Gruebner O, Van Der Linden S (2010) Mapping megacity growth with multi-sensor data. Remote Sens Environ 114–439

McDonald RI (2008) Global urbanization: can ecologists identify a sustainable way forward? Front Ecol Environ 6:99–104 (Google Scholar)

McDonald RI, Kareiva P, Forman R (2008) The implications of urban growth for global protected areas and biodiversity conservation. Biol Conserv 141:1695–1703 (Google Scholar)

Patkar VN (2003) Directions for GIS in urban planning. Map Asia conference, urban planning. GIS@development. www.gisdevelopment.net/application/urban/overview/urban0042p1/htm

Subudhi AP et al (1998) Modelling urban sprwal and future population prediction using remote sensing and GIS techniques J Potonirwachakpp 125–129

Theobald DM (2001) Land-use dynamics beyond the American urban fringes. Geogr Rev 91:544–564 (Google Scholar)

Theobald DM (2005) Landscape patterns of exurban growth in the USA from 1980 to 2020. Ecol Soc 10 (Google Scholar)

Xu H (2008) A new index for delineating built-up land features in satellite imagery. Int J Remote Sens 29(14):4269–4276

Chapter 9
The Meta-Analysis of Studies on Urban Sprawl

Rostam Saberifar, Muslim Nouri, and Prabuddh Kumar Mishra

Abstract Descriptive and analytical methods were used to assess and judge the research related to the topic of sprawl, and data were obtained from 110 studies that have been carried out since 1997 and were available. Before the study was completed, a list was prepared in order to perform more precise and comprehensive data collection. To analyze the information gathered, the analysis technique based on the interpretation, classification, and analysis of the research construct consistency was used. The results indicated that in addition to non-observance of the principles of the compilation of scientific and research articles, methodological ambiguities, poor quality, obvious results, neglect of the theoretical and analytical frameworks of the research, and as a result, the lack of necessary conditions for the production, can be considered as the characteristic of most studies reviewed. Indeed, the researchers have sufficed to the first step of the three stages of description, interpretation, and explanation, and have not considered the next steps. Accordingly, encouraging and stimulating scholars to familiarize themselves with the principles of scientific and technical writing, to identify the breakpoints and discontinuities of studies, and to find ways of joining them together can be considered as a strategic recommendation.

Keywords Meta-analysis · Urban sprawl · Indigenous knowledge · Theoretical foundation · Content analysis

9.1 Introduction

The scale of sprawl consequences has brought many experts to this area, so that now, with the advent of sustainable development, few specialized areas can be considered as not entering this realm. Even in some cases, specialized fields such as agriculture,

---

R. Saberifar · M. Nouri
Department of Geography, Payam Noor University, Tehran, Iran

P. K. Mishra (✉)
Department of Geography, Shivaji College, University of Delhi, Delhi, India
e-mail: prabuddh@shivaji.du.ac.in

P. Sharma (ed.), *Geospatial Technology and Smart Cities*, The Urban Book Series,
https://doi.org/10.1007/978-3-030-71945-6_9

as well as environment and health have entered into this area and carried out significant research (Seabrook et al. 2011). However, the exact definition of this category has not yet been provided, and everyone "*understands it based on his understanding*" (Galster et al. 2001, 690). This issue has led to fundamental challenges for the method analysis, findings, and the results of such research.

Issues associated with sprawl in Iran have attracted a wide range of researchers and scholars, while an enormous amount of research has been carried out in the field of geosciences and urban planning. The diversity and volume of researches in this area indicate that the sensitivity of the people and authorities in this connection is very high. However, conducting various studies in this area has been accompanied by several problems, including the variety of methods, samples, and finally, obtaining different and sometimes contradictory results, so that sprawl, variety, and different results have led to the misunderstanding that research work is "*basically waste of time and cost, or is at least costly and useless*" (Wolf 2009, 18).

To overcome this misunderstanding and to be able to connect the gaps of the studies to provide a good ground for achieving native knowledge in addition to use and apply the results of the plans, this study was carried out in the framework of the meta-analysis research on the urban sprawl.

9.2 Background and Theoretical Foundations

Although studies performed utilizing meta-analysis method are not numerous and abundant, they have been carried out with different attitudes, points of views, and various interpretations have been made from them. For example, in many studies, the purpose of the meta-analysis was to evaluate the method and the results of the research. However, in some cases, especially in medical and psychological fields, the purpose of this type of research is to obtain the overall and unite results of research carried out in a particular realm. In general, in a meta-analysis method, the researcher prepares the results by recording the features and findings of a mass of research in the form of "*quantitative concepts using powerful statistical methods*" (Mousavi Chalk et al. 2016, 2018). In Glass's view, meta-analysis is the statistical analysis of many individual studies and research in a specific domain, in order to "*integrate and unify their results*" (Safari 2004, 20). From the standpoint of Sediq Sarvestani (2000), in the existing meta-analysis research framework, the emphasis is "*placed on integrating their results for scientific and applied applications*" (Sediq Sarvestani 2000, 68). Therefore, meta-analysis provides a comparison of the results of others' research as well. The meta-analysis also includes all stages of the research, including "*theory, method, results,* etc." (Qasemi Ardahi 2006, 60).

Others believe that meta-analysis is a method for analyzing and combining analytic units to gain knowledge of "*a coherent or non-coherent set of scientific foundations and research structures*" (Doyle 2003, 323). Accordingly, the unit of analysis in the meta-analysis method is "*studies and research related to a particular topic*" (Ghazi Tabataba'i and Vedahir 2010, 39). The study of results, findings, and theories with "*a mainly critical approach is one of the most prominent features of this method*" (Vedahir 2010, 37). A meta-analysis is a tool for combining the results of scattered research and achieving a new approach to "*extending the boundaries of knowledge*" (Shafia et al. 2013, 21). While considering meta-analysis as a method, it tries to deduce the differences existing in the previous research and draw a series of general and practical outcomes (Khalatbari 2008). According to Reza'ian (2005), meta-analysis begins with a systematic review of resources to find, evaluate, combine, and if necessary, compile statistical data. The most important advantage of meta-analysis studies is that by integrating the results of various studies, the power of the study increases to find meaningful results. This has led to a significant increase in the number of articles related to meta-analysis in recent years (Reza'ian 2005).

Generally, a meta-analysis with an intensive view provides a general look at research activity. In other words, meta-analysis allows researchers to combine data from several studies and research works. The method of meta-analysis differs from that of literature review and goes beyond it. The review of literature is more descriptive and narrative, but the meta-analysis has an inferential and comprehensive perspective which extends beyond the mere review of literature and backgrounds through the use of statistical methods (Azadi Ahmadabadi 2013). Meta-analyses are carried out in two quantitative and qualitative ways. Most research carried out in Iran has emphasized qualitative aspects. In the qualitative meta-analysis, descriptive statistics are mainly used. Determining the frequency, frequency percentage, cumulative frequency and cumulative frequency percentage, plotting the bar and circle graphs for displaying the results, identifying the fashion, and so on are "*the main pillars of this analysis*" (Salimi and Maknun 2018, 9). This study also mainly focuses on the qualitative approach to meta-analysis in the field of urban sprawl.

9.3 Methodology

This research has been conducted using a descriptive and analytical method, in which library documents have been investigated through content analysis to obtain the required data (Fig. 9.1). In evaluating the research conducted in the field of sprawl in Iran, the required information has been collected according to subject categorization, the methodology of findings, and suggestions.

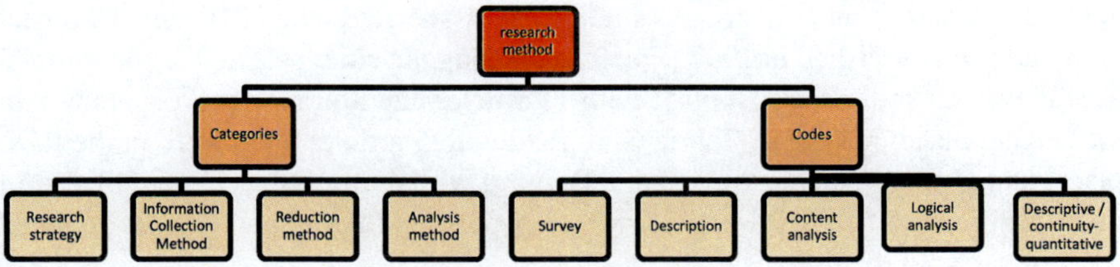

Fig. 9.1 Details of the research method

The statistical population of this study included all the projects, theses, and academic treatises whose results were presented in paper form and published in the period from 1997 to late 2018. To prepare this list, firstly, well-documented internal studies were compiled by searching in the internal online databases such as the comprehensive humanities portal, the scientific information database of the University Jihad, the websites of Magiran, Noormags, Civilica, and Iran Doc (Thesis Database of the country), as well as Google's search engine using keywords such as "sprawl, smart growth, horizontal development, urban development, congested city". According to this survey, it was found that nearly 130 titles of this type have been published, only 110 of which are available in scientific journals with their full text. Before the study was completed, a list was prepared in order to conduct a more precise and comprehensive data collection. To ensure the validity of the list, this form was distributed among 20 experts in the field of research methodology and geography along with the objectives of the research. By applying experts' opinions and final review, the formal validity of the comprehensive form was confirmed.

To analyze the collected data, the analysis technique was based on interpretation and classification and the analyses, while analysis was performed according to the harmony of the research structure so that all the required information was acquired and categorized using systematic review through the Marshall and Ramson model and open coding. In fact, a systematic review is of great importance due to summarizing available texts on a subject, giving meaning to a large amount of information, providing an analysis of existing texts, and making readers needless of all relevant documents, "*providing a complete picture of the subject, and achieving new discoveries*" (Vafaiyan and Mansourian 2014, 86). Open coding mainly seeks to reduce information and provide a detailed description of an issue, gradually summarizing the information to eventually reach "*the main concepts and implications related to the subject matter*" (Tabrizi 2014, 123). This type of encoding is the same as the first stage of data encoding in data grounded theory.

9.4 Discussion

9.4.1 Analysis of Findings

In this study, 110 published research papers from 68 authors in this field were evaluated. The largest number of articles studied was associated with 2016, and the oldest ones had been published in 1997 (Fig. 9.2). Among the cities selected as the sample, Tehran was ranked in the first place with 11 articles and Shiraz with 7 papers was in the second place. Mashhad, Tabriz, and Yazd with 5 articles each were in the third place, and Urmia, Isfahan, Ahvaz, and Bojnord with 4 articles each were placed in the fourth rank. Sari and Maragheh with 3 papers, and Amol and Chalus assigned the next ranks to themselves.

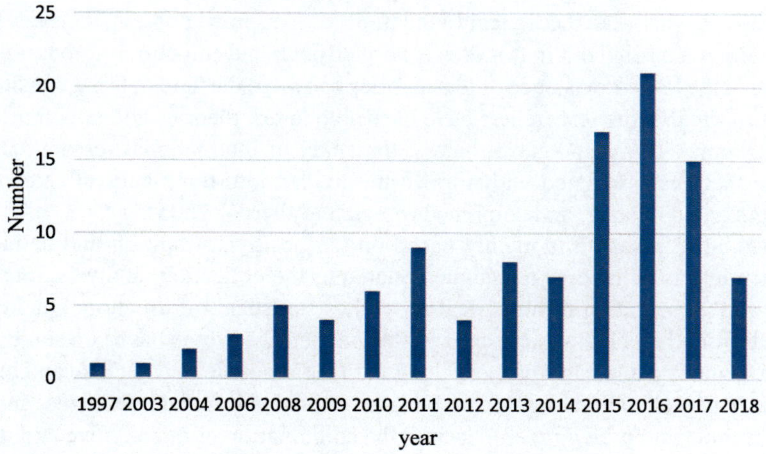

Fig. 9.2 Distribution of researches by year

According to the data collected, the leading 20 authors performed nearly 44% of the research, while 48 others accounted for 56% of the rest. The most active writer in this field was Dadashpour, who appeared in nine papers as the first author and after whom Mirnajaf Mousavi appeared in four papers as the first author. Of course, in this list, Zebardast, Pur Ahmad, and Ziyari also ranked in the first to fifth places due to the preparation of at least four research titles. After this group, Ahmadi, Pourmohammadi, Khairuddin, Azizi, and Rahnama were first to third authors with at least three papers each. Among the remaining 68 authors, at least nine of them had contributed to the compilation of two papers (Fig. 9.3).

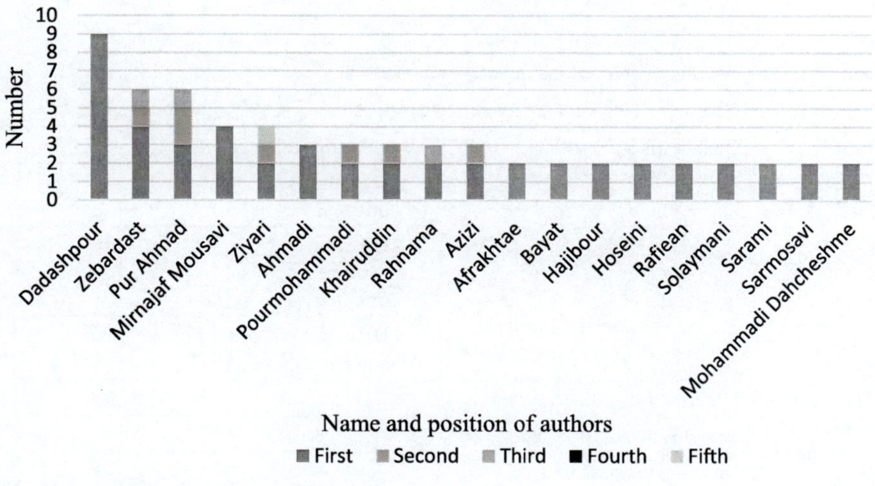

Fig. 9.3 The most active writer on Urban Sprawl

Because of the weak theoretical foundations of the sprawl domain, the evaluation of the research carried out in this area is very difficult and sometimes impossible. In the time of the lack or existence of few theories known in the field, it is not considered serious to say that the researchers have been able to test theories and obtain proportional or opposite results. Nevertheless, the study of international research shows that the theories associated with sprawl are divided into three general categories: quantitative, qualitative, and combined research (Saberifar 2005).

Quantitative research is mainly based on the positivist tradition and deductive approach and is defined as techniques related to the collection, analysis, interpretation, and presentation of numerical as well as statistical information. Qualitative research from the philosophical and intellectual perspectives is based on subjective, relativist, and semantic approaches, along with the rationale of deductive and inductive reasoning. Combined research relying on philosophical and intellectual support of the pragmatism paradigm emphasizes the combination of quantitative and qualitative research methodologies at all stages of the research and avoids the traditional quantitative–qualitative methodological opposition (Iman 2011; Mohammadpour 2009).

Of the research under investigation, 75% lacked theoretical foundations. However, this proportion is 65% in the experimental literature of research. This difference reflects the researchers' ignorance regarding previous studies. One of the major methodological weaknesses in the research under investigation is the lack of utilization of past scientific experiences. By summing up all the theories proposed in 110 research projects and classifying them in the proposed forms, the contribution of each of the theories in the whole study will be as follows: 8% biological theories, 15% sociological theories, and 77% spatial theories. Among sociological theories, the contribution of motivational theories was 47%, while cultural and control theories were equal to 26.5% each. Tables 9.1 and 9.2 are provided to give a general understanding of the main components and parameters of the research. These tables

Table 9.1 The overall structure of the investigated research

Categories	Codes	Number	Percentage	Codes	Number	Percentage
Type of research (110 items)	Quantitative-Qualitative	110	100	Question-oriented	12	11
				Hypothesis oriented	51	46
				Question and hypothesis oriented	20	18
				Lack of questions and hypotheses	27	25
The purpose of the research(110 items)	Describe the basic statistics				80	73
	Social Impact Assessment				7	6
	Describe the base text				13	12
	Exploration				10	9

Table 9.2 Details of the research method

Categories	Codes	Number	Percentage
Research strategy (110 items)	Survey	75	68
	Description	11	10
	Content analysis	8	7
	Logical analysis	9	8
	Action research	7	6
Information Collection Method (110 items)	Documentary research method	60	55
	Documentary- Survey	33	30
	Questionnaire	17	15
Reduction method (110 items)	Quantitative cluster analysis	19	17
	Indicator-quantitative	58	53
	Scale-quantitative	18	16
	Coding-qualitative	15	14
Analysis method (110 items)	Descriptive-quantitative	66	60
	Descriptive / continuity- quantitative	33	30
	Descriptive-qualitative	11	10

provide a basis for further analysis while providing generalizations and details of the methods used. As a result, in this section, the studies were classified according to the methodology, sampling techniques, information gathering techniques, study field, hypotheses, and nature of the research, after which the contribution of each of these categories in methodology was eventually calculated.

9.4.2 *Classification of Hypotheses*

Hypotheses are a fundamental tool for methodology in the project structure. In fact, the hypotheses relate the theoretical foundation of the research structure to its methodological basis and are of particular importance in this regard (Sediq Sarvestani 2000). However, the need for the hypothesis is not considered in some research structures. According to the survey, 75 percent of the projects have hypotheses and the remaining 25 percent have not used the hypothesis. Comparing these ratios indicates the use of the hypothesis in many projects, and this result is consistent with the prevalent use of the survey method in the studies under investigation (Saberifer 2005). The very important point is that the research hypotheses were not at one level; consequently, each of the hypotheses is evaluated in terms of the relation with various sprawl theories. These hypotheses have been evaluated based on the samples, being multivariate, being non-obvious, efficiency of the results of the hypotheses, and the expressiveness of its propositions along with their appropriateness.

In the meantime, some of the hypotheses have been relatively reasonably validated, despite the lack of desired attributes, due to specific features such as testability, transparency, and clarity, along with acceptable results. But a group of hypotheses did not have any of the conditions mentioned and are basically dumb, ineffective, and inexplicable. In fact, these hypotheses question the validity and value of the research and are basically not related to the subject of the research, while they do not need to be tested because they are obvious are essentially uncontrollable (Mohammadpour 2009; Iman 2011). A careful examination of the assumptions of the selected research suggests that 15% of the hypotheses have come out of sociological theories and 5% belong to the biologic theories and others are associated with the spatial theory.

The key and important variables on which the hypotheses rely are extracted and the contribution of each one is specified in the hypotheses. These variables included the role of government, urban managers, car ownership, land acquisition, and social systems, along with individual characteristics, such as age, gender, personal base, income, type of occupation, etc. Thirty-three percent of the total hypotheses have been based on the characteristics of the individuals, and then 62 percent of the hypotheses suggested the impact of the economic and social structure, while the rest of the hypotheses have emphasized on the role of other factors. In general, it can be said that the research has not been able to have a high theoretical level. These conditions are while the research questions and hypotheses are considered as the most important elements of qualitative and quantitative research. In fact, research is carried out in order to answer these questions, and "*confirm or reject the hypothesis*" (Bliki 2010, 40). Achieving such results implies that the researcher has not addressed the philosophical foundations and research approaches.

9.4.3 Combination and Categorization of the Results

Principally, meta-analysis is conducted in a research area where multiple and varied analyses are available in a scientific discipline. Therefore, meta-analysis is based on analyses in a scientific field (Tabatabaei Jabali et al. 2014). However, investigations on the sprawl indicate that the analyses carried out in a definite direction or path according to which their results can be evaluated. As many civil affairs administrators believe, the studies conducted in relation to sprawl have been inefficient or ineffective. Indeed, these studies have not been able to deal with the problems. Therefore, the fundamental question posed in this section is why the results have been weak? The most basic answer to this question is that the quality of the structure of most research works under study lacks the required strength and is apparently unsupportable. To assess the validity of this hypothetical response, the number of fundamental components must be examined, namely, how to design the subject, the purpose, theoretical and empirical foundations, methodology, results, and solutions, as well as the rational and practical connection of these elements with each other (Saberifer 2005).

9.4.4 *Epistemological Foundations*

All books on research methodology agree on the basic point that the selection of appropriate components such as title, problem design, theoretical and empirical foundations, methodology, results, and solutions can be considered as the precondition for accepting the results of scientific research. Without these main components, no research can be categorized as a scientific and targeted study. Therefore, the components mentioned in each research section should be considered by the researcher (Duas 1995). The mere existence of these components, although considered necessary, does not seem to be sufficient, and they should be mutually supportive and reinforcing. If research does not have these conditions, it will undoubtedly suffer from structural tensions and there will be no solid scientific results. The rational and methodological fit of each component with the other component and each element with all elements must be investigated scientifically and methodologically (Seddiq Sarvestani 2000). The logical criteria for assessing the fitness of the components are.

9.4.5 *Methodological Evaluation*

Based on the existing scores, 35% of the researches had a high score in terms of observing the criterion in their elements and had no methodological drawbacks. In 45% of other studies, the average criteria have not been met. Nevertheless, in the same cases, some criteria have been taken into account and the rest have been neglected. In the end, 20% of the research carried out had the lowest score, and almost the research structure of these projects lacked scientific and methodological strength.

9.4.6 *Methodological Meta-Analysis*

Typically, in books about the research method, there are sometimes different stages and processes. However, research can be based on common processes and stages that are usually shared. A survey conducted in relation to the research on sprawl indicates that in 55% of the research studied, the conventional methods of research have been observed. However, in 35% of studies, practices have not been observed logically and processes, as well as stages, have been confounded in many ways. For example, the interference of the research objectives with the problem design, the removal of theoretical foundations and empirical studies, non-observance of the priority of research objectives against the theoretical approach, the integration of research strategies with the results, and inadequate observance of the references can be mentioned. In another 10% of the studied projects, little attention has been paid to the stages of scientific research.

9.4.7 Elements and Components of the Research Structure

The elements and components of the research structure include objectives and methodology. For example, in structured research, the four main elements of the study design, samples, measurements, and the method of analysis (Saberifer 2005) in the research methodology section are presented in a way that other researchers able to reconstruct the study in the same way. In this section, passive verbs are usually used. A review of the research conducted in the field of sprawl shows that in both cases, conductors of 60% of the projects have been able to consider the scientific criteria of these sections well. In the context of the problem design, it is worth noting that in most of the projects the problem has not been accurately and clearly outlined. Logical extensions and spatial analyses of the problems are lacking in more than 53% of the projects. In the field of theoretical and empirical research, only 27% of the research has been carried out on acceptable bases. In 47% of the research, there are always weaknesses regarding the observance of methodological criteria. In fact, there has always been a non-spatial component in theoretical foundations that have usually had no effect on the direction of the project. Researchers have less clearly identified which theory or theories have been mainly applied in their analyses from a mass of theoretical foundations reported in their studies. One of the basic weaknesses in this section is that theories are less used for analysis and explanation. In the methodology section, although a better situation than other research elements can be observed, this section also has some basic weaknesses. For example, although the adherence to the hypothesis logic has been of interest to researchers, it has not been clear in some studies that which hypotheses have been approved and which ones have been rejected. In another part of the research, there are always hypotheses that have not been tested. Of course, sometimes the limited number of hypotheses of some studies prevents achieving applicable results.

The other weaknesses in the methodology of quantitative projects are the lack of utilization of optimal and advanced techniques for data analysis. In fact, in all researches, the same techniques of other studies have been applied, and fewer innovations in techniques are observed.

In the area of research results, only 39% of the projects presented reliable and valid results. The rest of the findings were either not based on scientific foundations or lacked theoretical insights. In many of the findings, less attention has been paid to the basic social variables through which the problems can be dealt with. Also, the researchers did not take the generalization of these projects into account while mentioning the results, and in some cases, their results have always been mentioned with extremes in generalization.

In the field of research solutions, only 23% of the studies have provided valid and feasible solutions. Many strategies are provided without considering the capabilities and facilities of the community. In addition, some of the strategies are so general that they question the necessity of conducting the research, and the scientific validity of solutions has been less widely considered. Finally, only 42% of the projects have considered the standard criteria for acceptable research components and elements. In sum, it can be said that research has not been in a good position in terms of observing the research criteria.

9.4.8 Coordination Among Structural Elements

In research, most researchers tend to choose topics that are more general than what they are studying (Duas 1995), while certain criteria are available for choosing a research title. Usually, some of the effects of an independent variable on a dependent variable in a given population are used as the title. A group also chooses to use a question for the title of the article. Some also answer the main question of the study in the title, and all of these cases have their own bottlenecks and problems. Accordingly, it should be noted that the title, first of all, needs to be simple, understandable, distinct, attractive, with exclusive and brief vocabulary, without additional words. Some also consider the exact and acceptable titles to have a maximum of thirty words. Surveys have shown that around 48% of the research has chosen unsuitable titles with the subject studied. Research objectives in 27% of the projects have not always been the same ones that the research topic has been looking for.

Theoretical foundations always contain elements that are incompatible with other elements of research. The large volume of theoretical and non-geographical foundations of some of the theories is one of the factors contributing to this incompatibility. In 53% of the projects, theoretical and empirical bases are considered to be a significant source of hypothesis for projects, while this alignment is minimized in another part of the projects. In fact, the present study indicates the separation of theoretical and empirical foundations from the hypothesis and ultimately the results of the research. The point that the language of the research results should be in line with the language of the analysis framework should also be considered. The results of the studies are poor in this regard, although it is only in the light of the relationship with the theoretical framework and the use of past experience studies that current research can reach the theoretical level (Sediq Sarvestani 2000). In only 18% of the projects, there is an alignment between theoretical and empirical foundations with the results of the research. The main subject of the research results is simply a plain description of hypothesis testing and a statistical table analysis. However, the results should be presented and explained in the theoretical framework, while going beyond the mere description of the data. Typically, in this section, researchers need to take very clear steps in order to provide their findings in line with scientific principles. As an example, in scientific research that is consistent with the accepted standards, the section of the findings should have the following structure: first, the description of the sample studied; second, the univariate analyses, which means that what information is true about what number of participants; third, the relation between the outcome variables (and how the description is, or the two-variable analysis); and fourth, the multivariate analyses, or the expression of the findings, taking into account the confounders and the modifiers of the effect. This is while most research studies have not paid attention to such a process. The most important consequence of this misunderstanding is that the research did not cope with the cumulative process.

There are also inconsistencies in solutions. Basically, researchers can provide precise and targeted strategies when they can correctly cover the topic of the discussion. In fact, in the discussion section, seven general principles are taken into consideration. At first, the answers to the questions raised in the research should be provided; further, the results should be defended and justified if there is a contradiction between the findings or there are some unexpected findings (Gustavi 2017). Then the limitations of the study are described and finally, after proposing the importance of the findings and emphasizing the novelty of the results, suggestions for future research are presented. But in the research of sprawl, such a process has not been followed, and as a result, some of the solutions are beyond what could be expected from the research subject. Some of them are not essentially related to the research problem, and others are vague or very general with no scientific use. The two sections of the research title and the research guidelines are similarly incompatible with other research elements since due to lack of scientific control over research, researchers are demanding exaggerated titles and providing general and non-scientific solutions.

In total, out of the total number of research studies, only 35% of the research has had a significant correlation between the research elements. The organic inconsistency between the elements has been observed in 35% of the research and 30% of the research, this organic disorder and inconsistency have been very uneven.

9.5 Conclusion

In the methodology section of the research, the findings of the article showed that researches in the field of sprawl on the urban scale did not pay attention to philosophical foundations and research approaches. In the meantime, most types of research have been devoted to quantitative research, and researchers have had the least attention to the use of qualitative and combined methods in urban sprawl studies. Looking at the other points, in most of the research studies, more attention is paid to applied research compared with fundamental research. In this regard, more than 70% of research has been formed with descriptive purposes based on quantitative statistics and the research strategy adopted in the advancement of research is a survey procedure. Therefore, quantitative methods such as indexing and analytical descriptive statistics, as well as correlation, have been used in the collection, reduction, and analysis of data. According to what is said, the components of the research methodology are in a particular form. These conditions indicate that the attention of researchers in the field of sprawl is more focused on the application of smart growth procedures and densely populated cities in the form of indexing models, to investigate the existing status of cities from the point of view of spatial and social regulation.

Some scholars have simplified the whole area of sprawl so that, in the end, the concept of this category along with its causes and solutions, is described in a linear and one-dimensional path, and without measuring the consequences of each solution to this massive problem. However, the city and human life have numerous and countless dimensions and the change of each component affects the function and position of the other components, making it is impossible to isolate the problems individually from others. Indeed, some scholars from the three-dimensional rings of description, interpretation, and explanation have only completed the first stage and did not consider the next steps. Accordingly, many of the results presented in the study can also be presented without conducting field studies. This trend brings more dimensions and angles to the unreliability of researchers and their studies that have been in our society for some years.

Regardless of the research method and the analysis of the research studies, the final results of the quantitative and qualitative studies show that the categories such as the description of the sprawl concept and the criteria for determining the extent of this phenomenon, the description of the congested city and its location in the country's planning system and comprehensive, as well as detailed plans, the explanation of the sprawl position in the provision of service infrastructures and the explanation of the spatial along with social inconsistencies in these studies have been given the main status. These not-so-valuable findings indicate the main weakness in the research that the researchers have not paid attention to the theoretical and analytical framework while explaining the results, therefore, this model is expected to have a certain meaningless and void theory. In this context, indigenous knowledge must be integrated constantly in promoting and emphasizing to expand and use in the development and progress of the country. But the reality is that speaking of native knowledge is meaningless as long as the research results cannot provide a proper theoretical framework. The fact is that in most cases, research has been a kind of addressing a task to make the author's scientific resume more attractive than seeking answers or solving a problem, since now the idea of "*propagation or degeneration*" (Habibi et al. 2008, 4) has taken all levels of the scientific and academic area within its domination. In this sense, those who act as book writers and authors remain in the scientific scene, and the rest will be isolated by others. Although this trend has been in the West and the United States for many years, and only articles that try to address the problem of community issues (Paradis 2018) can be published, problem-based studies have been taken into account only in recent years (Mehrshahi 2003). This is while international competition in innovation has become the core of market research and policymaking (Dodgson 2018), and for this reason, countries that do not move in this direction will not have a place among the science industry. But in Iran, subject-oriented research, without reference to the basic criteria for academic and scientific research, includes a wide range of studies. It seems that the effect and role of senior professors and top researchers in the early papers is much greater than the subsequent ones. In other words, after one or two primary articles, the paperwork is assigned to the students and the professors will only carry out general oversight, so that even without placing the articles in the evaluation models and accurate measurements with the standard criteria, the tone of the writing and the wording also clearly reveal that the author's vocabulary is very limited. For this reason, it is suggested that in further research, articles published in a particular scientific discipline or in a specific collection of publications should be evaluated from this point of view. One of the main limitations of this study was that it only focused on research in Iran, and external articles and reviews were not evaluated. Therefore, it is suggested that these considerations will be addressed in further studies.

References

Azadi AQ (2013) Metaphysical approach: capacities and gaps. Soc Sci 71(8):89–82

Bliki N (2010) Social research design. Ney, Tehran

Dodgson M (2018) The management of technological innovation: an international and strategic approach. Oxford University Press, Oxford

Doyle IH (2003) Synthesis throug meta-ethnography: paradoxes, enhaneements, and possibilities. Qual Res 3:321–345

Duas DE (1995) Scrolling in social research, translation by Hooshang Naibi. Nayan Rey, Tehran

Galster G, Hanson R, Ratcliffe MR, Wolman H, Coleman S, Freihage J (2001) Wrestling sprawl to the ground: defining and measuring an elusive concept. Hous Pol Debate 12:681–717

Ghazi Tabataba'i M, Vedahirm AB (2010) Meta-analysis in social and behavioral research. Publications of Sociologists, Tehran

Gustavi B (2017) How to write and illustrate scientific paper. Cambridge University Press, New York

Habibi A (2008) Principles of scientific writing. Knowledge Development Institute, Tehran

Iman MT (2011) Paradigm fundamentals of quantitative and qualitative research methods in the humanities. Institute of Hozeh and University, Tehran

Khalatbari J (2008) Statistics and research method. Publishing Process, Tehran

Mehrshahi D (2003) Analytical review of the progress of the structure of the writing of geographic articles quarterly journal of geographic research (2006–2004). Q J Geogr Res 7(68):151–145

Mohammadpour A (2009) Outcome: the philosophical and practical foundations of combined research in social sciences and behavioral sciences. Sociologists, Tehran

Mousavi Chalk A, Alaei Arani M, Salami M, Soheili F (2018) Meta-analysis of scientific researches on the basis of the use of scientific databases (Case study: internal research). Q J Inst Sci Technol Iran 34(1):112–189

Paradis JG (2018) The mite guide to science and engineering communication. The MIT Press, Cambridge

Qasemi AA (2006) Investigating the causes of immigration of villagers to cities in iran by meta-analysis of theses (1979–2003). Village Dev 23:80–51

Reza'ian M (2005) Descriptive glossary of meta-analyzes. Iran J Med Educ 5(2):145–143

Saberifar R (2005) An Introduction to the research method in geography. Noor Elm Publication, Hamadan

Safari H (2004) Methodological meta-analysis of social-research surveys in the ministry of the Interior. Tehran, Tehran University, unpublished PhD diss

Salimi J, Maknun R (2018) Qualitative meta-analysis of scientific research on the issue of governance in Iran. Governmental Adm 10(1):30–31

Seabrook L, Mcalpine C, Bowen M (2011) Restore, repair or reinvent: options for sustainable landscapes in a changing climate. Landsc Urban Plann 100:407–410

Sediq SR (2000) A meta-analysis of studies on social pathology in Iran. Soc Sci Lett 15:103–167
Shafia S, Shafia MA, Kazemian G (2013) Meta-analysis of the methods and results of research on the quality of urban life in Iran. Appl Sociol 50:40–21
Tabatabaei J, Delavar A, Borjali B (2014) Meta-analysis of the relationship between personality variables with job stress. Couns Psychother Cult 19:122–195
Tabrizi M (2014) Analyzing qualitative content in terms of deductive and inductive approaches. Soc Sci J 64:106–138
Vafaiyan A, Mansourian Y (2014) How to do a review? Moon Book 197:85–90
Vedahir A (2010) Further the results of qualitative analysis and cultural studies: reality or illusion. Barg Farhang 22:24–45
Wolf F (2009) Meta-analysis. Sage Publication, London

Chapter 10
Four-Dimensional Covid-19 Simulation in Slums Using Hologram Interferometry of Sentinel-1A—Satellite

Maged Marghany

Abstract This study is the first work in implementing four-dimensional (4-D) hologram interferometry to simulate COVID-19 rate variations in slums using microwave remote sensing technology. The study also proposed a new mathematical formula to simulate the rate of COVID-19 from 4-D hologram interferometry; termed as Marghany's 4-D hologram SAR interferometry. The most critical challenge of the world is pandemic COVID-19; which is spreading widely across slums. Slums are considered the main source of criminals and diseases owing to lack of perfect housing, unsanitary conditions, poor infrastructures and occupancy security. The meagre in the impenetrable urban slums are the furthermost susceptible to impurity because of (i) scarce and delimited admission to wellbeing drinking water and satisfactory extents of water for individual sanitation; (ii) the absence of elimination and handling of an excretory product; and (iii) the lack of removal of solid waste. In this view, urban slum requires the standard and accurate method to be identified automatically from remote sensing data. In this context, a remote sensing technique plays a tremendous role in monitoring land use spatial variations. The data are used that involved historical three Sentinel-1A data. The consequences present that the hologram Interferometric technique is being an admirable device for a disordered urban slum in place of it can differentiate between them from its adjacent setting. Reconstruction of a 4-D urban slum is delivered by hologram Interferometric phase unwrapping based on Particle Swarm Optimization (PSO), besides historical time variations of COVID-19 are allied with slum. Hologram interferometric reveals a countless imbrication of COVID percentage rate of 1.2% between high-class zone with the urban slum. The consequences disclose that urban slums, road networks, and infrastructures are effortlessly categorized. In conclusion, the hologram Interferometric based on Particle Swarm Optimization (PSO) is an appropriate algorithm for chaotic 4-D urban slum automatic detection in Sentinel-1A.

Keywords COVID-19 · Hologram interferometric · Particle swarm optimization (PSO) · Sentinel-1A · Slum · Fourth-dimensional

M. Marghany (✉)
Department of Informatics, Faculty of Mathematics and Natural Sciences, Universitas Syiah Kuala, Jln. Teuku Nyak Arief, Darussalam, Banda Aceh, Aceh 23111, Indonesia
e-mail: magedupm@hotmail.com

P. Sharma (ed.), *Geospatial Technology and Smart Cities*, The Urban Book Series,
https://doi.org/10.1007/978-3-030-71945-6_10

10.1 Introduction

The critical question now is: what is the relationship between COVID-19 rapid spreading and the slum cities. Consistent with the United Nations, approximately 1 billion peoples endure in so-known as slums. Abundant researches have revealed that this huge population is predominantly exposed to infective COVID-19 viruses. The recent COVID-19 plague, instigated by the innovative coronavirus SARS-CoV-2, categorically underneath appearances of this pandemic. The habitually-extraordinary-mass surviving residences tied with a bulky amount of peoples per lodging and the deficiency of passable cleanliness are bases why procedures to comprehend the COVID-19 plague merely operate to a restricted scope in shanty-towns. Besides, the obligation to hazard assemblies for severe passages of COVID-19 triggered by noncontagious viruses; for instance, cardiovascular diseases; which is not conceivable owing to derisory information accessibility. In this view, data on persons alive in slum areas and their health condition is either absent or merely be real for explicit zones. In this understanding, one of the supreme issues concerning the COVID-19 epidemic in the basis of slums in the developed countries is the shortage of information on the figure of persons, their livelihood status, and their health conditions.

Consistent with Corburn et al., (Corburn et al. 2020) slums and unceremonious communities are ailing organized to accomplish the plague and deliver the quick emergency responses to diminish the risk of the COVID-19. Therefore, this could not only an issue of curing the populations of slums on a range corresponding to the remainder of the high population density but correspondingly encompasses offering them with superior aid to tolerably pledge the dangers allied with their livelihood circumstances (Buckley 2020; Friesen and Pelz 2020).

Consequently, there are numerous developed proposals to restrict the spreading of the COVID-19 in slum areas towards the surrounding cities and the global countries. To this end, examines of slum inhabitants and their surroundings are required. It is required to recognize and categorize both individual houses and greater slum urban areas exploitation advanced technology worldwide. Remote sensing technology can be conducted in diverse places. Indeed, remote sensing technology is essential to create shared databases to assemble information on the inhabitants and spatial characteristics of slum settlements. Prior to implementing a novel advanced remote sensing to handle the COVID-19 pandemic, the perfect thoughtful of slum settlements are required.

10.1.1 What is Meant by Slum Settlements?

The initial understanding of slum settlements is an extremely massive population without perfect infrastructures and healthy accommodation in which accelerating the spreading of the COVID-19 virus rapidly. The critical question is then, what is the perfect identification of a slum city or urban?

UN-Habitat identified the slum cities as informal settlements or regions divested of access to secure water, tolerable cleanness, and secure housing; additionally, to being zones that are extremely overpopulated and shortage land tenancy safety (Kuffer et al. 2016). On the word of author thinking, slum cities are just similar as the permanent refugee camps with underprivileged compulsory facilities of live activities counting deprived of educational and health infrastructures, which mainly are the responsibilities of the worldwide governments. The bad urban planning due to corrupted governments leads to slum cities, which are the great factories for COVID-19virus spreading.

Figure 10.1 reveals unhealthy housing as there is no space for a person to sleep. Such circumstance is the keystone for spreading the infection of the COVID-19 virus. Social distances are required to prevent dangerous virus infections, which is not provided by such housing conditions. Moreover, to avoid the scattering of COVID-19 virus, a secure source of clean water is required too (Fig. 10.2).

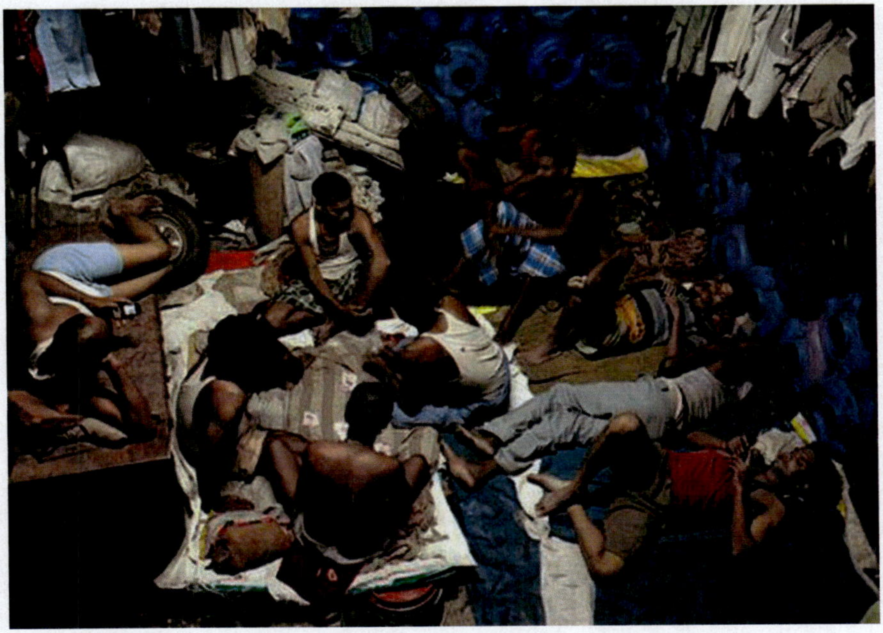

Fig. 10.1 Physical characteristics of urban slum people sharing the same room in urban slums. *Source* www.unmultimedia.org/photo/

Fig. 10.2 Polluted water resources in the urban slum. *Source* https://www.flickr.com/photos/worldbank/8775282532

10.1.2 Why do Urban Slums Trigger COVID-19?

Along with the above perspective, COVID-19 spreads rapidly in slum cities towards the surrounding cities. The question arises now why do slums trigger and spread the COVID-19? Slum dwellers generally have a high incidence of disease. Diseases reported in the slums include cholera, HIV/AIDS, measles, malaria, dengue, typhoid, drug-resistant TB, and other epidemics (Anand et al. 2007; Ross et al. 2020).

Therefore, factors attributed to the excessive rate of sickness transmission in the slums encompass excessive population density, terrible dwelling conditions, low vaccination rates, inadequate health-related data, and insufficient fitness services. Overcrowding consequences in an extra fast and sizable unfold of ailment due to housing shortages in the slums. Poor living prerequisites additionally make slum dwellers greater susceptible to some diseases. A clear instance of this is that negative water pleasant is the cause of many important diseases such as malaria, diarrhoea, and trachoma. Improving dwelling conditions, such as the introduction of better sanitation offerings and get admission to simple facilities, can mitigate the consequences of ailments such as cholera. In this understanding, historically, slums have been linked to epidemics and this trend has continued into the modern era. For instance, slums in West African countries like Liberia were paralyzed and also contributed to the 2014 outbreak and spread of the Ebola virus (Anand et al. 2007; Ross et al. 2020; Kangmennaang et al. 2020). Slums are a major public health problem and a potential breeding ground for drug-resistant diseases in the city (Kangmennaang et al. 2020; Ezeh et al. 2017; Lilford et al. 2017). The entire, nation, as well as the world community Consistent with this perspective, such COVID-19 virus would be spread rapidly through slums besides rapid scattering for the surrounding countries. In other words, the small scale of slums COVID-19 spreading turns to large-scale scattering in the surrounding countries.

10.1.3 How Do Remote Sensing Monitor the Slums?

Remote sensing can play a tremendous role in scrutinizing "space–time dynamics"; for instance, observing density and growth developments or serving device slum enhancement strategies. Besides, it tolerates relating the morphology of the slum city to social and economic criteria. Remote sensing; therefore, can determine the number of slums in high-risk zones or wide-ranging environmental circumstances that reveal a significant responsibility in urban health crusades. Slum Dwellers International (SDI), consequently, emphasized the advantage of slum mapping; for instance, maps are vital historical records in investigation cases, and they defend residents from illegal dislodgments. Supporting deprived decision-making, it is required to merge spatial information with community mapping to comprehend resident requests.

Consequently, slum identification procedures in remote sensing data involve: (i) visual clarification; (ii) object-based image analysis (OBIA); (iii) texture-based approaches; and (iv) community-based methods. In this regard, e object-based analysis involves the abstraction of urban objects; for example, slum roofs, area-based denotes the mining of homogeneous urban patches, correspondingly termed "systematic districts" signifying slum localities.

Accordingly, most of the investigations implemented automated or semi-automatic detection algorithms for distinguishing slums in low-resolution satellite data; for example, LANDSAT TM satellite data. In this regard, Ismail et al. (Lilford et al. 2017) suggested a new-fangled approach to distinguish urban slums, which is grounded on the Coexistent Urbanism technique. According to this view, they delivered a novel technique that is considered three constructions of factors to differentiate among the diverse slum dimensions. Accordingly, elementary and socioeconomic factors, urban analysis factors, and fabric morphological factors are entailed. Finally, they related space Syntax to the slum's network, then slums can be classified by an advanced normalized index termed Coexistence Potential index, which regulates productivity and deficiency for the interruption.

Lately, Marghany (Marghany 2014) announced a novel technique, which is founded on fuzzy B-spline with ENVISAT ASAR data to form 3-D urban slums that are existing in Cairo, Egypt. Along with Dekker (Dekker 2006), synthetic aperture radar (SAR) data are proper for studying urban land use. Numerous bands that involve L, C, or X bands; with a wide diversity of incidence angles between 35° and 50°; and signal radar polarization of HH and HV polarization (HV interesting to investigate urban land use or quad pol monitoring urban environments). Contrariwise, Dekker (Dekker 2006) designated that ENVISAT ASAR satellite data with 30 m resolution is impossible to imagine lesser object sizes. On the contrary, a comparable system can be cast-off in unification with higher-resolution satellites; for instance,

RADARSAT-2 SAR, TerraSAR X, and airborne SAR systems for monitoring land use. Consequently, Brook et al. (Dekker 2006) and Amarsaikhan et al. (Amarsaikhan et al. 2010), have correspondingly verified that the fusion of SAR and optical data can provide clear urban feature detections and perfections.

Innovative 3-D object perceptions have been industrialized newly by Marghany (Marghany 2018a) as announced like novel speculation based on hologram interferometry. Marghany (Marghany 2018a) inveterate that 2-D object is implied in the 1-D object, and 4-D would be implied in the 4-D object. In this understanding, Marghany (Marghany 2018b) verifies the potentials of hologram interferometry as a novel technique for radar usages for 4-D object detection and reconstruction.

10.1.4 Hypothesis of the Study

According to this view, the innovative remote sensing operation of fourth-dimensional (4-D) reconstruction or simulation can be valued when it can compromise the marvellous renovation of slums in satellite images with an amalgamation of hologram interferometry for microwave satellite data. Two hypotheses are examined:

(i) Hologram interferometry able to encode 2-D identified slum image into the 4-D image
(ii) 4-D slum image can involve access correlation with the rate of the infected population by COVID-19 virus.

The foremost aim of this book chapter is to restructure 4-D of urban slums with COVID-19 infection exploitation hologram interferometry for microwave remote sensing satellite data.

10.2 Synthetic Aperture Radar (SAR) Data

In this examination, the Sentinel radar images are used. They are satellites of the European Space Agency (ESA), which deliver a huge quantity of data and images for the Copernicus programme in Europe. Sentinel-1A is armed with bipolar orbiting satellites that work to offer geospatial data for global environmental security. Satellites activate day and night and exploit radar images to form an artificial aperture. Sentinel is also delivering long-term, consistent data archives for operation in the purpose ground on long time series.

Sentinel-1A is operated with C-band in which cannot be impassable by cloud cover or inadequate electromagnetic radiation. Moreover, it is fortified with an internal calibration system; which permits to convey pulses then being directed into the receiver to trace information. This process alleviates to monitor amplitude and phase to simplify elevated radiometric steadiness.

Four exclusive acquisition, modes are equipped in the Sentinel-1 A & 1 B: (i) stripmap (SM); (ii) interferometric wide swath (IW); (iii) extra-wide swath (EW); and (iv) Wave (WV). In these regards, stripmap, interferometric wide swath, and extra-wide swath modes operate on a 25-min cycle per orbit, while the wave mode works on a 75-min cycle per orbit. Consequently, the stripmap, interferometric wide swath and extra-wide swath modes exploit single polarisation; which is either in HH or VV, and dual polarization. Subsequently, these polarisations can be HH + HV and VV + VH. Besides, these modes are conceivable because of the operation of a transmission array and adjustment between dual analogous receiving arrays with both H and V polarisations. The WV modes are only supported by single polarisations in either HH or VV.

This Sentinel-1A has a spatial resolution of lower to 5 m and a swath of approximately up to 400 km. The Sentinel-1A constellation is on a sun-synchronous, near-polar (98.18°) orbit; which has a revisit cycle of 12-day (Marghany and Mansor 2017).

10.3 Fourth-Dimensional Using Hologram Interferometry

10.3.1 Definition of HOLOGRAM

The important interrogative is: what is meant by hologram and can it be obligated in remote sensing processing as a novel tool? The applicable explanation of a hologram is grounded in Greek. In this tendency, the Greek word ‘holos’ indicates a total, patch ‘gramma’ designates a message. It is outstanding as HOL-o-gram, which is a 3-D image view, formed by the photographic propulsion. Different from 3-D or virtual practicality on an n-dimensional machine assemblage, a hologram is an imitation three-dimensional and unfastened-status image that doesn’t emulate the spatial powerfulness or necessitates an inimitable observing tool.

In the view of physics, the coherent electromagnetic wave is the cornerstone of the hologram performance, as well as laser illuminates the object, then its image is chronicled. Certainly, if the laser is visible in the film, it is reflected from the object and to a direct beam of the laser. In this regard, the interference of the laser beam of coherent electromagnetic beams on the film illuminates the object and generates a 3-D image (Fig. 10.3) (Marghany 2018a; Eberhart and Kennedy 1995).

10.3.2 What is the Hologram Interferometry?

Holography is an exercise of the interference physical characteristics of the beam. By superimposing two sets of coloured and coherent wavefronts on a photographic plate, a microscopic interference pattern is produced. The developed plate, called a

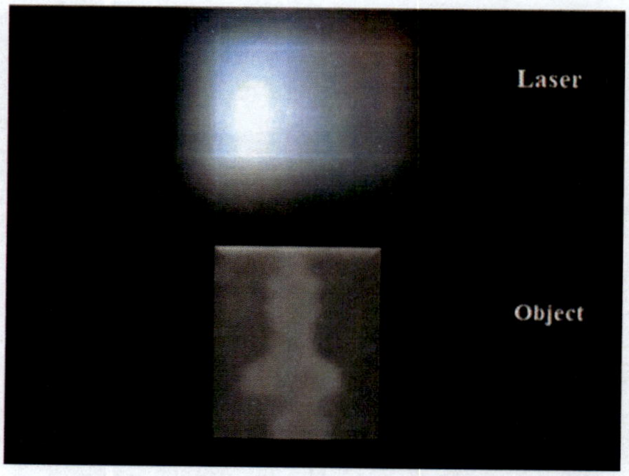

Fig. 10.3 3-D hologram of chess pieces is created by laser illumination (Marghany 2018a)

hologram, contains a permanent record of the interference pattern, i.e., it stores both amplitude and phase information. When the hologram is situated in an experience of the same seamless, coloured signalling of the diffracted fringes are recorded on the photo to produce a set of wavefronts that vary from those originally echoic from the objective. Consequently, when viewed, the diffracted wavefronts offer a remarkably realistic three-dimensional render of the object (Marghany 2018a, b; Holah et al. 2005; d'Ozouville et al. 2008; Zribi et al. 2005; Eberhart and Kennedy 1995; Meseery et al. 2009).

10.3.3 Generating Hologram Interferometry from SAR Images

Coherent signal sources such as radar signals are suitable in forming the holographic image. Similar to traditional interferometric synthetic aperture radar (InSAR), two radar complex signal is then exploited to construct holographic. In this view, an interference pattern consequences are encoded in either a 2-D or 3-D medium forms a hologram. The3-D space involves three dimensions; which are x,y,z. Further, time (t), is also a dimension. Therefore, space and time are not concepts that can be considered independently of one another when we are looking for change detection from satellite data (Marghany 2018a, b; Holah et al. 2005; d'Ozouville et al. 2008; Zribi et al. 2005; Eberhart and Kennedy 1995; Meseery et al. 2009).

In this understanding, the mathematical description of the complex two SAR hologram interferometry image $H(x, y, z, t)$ is formulated as

$$
\begin{aligned}
H(x, y, z, t) &= \varepsilon_0 c\left|S^2\right| = \varepsilon_0 c\left|(S_1+S_2)^2\right| = \varepsilon_0 c\left|S_1^2 + S_2^2 + 2S_1\cdot S_2\right| \\
&= \varepsilon_0 c\left(S_1^2 + S_2^2 + 2S_1 S_2 cos\vartheta_{12}\right) \\
&= \varepsilon_0 c\begin{bmatrix} S_{01}^2 cos^2(kz-\omega_1 t) + \dfrac{S_{02}^2}{r^2}cos^2(kr-\omega_2 t)+ \\ \dfrac{2S_{01}S_{02}}{r}cos(kz-\omega_1 t)cos(kr-\omega_2 t)cos(\vartheta_{12}) \end{bmatrix}
\end{aligned} \tag{10.1}
$$

Equation 10.1 says that the two complex radar images S_1 and S_2 are implemented to construct a 3-D hologram radar image. This hologram radar image involves the two signal wavenumber k, the baseline between both acquired SAR images r over the time of acquisition t and the radial frequency ω. Besides, Eq. 10.1 contains a part which just relies on both radar complex signal, plus an interferometry term involving the cosine of the angle ϑ_{12} between the two vector amplitudes (which may be a function of position *x,y,z*) (Abdul-Rahman et al. 2005; Marghany 2019a).

Generation of the realistic 3-D object in hologram radar images is then requested wrapped phase $\phi_i(x, y, z)$; which is given by

$$
\phi_i(x, y, z) = \arctan\left(\frac{H_0(x, y, z) - H_2(x, y, z)}{H_1(x, y, z) - H_3(x, y, z)}\right), \quad i = 1, 2, 3 \tag{10.2}
$$

Estimation of phase maps of different hologram radar images then can be achieved as

$$
\phi_{12}(x, y, z) = \arctan\left(\frac{\sin(\phi_2(x, y, z)) \times \cos(\phi_1(x, y, z)) - \cos(\phi_2(x, y, z)) \times \sin(\phi_1(x, y, z))}{\cos(\phi_2(x, y, z)) \times \cos(\phi_1(x, y, z)) + \sin(\phi_2(x, y, z)) \times \sin(\phi_1(x, y, z))}\right) \tag{10.3}
$$

$$
\phi_{23}(x, y, z) = \arctan\left(\frac{\sin(\phi_3(x, y, z)) \times \cos(\phi_2(x, y, z)) - \cos(\phi_3(x, y, z)) \times \sin(\phi_2(x, y, z))}{\cos(\phi_3(x, y, z)) \times \cos(\phi_2(x, y, z)) + \sin(\phi_3(x, y, z)) \times \sin(\phi_2(x, y, z))}\right) \tag{10.4}
$$

$$
\phi_{123}(x, y, z) = \arctan\left(\frac{\sin(\phi_{12}(x, y, z)) \times \cos(\phi_{23}(x, y, z)) - \cos(\phi_{12}(x, y, z)) \times \sin(\phi_{23}(x, y, z))}{\cos(\phi_{12}(x, y, z)) \times \cos(\phi_{23}(x, y, z)) + \sin(\phi_{12}(x, y, z)) \times \sin(\phi_{23}(x, y, z))}\right) \tag{10.5}
$$

Equations 10.3 to 10.4 demonstrate that $\phi_{123}(x, y, z)$ diverges from 0 to 2π without 2π discontinuities, and then the image of the wrapped phase can be an expression. Consequently, reconstruction of the 3-D surface directly from $\phi_{123}(x, y)$ cannot deliver excellent consequences owing to $\phi_{123}(x, y, z)$ does not reveal 2π discontinuities since SAR long-wavelength regularly rises in low quantity precision because of the nature of its coherence signal. In other words, the phase $\phi_1(x, y)$ has the wavelength ratio of three complex SAR images there are $\frac{S_{12}(\lambda)}{S_1(\lambda)}$ discontinuities with equal spacing $S_{12}(\lambda)$. In this circumstance, inside every, $S_{12}(\lambda)$ there are discontinuities of $\frac{S_{123}(\lambda)}{S_{12}(\lambda)}$ on $\phi_{12}(x, y, z)$. In other words, the conjugate of three complex SAR images $\frac{S_{123}(\lambda)}{S_{12}(\lambda)} \otimes \frac{S_{12}(\lambda)}{S_1(\lambda)}$ forms discontinuities on $\phi_1(x, y, z)$. In general, the hologram interferometry image can be expressed as

$$H_k(x, y, z) = \alpha(x, y, z) + \beta(x, y, z)\sin(\phi(x, y, z) + 0.5(\pi k)) + G(x, y, z). \tag{10.6}$$

Equation 10.6 demonstrates the occurrence of white Gaussian noise $G(x, y, z)$ since Eqs. 10.3 to 10.5 contain degradation in the computing of the absolute phased (Song et al. 2015).

10.3.4 4-D Phase Unwrapping Using Particle Swarm Optimization

The absolute phase cannot use to determine the 3-D object reconstruction in the complex SAR image in which is required involving the phase unwrapping technique. To this end, the 4-D phase unwrapping can be expressed as

$$\Psi_3(x, y, z, t) = \phi(x, y, z, t) + 2\pi \begin{pmatrix} \mathrm{int}\left(\frac{\phi_{12}(x, y, z, t)}{2\pi} \otimes \left(S_1(\lambda) \times S_2(\lambda)^{-1}\right) \otimes \left(S_2(\lambda) \times S_3(\lambda)^{-1}\right)\right) + \\ \mathrm{int}\left(\frac{\phi_{23}(x, y, z, t)}{2\pi}\right) \otimes \left(S_2(\lambda) \times S_3(\lambda)^{-1}\right) + G(\sigma) \end{pmatrix} \tag{10.7}$$

The novelty of Eq. 10.7 is based on the exchange of the absolute phase estimations of the three complex SAR images with different acquisition times. In other words, the novel equation of 4-D phase unwrapping is also considered the white Gaussian noise as a function of standard deviation σ. The aim of Eq. 10.7 is to reconstruct phase unwrapping in 4-D and also to remove a discontinuity.

10.3.5 Marghany's COVID-19 Procedure in 4-D Hologram Image

The simulation of the COVID-19 in Cairo slums is estimated based on the data are retrieved from https://covid19.who.int/region/emro/country/eg. In this regard, let us assume that the COVID-19 standard deviation $C_{123}(\sigma)$ over the three SAR image acquisitions is contained the 4-D phase unwrapping Ψ as $C_{123}(\sigma) \subset \Psi(x, y, z, t)$. In this understanding, the 4-D COVID variations into hologram interferometry can be expressed as

$$\begin{pmatrix} \Delta\Phi_1 \\ \Delta\Phi_2 \\ \Delta\Phi_3 \\ \Delta\Phi_4 \end{pmatrix} = \begin{pmatrix} \vec{C}_{1i} \; \vec{C}_{1j} \; \vec{C}_{1_k} \; \vec{C}_{1t} \\ \vec{C}_{2i} \; \vec{C}_{2j} \; \vec{C}_{2_k} \; \vec{C}_{2t} \\ \vec{C}_{3i} \; \vec{C}_{3j} \; \vec{C}_{3_k} \; \vec{C}_{3t} \\ \vec{C}_{4i} \; \vec{C}_{4j} \; \vec{C}_{4_k} \; \vec{C}_{4t} \end{pmatrix} \otimes \frac{2\pi}{\lambda} \begin{pmatrix} \vec{\Psi}_{1i} \; \vec{\Psi}_{1j} \; \vec{\Psi}_{1k} \; \vec{\Psi}_{1t} \\ \vec{\Psi}_{2i} \; \vec{\Psi}_{2j} \; \vec{\Psi}_{2k} \; \vec{\Psi}_{2t} \\ \vec{\Psi}_{3i} \; \vec{\Psi}_{3j} \; \vec{\Psi}_{3k} \; \vec{\Psi}_{3t} \\ \vec{\Psi}_{4i} \; \vec{\Psi}_{4j} \; \vec{\Psi}_{4k} \; \vec{\Psi}_{4t} \end{pmatrix} \otimes \begin{pmatrix} W^x_{i,j,k,t} \\ W^y_{i,j,k,t} \\ W^z_{i,j,k,t} \\ W^t_{i,j,k,t} \end{pmatrix} \tag{10.8}$$

Here W represents the user-defined weights, $i, j, k,$ and t cartesian coordinate of 4-D vector representation in hyperspace hologram interferometry. In this regard, this new formula is Marghany's 4-D hologram SAR interferometry.

10.3.6 Particle Swarm Optimization Algorithm

10.3.6.1 Optimization of 4-D Phase Unwrapping of Slum

Kennedy and Eberhart in 1995 insinuated Particle Swarm Optimization (PSO) as a population-based optimization method. In this investigation, PSO is utilized to optimize the phase unwrapping issue to simulate the spatial variation of COVID-19 spreading across the slum zones. In other words, every particle in PSO is considered as In PSO, each particle is a contender solution to the phase unwrapping and COVID-19. Consequently, particles approximate to the ideal result due to their existing velocity, i.e., their neighbours. Then the question is why do select PSO rather than other evolutionary computation techniques such as genetic algorithm, and multiobjective algorithms; for instance. PSO offers a quick ideal solution to any problem rather than other evolutionary computation techniques. PSO also delivers steady convergence properties, whereas it is straightforwardly executed.

The PSO algorithm pursuits in parallel exploitation a set of random particles. Separately particle in a swarm twins a candidate consequence to solve the 4-D phase unwrapping problem containing COVID-19. In this regard, particles in a swarm come close to the optimal result through their contemporary speed, their earlier practice, and the skill of their neighbours. In every initiation, every particle in a swarm is rationalized by dual superlative rates. The first one, therefore, is the finest result (best fitness) it has accomplished hitherto; which is termed as *Pbest*. Moreover, the alternative best rate that is traced by the particle swarm optimizer is the best fitness, attained hitherto by a little particle in the population set. In this view, the best fitness is a comprehensive best and termed as *gbest*. Consistent with each particle's flying experience and neighbour's flying experience, it moves its site in the search space along the 4-D phase unwrapping image and apprises its rapidity. In this understanding, the determination of the bbest and gbest is achieved by

$$V_{i,j,k,t}^{K+1} = \omega \times V_{i,j,k,t}^{K+1} + c_1 \times r_1 \times (Pbest_{i,j,k,t}^{K} - \Psi_{i,j,k,t}^{K}) + c_2 \times r_2 \times (gbest_{i,j,k,t}^{K} - \Psi_{i,j,k,y}^{K}) \tag{10.9}$$

Equation 10.9 reveals that the PSO particles move through 4-D in the phase unwrapping space image *i,j,k,t* within velocity V through iteration K; which regulates by the inertia weight factor ω. In this sense, c_1 and c_2 are the acceleration coefficients, r_1 and r_2 are positive random numbers between 0 and 1, whereas $Pbest_{i,j,k,t}^{k}$ is the best site of particle *i,j,k,t* in the phase unwrapping space image at iteration K; and $gbest_{i,j,k,t}^{K}$ is the best site of the set at iteration k. Consequently, the inertia weight ω is given by

$$\omega = \omega_{\max} - \frac{\omega_{\max} - \omega_{\min}}{K_{\max}} \times K \tag{10.10}$$

where ω_{min} and ω_{max} present the minimum and maximum value of inertia weight factor, respectively. $k_{\max}$ matches the maximum iteration number and k is the existing iteration number.

Following Marghany (Marghany 2019a, b) the initial swarm particles are initialized to contain 3000 facts of particles to be implemented in the 4-D quality map $Q_{i,j,,k,t}$ of phase unwrapping as

$$Q_{Vol,t} = \frac{1}{Vol \otimes t} \otimes \begin{bmatrix} \sqrt{\sum (\Delta\Psi^{x}_{i,j,k,V} - \overline{\Delta\Psi^{x}_{i,j,k,V}})^2} + \sqrt{\sum (\Delta\Psi^{y}_{i,j,k,V} - \overline{\Delta\Psi^{y}_{i,j,k,V}})^2} + \\ \sqrt{\sum (\Delta\Psi^{z}_{i,j,k,V} - \overline{\Delta\Psi^{z}_{i,j,k,V}})^2} + \\ \sqrt{\sum (\Delta\Psi^{t}_{i,j,k,V} - \overline{\Delta\Psi^{t}_{i,j,k,V}})^2} \end{bmatrix}. \tag{10.11}$$

here Vol is the 3-D volume of the hologram image to be encoded in 4-D, the time t is considered as the 4th dimension, where the spatial variation of COVID-19 grows extremely within the time.

10.3.6.2 Optimization of COVID-19

Determination of the COVID-19 in the 4-D slum model is required to initialize the variation of the virus location in spatial(space) and time dimensions. To this end, let us assume v_s is the virus spreading rate in a particular position $v_{i,j,k,t}$; which is mathematically expressed as

$$v_{i,j,k,t} = v_{i,j,k,t}(\min) + \left(v_{i,j,k,t}(\max) - v_{i+n,j+n,k+n,t+n}(\min)\right) \otimes rand(O, D) \tag{10.12}$$

$$v_{s_{i,j,k,t}} = v_{s_{i,j,k,t}}(\min) + (v_{s_{i+n,j+n,k+n,t+n}}(\max) - v_{s_{i,j,k,t}}(\min)) \otimes rand(O, D) \tag{10.13}$$

Both spreading and actual virus position is constrained to a minimum and maximum boundary in the circumstances of random virus variation of locations and dimension, i.e., respectively. Consequently, the COVID-19 new location $v_{i+1,j+1,k+1,t+1}$ is determined by

$$v_{i+1,j+1,k+1,t+1} = rand(O, 1) - D^n \tag{10.14}$$

Equation 10.13 reveals that random spreading of the COVID-19 over any dimension number *n*. In this regard, $\Psi \subset v_{i,j,k,t} \subset v_{s_{i,j,k,t}}, and \Psi \cup v_{i+1,j+1,k+1,t+1}$. In this understanding, the 4-D phase unwrapping slum image must be accomplished by random variations of COVID-19 spreading from one place to another in any space-dimension in the slum. This section is developed by the author in this recent study based on the work of Hudaib and Hwaitat (Hudaib and Hwaitat 2017).

10.4 Results and Discussion

Three Sentinel-1A complex images with different periods have been acquired along with the city of Cairo, Egypt, on September 3rd, 2020; November 12th, 2020; and January 11th, 2021; respectively. There is no much variation in the land covers reveals by the backscatter and coherence values. Urban zones have the highest backscatter of −5 dB than the surrounding environment. However, Nile river water has the lowest backscatter value of −30 dB. This leads to variation in the three images coherence. In this view, water dominates by the lowest coherence value of 0, while the urban has the highest coherence value of 1 (Fig. 10.4). The highest coherence value of urban zones indicates the density of building covers owing to the increment of the number of population which is approximately 100 million.

Figure 10.5 shows the Sentinel-1A selected zone of slums in Misr Al Qadima, Cairo, between 30° 03 and 30° 09′ N and 31° 09′E to 31° 15′E, respectively. These slums are nominated as a function of the criteria of miserable eminence construction, deprived structures, imbrication between high-class and low-class constructions, heavy-house density covers, risky setting, deprived ecological state, robustness. Figure 10.8 reveals that urban slum is subjugated by scanty roof material, nonappearance of streets, asymmetrical road networks, deficiency of vegetation covers, the

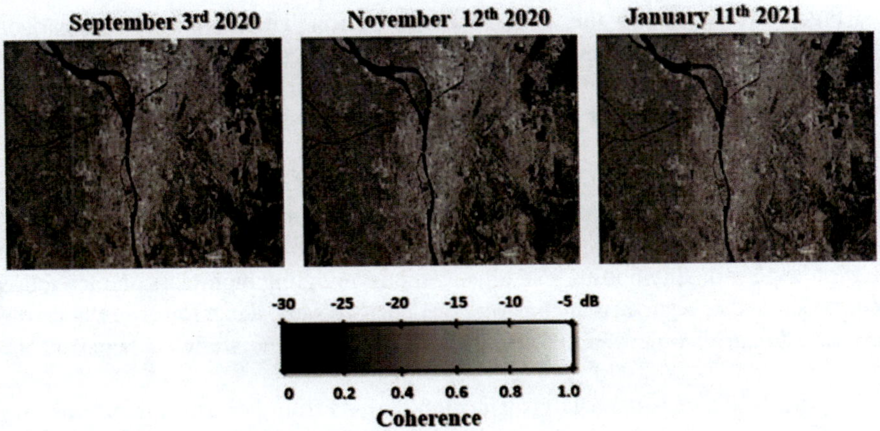

Fig. 10.4 Three Sentinel-1A images over Cairo, Egypt

Fig. 10.5 Selected location of the slum in Cairo, Egypt

asymmetrical outline of settlement link with the surrounding environment, texture, and vicinity. This approves the analysis of Kohli et al. (2012).

Figure 10.6 reveals the increment of the COVID-19 in Egypt, which retrieved from https://upload.wikimedia.org/wikipedia/commons/e/ed/COVID-19-Egypt-log.svg. It is obvious that the COVID-19 rate increases from February 2020 to January 2021. The infected reported COVID-19 victims are approximately 100,000 persons in Egypt.

Figure 10.7a shows the coherence SAR data. Therefore, the comparison between 2-D phase unwrapping (Fig. 10.7a); 3-D phase unwrapping(Fig. 10.7b); and 4-D phase unwrapping is shown in Fig. 10.7d. The hologram interferometry fringe patterns are more vibrant by using 4-D phase unwrapping as compared to other phase unwrapping dimensions i.e. 2-D and 3-D. This could be due to the rapid development and increment in the ratio of population in Cairo. This area is dominated by a heavy density of urban slums. Along with Marghany (Marghany 2018b), the foremost principles of an urban slum in Egypt is: poor-quality houses, bad organizations, mingling high-class and low-class constructions, heavy irregular construction density covers, dangerous sites, deprived conservational disorder, and resilience. This confirms the study of Ismail et al., (2013).

Figure 10.8 shows the 4-D visualization derived from hologram interferometry as a function of PSO. The 4-D visualization distinguishes between infrastructures

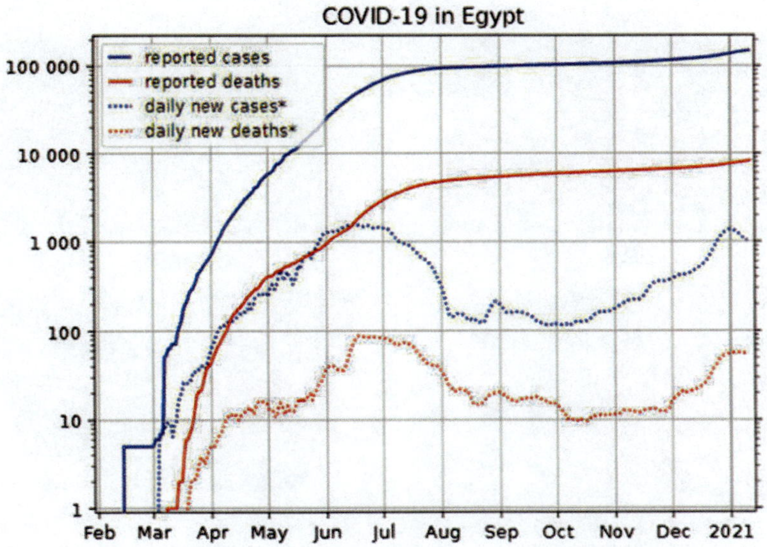

Fig. 10.6 COVID-19 rate in Egypt

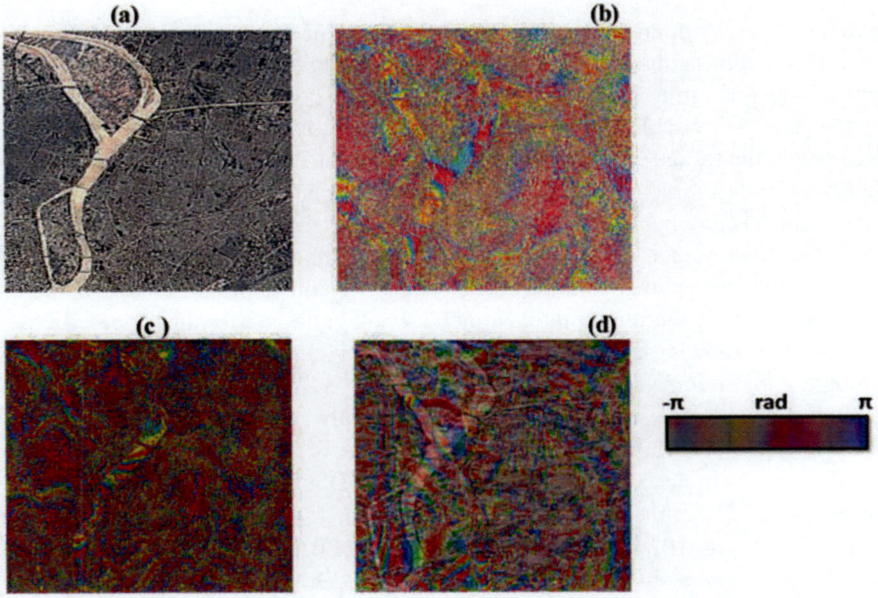

Fig. 10.7 Phase unwrapping of **a** coherence SAR image, **b** 2-D, **c** 3-D, and **d** 4-D

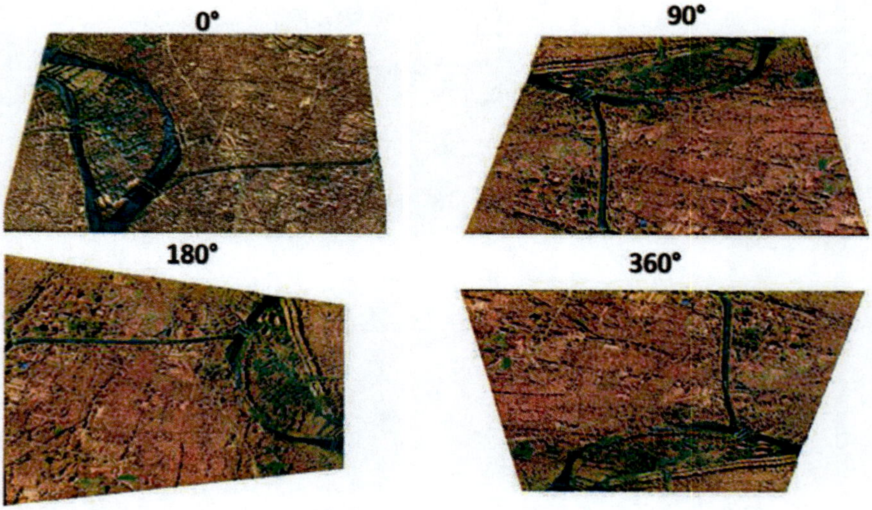

Fig. 10.8 4-D visualization produced by hologram interferometry in Sentinel-1A data with different angle of views

and buildings. With this regard, coding 2-D of Sentinel-1A data into 4-D can visualize much information about urban features, i.e., building, infrastructures, and roads despite the great irregular pattern of an urban area as one of the indicators of slum urban. The 4-D visualization discriminates between infrastructures and buildings. Roads, buildings, and infrastructures are displayed in Sentinel-1A. This is since the hologram interferometry based on PSO is well-thought-out as a deterministic set of rules, which is pronounced here to augment merely a triangulation domestic between dual dissimilar points. This corresponds to the feature of deterministic strategies of finding only sub-optimal solutions usually. The visualization of the infrastructures of urban slums is sharp with the Sentinel-1A hologram interferometry based PSO. This study is showing excellent promising for 4-D visualization that is derived from hologram interferometry.

Figure 10.9 demonstrates 4-D COVID-19 spatial variation from September 3rd, 2020 to January 11th 2021. At present, the population of Cairo is 20,901,000, whereas a 2.03% increase since 2019. Consequently, the 4-D hologram SAR interferometry reveals that 1.2% of the Cairo population is infected by COVID-19; which is found in the slums (Fig. 10.9). Needless to say that rate of 0.2% increment of COVID-19 infections occurrences in Cairo alone (Fig. 10.9). This could be attributed to low temperature and high humidity too besides high mass density population covers. In this understanding, COVID-19 is just a weak virus, as well as only 1.2% of the 20,901,000 of the Cairo population, are only infected. However, it should follow the health safety requirements to avoid more infections either in a short time or a long time.

The implementation of PSO with 4-D phase unwrapping assisted to determine the most fulfilling growth of COVID-19 region infection across the persevering with the

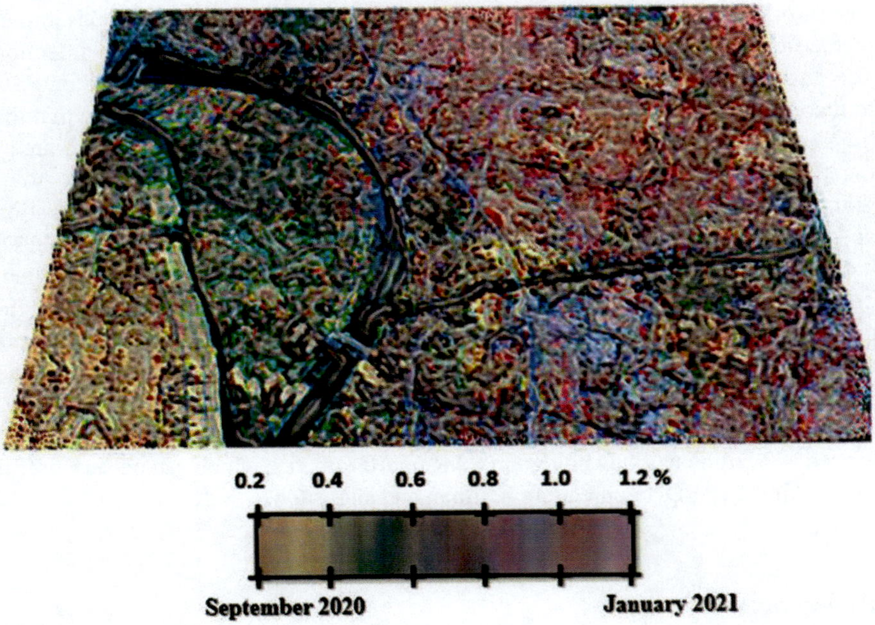

Fig. 10.9 4-D hologram interferometry of COVID-19 on slums

unwrapping of edges. In this view, PSO synchronized the voxels on each aspect of the area (Figs. 10.8 and 10.9). Besides, 4-D phase unwrapped algorithms constructed the discontinuity in quality order. This is fabulous in the excessive depth line or curve of fixed length and locally low curvature boundary is recognized to exist between edge factors and excessive noise ranges in Sentinel-1A data. PSO, conversely, optimizes the gaps between discontinuity slums edges and fulfilled the scenario of COVID-19 infections. In this regard, 4-D phase unwrapping-based PSO is the choicest search for actual partial COVID-19 and slums values, which are present at the boundary of slums and the optimization of 4-D section unwrapping in hypercube can reconstruct the 3-D object displacement with extra 4th dimension. Finally, 4-D phase unwrapping-based PSO approves for the dependable unwrapping of low signal to noise ratio (SNR). This finding out about ought to enhance of 3-D phase unwrapping proposed utilizing Peer et al., (Peer et al. 2003); Abdul-Rahman et al., (Abdul-Rahman et al. 2005); and Marghany (Marghany 2019a).

10.5 Conclusions

This study has demonstrated a new approach for simulation of the COVID-19 variations in Cairo slums. The novelty of the study was established a novel mathematical framework based on the 4-D hologram interferometry. To this end, the PSO is implemented and delivers new formula to determine the percentage of COVID-19 infection rate on slums. Then, the hologram interferometry was implemented based on PSO to reconstruct 4-D variations of the COVID-15 in slums. The results show that the high density of housing spatial distributions, inherent in many dense slum areas, together with the high level of infrastructure, can be observed. In conclusion, hologram interferometry based on PSO shows an excellent tool for 4-D visualization and discrimination between infrastructures and housing features in remotely sensed data, especially with Sentinel-1A satellite. The Sentinel-1A data shows an excellent promise in constructing 4-D hologram image and PSO used to determine the high-quality map of slums in such radar speckle images. Moreover, the COVID-19 rate is increased steadily with a rate of 1.2% from September 2020 to January 2021. The study concludes that the high rate of COVID-19 in Cairo is attributed to heavy population density variations. Besides, the novel formula proposed in this study can be termed as Marghany's 4-D hologram SAR interferometry; which can be very useful for tracking COVID-19 spreading in urban and slums too.

References

Abdul-Rahman H, Gdeisat M, Burton D, Lalor M (2005) Fast three-dimensional phase-unwrapping algorithm based on sorting by reliability following a non-continuous path. In: Optical measurement systems for industrial inspection IV, vol 5856. International Society for Optics and Photonics, pp 32–40. 2005 Jun 13

Amarsaikhan D, Blotevogel HH, Van Genderen JL, Ganzorig M, Gantuya R, Nergui B (2010 Mar 1) Fusing high-resolution SAR and optical imagery for improved urban land cover study and classification. Int J Image Data Fusion 1(1):83–97

Anand K, Shah B, Yadav K, Singh R, Mathur P, Paul E, Kapoor SK (2007 May 1) Are the urban poor vulnerable to non-communicable diseases? a survey of risk factors for non-communicable diseases in urban slums of faridabad. Natl Med J India 20(3):115

Buckley RM (2020 Jun) Targeting the world's slums as fat tails in the distribution of COVID-19 cases. J Urban Health: Bull N Y Acad Med 2:1

Corburn J, Vlahov D, Mberu B, Riley L, Caiaffa WT, Rashid SF, Ko A, Patel S, Jukur S, Martínez-Herrera E, Jayasinghe S (2020 Apr) Slum health: arresting COVID-19 and improving well-being in urban informal settlements. J Urban Health 24:1

d'Ozouville N, Deffontaines B, Benveniste J, Wegmüller U, Violette S, De Marsily G (2008) DEM generation using ASAR (ENVISAT) for addressing the lack of freshwater ecosystems management, Santa Cruz Island. Galapagos Remote Sens Environ 112(11):4131–4147

Dekker RJ (2006) Monitoring urban development using envisat ASAR. EARSeL, Munster

Eberhart R, Kennedy J (1995) Particle swarm optimization. In: Proceedings of the IEEE international conference on neural networks, vol 4, pp 1942–1948. 27 Nov 1995

El Meseery M, El Din MF, Mashali S, Fayek M, Darwish N (2009) Sketch recognition using particle swarm algorithms. In: 2009 16th IEEE international conference on image processing (ICIP). IEEE, pp 2017–2020. 7 Nov 2009

Ezeh A, Oyebode O, Satterthwaite D, Chen YF, Ndugwa R, Sartori J, Mberu B, Melendez-Torres GJ, Haregu T, Watson SI, Caiaffa W (2017) The history, geography, and sociology of slums and the health problems of people who live in slums. Lancet 389(10068):547–558

Friesen J, Pelz PF (2020) COVID-19 and slums: a pandemic highlights gaps in knowledge about urban poverty. JMIR Public Health Surveill 6(3):e19578

Holah N, Baghdadi N, Zribi M, Bruand A, King C (2005) Potential of ASAR/ENVISAT for the characterization of soil surface parameters over bare agricultural fields. Remote Sens Environ 96(1):78–86

Huang Y, van Genderen JL, van Veen BS (1997) The ITC filter: a new adaptive filter for SAR speckle reduction. In: Proceedings of the second international airborne remote sensing conference and exhibition

Hudaib AA, Hwaitat AK (2017) Movement particle swarm optimization algorithm. Mod Appl Sci 12(1):1–17

Ismail AM, Bakr H, Anas S (2013) A hybrid GIS space syntax methodology for prioritizing slums using coexistent urbanism. Coordinates IX, 44–50

Kangmennaang J, Bisung E, Elliott SJ (2020 Jan) 'We are drinking diseases': perception of water insecurity and emotional distress in urban slums in Accra, ghana. Int J Environ Res Public Health 17(3):890

Kohli D, Sliuzas R, Kerle N, Stein A (2012 Mar 1) An ontology of slums for image-based classification. Comput Environ Urban Syst 36(2):154–163

Kuffer M, Pfeffer K, Sliuzas R (2016 Jun) Slums from space—15 years of slum mapping using remote sensing. Remote Sens 8(6):455

Lilford RJ, Oyebode O, Satterthwaite D, Melendez-Torres GJ, Chen YF, Mberu B, Watson SI, Sartori J, Ndugwa R, Caiaffa W, Haregu T (2017) Improving the health and welfare of people who live in slums. Lancet 389(10068):559–570

Marghany M (2014) Fuzzy B-spline optimization for urban slum three-dimensional reconstruction using ENVISAT satellite data. In: IOP conference series: earth and environmental science, vol 20, no 1, p. 012036. IOP Publishing

Marghany M (2018) Advanced remote sensing technology for tsunami modelling and forecasting. CRC Press. 4 Jul 2018

Marghany M (2018) Four-dimensional water detection in mars using spline algorithm. Int J Hydro 2(5):607–611

Marghany M (2019) Four-dimensional of copper mineralization using tandem-X satellite data. Geosci Bull 1(1). 31 May 2019.

Marghany M (2019) Tandem-X satellite data application to four-dimensional of copper mineralization. Int Rob Auto J 5(2):38–41. https://doi.org/10.15406/iratj.2019.05.00170

Marghany M, Mansor S (2017) Three-dimensional Nepal earthquake displacement using hybrid genetic algorithm phase unwrapping from Sentinel-1A satellite. Earthq: Tecton, Hazard Risk Mitig, 139. 1 Feb 2017

Peer ES, van den Bergh F, Engelbrecht AP (2003) Using neighbourhoods with the guaranteed convergence PSO. In: Proceedings of the 2003 IEEE swarm intelligence symposium. SIS'03 (Cat. No. 03EX706). IEEE, pp 235–242. 26 Apr 2003

Ross AG, Rahman M, Alam M, Zaman K, Qadri F (2020 Mar) Can we 'WaSH'infectious diseases out of slums? Int J Infect Dis 1(92):130–132

Song L, Chang Y, Xi J, Guo Q, Zhu X, Li X (2015 Nov) Phase unwrapping method based on multiple fringe patterns without use of equivalent wavelengths. Opt Commun 15(355):213–224

Zribi M, Baghdadi N, Holah N, Fafin O (2005 Jun 30) New methodology for soil surface moisture estimation and its application to ENVISAT-ASAR multi-incidence data inversion. Remote Sens Environ 96(3–4):485–496

Chapter 11
Geospatial Technologies for Public Health Management System

Bhoop Singh, Ashok Kumar Singh, Shubha Pandey, and Mahak Garg

Abstract Geospatial technologies have made significant progress in data capturing, storing, analyzing, managing and presenting spatially referenced data in attending various applications. Many environmental conditions affect people's health. The distribution of wetland may affect the dispersion of Malaria, while the groundwater aquifer and the location of solid or waste dumping sites may impact the drinking water quality which, in turn, affects the residents' health. Thus, geospatial technologies and tools would facilitate public health policy formulation and planning, implementation and monitoring of appropriate interventions by regulatory authorities. In this technical paper, efforts are made to focus on the health care system in the patients. Geospatial analytical techniques such as proximity estimations and cluster analysis are built on statistical methods that incorporate distance and direction measurement to generate geospatially accurate maps and graphics reports. Disease clustering is classified as temporal clustering, spatial clustering, etc. These are examples of tools and technologies that can be made applicable to public health management. The authors have tried to address the following issues in this paper, where the use of geospatial technologies has a big role to provide a better health management system. The important issues are: Methods for disease and risk mapping; Spatial patterns of diseases; Hotspot detection of diseases; Spatial diffusion of disease outbreak; Roadmap for spatial epidemiological model; Geospatial analysis and visualization; Geospatial public health interoperability; Location-based hazard vulnerability assessment, etc. All the above issues have been properly carried out as part of the network project on

B. Singh · A. K. Singh · S. Pandey (✉) · M. Garg
Natural Resource Data Management System Division, Department of Science and Technology, Ministry of Science and Technology, Technology Bhawan, New Delhi, India
e-mail: shubha.p@nic.in

B. Singh
e-mail: bhoopsingh@nic.in

A. K. Singh
e-mail: ashokk.singh@nic.in

M. Garg
e-mail: mahak.garg@nic.in

P. Sharma (ed.), *Geospatial Technology and Smart Cities*, The Urban Book Series,
https://doi.org/10.1007/978-3-030-71945-6_11

the health GIS as one of the applications of geospatial technology under the NRDMS programme.

Keywords Heath GIS · Public Management System · Geospatial technologies · Disease clustering and risk mapping

Acronyms

CVD	Cardio Vascular Disease
DMHO	District Medical Health Offices
ERD	Entity Relationship diagram
GIS	Geographical Information System
IoT	Internet of Things
LST	Land Surface Temperature
NCDs	Non-Communicable Diseases
NDVI	Normalized Difference Vegetation Index
NRDMS	Natural Resources Data Management System
PMSSY	Pradhan Mantri Swasthya Suraksha Yojana
RSBY	Rashtriya Swasthya Bima Yojana
VBD	Vector Borne Disease
WHO	World Health Organization

11.1 Introduction

Natural Resources Data Management System (NRDMS)—a multi-disciplinary and multi-agency programme of the Department of Science & Technology is promoting research and development in geospatial science, data, solutions and applications to demonstrate the efficacy of geospatial technologies in good governance. As part of this, a number of initiatives have been taken to formulate and launch a variety of R&D projects. Health geospatial public management system is one of the important schemes of the NRDMS programme. Under this scheme, the emphasis has been given to utilize geospatial technologies in order to facilitate decision makers for implementing the health related services (Administrative Staff College of India 2018). Focus on managing of health for society has been described as public health management System (Hunter and Berman 1997). Public health management can be defined as the management of the health of the society by improvising the infrastructure, specific resources related health care services, which can ultimately lead to better health of the population over a large scale area. Therefore, public health management system in any country needs substantial government support (Hunter 2001). In general, public health differs from personal health in the following regards:

(i) focuses on the health of populations rather than of individuals, (ii) more emphasis on prevention than on treatment and (iii) mainly operates in governmental set-up (O' Carroll 2003).

The use of Geographical Information System (GIS) to solve public health issues has an exponential increase over the years due to advancement in geospatial technologies. Applying the fundamental geographic concepts of proximity, travel time, correlation and normalization by population and area to public health datasets can also target the source, and reveal transmission patterns of specific health problems/diseases. The use of these techniques using GIS, enables the exploration of a broad range of determinants (e.g., demographic, socio-economic, geographic, environmental) that influence disease risk, its transmission to the community (Blossom et al. 2011).

Considering the vast scope of geospatial technology in better management of public health management systems, efforts have been made to demonstrate this by undertaking specialized research in various sectors with real time data of various diseases spread all over the country. Increasingly, geospatial technology is bound to have scope for facilitating many tasks of governance, from facilities management to land records, natural resources and disaster management, etc. Further, it promotes E-governance which is a well liked expression at present by utilization of scientific tools and technologies for the transformation of government services. The following important themes have been covered while working on the scope of geospatial technology in health management system:

- Methods for disease and risk mapping
- Spatial patterns of diseases
- Hotspot detection of diseases
- Spatial diffusion of disease outbreak
- Road map of spatial epidemiological model
- Geospatial analysis and visualization
- Geospatial public health interoperability
- Location based hazard vulnerability assessment.

In India, there is an organized system of Health care levels and value chains which is depicted in Fig. 11.1. The detail of the Entity Relationship diagram (ERD) has been depicted in Fig. 11.2. Entity Relationship Diagram (ERD) represents data objects and their relationship. Basically, ERD is a high level conceptual data model illustrating how "entities" such as people, objects or concepts relate to each other within a system. In this paper, this concept has been adopted.

In this technical paper, two case studies of different urban areas, i.e. development of a statistical vector-borne disease risk prediction model and geospatial mapping of Diabetes Risk hot spots in Jhajjar, Haryana and North East Delhi have been discussed in detail.

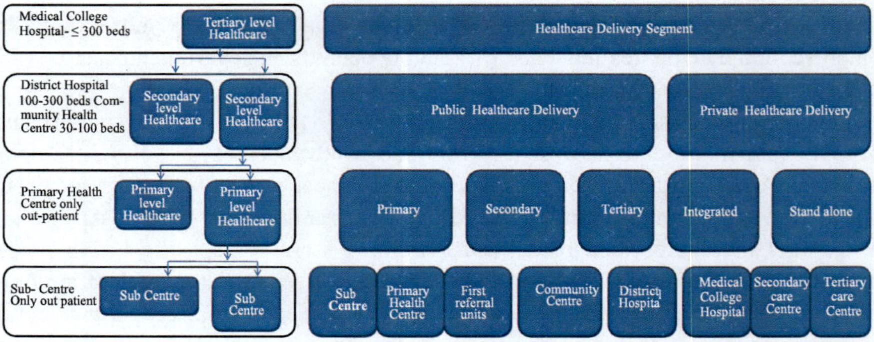

Fig. 11.1 Health care levels and value chains in India (Administrative Staff College of India 2018)

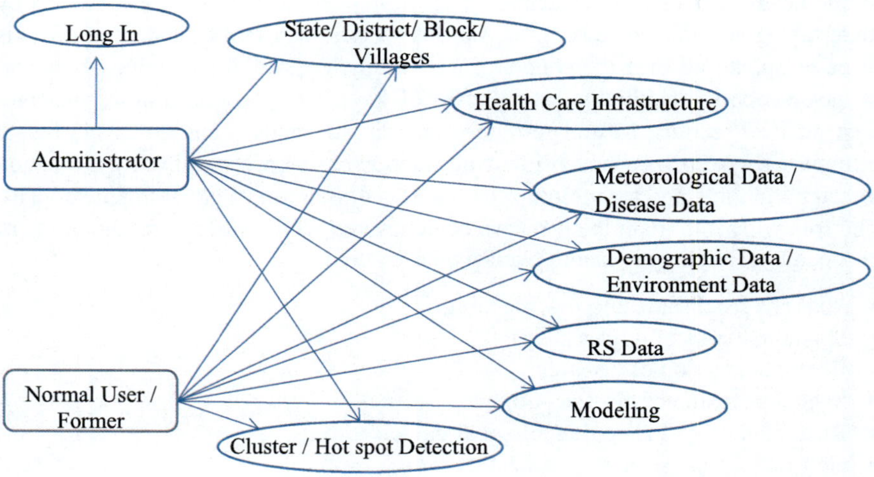

Fig. 11.2 Entity Relationship Diagram (ERD) (Administrative Staff College of India 2018)

11.2 Main Research Outputs of the Themes

In order to understand the use of geospatial technologies in handling the above issues, specific R&D projects were undertaken on each theme. While working on each of the specific themes, the following research outputs were envisaged:

(i) Disease mapping and developing community health surveillance to demonstrate the use of geospatial technologies in managing the health resources and population density, health and human services.
(ii) Overlaying the socio-economic, demographic and environmental issues helped in better management of health and human services.

(iii) Integration of geospatial analysis and modeling helped in strengthening the overall Health Management Information System through validation of operational public health GIS models.
(iv) Development of Web-based platforms for real time data collection also helped in the prediction of moment of vectors as well as vector-borne diseases.
(v) Hearing capabilities of school children and diseases of tribal people were carried out, where area specific data was collected to verify the hearing capabilities.
(vi) Prediction of mosquito diseases such as malaria and dengue were also studied using geospatial technologies.
(vii) Development of a querying system that facilitated multi-criteria queries for finding out spatial applications for dengue modeling, prediction and prevention using the mobile Apps.
(viii) Integration of Internet of Things (IoT) devices along with geospatial tools for diagnosis and monitoring of patients has helped in maintaining advanced health care services.
(ix) A Digital platform for public health comprising a scalable blockchain-based electronic health record was created to facilitate the use of data science, big data and data analytics.

11.3 Government of India—Health Management Priorities

For any country, the sound health of the people living over there is the top priority. All the developed countries have a very concrete programme on health care and public health management system. India with the priority of government is also making concerted efforts to develop a better health management system. Some of the important government schemes in the health sector are as under:

- Pradhan Mantri Swasthya Suraksha Yojana (PMSSY)
- National Rural/Urban Health Mission and Human Resources for Health and Medical Education
- Rashtriya Swasthya Bima Yojana (RSBY)

All these schemes related to health are very important to effectively penetrate among the people to ensure better health providing services. In this regard, geospatial science and technologies have a tremendous role to play by developing suitable spatial data management systems on the type of diseases and their spread over various parts of the country. With this, the analysis of demographic patterns with other attribute information like the quality of air, drinking water and other environmental information would provide a targeted area for providing better health services. Under the DST programme such efforts have been demonstrated by undertaking specialized projects in various conditions/areas in the country.

11.4 Public Health Awareness by the Government

In order to cover the maximum population in rural areas, as well as semi urban clusters, to ensure extending better health facilities and making the life of the people more happy and healthy, the government extended health insurance worth Rs.5.0 lakhs to each individual. This is the world's largest government funded health care programme. One lakh fifty thousand health and wellness centers for primary, secondary and tertiary health care were set up to provide free diagnosis and treatment. To ensure this, the provision of Rs.1200 crores has been made in the central budget of the government during the year 2018–19. Indian ecology supports the cultivation of highly specialized medicinal and aromatic plants. To support the organized cultivation of such medicinal plants, an allocation of Rs.200 crores was proposed. Therefore, the efforts by the government in this regard are worth appreciating.

11.5 Case Studies

The details of the two case studies are as follows.

11.5.1 Case Study 1

'Development of a statistical Vector-Borne Disease (VBD) risk prediction model based on various environmental variables'.

Efforts have been made to develop a statistical Vector-Borne Disease (VBD) risk prediction model based on various environmental variables such as soil, vegetation and water quality in an urban area Guntur, Andhra Pradesh. The main focus of the above-mentioned case study has been on Dengue only.

Methods such as hot spot analysis along with multi distance spatial clusters were mainly used to observe the change in the spatial patterns and hot spots of dengue fever and its vector. Various GISbased spatial analysis tools were utilized for integrating the dengue fever spread and female vector Aedes mosquitoes. The notifications were recorded at state/districts/block levels for identifying and visualizing the spatial patterns and hotspots of the Dengue in the area. In this study, the identified hot spots may predict the risk of dengue fever transmission and may explain the entire variance of the disease. Figure 11.3 represents the spatial distribution and clustering of dengue fever in Guntur District, Andhra Pradesh.

Based on the data collected from across the country for vector-borne diseases, a statistical model was developed by considering the parameters such as water bodies, dense urban vegetation, sparse vegetation/urban open lands and semi urban croplands in Guntur urban area, Andhra Pradesh. Following is the detail of the various activities carried out under this case study.

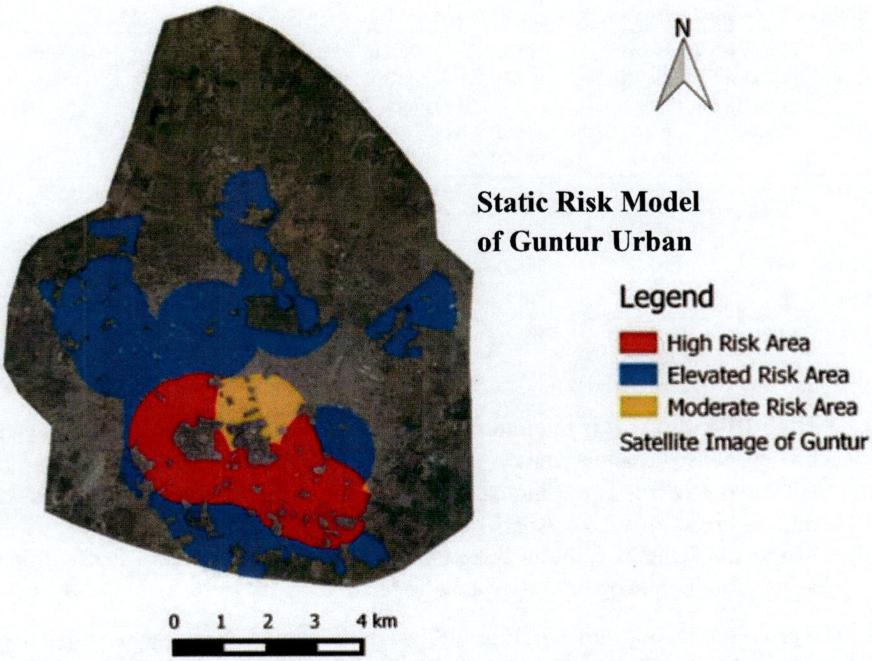

Fig. 11.3 Representation of high, elevated and moderate risk areas of the Guntur urban area (Administrative Staff College of India 2018)

Generation of Dengue Risk Map: This was map generated from taking into consideration of various environmental factors such as Land Use/Land Cover (LU/LC), Normalized Difference Vegetation Index (NDVI), Land Surface Temperature (LST) and Slope (High Land/Low Land areas), etc. The data sets that were originally acquired from LANDSAT 7 (ETM+) satellite, etc. data were classified as "Low", "Medium" and "High" risk potential zones having the dengue outbreak. The following algorithm was used to develop the dengue risk map from each environmental indicator:

$$\text{Dengue Risk Map} = \text{LULC(A)} + \text{NDVI(B)} + \text{LST(C)} + \text{SLOPE(D)}$$

Static Risk Model: Further, a static risk model was also developed by considering the various parameters such as water bodies, dense urban vegetation, sparse vegetation/urban open lands and semi urban croplands in Guntur urban area, Andhra Pradesh. The built-up area (urban area), water bodies, canals, sparse vegetation/open lands, dense urban vegetation and semi urban cropland maps were prepared by using high resolution images. This model for Guntur urban area was created by analyzing the mosquito habitat and conditions promoting mosquito breeding. Following three decision rules were used to develop the static risk map of the area (Fig. 11.3):

Table 11.1 Validation summary of the proposed model

Years	No. of high risk cases noted in the model	No. of elevated risk cases noted in the model	No. of moderate risk cases noted in the model	No. of cases outside the model	Total	No. of cases fit in the model	Validation accuracy of the model
2015	132	082	11	27	252	225	89.29
2014	141	109	11	24	285	261	91.58
2013	197	106	25	36	364	328	90.11
2012	206	72	28	37	343	206	89.21
2011	173	73	16	25	287	262	91.29

i. **High Risk Zone**: It is the intersection of 1 km buffer regions of urban water bodies, dense urban vegetation, Urban open lands/sparse vegetation.
ii. **Elevated Risk**: It is the intersection of 1 km buffer regions of urban water bodies, urban open lands/sparse vegetation.
iii. **Moderate Risk**: It is the intersection of 1 km buffer regions of dense urban vegetation, Urban open lands/sparse vegetation.

The proposed model shown in Fig. 11.3 was validated by using the retrospective data obtained from District Medical Health Offices (DMHO) of the years 2011–2015. After correlating both, it was observed that 173 (from the Year 2011) cases were noted in the high risk zone, 73 cases in the elevated risk zone, 16 cases in the moderate risk zone and 25 cases were fell outside of the model. The generation of this type of map is important as it provides the relationship between vector born disease incidence and distance from houses to breeding sites. Based upon this study, it was envisaged that 1 km distance would be considered as probable regions for Dengue occurrence. The validation summary of the proposed model has been shown in Table 11.1.

It was noticed that the proposed model showed the consistent validation accuracy for the consecutive five years from 2011 to 2015.

11.5.1.1 Significant Findings

As discussed earlier, this case study is mainly based upon the reported Dengue cases in Guntur Urban Area and their detailed analysis with respect to the optimum conditions for mosquito breeding such as water bodies, dense urban vegetation, urban open land, etc. The results obtained were divided into three categories such as high risk zone, elevated risk zone and moderate risk zone. This has also been depicted in a spatial pattern. The main output of the model could be validated with the standards of the District Medical Health Offices (DMHO).

11.5.1.2 Case Study 2: 'Geospatial Mapping of Diabetes Risk Hot Spots in Jhajjar, Haryana and North East Delhi

At present, demographic projections indicate a major increase in Cardio Vascular Disease (CVD) in the Country. Overall, CVDs alone are accounted for around one-fourth of all deaths in India up to 2008.[1] Till now this category of diseases is expected to be the fastest growing chronic illnesses that affect all ages of people. Both diabetes and CVD's are known to have associated risk factors leading to mortality and co-morbidity. Therefore, it is the need of the hour that diabetes and CVDs must be addressed together in terms of risk assessment. A number of interventions are required immediately required for the prevention and management of both. In the present case study, efforts have been made to study Diabetes, in the selected areas of Jhajjar, Haryana (Urban area) and North East Delhi considering various factors.

Behavioral, demographic and environmental risk factors assessment of Diabetes.

Various districts of North East Delhi (Delhi) and Jhajjar (Haryana) were taken as study areas. A cross-sectional, community-based survey of male/female (18 years and above) was carried out to estimate the common risk factors of the Non-Communicable Diseases (NCDs). The survey included household, demographic, behavioral and biological information.

Analysis of additional risk factors such as the use of tobacco, excess alcohol consumption, physical inactivity, overweight/obesity, hypertension and glucose levels.

(i) Tobacco Use: Distribution of tobacco users/ Non Users in Jhajjar and North East Delhi has been shown in Fig. 11.4a, b, respectively. Data represents that in North East Delhi, the percentage of tobacco users was less (24.6%) than that in Jhajjar, where 34.8% of tobacco product users were identified.

(ii) Alcohol Use: Alcohol consumption indicated a higher prevalence among men than women. It was observed that in both the districts, 23–24% of respondents

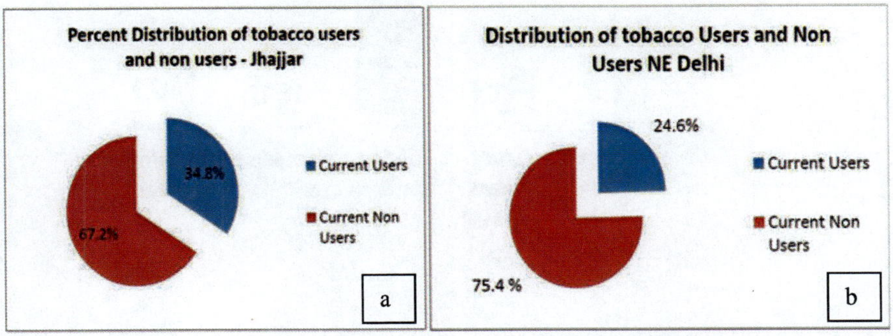

Fig. 11.4 Distribution of tobacco users and Non Users in Jhajjar (**a**) and North East Delhi (**b**) (Administrative Staff College of India 2018)

were current drinkers. It was also noted that in Jhajjar, the average % of respondents consuming alcohol every week was more than that of North East Delhi, where 22.8% out of total male respondents were current drinkers. The study showed that almost 94% of female respondents have never consumed alcohol in their lives. On an average, 2.72 standard drinks were consumed in Jhajjar compared to an average of 2.6 standard drinks in NE Delhi. In NE Delhi, those men who consumed alcohol daily or more than 4 times a week, took around 1 to 2 standard drinks. Figures 11.5 and 11.6 represent the distribution of alcohol consumption in Jhajjar and North East Delhi districts.

(iii) Diet: Consumption of fruits and vegetables is less in North East Delhi than in Jhajjar irrespective of gender. In both, the districts consumption of healthy diet of vegetables and fruits is lesser among women than men. Around 70–80% of all age groups both male and female consumed three or less number of servings of fruit and vegetables per day. The following Table 11.2 presents the percentage distribution of Respondents in North East Delhi and Jhajjar.

(iv) Physical Activity: Based on WHO criteria, individuals were classified as active and inactive. On comparison, significant differences were observed among both sexes. Jhajjar men (74%) were most active, followed by women (42%) out of the active respondents and in North East Delhi it was reversed. Women were more active (59%) compared to men (43%). Across the age groups, there were not much significant differences in the activity level in North East Delhi except in the age range 25–34 years (Figs. 11.7 and 11.8). In Jhajjar, 40–47%

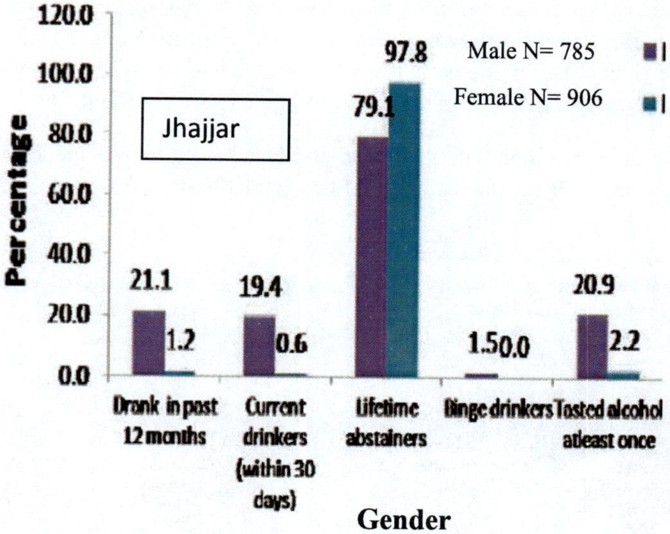

Fig. 11.5 Distribution of alcohol consumption in Jhajjar (Administrative Staff College of India 2018)

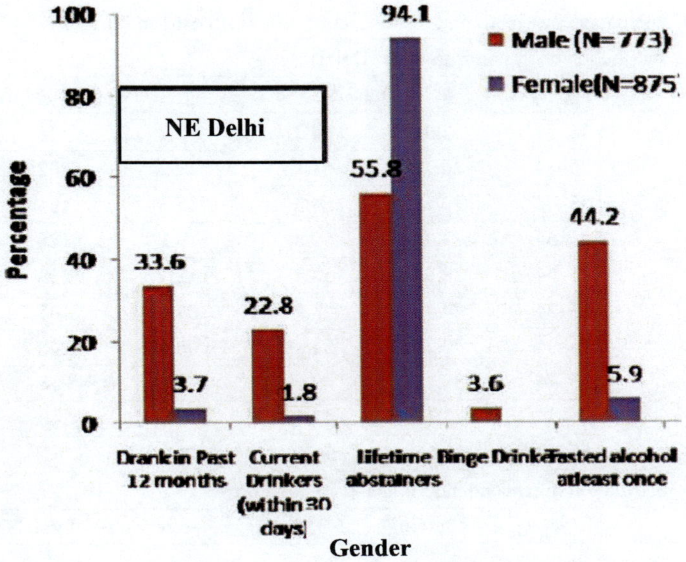

Fig. 11.6 Distribution of alcohol consumption in North East Delhi (Administrative Staff College of India 2018)

of respondents in each age group were physically active. In all the age groups, men were physically more active than women.

Distribution of Physical Activity: In addition to physical activity time spent in sedentary activities was surveyed. Among the various age groups, the propensity towards sedentary activities like was more among 35–44 age group in North East Delhi. In Jhajjar, women were found to have more sedentary behavior than men. The following Figs. 11.9 and 11.10 show the age wise distribution of respondents having sedentary activities in both Jhajjar and North East Delhi, respectively.

(v) Hypertension: In North East Delhi, a proportion of 28% ($N = 305$) had their blood pressure measured by a doctor or health care worker in the past 12 months, with 24% ($n = 84$) diagnosed as hypertensive; all of them were on prescribed anti-hypertensive medication. The systolic and diastolic BP was measured for all the respondents. Out of 1648 respondents, 23.4% of the total sample were found to have high systolic blood pressure > 140 mmHg and 37.3% had diastolic pressure > 70 mmHg. In Jhajjar, 22.2% of respondents were found to be hypertensive with 15% of men and 12% of women having a systolic BP > 140 mmHg and/or diastolic BP > 90 mmHg. Figure 11.11 shows the distribution of hypertension cases based on the age of respondents in North East Delhi and Jhajjar.

(vii) Elevated Glucose Level: Raised blood sugar or metabolic syndrome is equally a major risk factor for non-communicable diseases globally. Only 53.6 of the

Table 11.2 Percentage distribution of respondents North East Delhi and Jhajjar

Percentage distribution of respondents (New Delhi)					
Gender	Age (years)	≤2 servings	2–3 servings	≥ 3 serving	Total ($N = 1648$)
Male	15–24	27.0	64.3	08.7	115
	25–34	26.7	61.1	12.2	232
	35–44	30.4	54.2	15.3	161
	45–54	23.0	59.8	17.1	126
	55–64	28.8	58.3	12.9	139
Female	15–24	44.4	48.9	06.6	135
	25–34	40.2	52.8	06.9	246
	35–44	34.4	49.6	14.0	218
	45–54	42.1	47.4	10.4	140
	55–64	31.6	56.6	11.8	136
Percentage distribution of respondents (Jhajjar)					
Gender	Age (years)	≤2 servings	2–3 servings	≥3 serving	Total ($N = 1690$)
Male	15–24	70.3	22.3	07.4	0.87
	25–34	69.4	13.0	17.6	143
	35–44	63.0	12.9	24.1	154
	45–54	64.0	11.4	24.6	208
	55–64	63.0	12.1	24.9	192
Female	15–24	56.4	23.9	19.7	112
	25–34	53.7	24.6	20.7	253
	35–44	53.5	25.6	20.9	213
	45–54	56.1	25.5	18.5	165
	55–64	50.2	26.6	23.6	163

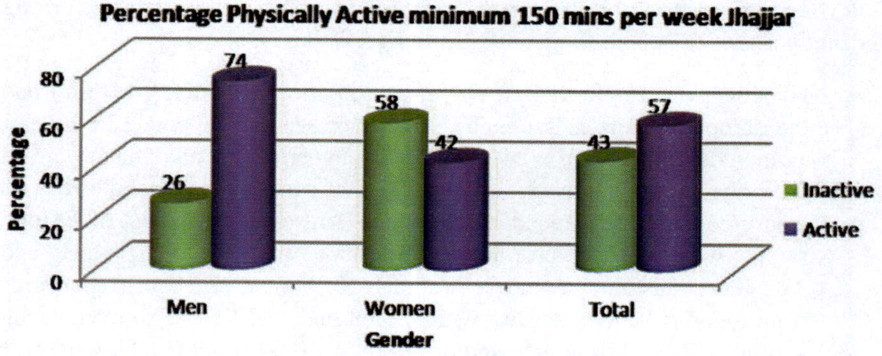

Fig. 11.7 Percentage physically active minimum 150 min/week in Jhajjar (Administrative Staff College of India 2018)

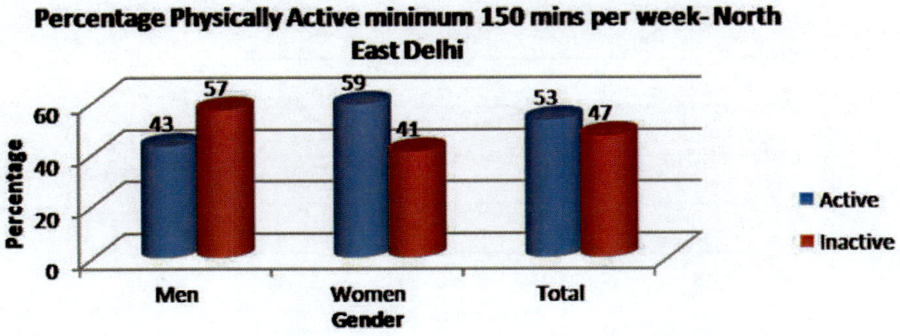

Fig. 11.8 Percentage physically active minimum 150 min/week in North East Delhi (Administrative Staff College of India 2018)

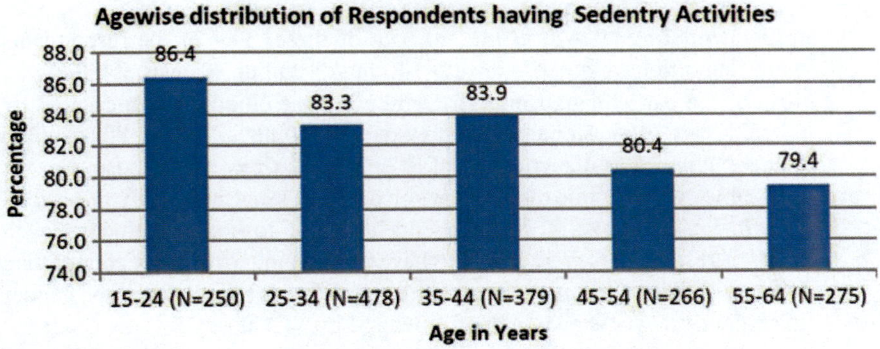

Fig. 11.9 Representing age wise distribution of respondents having sedentary activities in Jhajjar (Administrative Staff College of India 2018)

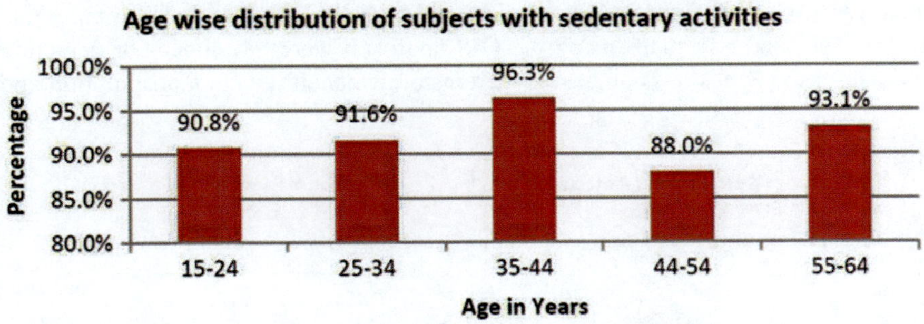

Fig. 11.10 Age wise distribution of respondents having Sedentary Activities in North East Delhi (Administrative Staff College of India 2018)

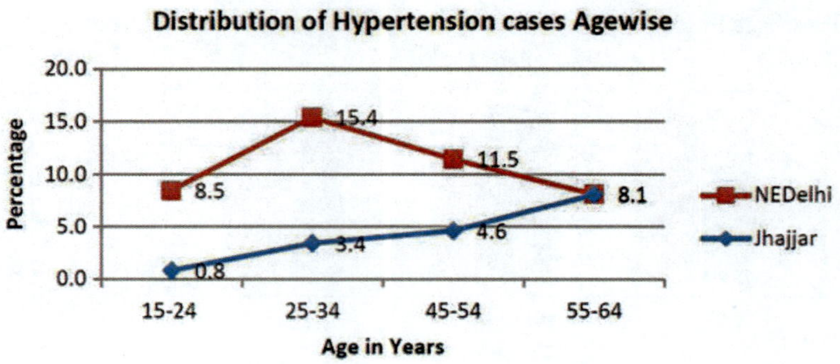

Fig. 11.11 Age wise distribution of hypertension cases (Administrative Staff College of India 2018)

respondents ($n = 926$) were tested. Out of these, 39.1 of the respondents were found to have glucose level > 140 mg/dl. In our analysis, 24.3% ($n =$ 401) of the participants from NE Delhi had their blood sugar measured by a health care professional in the previous 12 months, 22 ($n = 90$) of them being diagnosed as diabetic. Out of these, 33.3% ($n = 30$) of those were on injectable insulin, while the rest were on oral drugs and a specially prescribed diet. In Jhajjar, out of the 909 respondents tested for random glucose level, 6.6% of respondents were found to have random glucose levels greater than 140 mg/dl. The distribution of random glucose levels based on age and gender of respondents are shown in the following Fig. 11.12.

Mapping of the individual level risks factors

The risk factors were analyzed on the basis of gender, age and specific area. Selected area was also categorized as low, medium and high risk zones to that particular risk factor. The final risk maps using the GIS approach were developed for depicting the risk-based zones of Diabetes, Hypertension, Unhealthy Diet Consumption and

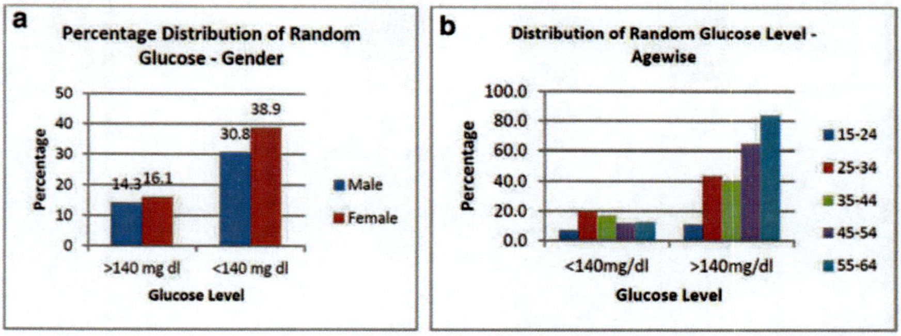

Fig. 11.12 Percentage distribution of random glucose level based on (**a**) gender and (**b**) age (Administrative Staff College of India 2018)

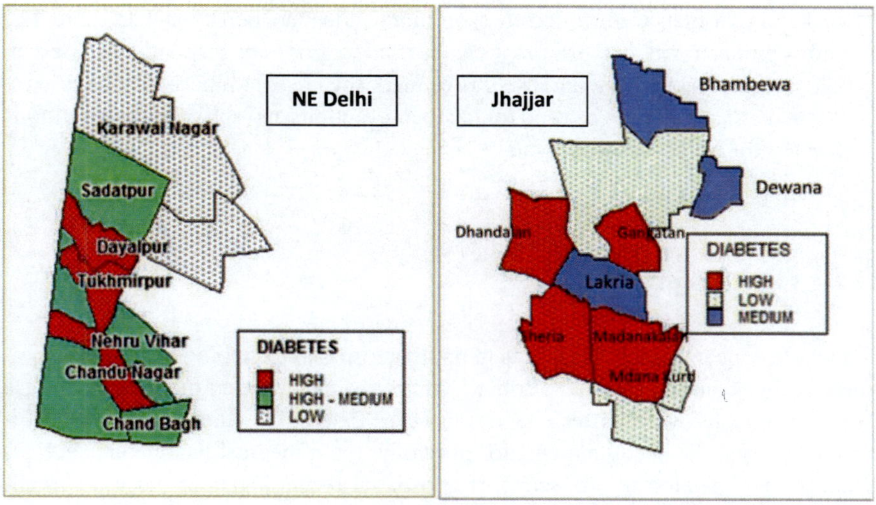

Fig. 11.13 GIS Map Indicating Diabetes risk-based zones in North East Delhi and Jhajjar (Administrative Staff College of India 2018)

Physical Inactivity in North East Delhi and Jhajjar, respectively. Figure 11.13 depicts the Diabetes risk-based zones in North East Delhi and Jhajjar.

Salient findings: The important findings of the study are as follows:

- The study was carried out in the various districts of North East Delhi (Delhi) and Jhajjar (Haryana), respectively. A cross-sectional, community-based survey of adults (both male and female), above the age group of 18 years was carried out to estimate the common risk factors of the NCDs—diabetes, and attempts were made to develop GISbased thematic maps based on the risk factors.
- It was observed that in North East Delhi 24.6% of respondents were using tobacco which was less than Jhajjar where 34.8% of respondents were tobacco users.
- AS per World Health Organization (WHO) adults aged 18–64 years should do at least 150 min of moderate intensity physical activity throughout the week, or do at least 75 min of vigorous-intensity physical activity throughout the week, or an equivalent combination of moderate- and vigorous-intensity activity. Based on WHO recommendations, individuals were classified as active/inactive along with spatial attributes. It was observed that Jhajjar men (74%) were most active, followed by women (42%) out of the active respondents. In North East Delhi reverse pattern was noticed. Women were more active (59%) compared to men (43%). On further investigation it was found that in Jhajjar the main occupation was agriculture and majority of the men worked, hence were more active than women. In addition to physical activity, time spent in sedentary activities was also surveyed. Among the various age groups, the propensity towards sedentary activities like was more among 35–44 age group in North East Delhi.

- In Jhajjar, women were found to have more sedentary behavior than men. The study demonstrated that GIS maps can be used to group the respondents based on risk category assessment and use this to mark the areas at block- or village-wise within districts, that can be used for health promotions and any other interventions targeted for NCDs at local levels.

11.6 Conclusion

The whole concept of this chapter is to apply geospatial technologies in data acquisition analysis with the related attribute data of the patients and the type of disease they were affected with. After integration of the data in the standardized format, it is easy to target the needy patients for providing them medical health care services in quick time. Therefore, the scope of geospatial technologies in disease surveillance and extending quick medical services has tremendous scope for decision makers/stakeholders to implement health care management schemes mostly in rural areas of the country.

Acknowledgements The figures and the tables used in this chapter are taken from the Project Completion Report of the Network programme supported by the Department of Science and Technology, Govt. of India. The authors would like to thank to Prof. I.V. Murli Krishna, National Coordinator of DST Networking programme on Geospatial Public Health Programme. The contribution made by all the Principal Investigators in undertaking various case studies of this networking programme, especially Dr. Valli Manickam and Prof. Anandhi Ramachandranan and their team from ASCI, Hyderabad and IIHMR, New Delhi are thankfully acknowledged.

References

Administrative Staff College of India (2018) Project completion report of national geospatial public health data and management system: a national networking project, Hyderabad

Blossom JC, Finkelstein JL, Guan WW, Burns B (2011) Applying GIS methods to Public Health Research at Harvard University. J Map Geogr Lib 7(3):349–376

Hunter DJ (2001) Professor of Health Policy and Management, University of Durham, England. Public Health Management World Health Organisation—University of Durham meeting report.

Hunter DJ, Berman PC (1997) Public Health Management. Eur J Pub Health 7(3):345–349

O' Carroll PW (2003) Introduction to public health informatics. In: Carroll PW, Yansoff YA, Ward ME, Ripp LH, Martin EL (eds) Public health informatics. Springer , New York, NY, pp 1–15

Chapter 12
Utilisation of Geo-Spatial Technology to Study the Variation in Access of Urban Health Care Centres in Kamrup Metropolitan, Assam, India

Namita Sharma, Jayanta Goswami, and Poonam Sharma

Abstract Nowadays, Geographic information system (GIS) has become a more efficient technology in applications of many fields in medical research. As health is always the primary concern of human beings, everyone wishes to access better health care for his family. As an emerging tool to solve every complex query in various fields of earth sciences, social sciences as well as in medical sciences. GIS plays a vital role in storing, monitoring, mapping, managing and developing huge amounts of spatial and non-spatial data. Recent increases in the availability of GIS and its modelling approaches have provided a base for the planning of different public services in a metropolitan context. Kamrup Metropolitan of Assam has been selected as the area of interest for the study. The present study aims to show the distribution pattern of all health centres, hospitals including government and private sector, accessibility to the health care centre from different population concentration zones. The study also identifies those locations which are vulnerable with respect to having fewer number of hospitals using geospatial technology. The study aims to the planning and management of the city to avail of a better health facility for the residents, as well as outsiders, which will lead to the smartest development of the city. A Geodatabase has been created which includes the total numbers of government and private health centres along with spatial and non-spatial attribute data, population distribution of the district and road networks. The study involves the use of some basic GIS functions like: Network analysis, proximity analysis, vulnerability analysis, as well as statistical analysis, using ARC GIS, QGIS software. The vulnerability analysis will give us a general conception about the need of health care centres for the city, as the use of geospatial technology can solve complex public issues.

Keywords Geographic information system (GIS) · Health care centre · Population concentration Zone · Network analysis · Hotspot analysis

---

N. Sharma (✉) · J. Goswami
National Institute of Rural Development & Panchayati Raj-NERC, Guwahati, Assam, India

P. Sharma
Department of Geography, Shaheed Bhagat Singh College, University of Delhi, Delhi, India

P. Sharma (ed.), *Geospatial Technology and Smart Cities*, The Urban Book Series,
https://doi.org/10.1007/978-3-030-71945-6_12

Acronyms

GIS	Geographic information system
ISBT	Inter-State Bus Terminal
FCA	Floating Catchment Area
MSL	Mean Sea Level
PHC	Primary Health Care
RT	Response Time
V.I	Vulnerability Index

12.1 Introduction

Geographical Information Systems (GIS) became a very advanced technological system that has the capability to solve problems related to almost every field of application and study in recent days. As an emerging tool to solve every complex query in various fields of earth sciences, social sciences, as well as in medical sciences, GIS plays a vital role in storing, monitoring, mapping, managing and developing a huge amount of spatial and non-spatial data. Application of GIS has the greatest achievement and innovation in the fields of medical research including mapping and monitoring of the spatial and temporal distributions of vectors of infection, surveillance of infectious diseases, etc. GIS can help the professionals to make the decision better and faster for different public health planning issues in a spatial context (Hanchette 2003).

The implementation of GIS in the health care industry for improving health has become widespread all over the world now. Some of the potential application of GIS in medical research are: Mapping and monitoring the spatial and temporal distribution of vector-borne infection; Surveillance of infectious diseases, Accessing the health care facilities in the context of spatial location; Monitoring air pollution-related health effects, water and food bore diseases; Development of detection and early warning system for any vector-borne diseases risk for any particular region; Health vulnerability analysis in terms of total income, hospital infrastructure, remote areas, climatic condition, etc.

Network refers to a system of elements, i.e. lines, points that are interconnected and represent routes from one location to another location. Hence, Network analysis is used to deal with network related activities. Network analysis is meant for evaluating, planning, ordering and installing any kind of communication equipment that a business needs.

GIS-based network analyst is a powerful extension tool which serves network-based spatial analysis such as service area analysis, closest facility, shortest path analysis, travel time, travel distance, location finding, routing, etc. As we all know that the transportation system of a country is a key indicator of development, so the focus should be given to transport and communication systems equally for the overall growth of the country. In recent trends of technology, GIS became a well known technology for any kind of network related analysis. A network data model can be built using GIS data on ArcGIS environment. As already mentioned above ArcGIS-based Network analysis provides multiple solutions for various problems pertaining to spatial networks such as finding the closest travel route, travel time, directions, etc.

GIS can relate unrelated information by using location as the key index variable hence it is very useful in urban health studies. The GIS is used in almost every facet of our daily lives, from earth science and other physical sciences to finance and management (Sadiq et al. 2013). Variation in Geographically distributed population of any region is the key to analysis and planning for health services to develop the region. People of different age, gender, economic status have different needs for their health care. The distance from a village to a good hospital located in the nearest city, travel time, cost-effect analysis, Road network, types of services available in the health care, vulnerability analysis, etc. are very important aspects to explore by using geospatial technology which will support the better health service facilities. Access to health care can be measured using the straight line or road distance for evaluating existing and future optimal service location (Murad 2018).

12.2 Review of Literature

A GIS-based study using drive-time analysis approach to determine the geographical access to health centres using ArcGIS Network Analyst and overlay analyses were used as the analysis tool for this paper. (Murad 2018). Mokgalaka (2015) used the population data, network data and facility data to access the primary health centres with straight-line distance approach used to analyse travel distance to closest primary health cares, used modelling approach to create facility catchment area irrespective of the distance, health centres on high demand. Bagheri et al. (2005) applied the drive-time analysis approach for estimation of shortest time through road networks between any pair of population and locations of health centres and calculated spatial accessibility to primary health care (PHC) services. A significant case study on measuring access to urban health services was evaluated by using Floating Catchment Area (FCA), minimum distance methods and Response Time (RT) accessibility technique (Masoodi et al. 2015). Implementation of GIS application in health sector became useful as it is a cost-effective tool (Dermatis et al. 2016). Reviews about the various GIS applications in the field of medicine, epidemiological or social research and health care were discussed by Fradelos et al. (2014). Potential use of network analysis in defining the optimal service area of different services such as hospitals, schools and fire stations of Chandigarh city was carried out by Kumar et al. (2016). Dabhade et al. (2015) discussed the Dijkstra's algorithm application to provide the shortest path and closest facility from one location to another in terms of hospital information. The Network Analysis module of ArcGIS was used to evaluate the best routes and closest facility on a route network (Kharel et al. 2018). Ahmed et al.

(2017) recommended that the GIS best route algorithm is better than the shortest route algorithm in emergency situations, especially in a crowded city like GCR.

12.3 Objectives

The aim of the study is to evaluate the shortest path and closet facility from the. ISBT, railway station and airport to the multi speciality hospital using geospatial techniques.

The following are the objectives of the studys:

1. To create a Geodatabase for roads and health care centres.
2. To apply the network analysis as shortest path and closet facility to access the health care centres
3. To generate vulnerability index for Kamrup Metropolitan district

12.4 Study Area

Kamrup metropolitan is a district of Assam, India, bounded by the mighty Brahmaputra river and Darrang district on the north, Meghalaya on the south, Kamrup rural on the west and Morigaon and Meghalaya on the east. The Kamrup Metropolitan district lies between 25°43′ N to 26°51′ N latitude and 91°30′ E to 92°12′ E longitudes. The district Kamrup Metropolitan was created by bifurcating the old Kamrup district on 3rd February 2003. A location map of the study area has been shown in Fig. 12.1.

The area of the Kamrup metropolitan district is 1527.84 sq. km and the altitude is 55–51 meters from mean sea level (MSL). The district is further divided into six revenue circles, three Community Development blocks, fourteen Gaon Panchayat (village council) and 316 villages. The six revenue circles (RC) are Guwahati RC, Sonapur RC, Chandrapur RC, Dispur RC, North Guwahati RC and Azara RC. Revenue circles or revenue blocks are the local revenue sub-division of each district of the states of India, which helps in local administration and governed by a revenue inspector. Total population of the district is 12, 53,938, out of which 2, 16,927 is under rural and 10, 37,011 is urban population. The Literacy rate is 88.66%. The Dispur is the Capital of Assam is in Kamrup Metropolitan as well as the headquarter of the district.

The climate is sub-tropical with semi-dry summer and cold in winter. Annual rainfall ranges between 1500 mm to 2600 mm, average humidity is 76% and temperature varies from maximum 37–39° C to 6–7° C. The Brahmaputra is the major river of the district. The languages spoken by the local people are Assamese, Bengali and Bodo.

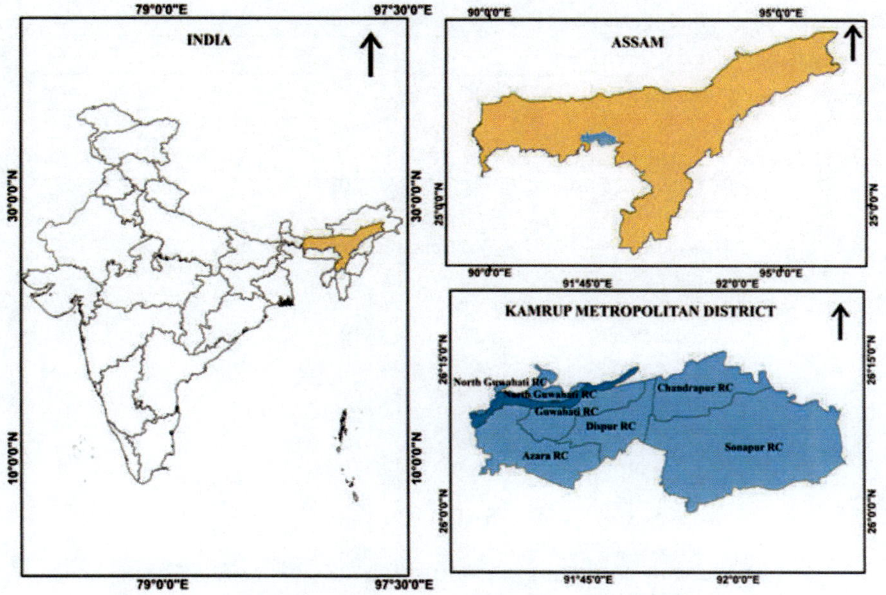

Fig. 12.1 Map of the study area

12.5 Database and Methodology

Study Area boundary: The study area boundary has been created in ArcGIS by digitizing the georeferenced image (Census of India 2011) of Kamrup metropolitan districts. There are six revenue circles in Kamrup metropolitan district.

Health Centre Data: Geodatabase of Health Centres in Kamrup Metropolitan has been created in Arc Map 10.5. Reference of the names of the Govt. Hospitals, District Hospitals, CHC, PHC, SC and Private Hospitals have taken from two departments, i.e. Health and Welfare Department, Govt. of Assam & Assam State Disaster Management Authority and exact spatial locations of health Centres has collected using GPS Device. The Name of all health centres with their location is given in Appendix 1.

Population Data: Population data has been taken from the Census of India (2011). The quantitative data of the population are joining into the study area shapefile for further analysis using ArcGIS software. From the data, population distribution, population density has been created.

Network Data: Road network data has been prepared by on-screen digitization method from Sentinel 2 imagery and verified from Google Earth and for network dataset has been created using The Network Analysis tool on ArcGIS environment. Network Analysis tool is used for determining the closest health centres facility and shortest path based on the road network.

A description about the database with the source of data is briefly mentioned in Table 12.1.

Table 12.1 Description of database

Data used	Description of data			Source
Primary data	Field data: Location of health centres			GPS Survey, Observation
Secondary data	• Satellite Imagery: Sentinel 2B • Date of acquisition: 7/01/2019, 11/02/2019 • Orbit: 133, 90			https://earthexplorer.usgs.gov/
	• Bands:	Wavelengths	Resolution (m)	
	Band 1-Coastal	0.443	60	
	Aerosol	0.490	10	
	Band 2-Blue	0.560	10	
	Band 3-Green	0.665	10	
	Band 4-Red	0.705	20	
	Band 5-RE	0.740	20	
	Band 6-VRE	0.783	20	
	Band 7-VRE	0.842	10	
	Band 8-NIR	0.865	20	
	Band 8A-VRE	0.945	60	
	Band 9-Water Vapour	1.1375	60	
	Band 10-WIR-Cirrus	1.1610	20	
	Band 11-WIR	2.190	20	
	Band 12-SWIR			
	• Shapefile of the study area			Census of India, 2011
	• Road Network Data			Satellite Imagery as well as Google Earth
	• Population data			Census of India (2011)
	• Background of Study area			Books, journals, Statistical handbook, assam, 2011

To fulfil the present work, both primary and secondary data are used. The methodology is the backbone of research. The observation methods, GPS surveys are used for primary data collection for the present study. Locations of Health centres are collected through a GPS survey by a practical visit to the study area to substantiate the objectives of the study. As shown in the Flow chart of methodology i.e. Fig. 12.2, after completion of data collection they are processed on GIS environment using geospatial technologies. All the data are kept in Geodatabase for further processing. After creating the Road network, Topology has been created to validate the dataset to prepare for network analysis, i.e. shortest path and closest facility analysis. Based on health centre concentration, the vulnerability Index has been calculated by taking a 2*2 km fishnet grid layer clipped for the study area. All the Data are prepared in shapefile format for Generations of maps for the study. As shown in the flow chart all steps have been followed, respectively, to get the final results and findings.

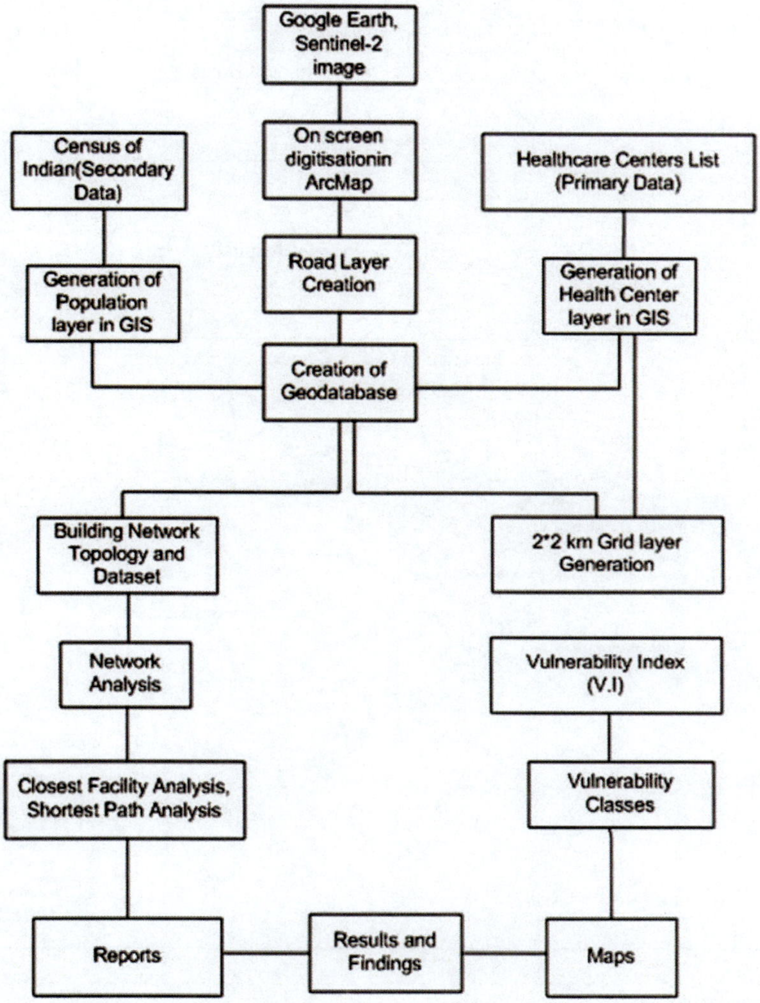

Fig. 12.2 Flowchart of methodology

12.6 Results and Discussion

a. Heath centre map

After the generation of the Geodatabase of the health centres, the main issue was to gain a clear picture of the current condition. In the whole district, 131 nos. of total health centres are present as shown in Table 12.2.

From the map as shown in Fig. 12.3: we can understand the picture that maximum nos. of health centres are distributed in the Guwahati city. And, we can see that all hospitals are in Guwahati only. The other parts are covered by only HCs like

Table 12.2 Different types of health centres

Sl. No.	Type	Nos
1	Government Hospital	8
2	District Hospital	1
3	CHC	3
4	PHC	26
5	SC	58
6	Private Hospital	35

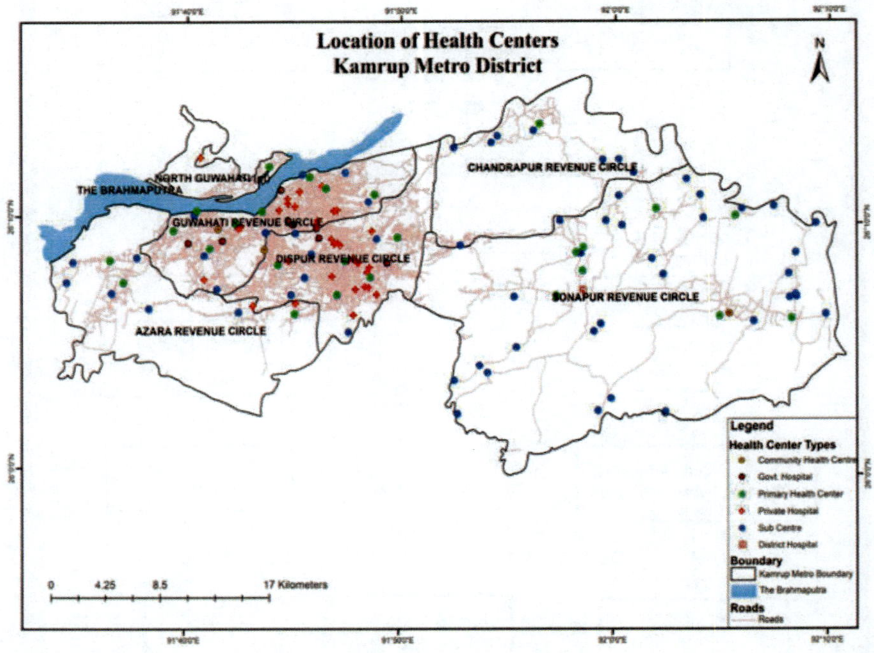

Fig. 12.3 Health centres location with road network

PHCs, SCs and CHCs. The location of health centres is also verified physically. After observing the scenario of the available health centres in this district, there are requirements of Govt hospital or private hospital in the right-hand side of the district.

b. Population Density map

As we all know that, Population density is a measurement of population per unit area or it is a quantity of type number density. Population density is frequently applied to living organisms, mostly to humans. To fulfil the purpose of the study, population density map has been prepared using the Point density tool on Arc Map 10.5. (Fig. 12.4). According to Person per Sq.km, the population density of the area has been calculated. The point density calculates the density of point features around

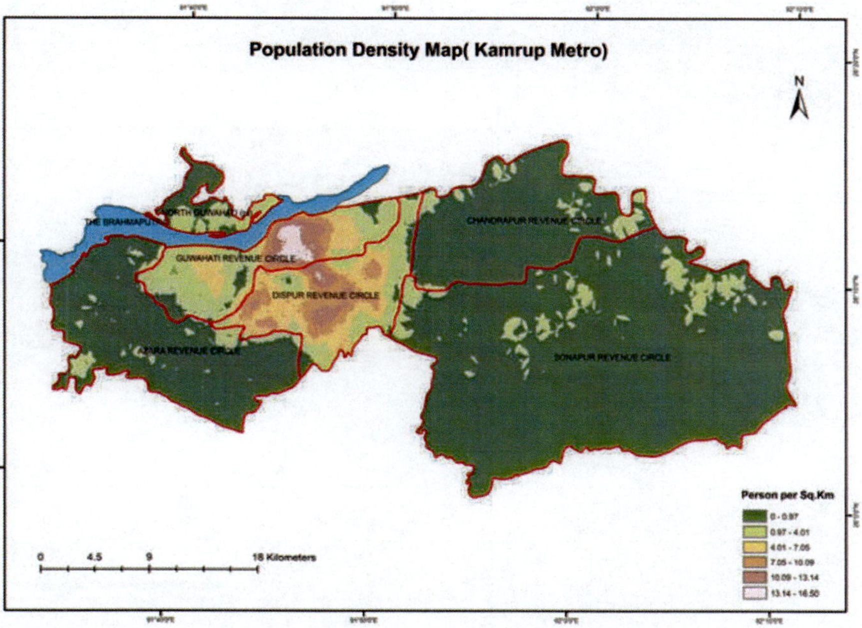

Fig. 12.4 Population density map

each output raster cell. Population density is high in Guwahati and Dispur Revenue Circle than the other part. In these both revenue circles, density per sq. km lies from 4–16.5. According to these, the facility of health centres is reasonable in Guwahati and Dispur Revenue Circle. The total population of Kamrup Metropolitan is almost 1.3 million (Census 2011).

c. **Network Analysis: Shortest Path and closest facility map**

ArcGIS Network Analyst enables users to dynamically model realistic network conditions, including turn restrictions, speed limits, height restrictions, and traffic conditions at different times of the day (Dabhade et al. 2015). For this study, we have used Dijkstra's algorithm. Dijkstra's algorithm is a graph search algorithm which solves the single source shortest path problem for a graph with non-negative edge path costs, producing a shortest path tree. This algorithm is often used in routing and as a subroutine in other graph algorithms. For a given source vertex (node) in the graph, the algorithm finds the path with the lowest cost (i.e. the shortest path) between that vertex and every other vertex (Dabhade et al. 2015).

The following assumptions were made when the shortest path and closest facility calculated for the study area:

(i) Traffic congestion not considered
(ii) State of the road not considered.
(iii) Calculations were based on road distances.

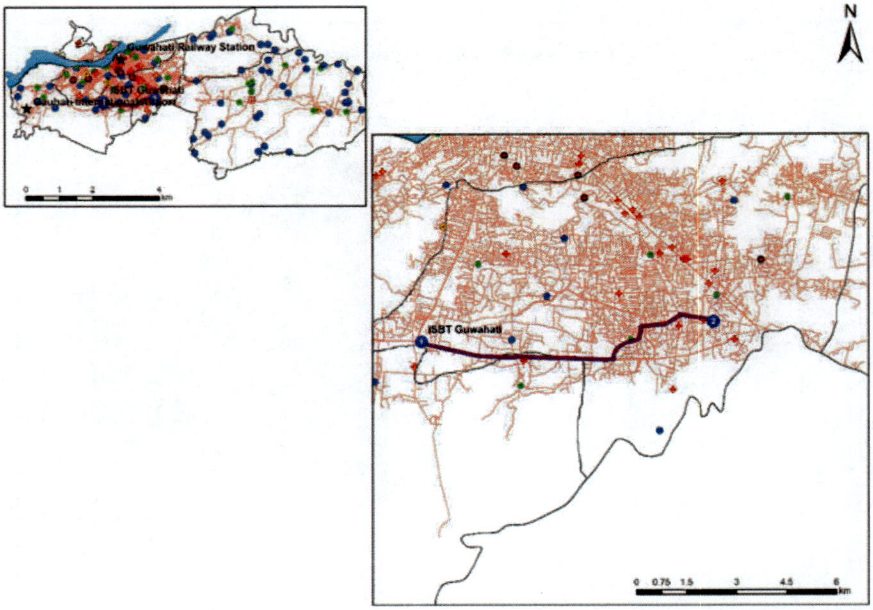

Fig. 12.5 Shortest path from ISBT (1)

Shortest path Analysis from the Airport, Railway Junction and ISBT to nearest health centre location (Figs. 12.5, 12.6, 12.7 and 12.8):

When a patient decides the hospital, the network model finds and calculates the distance from different routes and it takes minimum distance and shows the route in the map (Dabhade et al. 2015). In this study, the Shortest Path from three major locations of the city, i.e. The railway Junction, The Inter-State Bus Terminal (ISBT) and The International Airport has been calculated to the nearest Super speciality hospitals and Government Hospitals where the all facilities are available for different types of patient. The results have been calculated using the *Create network location* tool of Network analysis datasets from point 'A' to Point 'B', as for example, From the Guwahati Railway Junction to the nearest super speciality hospital or the government hospital.

Table 12.3 showing the calculated shortest path Location Name, i.e. from point A to point B and distance between the locations. In shortest path analysis, the model calculates the minimum possible distance between the starting and ending points and displays the shorter route on the digital platform. Likewise the above figures, i.e. Figure 12.5 to 12.8 showing the possible shortest path from the respective landmark location of the city to the nearest health care facilities. The maximum distance is found between Guwahati Airport to its nearest Excel Care super speciality Hospital, that is, 16 km. and the minimum is From Inter-State Bus Terminal (ISBT) to its nearest health care Ayursundra super speciality hospital.

Fig. 12.6 Shortest path from ISBT (2)

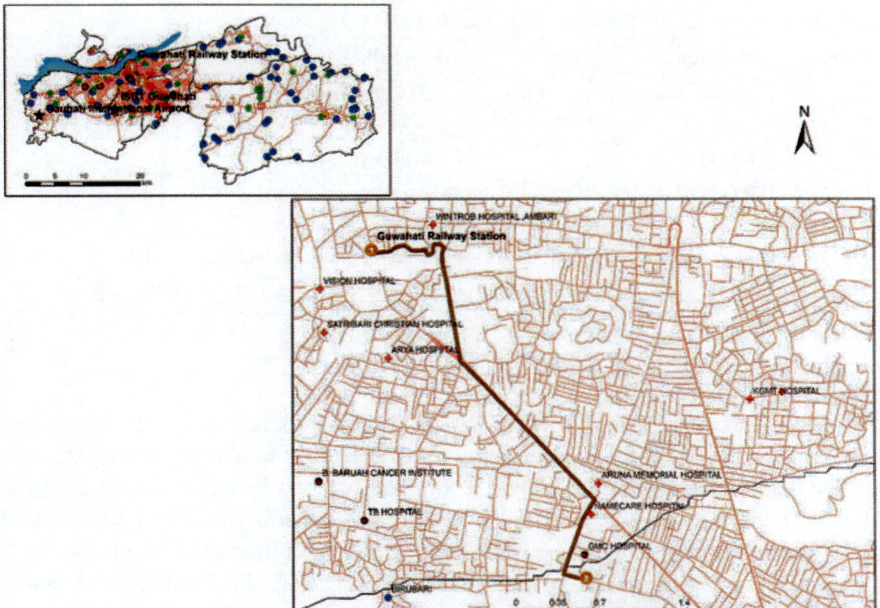

Fig. 12.7 Shortest path from Guwahati Railway Station

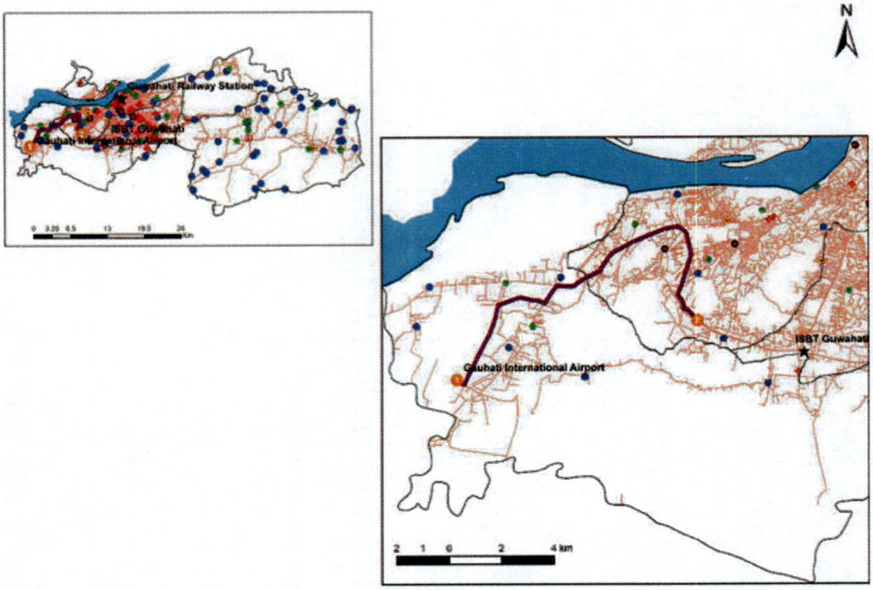

Fig. 12.8 Shortest path from Guwahati International Airport

Table 12.3 Distance from the locations to the Hospitals (Shortest path)

Sl. No.	From point A	To point B	Distance in Km
1	Guwahati International Airport	Excel Care Super speciality hospital	16
2	ISBT(1)	Esic model Hospital	9.8
3	ISBT(2)	Ayursundra super speciality hospital	3.11
4	Guwahati Railway Junction	Guwahati medical college & hospital	4.23

Closest Facility Analysis

From the population density map considering the high population concentration zone, two locations have been selected randomly to perform the closest facility analysis for the study area. Hengerabari and Gandhibasti are two locations from which the closest facility analysis has been calculated for health care. The closest facility tool measures the cost of travelling between incidents and facilities and determines which are nearest to one another. When a user has multiple options, i.e. hospitals, this model will help to find the closest hospital nearest to him. Figures 12.9 and 12.10 show the closest hospital nearest to the user out of many hospitals.

While calculating the closest facility multi-speciality hospital and two high population concentration zone are considered named, Hengerabari and Gandhibasti for

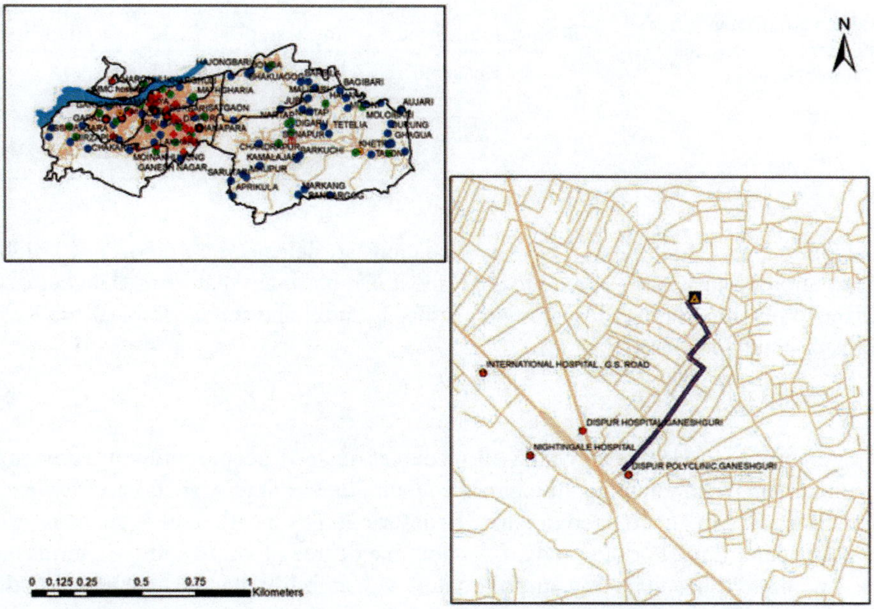

Fig. 12.9 Closest Facility (to Hospital) from one of the highest population concentrated zones (Hengerabari)

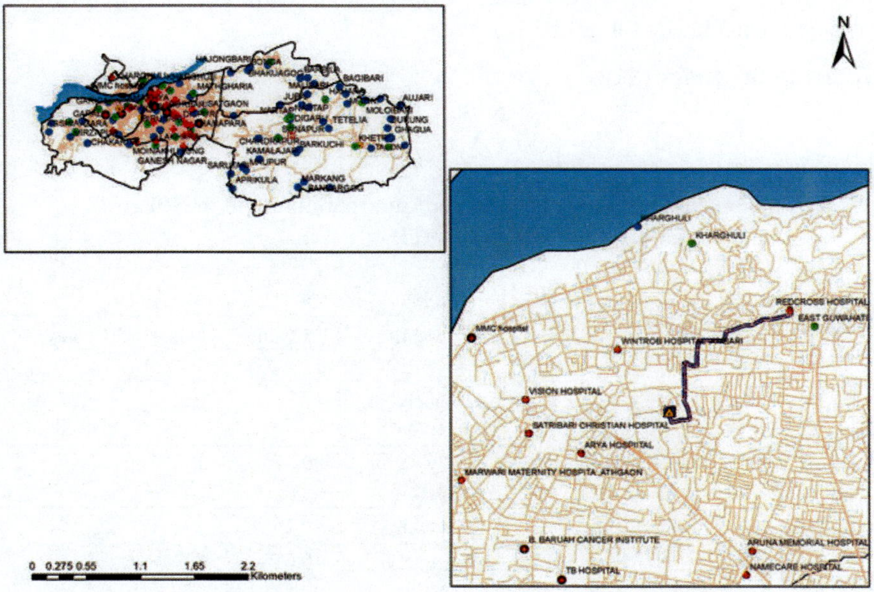

Fig. 12.10 Closest Facility (to Hospital) from one of the highest population concentrated zones (Gandhibasti)

Table 12.4 Distance from the user to the Hospitals (Closest Facility)

Sl.No	From location	To Closest HC	Distance In Km
1	Hengerabari	Dispur Polyclinic Hospital	1.044
2	Gandhibasti	Red Cross Hospital	2.36

the Study area. (Table 12.4). These two locations are selected on the basis of a high population density zone. From the analysis, it can be stated that form Hengerabari Dispur Polyclinic is the closest multi-facility hospital and to Gandhibasti the Red Cross Hospital is closer.

d. Vulnerability map

Vulnerability is a degree to which all resources such as people, property, government systems, economic conditions, surrounding environment, social activity, etc. are susceptible to get affected or harm, or destruction by a particular agent or factor at a particular time. For this study to find out the degree of vulnerability in terms of the numbers of hospitals in a spatial context vulnerability map has been prepared. The Vulnerability map according to the concentration of the health centres in the city has been calculated. The Fishnet Grid of 2*2 km has been prepared for the entire district. Then the calculation of health centres in each grid has been done in ArcGIS. Then by using the Vulnerability Index (V.I) algorithm (Phule et al. 2015), VI values have been calculated (Table 12.5).

Vulnerability Index (VI):

$$\mathbf{VI_i} = (\mathbf{V_i} - \mathbf{V_{max}})/(\mathbf{V_{max}} - \mathbf{V_{min}})$$

- VI_i is the vulnerability index for the indicator pertaining to the *i*th grid,
- V_i is the vulnerability value of the *i*th grid.

Table 12.5 Vulnerability Index

Sl no	Vulnerability classes	V.I Value	Nos. Health centres
1	Extremely Vulnerable	−1	0
2	Highly Vulnerable	−0.875	1
3	Vulnerable	−0.75	2
4	Moderately vulnerable	−0.625	3
5	Very Low Vulnerable	−0.50	4
6	Non Vulnerable	0	8

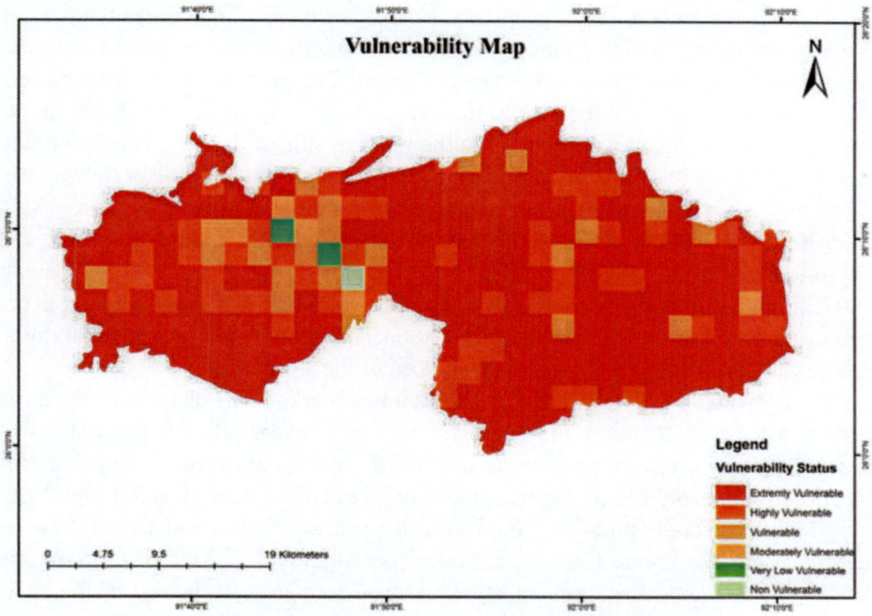

Fig. 12.11 Vulnerability mapping

- V_{max} and V_{min} are the highest nos. of the health centres and lowest nos. of health centres of all the grid cells (Phule et al. 2015).

In case of any particular hazard and disaster, resources, when it is more exposed the risk, are also high to get affected by that hazard or disaster or the resources became more vulnerable. But in this study, the number of Hospitals and vulnerability is inversely proportional, as fewer number of health centres increasing the more risk hence the area became more vulnerable. The VI value is coming in negative numbers because of the inverse relationship of risk factors and health care numbers. The Vulnerability map shows that the area is still more vulnerable in terms of the number. of health centres. The Number of health centres should be more for the area to access the better healthcare facility. From Fig. 12.11, it can be seen that a very less portion of the study area comes under very low to non-vulnerable Maximum number of grids are occupied by an extremely and highly vulnerable class that means the district must be developed in terms of more health care centres in the highly vulnerable areas.

12.7 Recommendations

From the above analysis and discussion, an overview of health care facilities that are available in the study area, about the population of the area, and network availability can be achieved. A clear picture of all health care facility maps including all types, i.e. Primary health centre. Sub-centre, government hospitals, private hospitals, etc. with latitude and longitude information are generated. The map shows that the eastern side of the city has less number of health centres with only one district hospital, i.e. Sonapur District Hospital comparing to the western side of the city. The Maximum number of Private hospitals are concentrated in Guwahati and Dispur Revenue Circle. It can be suggested to open more Government hospitals, as well as Private, within other Revenue Circles so that the people get equal benefits as it will be time and cost-effective too.

The population density map reflects the clusters of different population concentration zone, where the minimum population density is 0.97 and the maximum population density is 16.50 per person sq. km. of the study area.

Shortest path analysis from three important landmarks of the city allows the users to decide the stop location and nearest health care, especially for the visitors or patients coming from other districts or outside of the city. From Inter-State Bus Terminal two hospital is nearer than the other, i.e. Esic model Hospital and Ayursundra super specialty hospital according to the shortest path calculation. Likewise, there is a shorter route to Excel Care Super specialty hospital from the Airport and from Guwahati railway station Guwahati medical college & hospital is near.

According to the closest facility analysis from the two highly population concentrated locations which is Hengerabari and Gandhibasti, Dispur Polyclinic Hospital and Red Cross Hospital are the closest health care facilities found in the city. The positive side of the closest facility analysis is if any patient or user wanted to visit any of the above-mentioned hospitals they can directly go to the nearest junction or location.

The vulnerability map is a good example to show that the area comes under a vulnerable zone in terms of healthcare facilities. The coverage of the health care facilities should be 100% for any region so that the people get the ultimate benefits from it.

12.8 Conclusion

In Assam, Kamrup Metropolitan is the highest populated city. The main aim of the study is to show the use of Geospatial technology to access the urban health care centres to utilize the resources and to observe the vulnerability for the study area for the growth and development of the region. The adoption of GIS technologies for the on-going health challenges of the public to manage a large population is the key indicator towards the development. In the study, Dijkstra's algorithm was used to get the results of Network analysis from the digitized road network dataset. As already mentioned above, the vulnerability map shows that the maximum part of the city is still lacking behind with respect to the good and super speciality health care facilities. To store the health care data, medical records in a digital environment helps in better utilization of health services, temporal case studies and accessing the information of health services in the digital platform. For the Study area mapping of health care centres and all other observation that has been done for the project, is a positive footstep towards the development of the region.

Acknowledgements Authors would like to acknowledge and extend our heartfelt gratitude to Dr. Poonam Sharma, Associate Professor of Geography Department, Shaheed Bhagat Singh College, University of Delhi for giving me an opportunity to work. And we also like to thank Dr. A. Simhachalam, Assistant Professor of CGARD, NIRDPR-NERC for encouraging research work.

Appendix 1: Name and Location Information of Health Centres

S N	Type	Name	Longitude	Latitude
1	Community Health Centre	Maternity and child welfare hospital, Dhirenpara	91.72762	26.14643
2	Community Health Centre	Khetri CHC	92.08957	26.10675
3	Community Health Centre	Pandu FRU	91.69141	26.15977
4	District Hospital	Sonapur DISTRICT HOSPITAL	91.97592	26.12143
5	Govt. Hospital	MMC hospital	91.74019	26.18601
6	Govt. Hospital	Govt AYURVEDIC College	91.66811	26.15024
7	Govt. Hospital	GOVT HOMEO MEDICAL COLLEGE	91.82291	26.13827
8	Govt. Hospital	GMC HOSPITAL	91.76790	26.16075
9	Govt. Hospital	RAILWAY CENTRAL HOSPITAL	91.69457	26.15208
10	Govt. Hospital	B. BARUAH CANCER INSTITUTE	91.74578	26.16598
11	Govt. Hospital	TB HOSPITAL	91.74958	26.16312
12	Govt. Hospital	GUWAHATI MEDICAL COLLEGE	91.76986	26.15456
13	Primary Health Centre	PANIKHAITI	91.94091	26.23181
14	Primary Health Centre	GOTANAGAR	91.68480	26.14673
15	Primary Health Centre	MOLOIBARI	92.09389	26.17162
16	Primary Health Centre	EAST GUWAHATI	91.77519	26.18737
17	Primary Health centre	KHARGHULI	91.76263	26.19493
18	Primary Health centre	GARAL	91.60764	26.13825
19	Primary Health centre	AZARA	91.61805	26.12363
20	Primary Health centre	GARIGAON	91.65664	26.15845
21	Primary Health centre	NORTH GUWAHATI OPD	91.73127	26.20187

(continued)

(continued)

S N	Type	Name	Longitude	Latitude
22	Primary Health centre	KAMAKHYA	91.70489	26.16358
23	Primary Health centre	WEST GUWAHATI	91.72569	26.17172
24	Primary Health centre	ODALBAKRA	91.73821	26.13644
25	Primary Health centre	LAKHARA	91.75129	26.10424
26	Primary Health centre	BHETAPARA	91.78407	26.11689
27	Primary Health centre	DISPUR	91.78997	26.13935
28	Primary Health centre	KHANAPARA	91.80977	26.12869
29	Primary Health centre	SATGAON	91.83090	26.15534
30	Primary Health centre	SONAPUR	91.95466	26.11751
31	Primary Health centre	DIGARU	91.97521	26.13430
32	Primary Health centre	HAHARA	92.03223	26.17585
33	Primary Health centre	KHETRI	92.08196	26.10513
34	Primary Health Centre	DIMARIA	92.13810	26.10402
35	Primary Health centre	NARTAP	91.97565	26.14977
36	Primary Health centre	FERRY GHAT	91.67464	26.17179
37	Primary Health centre	HATIBAGARA	91.97011	26.14634
38	Primary Health centre	MATHGHARIA	91.81300	26.18366
39	Private Hospitals	MARWARI MATERNITY HOSPITA,ATHGAON	91.73930	26.17245
40	Private Hospitals	ARYA HOSPIITAL	91.75149	26.17511
41	Private Hospitals	REDCROSS HOSPITAL	91.77262	26.18882
42	Private Hospitals	WINTROB HOSPITAL,AMBARI	91.75510	26.18501
43	Private Hospitals	SRIMANTA SANKARDEV NETRALAYA,BELTOLA	91.79869	26.12048
44	Private Hospitals	MILITARY HOSPITAL,BASISTHA	91.79710	26.10354
45	Private Hospitals	GNRC HOSPITAL,SIXTH MILE	91.80757	26.13171
46	Private Hospitals	PRATIKSHA INFERTILITY HOSPITAL	91.81137	26.15931
47	Private Hospitals	GNRC HOSPITAL,DISPUR	91.79291	26.13971
48	Private Hospitals	GOODHEALTH HOSPITAL,DISPUR	91.79652	26.14142
49	Private Hospitals	INTERNATIONAL HOSPITAL, G.S. ROAD	91.77996	26.15379
50	Private Hospitals	DOWNTOEN HOSPITAL	91.79995	26.13818
51	Private Hospitals	DISPUR HOSPITAL,GANESHGURI	91.78457	26.15143

(continued)

(continued)

S N	Type	Name	Longitude	Latitude
52	Private Hospitals	DISPUR POLYCLINIC,GANESHGURI	91.78668	26.14962
53	Private Hospitals	NARAYANA SUPERSPECIALITY HOSPITAL	91.67766	26.20737
54	Private Hospitals	SANJEEVANI HOSPITAL	91.70941	26.16125
55	Private Hospitals	SWAGAT HOSPITAL	91.70744	26.16029
56	Private Hospitals	VISION HOSPITAL	91.74576	26.18017
57	Private Hospitals	SATRIBARI CHRISTIAN HOSPITAL	91.74613	26.17698
58	Private Hospitals	ARUNA MEMORIAL HOSPITAL	91.76901	26.16603
59	Private Hospitals	NAMECARE HOSPITAL	91.76841	26.16373
60	Private Hospitals	KGMT HOSPITAL	91.78154	26.17235
61	Private Hospitals	PRAGJYOTI EYE CARE CENTER	91.78421	26.17293
62	Private Hospitals	NIGHTINGALE HOSPITAL	91.78217	26.15036
63	Private Hospitals	HAYAT HOSPITAL	91.74668	26.13918
64	Private Hospitals	EXCEL CARE HOSPITAL	91.68084	26.12603
65	Private Hospitals	AYURSUNDRA SUPERSPECIALITY HOSPITAL	91.71935	26.10893
66	Private Hospitals	CRITICAL CARE HOSPITAL	91.75225	26.11085
67	Private Hospitals	GLOBAL HOSPITAL OF SURGERY	91.78051	26.12886
68	Private Hospitals	HEALTH CITY HOSPITAL	91.81529	26.11708
69	Private Hospitals	ESIC MODEL HOSPITAL	91.80898	26.12182
70	Private Hospitals	AGILE HOSPITAL	91.80598	26.12220
71	Private Hospitals	RAHMAN HOSPITAL	91.80948	26.13507
72	Private Hospitals	SUN VALLEY HOSPITAL	91.80107	26.13823
73	Private Hospitals	ASG EYE HOSPITAL	91.80047	26.13817
74	Sub Centre	KALANGPUR	92.04024	26.04151
75	Sub Centre	BANDARGOG	91.99778	26.05000
76	Sub Centre	MARKANG	91.98779	26.04191
77	Sub Centre	TOPATULI	92.16450	26.10684
78	Sub Centre	TALONI	92.10857	26.10196
79	Sub Centre	GHAGUA	92.14193	26.11804
80	Sub Centre	MURKATA	92.13650	26.11742
81	Sub Centre	DURUNG	92.13573	26.13331

(continued)

(continued)

S N	Type	Name	Longitude	Latitude
82	Sub Centre	KAMARKUCHI	92.14105	26.14712
83	Sub Centre	AUJARI	92.15663	26.16685
84	Sub Centre	PUB-MALOIBARI	92.12367	26.17798
85	Sub Centre	DAPATA TINIALI	92.06896	26.16980
86	Sub Centre	MITONI	92.06635	26.18501
87	Sub Centre	BAMANIJURABAT	92.02914	26.14266
88	Sub Centre	TETELIA	92.03799	26.13227
89	Sub Centre	GUMARIA	92.00563	26.16459
90	Sub Centre	MALIBARI	92.00315	26.18414
91	Sub Centre	KASUTALI	91.99320	26.16764
92	Sub Centre	NARTAP	91.97403	26.14571
93	Sub Centre	BARKUCHI	91.98428	26.09448
94	Sub Centre	BAGIBARI	92.05577	26.19560
95	Sub Centre	KALANGPAR	92.01456	26.19974
96	Sub Centre	DHANKHUNDA	92.00297	26.20850
97	Sub Centre	BARBILA	91.99066	26.20817
98	Sub Centre	JUBAI	91.95743	26.16744
99	Sub Centre	BHAKUAGOG	91.93649	26.22742
100	Sub Centre	BONDA	91.90351	26.21914
101	Sub Centre	KAMALAJARI	91.92400	26.08399
102	Sub Centre	MARAGDOLA	91.89535	26.07138
103	Sub Centre	MAUPUR	91.90152	26.06650
104	Sub Centre	SARUTARI	91.87578	26.06108
105	Sub Centre	CHANDRAPUR FW	91.92205	26.11652
106	Sub Centre	AMSING,JORABAT	91.87996	26.15041
107	Sub Centre	PANIKHAITI	91.87462	26.21556
108	Sub Centre	GONESH NAGAR	91.79007	26.19822
109	Sub Centre	MATHGHARIA	91.80822	26.17870
110	Sub Centre	BORBARI	91.81484	26.15423
111	Sub Centre	GANESH NAGAR	91.79313	26.09264
112	Sub Centre	BETKUCHI	91.74856	26.11651
113	Sub Centre	DAKHIN GAON	91.75889	26.12807
114	Sub Centre	NARKASUR	91.76402	26.14364
115	Sub Centre	BIRUBARI	91.75160	26.15741
116	Sub Centre	FATASIL,AMBARI	91.72845	26.15782
117	Sub Centre	MOINAKHURUNG	91.70757	26.10496
118	Sub Centre	PUB-BORGAON	91.69058	26.11995

(continued)

(continued)

S N	Type	Name	Longitude	Latitude
119	Sub Centre	GOTANAGAR	91.68061	26.14196
120	Sub Centre	CHAKARDA	91.63793	26.10660
121	Sub Centre	DHARAPUR	91.62848	26.14026
122	Sub Centre	MIRZAPUR	91.60898	26.11631
123	Sub Centre	GARIGAON SADILAPUR	91.67276	26.16882
124	Sub Centre	AGSHIA	91.57366	26.12322
125	Sub Centre	MAZIRGAON	91.57850	26.13667
126	Sub Centre	APRIKULA	91.87834	26.03891
127	Sub Centre	PANBARI	92.14099	26.11957
128	Sub Centre	CHANDRAPUR	91.98950	26.09939
129	Sub Centre	HAJONGBARI	91.90819	26.22353
130	Sub Centre	MOLOIBARI	92.09887	26.17614
131	Sub Centre	KHARGHULI	91.75706	26.19651

References

Ahmed S, Ibrahim RF, et al (2017) GIS-based network analysis for the roads network of the greater Cairo area. In: Applied research in computer science and engineering ICAR'17. Lebanon, CEUR–WS.org, vol 2144, pp 1–9

Bagheri N, Benwell GL, et al (2005) Measuring spatial accessibility to primary health care. In: The 17th annual colloquium of the spatial information research centre university of Otago. Dunedin, New Zealand, pp 2–7

Dabhade A, Kale KV et al (2015) Network analysis for finding shortest path in hospital information system. IJARCSSE 5(7)

Dermatis Z, Tsoromokos D, Gozadinos F, Lazakidou A (2016) The utilization of geographic information systems in healthcare. Int J Health Research and Innov. 4:39–57

District Census Handbook, Kamrup Metropolitian, Assam, Series-19, Part XII-A, Census of India (2011)

Fradelos E, Papathanasiou I, Mitsi D, Tsaras K, Kleisiaris C, Kourkouta L (2014) Health based geographic information systems (GIS) and their applications. Acta Informatica Medica. 22:402–405. https://doi.org/10.5455/aim.2014.22.402-405

Hanchette CL (2003) Geographic information systems. In: O'Carroll PW, Ripp LH, Yasnoff WA, Ward ME, Martin EL (eds) Public health informatics and information systems. Health Informatics Springer, New York, pp 431–466. https://doi.org/10.1007/0-387-22745-8_21

Kharel S, Shivananda DP et al (2018) Transportation network model for route and closest facility analysis in central Bengaluru. IJAIEM 7(4):58–62

Kumar P, Kumar D (2016) Network analysis using GIS techniques: a case of Chandigarh city. IJSR 5(2):409–411

Mokgalaka H (2014) Measuring access to primary health care: use of a GIS-based accessibility analysis. In: Planning Africa 2014 conference, at international convention centre. Durban, South Africa, pp 2–26

Mokgalaka H (2015) GIS-based analysis of spatial accessibility: an approach to determine public primary healthcfare demand in metropolitan areas. In: Master's thesis University of Cape Town
Phule RR, et al (2015) Vulnerability mapping for disaster assessment using ArcGIS tools and techniques for Mumbai city, India. In: 16th ESRI-India user conference, pp 1–12
Sadiq A et al (2013) Geospatial Mapping of Health Facilities in Yola, Nigeria. IOSR-JESTFT 7(3):79–85
Masoodi M, Rahimzadeh M (2015) Measuring access to urban health services using geographical information system (GIS): A Case Study of Health Service Management in Bandar Abbas, Iran. Int J Health Policy Manage 4:439–445. https://doi.org/10.15171/ijhpm.2015.23
Murad A (2018) Using GIS for Determining variations in health access in jeddah city, Saudi Arabia. ISPRS Int J Geo-Inf 7(7):254. https://doi.org/10.3390/ijgi7070254

Chapter 13
Geo-Spatial Analysis of Health Care Service Centres for Smart Cities: A Study of South-East District, Delhi-India

Mohammad Tayyab, Babita Kumari, Shahfahad, Asif, Hoang Thi Hang, Safraj Shahul Hameed, and Atiqur Rahman

Abstract The provision of Health Care Service Centres (HCSCs) is one of the biggest challenges in developing countries like India. Proper and equitable distribution and provision of health care services is a major challenge in a country like India especially when we talk about concept like 'Smart cities'. The primary objective of this study was to analyze the distribution of health care facilities in making smart cities: a case study of SE Delhi. We need to pay special attention to spatial arrangement of the health care services to make city a smart city in a country like India. The design of the study is such that satellite data together with population data has been modelled to see the spatial distribution of health care services in GIS domain. The availability of satellite data e.g. IKONOS, Quick-bird, Geo-Eye, World-View I and II and even Google Earth in combination with Geographical Information System (GIS) has made added advantage in this context. Health care data were taken from Google Maps and Google Earth, population data from Census of India-2011 and other sources. GIS is used to assess, examine and map the distribution of health care services. Therefore, in this paper, in the light of the above discussion we have tried to assess the spatial distribution of health care services, to relate it with population distribution and if HCSCs are sufficient and equitably distributed or not using spatial data in GIS domain in South-East District of national capital of India, Delhi. As a part of this study, we geocoded 197 health care centres in the 36 ward of SE Delhi. Study shows that healthcare service centres are maximum in the wards where total population, density of population, household density, no. of households, no. of beds are less and moderate, while maximum wards have low spaces to build the healthcare facilities. Density of population and healthcare facilities are less in commercial and industrial area. Hospitals have more than 300 beds such as Indraprastha Apollo

---

M. Tayyab · B. Kumari · Shahfahad · Asif · H. T. Hang · A. Rahman (✉)
Department of Geography, Faculty of Natural Sciences, Jamia Millia Islamia, New Delhi, India

M. Tayyab
DDA, New Delhi, India

S. S. Hameed
CCDC, New Delhi, India

B. Kumari
GIS Analyst at GeoSpectrum Technologies Pvt. Ltd., Bangalore, India

P. Sharma (ed.), *Geospatial Technology and Smart Cities*, The Urban Book Series,
https://doi.org/10.1007/978-3-030-71945-6_13

are located nearby very dense populated area such as Okhla, Madanpurkhadar and Sarita Vihar and Batra Hospital & Medical Research Centre are located near Sangam Vihar, Badarpur and Jaipur. In order to make Delhi a Smart city, Governments and healthcare planners need to provide healthcare facilities in the wards where it is low or not available.

Keywords Health care service centres · Smart Cities · Satellite data · GIS · Delhi-India

Acronyms

GDP	Gross Domestic Product
GIS	Geographical Information System
HCSCs	Health Care Service Centres
HHs	Households
MCD	Municipal Corporation of Delhi
NCT	National Capital Territory
NDMC	New Delhi Municipal Council
SE	South East
UP	Uttar Pradesh
UTM	Universal Transverse Mercator
WHO	World Health Organization
WGS 84	World Geodetic System 1984

13.1 Introduction

World Health Assembly adopted the global goal of Health for all by the year 2000 way back in 1977 (WHO 1979). And later in 1998, the World Health Assembly adopted a revised strategy for the twenty-first century that continues to emphasize availability, accessibility and quality of care (WHO 1998). But even today forget about health, the provision of Health Care Service Centre (HCSC) is one of the biggest challenges in developing countries like India. Further, the accessibility of health care institutions is one of the most important factors in constituting healthy communities (Fatih and Egresi 2013). The degree of accessibility of health care institutions is one of the most significant indicators for measuring the efficiency of the health care system (Gatrell and Elliott 2014). The access of public to health care services could be seriously restricted by distance (Black et al. 2004). Longer distances may affect especially the access of children, elderly and physically impaired people to health care. In general, longer the distance to health care services higher the risk of fatalities (Hare and Barcus 2007). The study of accessibility to health care has long been of interest to public health researchers, medical geographers and other social scientists (Joseph and Bantock 1984). Studies on health care accessibility have been done in the more developed countries (Lin et al. 2002; Kalogirou and Foley 2006; Pearce et al. 2006; Ohta et al. 2007) and also in the developing countries too (Okafor 1990; Perry and Gesler 2000; Murad 2004). But still there is hardly any study to assess the spatial distribution of health care centres in urban areas in India.

There is a mass migration of population to cities in 2014, more than half of the World (54%) has become urban and about 31% in India (Census of India 2011). Globally, one million people are moving into cities every week (UK Trade and Investment 2015). In India alone, every minute, 30 country dwellers move permanently to a city. This has led to an active demand for 'Smart City' approaches to solve challenges of urbanization, which if not met will threaten to strangle city life. The limitations include choking traffic, generating pollution, which we witness these days in all big Indian cities, and poor health services, etc. The utilization levels and patterns of health care facilities indicate the awareness and attitude of people towards their health (Prakasam 1995). Education, economy, sex and social status, etc. are major influencing factors for utilization of health care facilities in India. An educated person is more careful about his health than a non-educated. The geography of health care comprises the analysis of spatial organization (number, sizes, types and locations) of health care services, how and why spatial organization changes over time, how people can get access to health care services and the impacts on health (Fortney et al. 1999).

The World Health Organization (WHO) included the effects of urbanization and associated living conditions in their conceptual framework as one of nine overarching social determinants integral to health and well being (Solar and Irwin 2010). This followed the concept of 'Healthy Cities' initiative launched by the WHO in recognition in 1986 of the fact that cities themselves are 'habitats' whose conditions can be meliorated with the goal of health promotion (Ashton et al. 1986; Leeuw et al. 2014). On similar lines in recent days, the concept of making 100 smart cities in India has gained momentum and today its buzz word. What is the most important in this context a city cannot be called a smart city merely by pumping money and accelerating economic growth? But two most important components would be healthy people and healthy cities and then it can become Smart city. It is strongly believed that for the healthy people we need to have good health care services and how they distributed in a space. So there is a need to assess how best people are being served with these kinds of services which be reflected in the spatial arrangement and distribution.

Proper and equitable distribution and provision of health care services is a major challenge in a country like India especially when we talk about the Smart cities. Health care service providers are opening and closing, new forms of health care services are coming up, and the persistently high costs of health care services are raising concerns about quality, effectiveness, and access especially (Rai and Nathawat 2014) in the big cities like Delhi. Utilization of health care services is a complex phenomenon which, on the hand, is influenced by the perception by an individual of the need for services thereby, promoting him to take a decision to utilize them and, on the other hand, by the availability, accessibility and organizational aspects of health services itself (Murali 1981). Besides, successful utilization of health care services depends on reliability, awareness, motivation and on the perception of the people about the services and the need about a particular health care service (Rai and Nathawat 2014).

It is well-known fact that Cities have become the engines of economic growth in all parts of the World and the same are true for India as well. Today about 31% of India's population lives in urban areas and contributes about 63% to the India's GDP (Census of India 2011). With increasing urbanization, growth of urban areas it is expected that about 40% of India's population will live in urban areas and may contribute 75% of India's GDP by 2030. For this, there is a need for precise and comprehensive development of physical, economic institutional, social and the most important health infrastructure. Therefore, Ministry of Urban Development, Govt. of India has started the plan to develop 100 Smart Cities India. Now the big question is that do we really need smart cities that will solve our problem? Or first we need to address the major problems of the existing cities so that the cities may look like cities and then can think of smart cities. If we straight away jump to develop smart cities then we need to address the health and health care facilities seriously without that we cannot make Indian cities as smart cities.

In the draft document of Smart Cities: Mission Transformation; health is the key element in it (Ministry of Urban Development, Govt. of India 2015). Therefore, to make city a Smart city, we need to pay special attention to spatial arrangement of the health care services in India. During the last two decades or so public health scientist, Geographers and other researchers have started using GIS to assess and examine the distribution of health care services. Later on in recent time the availability of satellite data e.g. IKONOS, Quick-bird, Geo-Eye, World-View I and II and even Google Earth in combination with GIS has made added advantage in this context. Therefore, in this paper, in the light of the above discussion we have tried to assess the spatial distribution health care services centre, to relate it with various population and its related characteristics and if HCSC are sufficient and equitably distributed or not using spatial data in GIS domain in South-East District of national capital of India, Delhi.

South-East district Delhi is one of the eleven districts of National Capital Territory (NCT) of Delhi which came into existence in September 2012 when two new districts were created by altering and modifying the limits of the sub-divisions. The SE district Delhi occupies an approx. area of 44.34 km^2 which is 3% of NCT of Delhi and has a population of 19, 34,359 as per Census of India, 2011. South East district will have Defence Colony, Kalkaji and Sarita Vihar as sub-divisions (www.mapofindia.com). It is surrounded by the states of Uttar Pradesh (UP) and Haryana. On its southern side lie the Gurgaon and Faridabad districts of Haryana and on South Eastern side is located the Gautam Budh Nagar district of UP. South district on its eastern side is flanked by South and East Districts of Delhi while on northern side lies New Delhi district (Fig. 13.1).

Numbers of agencies are being provided health care centres in NCT of Delhi such as Govt. of NCT of Delhi, local bodies, MCD, NDMC, Delhi Cantonment Board,

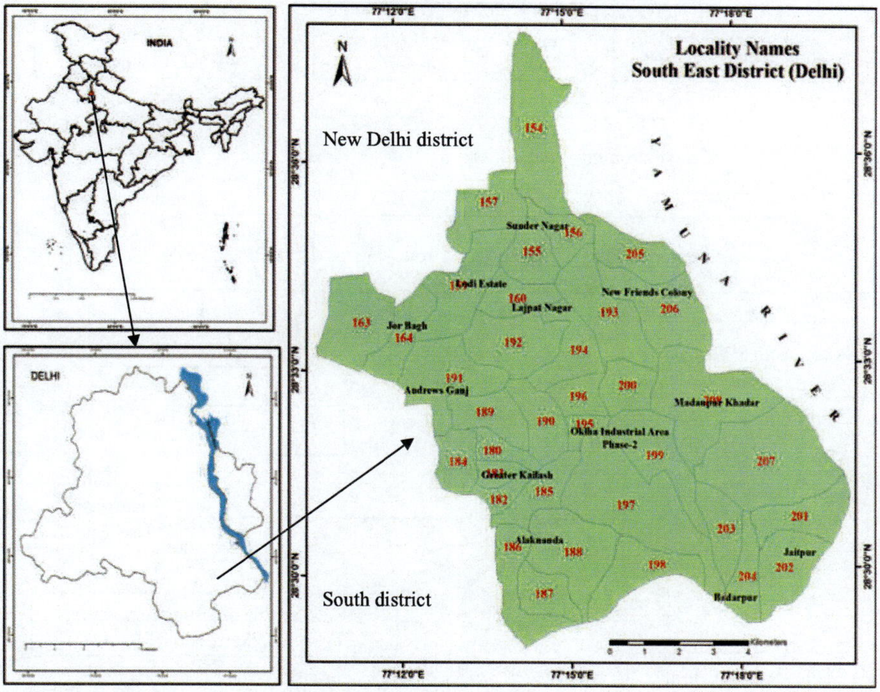

Fig. 13.1 Locational aspect of the study area

Ministry of Health and Family Welfare so on. Public sector is also contributing to provision of health care centres in Delhi. There are 509 primary and 38 secondary and tertiary health care services being provided by Allopathic, Ayurvedic & Unaniand Homeopathic hospitals by Govt. of Delhi (Medical & Public Health, Govt. of Delhi).

13.2 Data Base and Methods

The satellite-based spatial data for the analysis of health care is important in making and designing the smart cities in India. Further, these spatial data can be used in Geographical Information System (GIS) as a tool to analyze the spatial relationship, pattern and distribution of the health care services. Therefore, in this study, we used GIS technique for the analysis of healthcare centres in SE district Delhi. This study used both spatial and non-spatial data, (i) spatial data, i.e. ward/district maps of South East, Delhi (SE, Delhi) from www.mapofIndia.com, non-spatial data from Census of India 2011 and (iii) healthcare data captured latitude/longitude of health care service centres by using Google Maps/Earth and primary survey and after that these data were converted into shape files (Fig. 13.2). The ward/district maps of SE Delhi

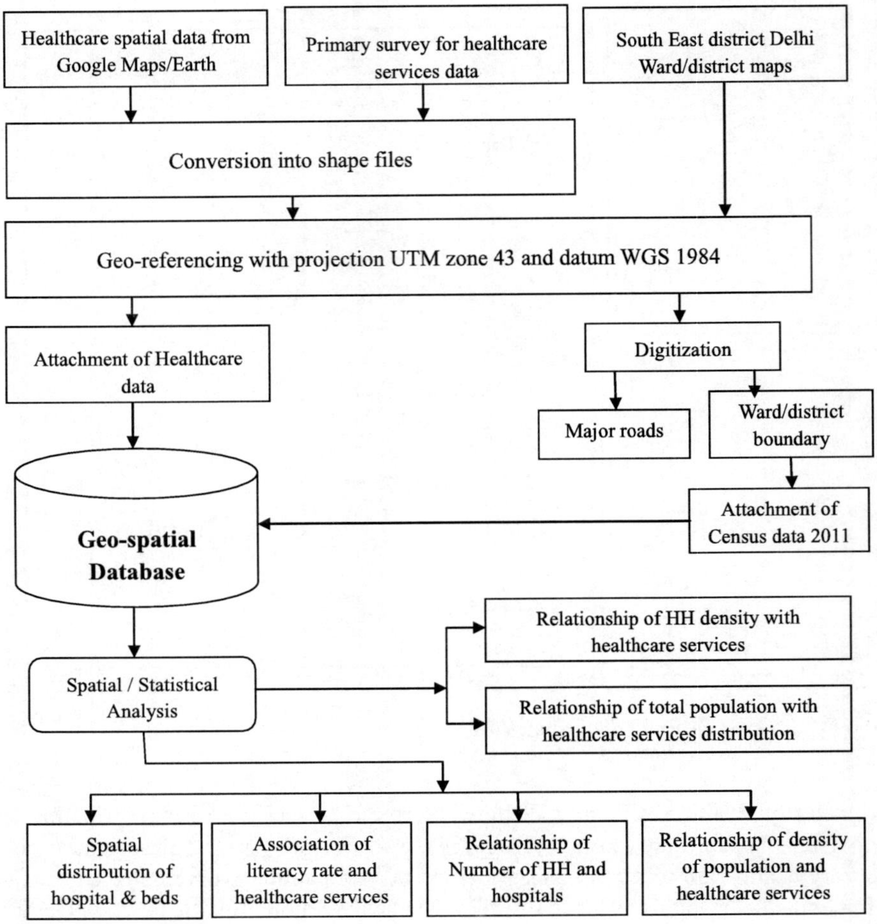

Fig. 13.2 Flow chart of the methodology

were geo-referenced with projection UTM Zone 43N and Datum WGS 84. The geo-referenced maps were used to digitize the district and ward boundary of the study area. Total population and Households (HHs) data have been taken Census of India and then population density and HHs density were calculated using these data using the ward maps of the SE district in the GIS environment. Healthcare service data like number of beds was linked with spatial data of health care centres like hospitals which are generated in point shape file. Thereafter, Census data of 2011 was attached with ward/district boundary of SE Delhi. This is done to draw a relationship among various attributes the detailed methodology is given in Fig. 13.2.

13.3 Results and Discussion

Providing smart and affordable health care services is an increasingly difficult challenge in city like Delhi that is may be due to the complexities of health care services, costs, quality, accessibility and delivery making decisions about health-related issues (Hughes 2008). There are about 197 healthcare service centres which include hospital, nursing home, clinic, and lab of various sizes in SE district of Delhi (Fig. 13.3). Now the big question is that if these service centres are sufficient enough to meet out the demand and need of the population in the present time or not? Therefore, in order to assess this spatial analysis has been done using Census of India data of 2011 and Google Earth satellite data under the following heads.

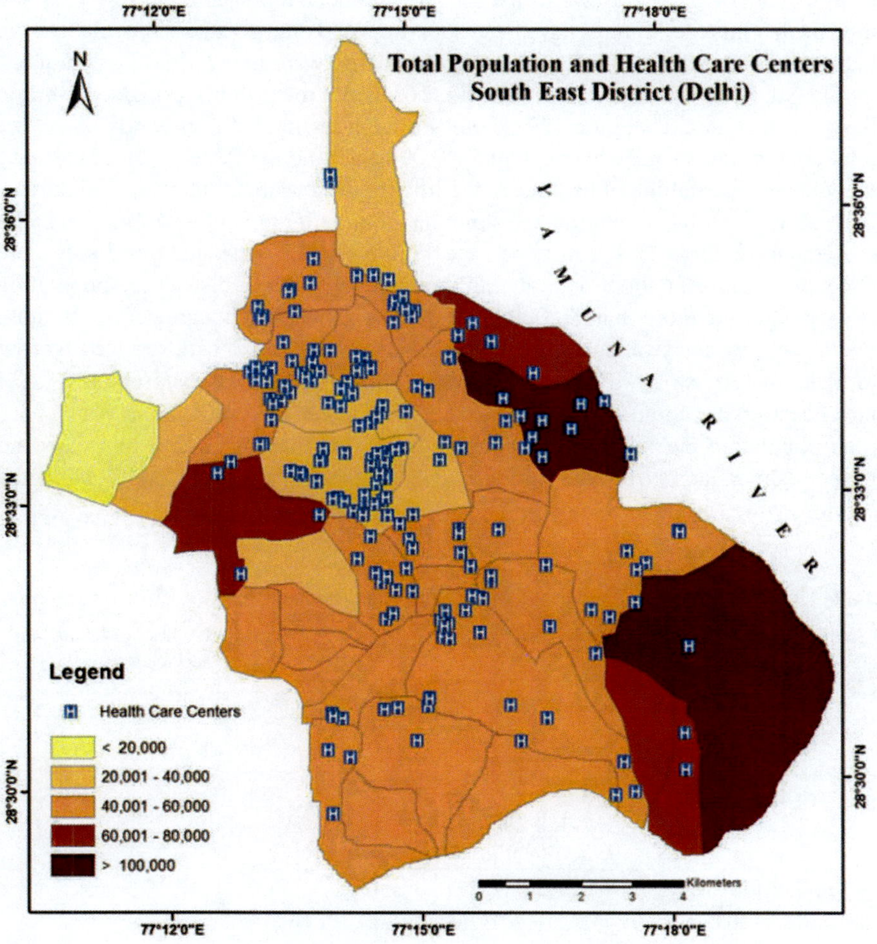

Fig. 13.3 Spatial distribution of population and Health care centres, South-East district (Delhi)

13.3.1 Total Population and Heath Care Service Centres

Health care service centres includes hospitals both public and private, individual clinics that are being run by qualified and non-qualified doctors in lanes and by lanes of the cities in India, personal, i.e. number of nurses and physicians and lastly technology that creates the capacity to provide health services to the people. The accreditation of health care facilities, e.g. hospitals, nursing homes is the review of the adequacy of structural characteristics, including staffing, on-call doctors, technology, and support services, i.e. laboratory, pharmacy, radiology etc. (Hughes 2008). In this context, it has been tried to assess the total number of hospital/healthcare service centres with respect to the total population in SE Delhi. Total population has been classified into four groups, i.e. <20,000, 20,001 to 40,000 and so on using Census of India data of 2011 (Table 13.1). The same has been represented in Fig. 13.3 and then the health care services have been superimposed on it so assess the relationship between population and the hospitals in a particular ward using of GIS technique.

The study shows that in the ward no. 163 is the most thinly populated and do have any hospital. There are 120 healthcare facilities in about 20 wards where the total population is somewhere between 40,000 and 60,000 (Table 13.1). This shows that uneven distribution of healthcare facilities in these wards. In terms of healthcare centres and the total population, the situation looks better in South-East district Delhi if we analyze Table 13.1. But once we examine Fig. 13.3 the situation looks quite different because in about 5 wards in the eastern side of SE district the population is very high, i.e. more than 80,000 but there is only one healthcare centre. In these wards, people face problems in terms of getting proper health care service. In order to make the city smart, the planners need to look into such issues/problems, i.e. if new hospitals are come up then it has to come up in these potential areas which have high population but fewer healthcare services. This map can be an important aid to the planners and government official responsible for providing smart healthcare services which an important component of the smart cities.

Table 13.1 Total population and health care centres in SE district, Delhi

S. no.	Total population	Wards	Number of wards	Number of healthcare service centres
1	<20,000	163	1	0
2	20,001–40,000	164, 154, 189, 194, 192, 160	6	57
3	40,001–60,000	155–157, 159, 180, 182–188, 190, 193 & 195–200	21	120
4	60,001–80,000	204, 203, 191, 205	4	10
5	>80,000	202, 201, 207, 206	4	10
	Total		36	197

13.3.2 Population Density and Health Care Service Centres

In densely populated areas the feasibility to deliver healthcare service centres should be more than the areas with dispersed population. Some study shows that population density positively affects coverage rates of health care services. The study shows that there are about 100 healthcare centres in 17 wards where density of population is 20,000–30,000 and 83 health care centres are in 12 wards where population density is between 10,000 and 20,000 (Table 13.2). Figure 13.4 shows that most of the healthcare service centres are concentrated in the central part of the district where

Table 13.2 Population density and health care service centres in SE district, Delhi

S. no.	Population density	Wards	Number of wards	Number of healthcare service centres
1	<10,000	163	1	0
2	10,000–20,000	154, 155, 157, 160, 164, 182, 184, 189, 192, 194, 199, 200	12	83
3	20,000–30,000	156, 159, 180, 181, 183, 185–188, 190, 191, 193, 195–198, 204, 208	17	100
4	30,000–40,000	202, 203, 205 and 206	4	13
5	>40,000	201 and 207	2	1
	Total		36	197

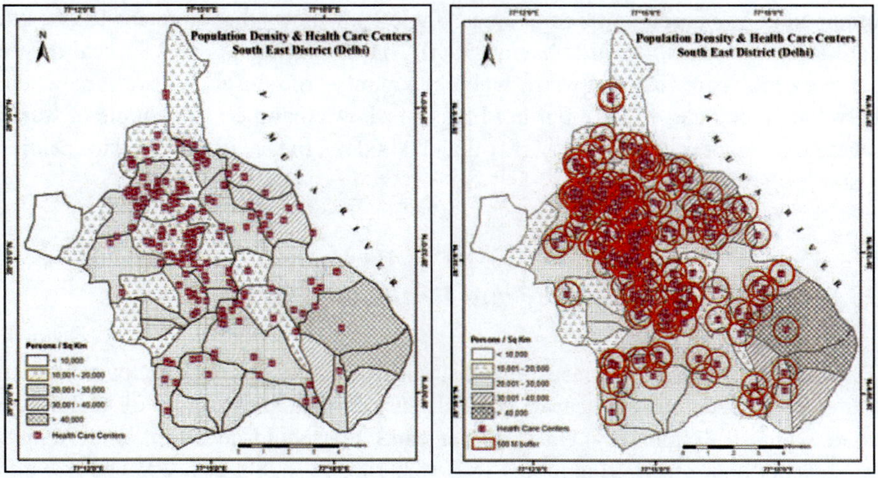

(a) (b)

Fig. 13.4 Spatial pattern of population density and healthcare centres, South-East district (Delhi)

Table 13.3 Households and health care centres in SE district, Delhi

S. no.	Number of households	Wards	Number of wards	Number of healthcare service centres
1	<5000	163	1	0
2	5000–10,000	156 & 159 & 180 & 185 & 186 & 188 & 190 & 193 & 195–198 & 203–205 & 208	16	102
3	10,000–15,000	154 & 155 &157 & 160 & 164 182–184 & 187 & 189 & 192 & 194 & 199 & 200	14	83
4	15,000–20,000	191 & 202 & 206	3	11
5	>20,000	201 & 207	2	1
	Total		36	197

mostly low to moderate population density is seen. This is mainly because in these areas (Lajpat Nagar & New Friends colony, etc.) mostly high to very high people live therefore; good hospitals are located in these wards.

13.3.3 Households and Health Care Service Centres

In this part of the paper it has been tried to assess the number of Households (HHs) and health care service facilities. Access to health care facilities within a ward is closely related to the population and the available facility which is normally more in urban areas compared to the rural areas. Table 13.3 shows that there are 102 health care service centres in wards having 5000–10,000 households and 83 health care service centres are found in wards where the number of households are somewhere between 10,000 and 15,000. Further Fig. 13.5 shows that there are couple of wards where the numbers of HHs are higher than 20,000 but the health care service centres just a few.

13.3.4 Household Density and Healthcare Service Centres

The study shows that the household density is less than 4 persons per house in ward number 190 because it is near to Okhla Industrial area and number of health care centres are 10 (Table 13.4).On the other hand, household density is more than 5 persons per house in ward numbers 182, 183, 186, 187, 188, 201, 205 and 206 and number of health care centres are 22. This shows that in several wards the household density is more than 5 persons per household, the numbers of hospitals are less but

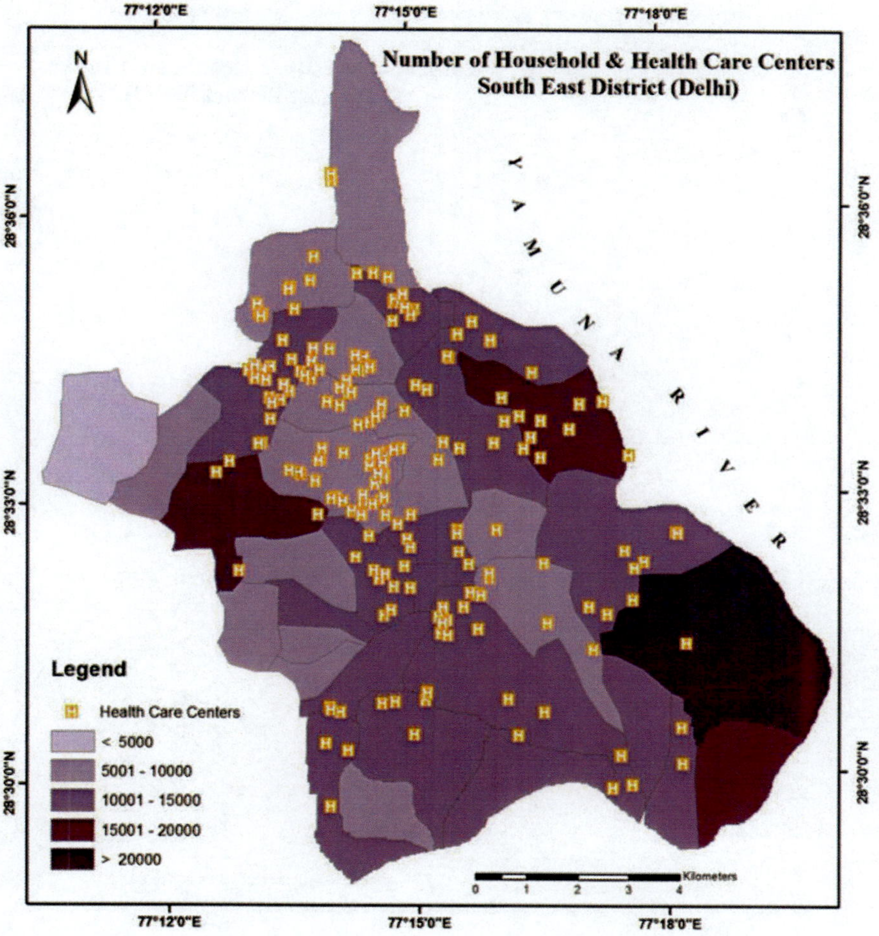

Fig. 13.5 Spatial distribution of households and healthcare centres, South-East district (Delhi)

Table 13.4 Household density and healthcare in SE district, Delhi

S. no.	Household density	Wards	Number of wards	Number of healthcare service centres
1	Less than 4.0	190	1	10
2	4.0–5.0	154–157, 159, 160, 163, 164, 180, 184, 185, 189, 200, 202–204, 207 & 208	27	165
3	More than 5.0	183, 182, 186, 187,188, 201, 206 & 205	8	22
	Total		36	197

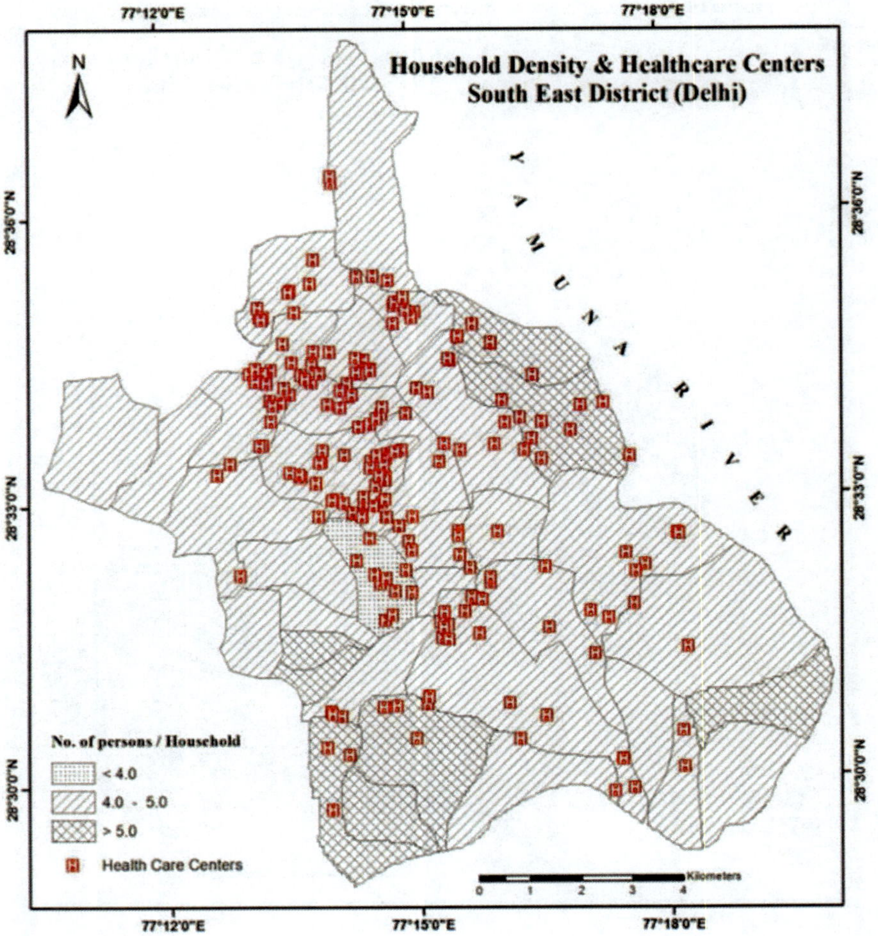

Fig. 13.6 Spatial pattern of household density and healthcare centres, South-East district (Delhi)

it should be more in a smart city. Figure 13.6 shows that most of the health care facilities are located in the central and NE part of the study area and outer wards have lesser rather do not have a health care facility that means people have to move to a certain distance even to get any health care services.

13.3.5 Literacy Rate and Healthcare Service Centres

An attempt has been made to relate the literacy rate with health care centres in SE district of Delhi because literacy is one of the key socio-economic factors which have direct relationship with the health and health care services. People who are highly

Table 13.5 Literacy rate and healthcare centres in SE district, Delhi

S. no.	Literacy rate (%)	Wards	Number of wards	Number of healthcare service centres
1	<70	188,197, 200 & 199, 203, 207	6	32
2	70.1–75	201, 198, 187, 186, 185, 180, 195, 193 & 205	9	20
3	75.1–80	182, 183, 204, 202, 208, 206, 155 & 154	8	32
4	80.1–85	163 & 184 & 191 & 196 & 160 & 157 & 156	7	40
5	>85	189 & 190 & 194 & 192 &164 & 159	6	73
	Total		36	197

literate there are more chances that they will be more concerned about their health and health-related issue. Data of 2011 on literacy rate taken from Census of India has been classified into five classes, i.e. <70%, 70.1–75% and so on (Table 13.5). It is seen that that only 6 ward has less than 70% literacy rate and number of health care centres are 32 and literacy rate more than 85% in about 7 wards and healthcare facilities are 73. Figure 13.7 shows that in the North-West part of the study area the literacy rate is quite high, i.e. more than 75% and in these areas there is high concentration of healthcare service facilities.

13.3.6 Spatial Distribution of Hospitals and Hospital Beds

The spatial distribution of hospitals and the services being offered, specialized doctors and the number of beds are some of the key features of any health care service centres. Of all these number of bed in a hospital is one of the most important features. This is because in the last few years it has been seen that during the summer and rainy seasons there is an outbreak of various diseases like dengue and swine flu, etc. and most of the hospitals do not have adequate beds to admit the patients and many deaths have been reported in such situation. Taking into account this aspect we tried to assess the number of beds per hospital. The study shows that out of 156 hospitals, 146 hospitals have less than 50 beds in most of the wards of study area (Table 13.6). While only 2 multi-specialist hospitals have more than 300 beds such as Indraprastha Apollo located close to very densely populated areas such as Okhla, Madanpur khaddar and Sarita vihar, etc. ward no. 208. The second is the Batra Hospital & Medical Research Centre which is again located in the thick populated areas like Sangam Vihar, Badarpur and Jaitpur, i.e. ward 188 (Fig. 13.8).

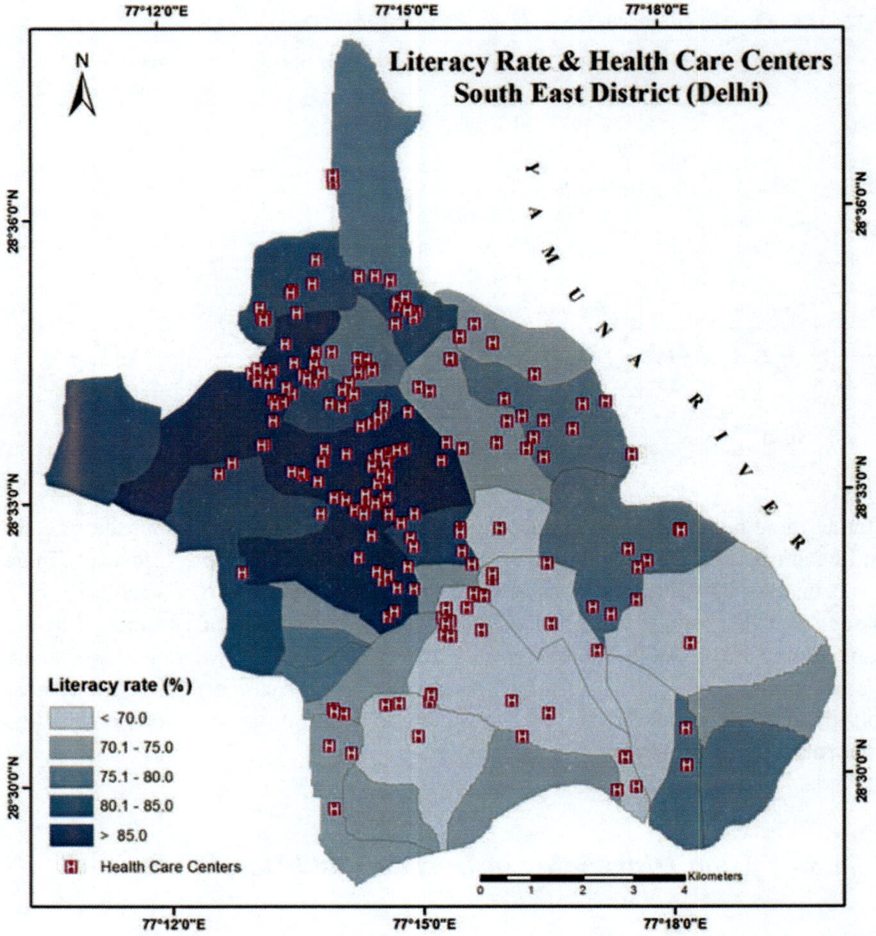

Fig. 13.7 Spatial pattern of literacy rate & healthcare centres, South-East district (Delhi)

Table 13.6 Spatial distribution of hospitals and hospital beds in SE district, Delhi

S. no.	Number of beds	Total number of hospitals	Ward numbers	Number of wards
1	<50	146	154–157, 159, 185, 186, 190, 194–196,198–200, 203, 204 & 207	17
2	51–100	1	193	1
3	101–200	3	197, 192 & 205	3
4	201–300	4	206, 160 & 191	3
5	>300	2	208 & 188	2
	Total	156		26[a]

[a]In10 wards do not have any hospital with beds

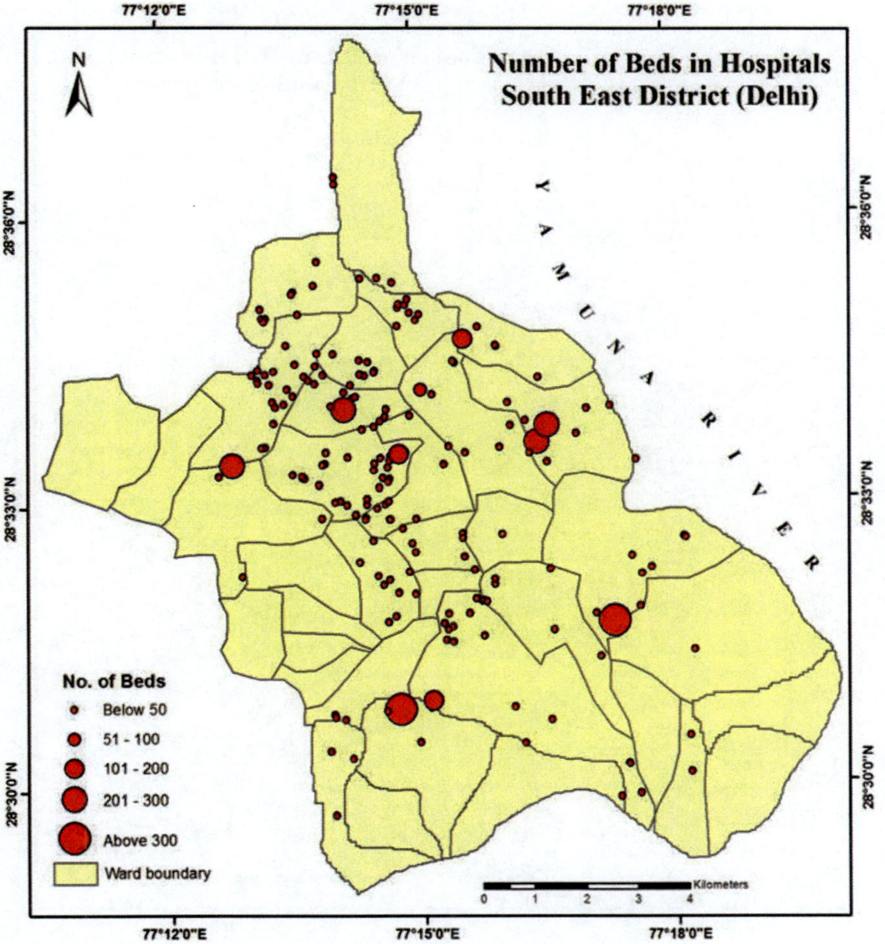

Fig. 13.8 Distribution of hospital beds, South-East district (Delhi)

13.3.7 Population Density and Ranking of Health Care Service Centres

In order to assess relationship between population densities with the hospitals, it has been tried to rank the health care centres, i.e. from largest and best hospitals based on the service which they provide. They are ranked 1 shown with big circles (Fig. 13.9) to the lowest which is given rank 5. The study shows that the low ranked hospitals which have fewer services are mostly in the high-density wards (Table 13.7) and vice versa. Ideally, it should not be like that either in urban or rural areas. This is more important in the context of smart cities planning process.

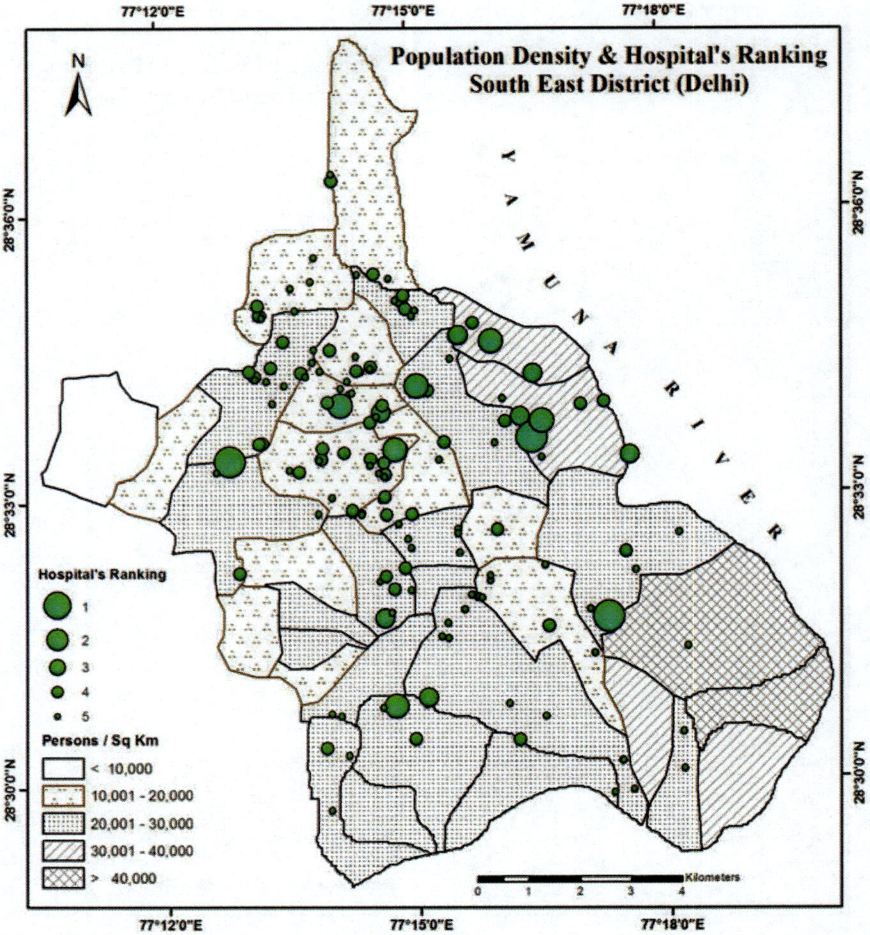

Fig. 13.9 Spatial pattern of population density & hospital's ranking, South-East district (Delhi)

An attempt is made to establish a relationship between various explanatory variables like total population, population density, no. of households, household density, literacy rate and no. of beds and spatial variability in health care service centres. Out of all these explanatory variables only household density and literacy come out as the significant explanatory variables in the linear regression models. Since analysis of variance return F value of 8.205 with 1 and 34 degrees of freedom which is significant at 0.007, it means the model is a good fit but R^2 value shows that only 19.4% spatial variability in health care centres is explained by household density at maximum and 17.1% spatial variable is explained by this model at minimum as indicated by adjusted R^2. Low household density indicates a population with small size of family; it means that household density signifies posh colonies (Fig. 13.10). In these areas income level is expected to be high, therefore, healthcare services

Table 13.7 Population density and ranking of health care service centres

S. no.	Population density	Ward numbers	Hospital's ranking	Ward numbers
1	<10,000	163	1	191, 206 & 208
2	10,000–20,000	154, 155, 157, 160, 164, 182, 184, 189, 192, 194, 199 & 200	2	160, 188, 192, 193, 205 & 206
3	20,000–30,000	156, 159, 180, 181, 183, 185–188, 190, 191, 193, 195–198, 204 & 208	3	160, 190, 197, 205 & 206
4	30,000–40,000	202, 203, 205 & 206	4	154–157, 159, 160, 188, 186, 190–194, 198–200, 205, 206 & 208
5	>40,000	201 & 207	5	154–157, 159, 160, 185, 186, 188, 190–194, 197, 198, 199, 204, 206, 207 & 208

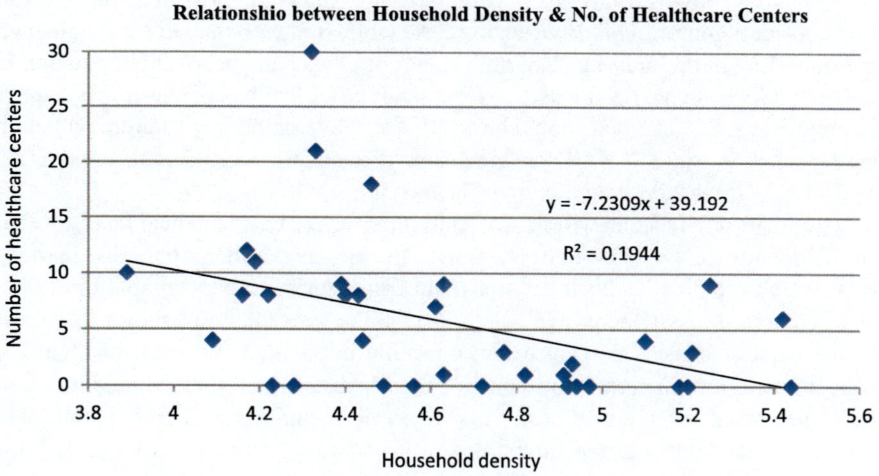

Fig. 13.10 Relationship between household density and number of healthcare centres

are concentrated there. However, it is in contrast to general principle of location of services which tend to concentrate in high-density areas.

In the case of literacy rate, the linear regression model fits well at a significance level of 0.031 and literacy rate explains at maximum 13.0 (R^2) % variation in distribution of healthcare services and at minimum 10.4 (adjusted R^2) % variability (Fig. 13.11). Literacy rate is high everywhere in Delhi. However, a higher literacy

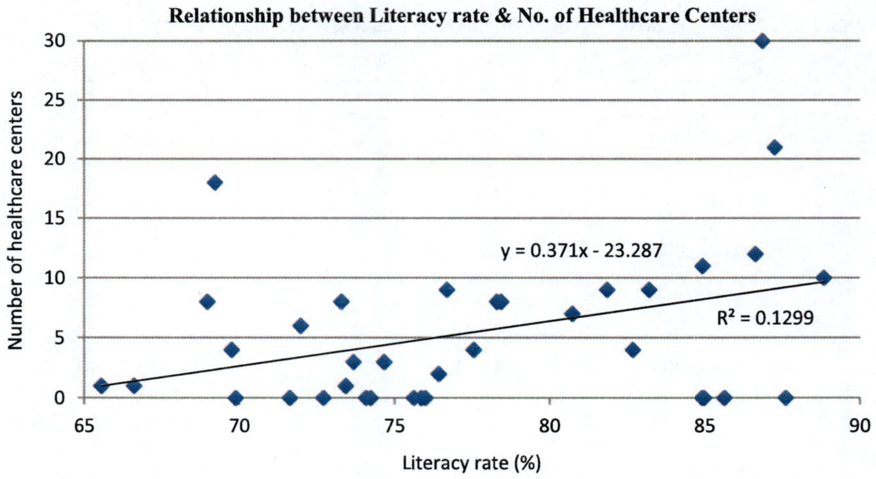

Fig. 13.11 Relationship between literacy rate and number of healthcare centres

rate in this scenario represents high educational level of the population which is more health conscious.

Table 13.8 shows various health parameters for Smart Cities that should be taken into consideration in India. If we analyze the table, it is seen that all the parameters are quite low in the present scenario, i.e. no. of physicians per 1000 population is only 0.7 which should be at least 2.0 for a smart city. Health expenditure per capital in India is just \$61 and that should be \$120. The Government expenditure on health care services is only 1% of GDP but that should be no less than 5% of the total GDP. This clearly shows the grim picture of health care services.

The smart city in India will first need to improve the basic medical facilities that should ideally be available in every part of the city as per the population distribution, which requires health institutions and dispensaries to be available to all and at affordable prices (Reddy and Singh 2015). The present study shows that most of the health care service centres are unevenly distributed and most of them are concentrated only in centre and north of the SE district and there should be more rationale spatial arrangement of health care service centres in a smart city. In Indian condition, the health-related infrastructure in the smart city should be ideally for

Table 13.8 Smart health parameters for smart cities in India (2011–2012)

S. no.	Parameters	Current scenario	Smart city measures
1	No. of physicians/1000 population	0.7	2.0–2.5
2	No. of hospitals beds/1000 population	0.7	2.0–2.5
3	No. of nurses and midwives/1000 population	1.7	3.0
4	Health expenditure per capital	\$61	\$120

Source Reddy and Singh (2015)

every 1,00,000 of population at least one nursing home, child, welfare and maternity centre of about 25 to 30 beds one multi-speciality hospital of 200 beds; one general hospital of about 500 beds per 1,00,000 of population. Similarly, there should be an intermediate hospital of 200 beds per 1,00,000 of population, one dispensary per 15,000 residents and one diagnostic centre per 50,000 residents. Also, one veterinary hospital per 50,000 residents and the most important is the 30 minutes emergency response time. The existing situation, i.e. number of physicians per 1000 population is only 0.7 that should be at least 2.0 for a smart city. Health expenditure per capital in India is just $61 and that should be $120. The Government expenditure on health care services is only 1% of GDP but that should be no less than 5% of the total GDP. This clearly shows the grim picture of health care services in India. It seems that of all above standard parameters set for Smart Cities, in S E Delhi most of the areas it's quite low; therefore, it is the need of the hours that proper health care service should be provided to all the resident population in all cities across India rather than making smart cities.

13.4 Conclusion

The aim of this study was to carry out spatial analysis of healthcare Service Centres in the context of the Government of India's 'Smart Cities ' concept. We explored the use of satellite data and Google Earth in combination with GIS to carry out the Analysis. Our experience shows that GIS together with satellite data is a very useful tool in this regard and can help planners in designing truly smart and healthy Cities. In this study area, we related population and demographic characteristics with provision of Health Care Service Centres (HCSCs). Our study shows low availability of health care facilities in the study area. Resources available are not enough to meet out the demand of the population. There is mismatch between availability and need; also the spatial distribution of health care service centres is not uniform. Government needs to consider and include health as a major component of Smart Cities. The study shows that there will be a long way in terms of providing better Health Care Service Centres in a city to be called as smart city because in the present scenario most of the health parameters are quite low than the standard set for the smart cities.

Statements Section The author(s) affirm that there is no ethical approval involved in the study, also there no funding is received for the study and we do not have any competing interests amongst us.

References

Ashton J, Grey P, Barnard K (1986) Healthy cities—WHO's new public health initiative. Health Promot Int 1(3):319–324

Black M, Ebener S, Aguilar PN, Vidaurre M, El Morjani Z (2004) Using GIS to measure physical accessibility to health care. World Health Organ 3–4

De Leeuw E, Tsouros AD, Dyakova M, Green G (2014) Healthy cities, promoting health and equity-evidence for local policy and practice. WHO Regional Office for Europe

Fatih KA, Egresi IO (2013) Accessibility of health care institutions: a case study by using GIS. Int J 3(4):2305–2493

Fortney J, Rost K, Zhang M, Warren J (1999) The impact of geographic accessibility on the intensity and quality of depression treatment. Med Care 37(9):884–893

Gatrell AC, Elliott SJ (2014) Geographies of health: an introduction. Wiley

Hare TS, Barcus HR (2007) Geographical accessibility and Kentucky's heart-related hospital services. Appl Geogr 27(3–4):181–205

Hughes R (ed) (2008) Patient safety and quality: an evidence-based handbook for nurses. Rockville, MD, Agency for Healthcare Research and Quality

Joseph AE, Bantock PR (1984) Rural accessibility of general practitioners: the case of Bruce and Grey counties, Ontario, 1901–1981. Can Geogr/Le Géographecanadien. 28(3):226–239

Kalogirou S, Foley R (2006) Health, Place and Hanly: modelling accessibility to hospitals in Ireland. Ir Geogr 39(1):52–68

Lin G, Allan DE, Penning MJ (2002) Examining distance effects on hospitalizations using GIS: a study of three health regions in British Columbia. Can Environ Planning 34(11):2037–2053

Ministry of Urban Development (2015) Smart cities: mission transformation; mission statement & guidelines, ministry of urban development, Govt. of India, June 2015, https://smartcities.gov.in/

Murad AA (2004) Creating a GIS application for local health care planning in Saudi Arabia. Int J Environ Health Res 14(3):185–199

Murali I (1981) Pattern of utilization of health care facilities by a selected urban community in New Delhi. Indian J Prev Soc Med 12(2):117–125

Ohta K, Kobashi G, Takano S, Kagaya S, Yamada H, Minakami H, Yamamura E (2007) Analysis of the geographical accessibility of neurosurgical emergency hospitals in Sapporo city using GIS and AHP. Int J Geogr Inf Sci 21(6):687–698

Okafor FC (1990) The spatial dimensions of accessibility to general hospitals in rural Nigeria. Socioecon Planning Sci 24(4):295–306

Pearce J, Witten K, Bartie P (2006) Neighbourhoods and health: a GIS approach to measuring community resource accessibility. J Epidemiol Community Health 60(5):389–395

Perry B, Gesler W (2000) Physical access to primary health care in Andean Bolivia. Soc Sci Med 50(9):1177–1188

Prakasam CP (1995) Utilization of primary health care services in a rural setup. Health Popul Perspect Issues 18(2):72–80

Rai PK, Nathawat MS (2014) Utilization of health care services in Varanasi District: a geographical analysis, GEOGRAFIA. Malaysian J Soc Space (Online). 10(2):14–33

Reddy GRK, Singh SP (2015) Smart and Human: building cities of wisdom. Hoper Collins Publishers India

Registrar GI (2011) Census of India 2011: provisional population totals-India data sheet. India. Indian Census Bureau, Office of the Registrar General Census Commissioner

Solar O, Irwin A (2010) A conceptual framework for action on the social determinants of health

UK Trade and Investment (2015) Indian smart city programme. The UK Offer to Build Together. https://www.gov.uk/government/uploads/system/uploads/attachment_data/file/460151/UKTI_-_The_UK_offer_to_build_together__1_.pdf

World Health Organization (1979) Coordination of activities with other organizations of the United Nations system and attainment of health for all by the year 2000. Geneva, Resolution WHA32.24, Thirty-second World Health Assembly, 7–25 May 1979. https://www.who.int/governance/en/

World Health Organization (1998) Health-for-all policy for the twenty-first century. Geneva, Resolution WHA51.7. Fifty-first World Health Assembly, 11–16 May 1998. https://www.who.int/governance/en/

www.mapofindia.com

Chapter 14
Usage of Transport Apps by Indian Commuters: An Empirical Investigation

Piali Haldar and Pooja Goel

Abstract The present study with an objective to make an original contribution in the literature of technology adoption by developing a conceptual framework based on Technology Acceptance Model (TAM) in the context of usage transport apps. The current study examined the relationships among perceived ease of use (PEU), perceived usefulness (PU), habit, collaboration, and usage of transport apps. The model was tested empirically using AMOS 21 by collecting data using convenient sampling method from 388 respondents by means of structured questionnaire. The results exhibited that PU, PEU, habit, and collaboration had a significant positive relationship with usage of transport apps. Overall, the extended TAM model contributed valuable insights on Indian commuters' usage behaviour of transport apps. Further, this study provided inputs to policymakers, marketers, and app developers to redesign their strategies to make transport system technologically driven and sustainable.

Keywords Transport apps · TAM · Smart cities · Urban mobility · Usage behaviour · Sustainability

Acronyms

EPI	Environmental Performance Index
NCR	National Capital Region
PEU	Perceived Ease of Use
PU	Perceived Usefulness
TAM	Technology Acceptance Model
VIF	Variance Inflation Factor

P. Haldar
School of Business, Sharda University, Greater Noida, India

P. Goel (✉)
Shaheed Bhagat Singh College, University of Delhi, Delhi, India

P. Sharma (ed.), *Geospatial Technology and Smart Cities*, The Urban Book Series,
https://doi.org/10.1007/978-3-030-71945-6_14

14.1 Introduction

India ranked 177 out of 180 countries in the Environmental Performance Index (EPI) 2018 on account of poor performance in the environmental health policy and air pollution categories. Emission from motor vehicles is one of the major reasons which is responsible for degrading the air quality in the country. One way to mitigate the present air pollution level is to use the smart technology applications (apps) related to transportation, such as route planning apps, car sharing apps, ridesharing apps, and parking information apps. Globally, these apps are contemplated in reducing travelling time, road congestion, economical, safe, and environment-friendly.

With the advent of Smart Era, everything around us is converting itself into smart; be its smart watch, smart phones, smart cards, smart cities, etc. The smart devices backed with smart technologies are helping us in making our life easier, comfortable, and better. In the same vein, mobile technology applications especially in the field of transportation have transformed the way of travelling. Several apps are available related to transport like route planning, trip planning, car sharing, carpooling, traffic safety, and parking information. In India, apps related to travel are seventh most popular app categories downloaded in smartphones (Statista 2015).

Advocates of smart urban mobility suggest that transport apps will play important role in regulating interactions among the commuters using transport apps for navigation information, real-time booking, journey planning apps, etc. The application of transport apps in urban mobility system will be indispensable for managing the growing mobility demand (Docherty et al. 2018; Cohen and Sundararajan 2017). Also, these apps will be instrumental in other issues related to transport like reducing road congestion, reducing number of vehicles on the roads, route planning, and reducing carbon emission, etc.

Interestingly, majority of the previous studies have focused primarily on shared mobility transport apps (Vibhanshu 2018; Zhou et al. 2017; Ballus-Armet et al. 2014; Baptista et al. 2014). Additionally, to the best of our knowledge, no study till now has investigated the usage behaviour of transport apps among the Indian commuters' especially big cities. Hence, this gives us an opportunity to study the key factors influencing usage behaviour of transport apps in Indian urban setting. Thus, the specific purpose of the present paper is twofold; first to identify the important factors responsible for usage of transport apps and second to measure the relative effect of these factors on usage of transport apps and develop a robust model on transport apps usage behaviour.

The rest of the paper proceeds as follows. Section two presents a review of the literature about the growth of transport apps and its contribution in urban mobility. Section three describes the research framework of the study and the hypotheses development. Section four details the research methodology and data collection. In subsequent sections, results, conclusion, and implications are presented.

14.2 Review of Literature

14.2.1 Brief History of Mobile Apps

India has become the global leader in mobile apps and also named as app-economy with an average download of 40 apps per month per person (Mannan 2018). The journey of mobile apps is dated back from 2008, when Apple opened its IOS app store and enabled other mobile companies to distribute mobile applications by setting standards. These standardized apps were used to surf websites by the consumers on their smartphones. This scenario led to the proliferation of mobile apps so much that American Dialect Society (2010) listed the word 'app' as the 'the word of the year'. E-mail, contacts, calendar, weather information, etc. are the few oldest mobile apps followed by mobile games, banking, and ticket purchasing. Broadly, mobile apps can be categorized into six main categories namely lifestyle apps, utility mobile apps, social media apps, entertainment apps, productivity apps, and news information apps (Matteo 2018).

14.2.2 Transportation Apps and Urban Mobility

Transport apps are helpful in solving the traffic-related problems and making commuting more flexible, sustainable, and rewarding. Earlier, transport apps were confined to navigation and location-based services only. As of now, there is a varied number of transport related apps such as logistic apps, parking apps, route planning apps, navigation apps, travel related apps, car sharing apps, ride sharing and ride-hailing apps, and carpooling apps. In an interesting study, Dio et al. (2018) showed that commuters' behaviour could be altered positively towards environment through a mobile apps. Similarly, Sospedra et al. (2015) implemented two end-user mobile applications without utilizing any external infrastructure in an university and confirmed the usability of the apps. However, Ma et al. (2018) identified four pitfalls, namely, information insecurity, privacy leakage, information islands, and digital divide of using mobile apps for providing real-time parking information. Gebresselassie and Sanchez (2018) found that transport apps failed to address the urban challenges with special focus on sustainability and after reviewing sixty transport apps. They established that transport apps do not cater to language barriers, digital divide, and people with physical disabilities. Further, Siuhi and Mwakalonge (2016) pointed out that transport apps could manually, visually, or cognitively distract the commuters while driving or walking.

14.3 Research Framework

Originally TAM was developed to investigate the causal relationship between beliefs and attitude to adopt computer-related environment. Later on, TAM was successfully adopted across various domains related to technology adoption such as mobile shopping (Tak and Panwar 2017), online banking (Patel and Patel 2018), social media usage (Rauniar et al. 2014), and digital commerce (Akhlaq and Ahmed 2015). However, few studies have shown skeptism in using TAM model because of its lack of predictive explanation power, non-inclusion of social and community influence, and oversimplified constructs (Rauniar et al. 2014; Raaji and Schepers 2008). Hence, many researchers either extended original TAM model or integrated different theories such as innovative diffusion theory, theory of planned behaviour, and reported improved results.

Accordingly, the present research paper has developed a theoretical model grounding on technology acceptance model (TAM) suggested by Davis (1989) by adding two additional determinants that is collaboration and habit. The purpose of this study is to formulate a robust model to analyse Indian commuters' transport apps usage behaviour. It has been observed that recommendations, suggestions, and reviews from the members of groups collaborating resources influence the attitude and behavioural intention of the individual (Hu and Zhang 2016). Further, the existing literature has also demonstrated the relationship between collaboration and adoption or usage of information technology (Chopdar et al. 2018; Ng 2016; Amin 2008). Venkatesh et al. (2012) established that individual develops habits which are acquired either directly or indirectly via behavioural intention linked with actual usage. The urban Indian commuters are using transport apps for route planning or booking rides for last few years, which is viewed as prior behaviour (Chen and Chao 2011; Kim and Malhotra 2005). Since the adoption rate of mobile apps is high in India, it is expected that they will use transport apps in future to save their time and money while travelling. Therefore, in addition to PEU and PU, collaboration and habits have been incorporated to understand usage behaviour of transport apps among the Indian urban commuters. Thus, the present study has an objective to establish the relationship between collaboration and transport apps usage behaviour among Indian commuters.

The underlying extended technology acceptance model (with two additional variables namely collaboration and habit) is presented in Fig. 14.1.

Perceived Usefulness and Usage of Transport Apps

Perceived usefulness (PU) is the first belief of TAM framework which states that there is a high probability of using a specific technology by a user if s/he finds it useful (Davis 1989). In the context of transport, route planning, traffic information, mobile ticketing, schedule information of public transits, multi-modal comparison, on-demand mobility, etc. are some of the benefits of transport apps. Further, commuters will use any transport app only after evaluating the benefits and costs involved in it. Several studies have also established the positive relationship between

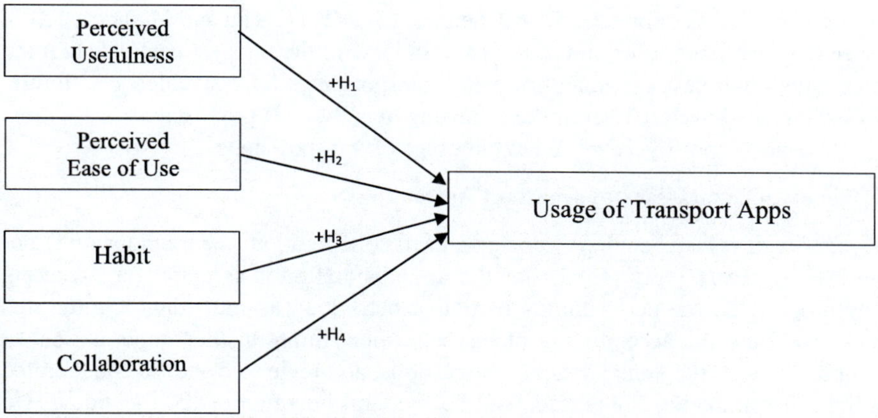

Fig. 14.1 Research framework of the study. *Source* Authors' Work

PU and usage of apps (Niemand and Chauke 2017; Schmitz et al. 2016). Thus, we also hypothesize that

H1: PU is positively related to usage of transport apps.

Perceived Ease of use and Usage of Transport Apps

Perceive ease of use (PEU) is the second belief of TAM framework which states that using a given technology largely depends on the degree of user-friendliness or effortlessness (Davis 1989). Because of the emergence of new technologies, users can find the desired resources anytime anywhere which further helps them in timely completion of their task with ease (Li and Wen 2019). In transport apps context, the adoption of different transport apps facilitates the daily life of the users by reducing the cost of searching transport or transport related information. Such PEU is likely to have positive significant impact on usage of transport apps. Hence, the study hypothesizes that

H2: PEU is positively related to usage of transport apps.

Habit and Usage of Transport Apps

The frequently performed behaviour becomes a habit and is seen as automatic response to a given situation. Previous studies have identified the accumulation of knowledge, experience, and skill acquired over a period of time as the antecedents of habit (Limayem et al. 2007). In various disciplines such as health sciences, consumer behaviour, and organizational behaviour. 'Habit' has been the part of the research model but information system research was included much later (Limayem and Cheung 2011). Venkatesh et al. (2012) found that habit is either directly or indirectly acquired via behavioural intention linked with actual usage while Triandis (1980) found the insignificant influence of behavioural intention on usage behaviour after transformation of behaviour in the habit. In the present technology era, people have already developed a habit of using apps for route planning or booking rides, which is viewed as prior behaviour (Chen and Chao 2011; Kim and Malhotra 2005), since they are using it for quite few years. In view of the fact, we postulate that the commuters will develop a habit of using transport apps for convenience in future. Based on the above arguments, the following hypothesis is proposed:

H3: Habit is positively related to the usage of transport apps.

Collaboration and Usage of Transport Apps

A number of researchers have conceptualized collaboration as a relationship structure (Gray 1989). Gray (1989) stated that collaboration is essential for executing any process. Before performing a particular behaviour, an individual is interested to know about the acceptability of his behaviour among his/her important set of people. Further, recommendations, suggestions, and reviews from the collaborated groups also influence the attitude and behavioural intention of the individual (Hu and Zhang 2016). Also, collaboration helps the members of group to interact and gain new knowledge and information. Existing studies have identified the influential relationship of collaboration on usage behaviour of information systems (Chopdar et al. 2018; Ng 2016; Amin 2008). Due to the processes of internationalization and identification, technology users are more likely to use transport apps too which leads to the following hypothesis:

H4: Collaboration is positively related to the usage of transport apps.

14.4 Research Methodology

14.4.1 Questionnaire Development

The role of TAM has been examined in transport related studies like booking rides, car sharing, etc. However, there is hardly any study which has used TAM to explore on usage of transport apps in general. For the present study, we have adopted TAM as a base model and extended it by adding two new constructs that is habit and collaboration. The survey was carried out using structured questionnaire for measuring the relationship between exogenous and endogenous constructs. The opening question of the questionnaire was a screening question seeking to explore whether the respondent has used any transport app (Table 14.1) in the past 1 month. Hence, the screening question helped in gathering data only from the actual users of the transport apps. Further, the statements for measuring PU, PEU, habit, collaboration, and usage of transport apps were adapted from previous studies (Chopdar et al. 2018; Ahuja and Khazanchi 2016; Chen and Chao 2011; Davis 1989). In total, questionnaire consisted of 16 statements. All the statements have been earlier used in different mobile apps context and to make the questionnaire contextual the questions were modified. Each item of the questionnaire was measured on a five-point Likert scale ranging from 1 (strongly agree) to 5 (strongly disagree). Further, to maintain the anonymity of

Table 14.1 Select transport apps operating in India

Name of the app	Name of the app
Google Maps, Waze, Map My India	Provides real-time information about traffic, construction, crashes, GPS tracking, navigation service
24 Response, Himmat App assistance	Provides quick response and immediate for any real-time situation involving personal safety service
Delhi Transit arrival Bus Info., One Delhi	Helps users to get vital information such as buses, bus routes, track location of a bus, multi-modal journey planner
Uber, Ola, Bla-Bla	Mobility related apps for Ridesharing, Car sharing, Carpooling

Source Authors' compilation

the respondents, limited demographic information such as gender, age, educational qualification, and occupation was asked from the participants.

14.4.2 Data Collection

Before conducted the final survey, the questionnaire was pre-tested to ensure the validity. By using convenience sampling method, data was collected from 338 respondents (72% response rate) located in Delhi and National Capital Region (NCR) via both online and offline methods of data collection. Table 14.2 presents the demographic profile of the respondents using transport apps. 53.6% of the respondents were male and majority of the respondents were from the age group of 21–30 years (40.2%). Two-third of the respondents were graduates and above. 54.4% of the respondents were self-employed while 24.3% were in the service sector.

14.5 Data Analysis

14.5.1 Evaluation of the Measurement Model

Both confirmatory factor analysis and structural equation modelling were applied on the proposed model. The psychometric properties of the questionnaire in terms of reliability, convergent, and discriminant validity were ensured before proceeding further. Table 14.3 shows that Cronbach alpha value of each construct is above the threshold limit of 0.7 (Hair et al. 2015) and composite reliability which is a measure of internal consistency of the constructs also depicting good values ($\geq$0.7 for each construct, Hair et al. 2015).

Table 14.2 Descriptive statistics of respondents' characteristics

Measure	Value	Frequency	Percentage
Gender	Male	181	53.6
	Female	157	46.4
	Total	338	100.0
Age	21–30	136	40.2
	31–40	84	24.8
	41–50	53	15.7
	51 and above	65	19.3
	Total	388	100.0
Qualification	Undergraduate	82	24.3
	Graduate	143	42.3
	PG	113	33.4
	Total	338	100.0
Occupation	Student	72	21.3
	Self-employed	184	54.4
	Service	82	24.3
	Total	338	100.0

Source Authors' compilation

Table 14.4 depicts the covergent and validity measures and also shows the correlation matrix. The values of AVE were greater than 0.5 which showed the convergent validity of the items within a construct while all the values of MSV and ASV are less than AVE suggesting the discriminant validity among the constructs.

Further, Variance inflation factor (VIF) values were also obtained for each construct. The VIF values ranging from 1.81 to 1.97 which was below than the threshold of 2, indicating no multicollinearity issue. Furthermore, Table 14.5 shows the measurement models' estimates which were well within the desirable limits of values suggested by Hair et al. (2015).

14.5.2 Evaluation of the Structural Model

Establishing the Model fit

To establish the relationships among variables, we used structural equation modelling by applying using AMOS 21. All the four exogenous constructs that is PU, PEU, habit, and collaboration, and one endogenous construct that is usage of transport apps were taken together to study the hypothesized model. Table 14.6 gives a brief summary of the output of second-order analysis of structural model suggesting the good model fit as all the indices were falling within the prescribed limits.

Table 14.3 Reliability measures

Statements	STD. loading[a]	Alpha Val	CR
Perceived usefulness		0.718	0.716
Transport apps help me in hassle free travelling	0.632**		
Transport apps help in saving my time while travelling	0.686**		
Transport apps made my life more comfortable	0.645**		
Transport apps help me in making better travelling decisions	0.723[n.a]		
Perceived ease of use		0.825	0.811
I can easily use transport apps	0.827***		
I think transport apps are easy to operate	0.718***		
I find transport apps easy to understand	0.731[n.a]		
Perceived ease of use		0.751	0.756
I can easily use transport apps	0.783***		
I think transport apps are easy to operate	0.843***		
I find transport apps easy to understand	0.647[n.a]		
Collaboration		0.751	0.756
My peer group influences me to use transport apps	0.783***		
People closer to me feel that I should use transport apps	0.843**		
I use transport apps because many people are using it	0.647[n.a]		
Habit		0.701	0.754
The usage of transport apps has become habit for me	0.803***		
I can't travel without using transport apps	0.731**		
I am accustomed to use transport apps	0.811[n.a]		
Usage of transport apps		0.730	0.745
I am using different kinds of transport apps at least for past 6 months	0.812[n.a]		
I would recommend others to use transport apps	0.791***		
I will continue to use transport apps in the future	0.731***		

Source Authors' work
Note [a]significant at $p \leq 0.05^{**}$, $p \leq 0.001^{***}$; n.a not applicable

Table 14.4 Measures of validity and correlation matrix

	AVE	MSV	ASV	PU	PEU	Habit	Collaboration	Usage
PU	0.637	0.371	0.272	1.000				
PEU	0.624	0.369	0.283	0.518	1.000			
Habit	0.679	0.388	0.252	0.536	0.506	1.000		
Collaboration	0.602	0.365	0.253	0.194	0.216	0.232	1.000	
Usage	0.625	0.361	0.322	0.684	0.361	0.454	0.244	1.000

Source Authors' own work

Table 14.5 Measurement model estimates

Model REMSEA	χ2	d.f	χ2/d.f	GFI	TLI	CFI
0.058	619.20	215	2.880	0.895	0.813	0.816

Source Authors' work

Table 14.6 Structural model estimates

Model REMSEA	χ2	d.f	χ2/d.f	GFI	TLI	CFI
0.033	13.572	4	3.393	0.911	0.915	0.898

Source Authors' work

Hypothesis testing

After achieving the model fit, the next step was to test the hypothesized theoretical relationships. The structural model with standardized regression weight is depicted in Figure 14.2. PU has found to have positive significant relationship with usage of transport apps ($\beta = 0.36$, $p < 0.001$) hence accepting H_1. Similarly, PEU and ($\beta = 0.17$, $p < 0.01$) and habit ($\beta = 0.23$, $p < 0.05$) have significant positive effect on usage of transport apps, resulting in acceptance of H_2 and H_3, respectively. Further, collaboration and usage of transport apps have been significantly positively related ($\beta = 0.34$, $p < 0.001$) thereby supporting H4 (Table 14.7). In sum, all the four hypotheses framed for the study have been accepted. Figure 14.2 also shows that altogether all the exogenous constructs undertaken in proposed model are explaining 63.3% ($R^2 = 0.633$) of usage of transport apps.

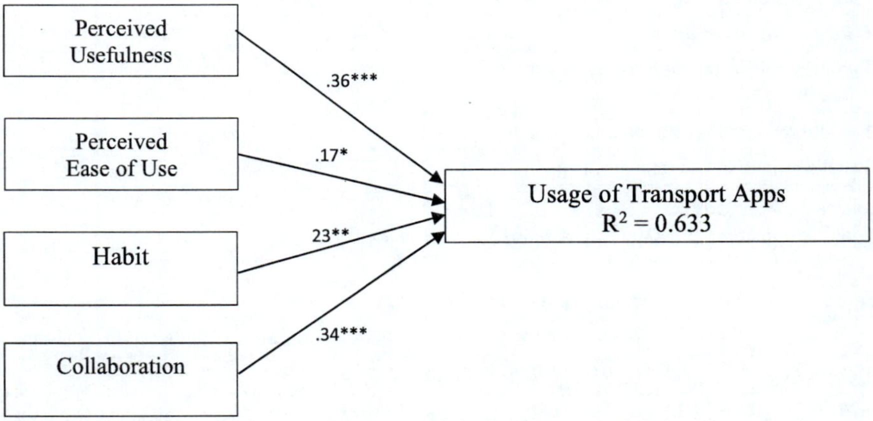

Fig. 14.2 Structural model. *Source* Authors' work

Table 14.7 Hypothesis testing results

Hypothesis	Standardized co-efficient (β)	P-value	Results
H1. PU ——> Usage	0.36^{***}	0.001	Accepted
H2. PEOU ——> Usage	0.17^{*}	0.23	Accepted
H3. Habit ——> Usage	0.23^{**}	0.049	Accepted
H4. Collaboration ——> Usage	0.34^{***}	0.001	Accepted

Source Authors' work
Note Significant at the $p < 0.001^{***}$, $p < 0.05^{**}$, $p < 0.01^{*}$

14.6 Discussion

In the present study, we have investigated the influence of PU, PEU, habit, and collaboration on usage of transport apps such as route planning, travel apps, and mobility apps among Indian urban commuters. The proposed model has adopted PU and PEU from TAM and two new constructs that is habit and collaboration have been added to the proposed model to make it more robust model in understanding the commuters' usage of transport apps. The results exhibit that PU, PEU, habit, and collaboration have significant positive impact on usage of transport apps.

To begin with, result reveals that PU is the strongest predictor of the usage of transport apps. The significant influence of PU is reported in different studies in different countries for various other apps, for example, m-health adoption in Bangladesh (Hoque 2016), usage of library apps in China (Hu and Zhang 2016), shopping apps in India (Tak and Panwar 2017), mobile apps for information purchasing and sharing in USA (Taylor and Levin 2014), and many others. Commuters are using transport apps because they feel that it saves their travel time by providing information about the busy routes and suggest the best route for reaching their destination. It helps them to book a parking space, travel tickets, or a vehicle for travelling. It also helps in finding the route of unknown destinations without human interventions. Commuters feel that with the help of transport apps, they have become more confident and comfortable in taking travel related decisions.

Moving further, the present study suggested the significant positive but comparatively weak relationship of PEU with usage of transport apps. Nowadays, commuters are using apps for various services, such as for health and fitness, shopping, and making payments. Therefore, they are well versed with the functioning of the apps. Also, the regular usage of these apps has helped them in attaining the technological self-efficacy. Commuters feel that the transport apps are easy to use and understand, hence they are able to use these apps very easily. Hence, the results of the current study are in line with previous studies suggesting the contribution of PEU in predicting app usage behaviour (Singh et al. 2017; Kim et al. 2016).

This study has also discovered that habit is another important factor that influences usage of transport apps. There can be several reasons for this result. First, people find transport apps useful and if anything useful will be used regularly, which further leads to habit formation. Verplanken et al. (1998) also stated that when we experience something useful or beneficial, there is a possibility to develop habitual pattern towards that phenomenon. Second, nowadays we are living in app-economy and people are using different types of apps on regular basis hence they look forward towards different apps which makes their life easy and comfortable. Third, usage of apps has a negligible cost and it is freely and easily downloadable. Further, these apps do not acquire much space in the phone also. Lastly, people perceive that apps are easy to use and save lot of surfing time in comparison to using browser by directly providing required information. Also, it saves all necessary information and search history Therefore, apps work really fast than any other digital platform of providing information.

This study validates collaboration as a significant predictor and suggests that commuters tend to get influenced by their peer groups, family, and friends to use transport apps. Thus, the behaviour of a commuters is largely influenced by the endorsement of their peer groups or people close to them. The previous studies also support that people are influenced by the suggestion, reccomendation, and opinion of the people whom they think they should flow, thus a shared goal is a key element of collaboration (Wood and Gray 1991; Bedwell et al. 2012).

14.7 Conclusion and Implications

While several other studies have addressed the consumers attitude and adoption behaviour related to apps related to e-commerce, but no previous study so far studied the commuters' willingness to use transport apps available in India. Thus, the present study has not only contributed in the literature of travel behaviour but also provided insights for managerial implications and policy measures. The theoretical contribution of the current study is stated in the terms of identifying the key factors which are responsible for the usage of transport apps and assessing their relative importance for the commuters. These insights can be used by the service managers and government agencies to understand the commuters' behaviour with respect to usage of transport apps.

In terms of practical contribution, several suggestions are given. To begin with, as the study established that commuters find transport apps useful because it helps them in many ways such as for navigation, searching location, traffic data collection, travel information, route planning, and ridesharing. Therefore, companies should develop more user-friendly, easy to use, interactive, useful, and highly reliable transport apps for optimal response. Also, the app developers should design the apps keeping in mind the usefulness of the products as the performance of the apps determine the adoption of transport apps among the commuters. Lastly, the commuters will develop the habit of using apps only when they find it useful and ease to use. They will further recommend these apps to others for making their travel and journey easy and comfortable. Thus, the marketers should enhance the value proposition of the transport apps.

Lastly, the insights developed from the study can be used by the city planners and civic authorities in taking several proactive steps. First, the government can utilize the data generated through smartphone usage in making several policies. For example, government can encourage and turn the commuters into prosumers (producer and consumers) as they are doing for electricity generation by understanding the travel needs and pattern of the city dewellers. It will also provide a sustainable solution for the increasing travel demands of the big cities. Second, the government can use the apps in marketing the cities especially for the tourism purpose. These apps can work as a travel guide for all those people who are either visiting these cities for official or leisure purpose or living in the city. Finally, the government can enhance the personal safety of the people by using the locational information of the commuters in particular of female commuters.

To conclude, the study on factors influencing the usage of transport apps among the Indian commuters will help the government policymakers, transport sector, and service providers to design policy and product to inspire the commuters to use transport apps, which in turn will reduce the problems of the commuters. There is no doubt about the fact that proliferation of apps related to transport in India is low and far behind compare to other developed countries. This study will throw some light on commuters' usage behaviour of transport apps, and there is an urgent need for an extensive study to get a better picture on usage of transport apps in India.

14.8 Limitation and Future Research

Some limitations concerning the generalization of the present study are needed to be addressed. First, this study is limited to Delhi and NCR commuters only; it would be better to see the replica of the proposed model in different parts of India and may be in other developing countries. Second, the adoption of transport apps seems to be beneficial for society and country as a whole, but it did not include issues related to hindrance in adoption. The present study did not consider any factor like social security, perceived risk, or physiological barrier which may affect the adoption of transport apps. Third, it is important to research each and every transport apps separately to understand how they are helping the commuters and how they can make it more beneficial for them in the future. Fourth, the transport apps have opened up a new paradigm in the transport sector, it is vital to study the impact of these apps on commuters and everyone directly or indirectly associated to transport sector. Lastly, the future researchers may use longitudinal study to compare the commuters' behavioural changes towards usage of transport apps.

References

"App" voted 2010 word of the year by the American Dialect Society (2010). Retrieved from www.americandialect.org: https://www.americandialect.org/app-voted-2010-word-of-the-year-by-the-american-dialect-society-updated. Retrieved from www.mapsofindia.com: https://www.mapsofindia.com/top-ten-cities-of-india/top-ten-highest-no-of-car-ownership.html (2016)

Ahuja V, Khazanchi D (2016) Creation of a conceptual model for adoption of mobile apps for shopping from e-commerce sites—an Indian Context. Procedia Comput Sci 91:609–616

Akhlaq A, Ahmed E (2015) Digital commerce in emerging economies: factors associated with online shopping intentions in Pakistan. Int J Emerg Mark 10(4):634–647

Amin H (2008) Factors affecting the intentions of customers in Malaysia to use mobile phone credit cards. Manage Res News 31(7):493–503

Ballus-Armet I, Shaheen SA, Clonts K, Weinzimmer D (2014) Peer-to-Peer carsharing: exploring public perception and market characteristics in the San Francisco Bay Area. Transp Res Rec J Transp Res Board 2416(1):27–36

Baptista P, Melo S, Rolim C (2014) Energy, environmental and mobility impacts of car-sharing systems. Empirical results from Lisbon. Portugal. Procedia-Soc Behav Sci 111:28–37

Bedwell W, Wildman J, DiazGranados D, Salazar M, Kramer W, Salas E (2012) Collaboration at work: an integrative multilevel conceptualization. Hum Resour Manag Rev 22(2):128–145

Chen CF, Chao WH (2011) Habitual or reasoned? Using the theory of planned behaviour, technology acceptance model, and habit to examine switching intentions toward public transit. Transp Res 14(2):128–137

Chopdar P, Korfiatis N, Lytras M, Sivakumar VJ (2018) Mobile shopping apps adoption and perceived risks: a cross country perspective utilizing the unified theory of acceptance and use of technology. Comput Hum Behav 86:109–128

Cohen M, Sundararajan A (2017) Self-regulation and innovation in the peer-to-peer sharing economy. Univ Chicago Law Rev Online 82(1):116–134

Davis FD (1989) Perceived usefulness, perceived ease of use, and user acceptance of information technology. MIS Q 13(3):319–340

Dio S, Gennusa M, Peri G, Rizzo G, Vinci I (2018) Involving people in the building up of smart and sustainable cities: how to influence commuters' behaviours through a mobile app game. Sustain Cities Soc 42:325–336

Docherty I, Marsden G, Anable J (2018) The governance of smart mobility. Transp Res Part A 115:114–125

Gebresselassie M, Sanchez TW (2018) "Smart" tools for socially sustainable transport: a review of mobility apps. Urban Sci 45(2(2)):1–10

Gray B (1989) Collaborating: finding common ground for multiparty problems. Jossey-Bass Publishers, San Francisco

Hair JF, Black WC, Babin BJ, Anderson RE, Tatham RL (2015) Multivariate data analysis. Pearson Education, New Delhi

Hoque MR (2016) An empirical study of mhealth adoption in a developing country: the moderating effect of gender concern. BMC Med Inf Decis Making 1–10.

Hu J, Zhang Y (2016) Chinese students' behaviour intention to use mobile library apps and effects of education level and discipline. Library Hi Tech 34(4):639–656

Kim SC, Yoon D, Han EK (2016) Antecedents of mobile app usage among smartphone users. J Market Commun 22(6):653–670

Kim SS, Malhotra NK (2005) A longitudinal model of continued IS use: an integrative view of four mechanism underlying post-adoption phenomena. Manage Sci 51(1):741–755

Li H, Wen H (2019) How is motivation generated in collaborative consumption: mediation effect in extrinsic and intrinsic motivation. Sustainability 11(3):640–653

Limayem M, Cheung CM (2011) Predicting the continued use of internet-based learning technologies: the role of habit. Behav Inf Technol 30(1):91–99

Limayem M, Hirt S, Cheung C (2007) How habit limits the predictive power of intention: the case of information systems continuance. MIS Q 31(4):705–737

Ma R, Lam P, Leung C (2018) Potential pitfall of smart city development: a study on parking mobile applications (apps) in HongKong. Telematics Inform 35(6):1580–1592

Mannan L (2018) With over 1 B installs, and $400 M in-app revenue, India has a booming app economy. Retrieved from yourstory.com: https://yourstory.com/2018/03/1-b-installs-400-m-app-revenue-india-booming-app-economy/

Matteo (2018) What are the different types of mobile apps? breaking down industries and functionalities. Retrieved from duckma.com: https://duckma.com/types-of-mobile-apps/

Ng M (2016) Examining factors affecting mobile commerce adoption of Chinese consumers' . Int J Electron Bus 13(1):98–115

Niemand C, Chauke H (2017) The use of mobile applications by public transport commuters in Gauteng, South Africa. S Afr J Inf Manage 19(1):1–10

Patel KJ, Patel HJ (2018) Adoption of internet banking services in Gujarat. Int J Bank Mark 36(1):147–169

Raaji E, Schepers J (2008) The acceptance and use of a virtual learning environment in China. ScienceDirect 50:838–852

Rauniar R, Rawski G, Yang J, Johnson B (2014) Technology acceptance model (TAM) and social media usage: an empirical study on Facebook. J Enterp Inf Manage 27(1):6–30

Schmitz C, Bartsch S, Meyer A (2016) Mobile app usage and its implication for service management-empirical findings from German public transport. Procedia 224:230–237

Singh N, Srivastava S, Sinha N (2017) Consumer preference and satisfaction of M-wallets: a study on North Indian consumers. Int J Bank Mark 35(6):944–965

Siuhi S, Mwakalonge J (2016) Opportunities and challeneges of smart mobile applications in transportation. J Traffic Transp Eng 3(6):582–592

Sospedra J, Avariento J, Rambla D, Montoliu R, Casteleyn S, Bordonau M, Huerta J (2015) Enhancing integrated indoor/outdoor mobility in a smart campus. Int J Geogr Inf Sci 29(11):1955–1968

Statista (2015) India most installed app categories 2015. Retrieved from www.statista.com: https://www.statista.com/statistics/509652/most-installed-app-categories-india/

Tak P, Panwar S (2017) Using UTAUT 2 model to predict mobile app based shopping: evidences from India. J Ind Bus Res 9(3):248–264

Taylor DG, Levin M (2014) Predicting mobile app usage for purchasing and information sharing. Int J Retail Distrib Manage 42(8):759–774

Triandis HC (1980) Values, attitudes, and interpersonal behaviour. In: Howe H (ed) Nebraska symposium on motivation, 1979: Beliefs, attitudes, and values. University of Nebraska Press, Lincoln, pp 195–259

Venkatesh V, Thong JL, Xin X (2012) Consumer acceptance and use of information technology: extending the unified theory of acceptance and use of technology. MIS Q 36(1):157–178

Verplanken B, Aarts H, Knippenberg A, Moonen A (1998) Habit versus planned behaviour: a field experiment. Br J Soc Psychol 37(1):111–128

Vibhanshu A (2018) Manufacturers can also win in the sharing economy. MIT Sloan Manage Rev 59(2):1–5

Wood DJ, Gray B (1991) Toward a comprehensive theory of collaboration. J App Behav Sci 27(2):139–162

Zhou F, Zheng Z, Whitehead J, Perrons R, Page L (2017) Projected prevalence of car-sharing in four Asia-Pacific countries in 2030: what the expert think. Transp Res Part C 84:158–177

Chapter 15
Parking Maximums and Work Place Levies: Time to Adopt New Paradigms in India, the Case of Kochi

Paulose N. Kuriakose and Suraj P. Rajeendran

Abstract The objective of this study is to understand how and to what extent Indian Cities are prepared to implement contemporary parking management strategies. India follows decades-old generic minimum parking standards and parking charges are one of the lowest in the world. But in a recent decade, India has invested a considerable amount of money in improving public transport. There are close to 180 cities with bus-based public transport, 10 cities with BRTS, 8 cities with metro rail and 1 city with monorail facility. India has 515 km of operational metro lines and 381 stations in 11 cities. It is the right time to initiate the adoption of innovations in parking management to give support to public transport development using parking as a travel demand management tool. The city of Kochi is taken to study the feasibility of implementing parking maximums and parking levies. Kochi has 51% of public transport ridership at present with three different types of public transport modes like ferry, buses, and metro. Accessibility assessment and zone delineation are conducted to make parking capping zones. It is possible to implement parking control zones and parking maximums in Kochi, because of the high accessibility to public transport. It can have an impact on public transport supply and optimal use of parking spaces.

Keywords Parking maximums · Parking caps · Workplace levy · Travel demand management

Acronyms

AWT	Average Waiting Time
ECS	Equivalent Car Space
EDF	Equivalent Doorstep Frequency
ETB	Electric trolley bus
HCBS	High capacity bus system

P. N. Kuriakose (✉) · S. P. Rajeendran
Department of Urban and Regional Planning, School of Planning and Architecture, Bhopal, India
e-mail: paulosenk@spabhopal.ac.in

P. Sharma (ed.), *Geospatial Technology and Smart Cities*, The Urban Book Series,
https://doi.org/10.1007/978-3-030-71945-6_15

KCR	Kochi City Region
KMBR	Kerala Municipal Building Regulations
KSWTD	Kerala State Water Transport Department
NUTP	National Urban Transport Policy
PTAL	Public Transport Access Level
POI	Points of interest
SAP	Service access points
SWT	Scheduled Waiting Time
TAT	Total Access Time
TOD	Transit-Oriented Development
UMTA	United Metropolitan Transport Authorities

15.1 Introduction

Regardless of how fuel efficient the cars are or how little pollution they emit, cars need to be parked somewhere and generally, a car spends about 95% of its life parked and uses several parking spaces each week (Marsden et al. 2011). The way car parking is managed has many ramifications on travel behaviour and overall quality of accessibility in urban areas. In the 1920s when the car usage increased, American urban planners came up with the solution of minimum parking requirements for every type of building use to reduce the pressure of parking demand on streets. Minimum-based parking standards are followed in most of the countries in the world (Shoup 1999). It was perceived as a panacea for the parking issues in urban areas, but slowly researches have proved that minimum-based parking standards have larger negative impacts on the urban fabric, land use and on the travel characteristics. Generic minimum-based parking supply keeps on adding parking spaces with an increase in car ownership and parking space demand. It does not consider other factors like mixed land use, transit accessibility, employment density, income and pricing, etc. that influence the parking demand. Generic minimum suggest the 'one size fit for all' parking standard for all areas of the city. Parking management falls into bigger trouble even after providing minimum parking spaces with free parking or suboptimal parking prices. Generic minimum parking indirectly pushes urban sprawl, increases the trip length, increases the price of real estate and creates a vicious cycle of many problems. It also causes a reduction in the availability of affordable housing (McDonnell et al. 2011).

By realizing the ill effects of generic minimum parking urban planners in different countries have come up with different modifications in parking standards and pricing. It has been recognized that parking management can be used as a strategy to reduce the negative externalities of urban transport. Many studies have concluded that parking norms affect parking supply, built-up space utilization, size and location of the development. Impact of parking standard goes beyond the parking market and it affects the price demand equations of housing and commercial built-ups development. Several cities have modified parking by adopting parking maximums, parking caps, areas specific parking norms or flexible parking standards (Kodrasky and Hermann 2011). Sections 15.2 and 15.3 detail out the maximum-based parking standards and work-place levy or employer cash-out scheme in parking pricing. Implementations of these two parking management strategies have impacted positively to improve modal shift, reduction in the ownership of cars and many other indirect co-benefits.

15.2 Maximum-Based Parking Supply

Contrary to the generic parking minimum, maximum parking norms prevent builders from constructing extra parking spaces than the given parking cap per built-up area. It helps to reduce the parking supply in areas where considerable accessibility levels are achieved. In addition to the site level parking maximums, it can be further extended to wide-area parking caps. Parking maximums are implemented based on the accessibility, nature of the road inventory, land use density, etc. Based on these factors, the city will be divided into different parking districts or precincts with different parking standards and caps (Joshua and Dylan 2010). Between the year 2000 and 2001 London made parking reforms that are pushed by national policies to give a fillip to sustainable modes of transport and restrict personal vehicle use. It removed street parking and established Controlled Parking Zones (Li and Guo 2014). In comparison to the pre-parking standard reform years, there was a reduction of 49% in parking supply and on an average 0.76 parking spaces were reduced per dwelling unit. Modification of parking through market approach found to play a key role in parking supply reduction, commuting by car and car ownership (Li and Guo 2014). Guo and Ren (2013) also found that elimination of the generic minimum parking norm in London has lead to reduction parking supply, but to achieve the full potential of its parking maximum policy has to be implemented in a combination of other parking management strategies. Parking maximums and supply caps are implemented in many European cities like Zurich, Hamburg, and Budapest. A policy called the Historischer Parkplatz Kompromiss—literally the historic parking compromise—was implemented in 1996 to establish an upper limit for the parking space availability. Through this policy, every off-street parking space addition has lead to the removal of an on-street parking space. In 1976 Hamburg introduced a cap of 30,000 spaces in the city centre (Kodrasky and Hermann 2011).

15.3 Workplace Levy

Generally, employers offer free parking for their employees, but this distorts averaged generalized cost of car users and skews the trip cost of other modes of transport and lead to inefficiencies in urban transport. To remove this average cost manipulation parking cash-out or workplace levies are suggested as an effective parking management strategy (Evangelinos et al. 2018). Workplace levy is a parking strategy that makes the employer offer employees the option to choose cash instead of free parking space or parking subsidy offered (Shoup 1997). Instead of pricing the employer-provided parking, parking cash-out scheme rewards the people use alternate modes. It would indirectly make the employee aware of the opportunity cost of workplace parking (Evangelinos et al. 2018). Parking cash-out schemes help the employers to pay lower rent due to the fewer requirements of parking spaces.

In 1992 California created an Act to implement employer-paid parking. A Study conducted by Shoup (1997) found that the number of solo divers reduced by 17 percentage after the implementation of the Law. There was a 64% increase in the carpooling and a 50% increase in the public transport ridership. The study summarized that it benefited employees, firms, and taxpayers. In 2012 the City Council of Nottingham the UK implemented workplace levy to reduce congestion and to mobilize money for public transport infrastructure (Dale et al. 2017). Post-policy implementation studies prove that parking cash-out has a conspicuous impact on personal vehicle trips. It has helped to reduce congestion (Dale et al. 2017).

Factors that influence parking demand, such as the popularity of a particular establishment, transit proximity, walkability, land use density, parking management practices, pricing, and availability of public lots, are considered for arriving at the parking demand (Litman 2016). Nearby transit service frequency and quality help reduce requirements of parking by 10% for housing and employment within ¼ mile of frequent bus service, and 20% for housing and employment within ¼ mile of a rail transit station. If the site is located near the car-sharing service facility, parking spaces for residential requirements would be reduced by 5–10% (Litman 2016).

15.4 Problem Statement: Urbanization, Mobility Demand, Pollution and Road Accidents

With the realization of the evolving urban development and demand for travel Government of India had launched the National Urban Transport Policy in 2006 (MoUD 2006). NUTP recognizes the increasing trend of personal vehicles and their negative impact on the quality of life and environment. It stresses the need for charging the parking spaces based on the real estate value of the property on which the vehicle is parked. Post-NUTP many programmes and policies are implemented to improve public transport in urban areas. A national urban renewal mission was started in 2007 to improve the urban infrastructure, especially public transport infrastructure. National Metro policy and Transit-Oriented Policy has been implemented in 2017. Currently, Smart Cities Programme and AMRUT programmes are focusing on infrastructure development in urban areas. There are close to 180 cities with bus-based public transport, 10 cities with BRTS, 8 cities with metro rail and 1 city with monorail facility. India has 515 km of operational metro lines and 381 stations in 11 cities (MoHUA 2018). A total length of 537 km of metro corridor is under constriction stage and planning and appraisal is under process for another 595 km of metro rail. There are 11 metro cities with at least three different modes of public transport available. Close to 400 km of BRT service available in different cities of India. It does not mean that India has a flawless public transport available, but it is on a sustainable path to achieving inclusive accessibility for everyone.

15.5 Methodology

A detailed literature review was conducted to understand various strategies on parking management. In the second stage, the existing situation of parking management was studied with the help of reports from various sources and government regulations. In the third stage parking management and public transport infrastructure of the Kochi city are analyzed. As part of the third stage detailed assessment of public transport, accessibility has been conducted using secondary data available from Road Transport Office, Government of Kerala. To assess the public transport accessibility levels throughout the city, the London Public Transport Access Level (PTAL) method was adopted. Connectivity by public transport can be measured using PTAL, which has been used for various planning processes throughout London. It would show how well a place is connected to the public transit services for any place selected. PTAL values range from '0' to '6', where '6' represents the best connectivity and '0' represents the worst connectivity (Transport for London 2015).

The approach adopted in this study was to derive the PTAL for a grid of points covering the Kochi City Region. Since the study area was large, a grid size of 500 m^2 has been used. The detailed calculation of PTAL for a single point of interest or location can be broken down to a series of stages:

(i) **Defining the points of interest**
In the London methodology the built developments constituted the points of interest (POI) but due to the limitations in data availability, this study has taken an alternate approach by dividing the study area into 1,679 500 m^2 grid cells. The centroids of each cell were considered as the points of interest for the calculation of PTAL of that cell.

(ii) **Calculation of walk time to service access points**
There are approximately 551 service access points (SAP) (491 Bus stops, 38 Boat Jetties and 22 Metro stations) in the study area. The average walking speed was assumed as 4.8 KMPH and the calculation assumes that a commuter will walk up to 500 m to bus service, 800 m to a metro service and 1,000 m to a ferry service. The walk access distance from the POI to SAP is measured in ArcMap software using Shortest Path Method algorithm over the road network layer.

(iii) **Calculation of Scheduled Waiting Time (SWT) at each SAP**

Routes are identified for each SAP and the service frequency for the peak hour is derived. It is assumed that the passengers arrive at the SAP at random without adjusting their arrival to the service timings. The SWT is estimated as half the time interval between the arrival of services.
i.e., SWT (in minutes) $= 0.5 \times (60/\text{frequency})$.

(iv) **Calculation of Average Waiting Time (AWT) at each SAP**
The average waiting time is the sum of scheduled waiting time and reliability factor. The reliability factor reflects the possible uncertainties in the delivery of the service and it is different for each mode of service. A reliability factor of 2 min is used for bus services; 1 min for ferry services and 30 s for metro services have been used to calculate the AWT in this study.

(v) **Calculation of Total Access Time (TAT)at each SAP**
The Total Access Time is the sum of Walk time to the SAP and the AWT at the SAP.
i.e. TAT = Walk Time + AWT.

(vi) **Calculation of Equivalent Doorstep Frequency (EDF)at each SAP**
The Equivalent Doorstep Frequency converts TAT back into units of frequency.
That is, $\text{EDF} = 0.5 \times (60/\text{TAT})$.
This estimates the service access time as a notional average waiting time by assuming that the service was available at the doorstep of the POI i.e., the service was available without any walking time.

(vii) **Calculation of Accessibility Index**
The PTAL method gives higher weightage to the single service with the highest EDF at each POI within the same mode of service. The Accessibility Index is calculated based on the EDFs at all SAPs that come within the acceptable walking distance but by giving a weight of one to the highest EDF and 0.5 to all other EDFs.
That is, AI = Largest EDF $+ 0.5 \times \Sigma$ (all other EDFs).
The AI for each mode is separately calculated this way and the total AI is calculated or a particular POI as the sum of the AIs of all modes.
That is, $AI_{total} = \Sigma\,(AI_{bus} + AI_{metro} + AI_{ferry})$.
The AI_{total} value obtained this way is converted to PTAL as per Table 15.1.
Different Access Index Zones are merged to create parking cap zones.

Major Source of the data used for this study is collected from three major reports (Comprehensive Mobility Plan 2035, Master Plan of Kochi City Region 2031 and Transit-Oriented Development Action Plan 2035) prepared for the urban and transportation planning of Kochi City Region.

Table 15.1 Conversion of access index to PTAL

PTAL	PTAL Access index range	Parking zones[a]
0 (worst)	0	Zone D
1a	0.01–2.50	
1b	2.51–5.0	
2	5.01–10.0	Zone C
3	10.01–15.0	
4	15.01–20.0	Zone B
5	20.01–25.0	
6a	25.01–40.0	Zone A
6b (best)	40.01+	

Source Transport for London (2015)
[a]Delineated by the authors for Parking Cap Zoning

15.6 Analysis and Discussion

15.6.1 Current Indian Parking Management

Around 400 million people are living in urban areas of India. According to the decadal census of 2011, 31% of the people are from urban areas. There are 7935 urban areas in India (Census of India 2011). The number of cities with more than one million inhabitants is 53 in 2011. After the liberalization of the Indian economy in 1991, the Indian economy has grown much faster especially in industrialization and manufacturing and information technology. Cities and considered as the engines of growth and it grows further with rural–urban migration, natural increasing of population and annexing of new areas to the urban boundary. With the increase in urban population, the demand for travel is also induced and many transport issues emerged. The total number of registered vehicles in India was 230 million in 2016. Two-wheelers (73.5%) constitute the highest share of registered vehicles, followed by cars, jeeps and taxis (13.1%). The number of registered buses is decreasing and the number of private passenger vehicles is fast growing (MoRTH 2017).

Indian cities regulate the supply of parking spaces by generic minimum parking requirements and parking prices. Parking standards are decided by the Indian Road Congress, National Building Codes and Development Control Regulations under urban development acts. IRC Standards of 1988 and 1997 contain provisions relating to parking while laying roads (NBC 2016). While the 1988 guidelines mention the requirements relating to off-street/building parking, the 1997 guidelines mention the road marking requirements for parking on the street. The Motor vehicle Act deals with traffic and parking area regulations and enforcement. It prohibits parking at or near road crossings, on a footpath, obstructing another vehicle, etc. Municipal Corporation and City Traffic Police are responsible for parking policy enforcement. Available public off-street parking spots and major street stretches would be leased out to private contractors for management. Lack of personnel to monitor parking areas, lack of resources, such as towing vehicles, and locks, and the inability to penalize for violations owing to various factors are some of the maladies that have been affecting. Town Planning wing of the municipal corporation is responsible for ensuring the provision of parking supply as per the building control regulations. All the standards presume that with the increasing population size of the city's vehicle ownership increases and less attention is paid to the fact that the share of public transport and proximity to Mass Rapid Transit System locations needs to be considered to fix caps on parking supply. None of the factors like transit accessibility, mixed land use, high residential or employment density is considered in deciding the parking standards. Residential parking standards normally relate to the floor space index (FSI) or total built-up area (Kuriakose 2015).

Parking charges in India are one of the lowest in the world. 98 percentage of on-street parking in arterial and sub-arterial roads are for free. Since on-street parking is free most of the Multi-Level Car Parking facilities are vacant. There is a strong protest going on in almost all metro cities against the parking fees charged by shopping malls. Commuters are claiming that it should be provided for free. There are many court cases going on in many consumer courts and high courts of India against the parking fee at shopping malls. The State Government of Telangana has issued a notification banning parking charges for the first half an hour and for 30 min to 1 h, it would be free if the visitor brings a bill of any amount after shopping. For above one-hour parking, charge cannot be levied if the visitor brings a bill worth value more than the parking charge. In effect, it prevents shopping mall authorities from collecting the parking charge. Gujarat High court has questioned the right of private malls for collecting the charges, According to the 74th constitution amendment act municipal corporations are the agencies that should ensure the earmarking regulation and supervision of parking places. Considering the cost of construction and the spiralling effect of free parking in commercial rental values, reducing the average generalized cost of private mode-based trips governments should make legal provisions for parking charge collection. With the introduction of Goods and Service Tax vendors who manage the parking spaces under the control of the municipal corporation and Indian Railway are required to pay 18% tax. Once the legal tools ensure the parking fee collection with Goods and Service Tax, it would add revenue to the government. Such revenue collected can be ring-fenced to use for the development of sustainable modes of transport. Mall managers can decide whether the cost of parking space from the retailer or the mall user. Generally, people expect parking to be free. There is a huge cost associated with the parking space construction in malls. If the government prevents malls from collecting parking fees, mall authorities will be forced to retrieve the construction cost of the parking space through the rent of the leasable space. It would create a situation in which a public transport user is indirectly paying for the parking space that they are not using.

15.6.2 Case Study of Kochi

Kochi is the third-largest city after Mumbai and Surat on the western coast of India (Murali and Kumar 2014). Kochi is situated in the estuary formed at the mouth of Vembanad Lagoon on the southwestern coast of Kerala (CMC 2010). At present, Kochi has its economic base in port activities and shipbuilding, tourism, spice export, and information technology-based industries. Kochi's economic growth has attracted many business ventures to the city. Such ventures are located in the special economic zone, international container transshipment terminal, software parks, and info city, as well as the Kochi metro corridor and, have brought various opportunities for growth and advancement. As a consequence, this has brought a significant population of migrants from many north and northeastern states of India. Also, Kochi receives the highest number of international and domestic tourists in Kerala, which stimulates high employment in the tourism industry.

Kochi City Region comprises Cochin Corporation and its immediate influence areas covering an area of 369.72 km^2 with a population of 12.23 lakhs in 2011. The population in the Kochi City Region increased from 11.28 lakhs in 2001 to 12.23 lakhs in 2011, with a decadal growth rate of 8% (Census of India 2011). Traffic analysis zone-wise population density is given in Fig. 15.1. Kochi City Region has a travel demand of around 1 million passenger trips per day. With a growing population and mega development plans coming up for this port city, travel demand is expected to grow steeply. The percentage of registered cars in the total vehicles registered has increased from 18% in 2007 to 22% in 2014. It is observed that 50.8% of the households own a two-wheeler and 9.3% own only car. 22.2% own both car and two-wheeler. Increased private vehicle registration (Table 15.2), will not only aggravate the congestion on the city roads but will also increase the pollution level. This section further details the public transport characteristics of Kochi city (CMP 2015).

It is observed that the share of two-wheelers is the highest at about 62.13%, followed by cars with 22.53%, whereas buses and auto-rickshaws contribute only 7% (Table 15.3). The sharp increase of two-wheelers and cars could be attributed to the improved economic status of people. The phenomenal increase of cars and the resulting demand for more road-space has resulted in a dense concentration of traffic on roads. This trend has to be kept checked, in terms of the cost it imposes on users demand after careful consideration. Based on the speed and delay survey, the average speed of private vehicles in the city is about 23.87 km/h and for buses is 28.17 km/h in peak hour. It is observed from the analysis that the majority of the roads have a speed between 20 and 30 km/h in peak hour (private vehicles). The average trip length of all passenger modes was is 6.65 km. The average trip length of different modes like cars, buses and two-wheelers were, 10.32 km, 9.5 km and 9.44 km, respectively. In 2014 there were 2634 accidents occurred in Kochi. Increase in private vehicle usage is causing road accidents; it is observed that almost 21% of the victims involved were cyclist or pedestrians in 2014 (CMP 2015).

Traffic volume count conducted for the preparation of the comprehensive mobility plan of Kochi shows that a major percentage of the vehicles are from private passenger

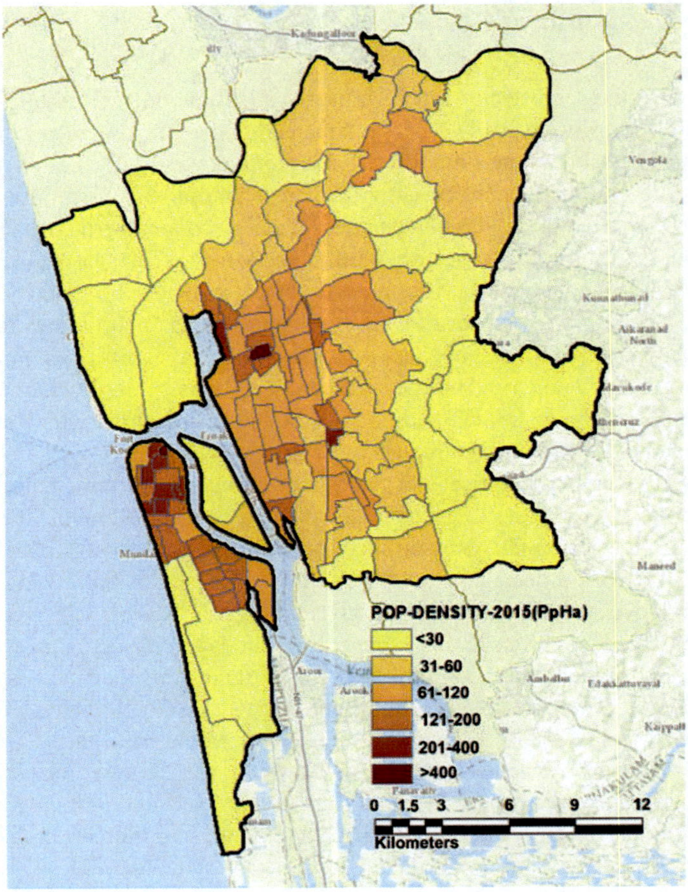

Fig. 15.1 Population density in Kochi City Region. *Source* Comprehensive Mobility Plan, 2015

Table 15.2 Total registered vehicles in Kochi

Vehicle type	2006–07	Percentage	2013–14	Percentage
Scooter/Bikes	422,473	63.10	911,423	62.13
Cars	121,503	18.15	330,524	22.53
Auto-rickshaws	41,894	6.26	87,906	5.99
Buses	21,823	3.26	23,350	1.59
Commercial vehicles	61,816	9.23	113,825	7.76
Total	669,509	100	1,467,028	100.00

Source Comprehensive Mobility Plan, 2015

Table 15.3 Growth rate of vehicles in Kochi

Vehicles	Share of vehicles		Increase (lakh units)	Growth rate (2006–2013) (%)
	2006–2007 (%)	2013–2014 (%)		
Motorcycle/scooter	63.10	62.13	4.89	53.63
Cars	18.10	22.53	2.09	63.24
Auto-rickshaws	6.30	5.99	0.46	52.34
Buses	3.20	1.59	0.02	6.54
Goods	8.20	5.97	0.33	37.42
Tractors	0.40	0.13	− 0.01	− 27.21
Other vehicles	0.70	1.66	0.20	81.22
Total	100.00	100.00	7.98	54.36

Source Comprehensive Mobility Plan, 2015

modes like two-wheelers and cars. On an average 43% of the vehicles are two-wheelers and 23. 49% of cars (Table 15.4). Two third of the road-space is used by personalized modes of transport that carry only 21% of the total trips per day.

15.6.3 Draft Kochi City Region (KCR) Master Plan 2031

The development strategies for traffic and transportation in Kochi are worked out with the aim to support the concept of making Kochi City Region a 'Global City'. It is estimated that the total population would grow from 1,164,225 of 2011 to 2,273,512, which includes migration and floating population components by 2031. In order to arrest urban sprawl in the region, it is proposed to develop a transit-oriented development corridor. The strategies also aim to ensure safe and economical commuting between place of origin and destination, convenient and quick access to all areas, reduction of pollution and congestion, energy efficiency and conservation, safety for all sections of the road and transport users. As part of promoting mass transport system, along with the narrow congested roads in the planning area, the use of minibuses shall be adopted. Simultaneously pedestrianization of internal roads and/or provision of pedestrian facilities shall also be promoted identifying such roads in a manner that mass transit corridors are accessible within walking distance. Use of personal vehicles shall be discouraged as well. New transit-oriented development corridors shall be opened up to stimulate the growth of other growth centres, to attract more population. KCR shall be considered as a single planning unit comprising of eleven planning divisions. The planning divisions should be delineated based on access to major transportation corridors that provide intra-city and or inter-city connectivity. Traffic generated from the planning divisions may flow into the sub-arterial and arterial roads which need to be developed as 'first-order mass transportation corridors'—opting for higher order of mass transport modes

Table 15.4 Traffic composition at screen lines

Survey points	Buses	Other buses	Minibuses	Car	Auto-rickshaw	2 Wheeler	Cycle	Commercial vehicles
Aroor Thopumpady	4.66	0.40	1.45	18.59	21.62	42.75	2.06	8.47
Kudanoor Willington Iland	0.79	0.33	0.70	22.32	12.70	53.74	0.36	9.06
Kizhvanna Road	0.96	0.16	0.22	27.06	20.75	44.50	1.21	5.14
SA Road	6.39	0.08	0.33	27.09	29.69	32.57	0.09	3.76
AL Jacob ROB	1.1	0.06	0.27	23.98	28.92	41.97	0.42	3.28
Pullepady ROB	0	0.00	0.05	9.82	31.52	53.83	0.36	4.42
Banarji Road	26.87	0.47	0.54	24.32	2.53	40.76	0.03	4.48
Chittoor road	0	0.00	0.65	12.47	17.30	61.80	1.99	5.79
Edapally ROB	2.91	0.09	0.60	24.21	11.45	51.53	0.17	9.04
Kalamaserry ROB	0.26	0.64	0.87	16.21	22.78	49.29	0.54	9.41
Aluva Munnar Road	17.43	0.28	0.11	18.02	23.56	27.86	0.07	12.67
Airport Road ROB	1.41	1.19	1.77	50.89	8.05	24.06	0.04	12.59
ROB near Vytilla	2.73	0.12	0.68	34.49	10.95	40.69	0.42	9.92
Vylopillil Road	1.53	0.00	0.13	12.10	27.84	55.32	0.75	2.33
ROB Kathrikadavu	2.62	0.00	0.28	26.31	24.46	41.67	0.26	4.40
Chitoor Road	5.9	0.00	0.10	22.99	9.96	51.43	1.05	8.57
MG Road	29.38	0.00	0.23	25.61	19.76	20.98	0.27	3.77
Park Ave Maharajas College	28.23	0.08	0.55	26.26	2.47	38.22	0.17	4.02
Average	7.40	0.22	0.53	23.49	18.13	42.94	0.53	6.73

Source Comprehensive Mobility Plan, 2014

(high capacity bus system (HCBS), electric trolley bus (ETB), electric tram, sky bus, metro rail, suburban rail or any other mode of mass transport based on feasibility and viability studies. Land zoning and permissible intensity of development within each planning division shall also be guided by the level to which the transit corridors would be developed. Land parcels which have direct access from the major transport corridors can be considered for a higher intensity of development. The intensity of development would depend on the width and grade of transport facility. This means that even within a planning division development regulations may vary (CMC 2010).

15.6.4 Kochi Transit-Oriented Action Plan 2034

With the implementation of Kochi Metro, a transit-oriented action plan has been prepared for the city. The influence area of transit corridor/stations on both sides shall extend to 500 m on either side. The influence area is identified at the corridor level as well as around the transit station (radius of 500 m). 500 m is the comfortable distance with a 5-min walk. The TOD influence area is divided into: (1) Direct Influence Zone or Mixed-Use Zone 1—MX1 (250 m radii from station location). MX1 is envisaged to be predominantly non-residential mixed-use zone in character with residential component essentially less than 30% of the consumed Floor Area Ratio (FAR). The rest of the FAR may be consumed by commercial retail, offices, entertainment uses non-polluting manufacturing, (2) Indirect influence Zone or Mixed-Use Zone 2—MX2 (from 250 to 500 m radii from station location) MX2 is envisaged to be predominantly residential mixed-use zone in character with residential component essentially equal to or more than 50% of the consumed FAR. Currently, the density all along the metro corridor is quite low, i.e., around 100 persons per hectare (pph) and there is scope for further densification. However, the basic premise of the Kochi Transit-Oriented Development (TOD) guidelines is that while the appropriate level of density for a given station will vary with its location, community setting, and function, development should be relatively dense and compact in the immediate station area, compared to its surroundings (KMRL 2016).

To achieve an appropriate TOD density for a given station area, a combination of density baselines, i.e., population and residential density and density bonuses are described. The density baseline would allow the greatest density in the core of the station area, i.e., MX1, area up to 250 m radii from the station location, immediately surrounding the station, and is proposed to undergo a transition downward towards the edges of the station area, where it meets the surrounding neighbourhoods or suburbs. Minimum population density for each station area is proposed to be 500 pph. However, most of the station areas are already developed more than 50% but are characterized with low densities, population as well as residential densities, a slow transition and transformation are expected over the period of time. Hence to guide the urban form and structure of the station area, a minimum residential density standard for all the station areas has been adopted as 200 dwelling units per hectare. Considering the above density norms, a total population of 672,656 is proposed to be accommodated in the identified 22 station areas by 2034. In the case of new development projects and redevelopment projects, the entire parking required for a project shall be provided within the project site as an 'unbundled' facility, i.e., parking spaces shall be sold/leased/rented independent of the dwelling unit or saleable floor space, so that the same space may be used by different users at different times of day (KMRL 2016).

15.6.5 Parking Management in Kochi

At present Kochi city follows the generic minimum parking requirements provided in the Kerala Municipal Building Regulations (KMBR) (KMBR 1999). Majority of off-street Pay and Parking facilities in Kochi are maintained by GCDA, Kochi Metro, and Southern Railway. Parking charges for spots under different agencies are given in Table 15.5. On-street parking is mainly maintained by the Kochi municipal corporation for areas within the corporation boundary. The parking charge for the car (2 h) is Rs. 10/- and for two-wheeler is Rs. 5/-. Street parking within the Kochi municipal corporation is tendered out every year for its operation and management. In addition, there are private off-street pay n park facilities in the city, which can be operated after taking licenses from municipal corporation. At present, there are 12 registered pay and park within the corporation area. The parking inside the railway premises is maintained by the self-help group known as 'Kudumbashree' at

Table 15.5 Parking charges in Kochi in rupees (One US Dollar is equal to Rs. 70)

Duration of parking	Private shopping mall (Lulu)		Kochi Metro		GCDA	
	Cars	2 Wheelers	Cars	2 Wheelers	Cars	2 Wheelers
0–15 min	Free	Free	25	10	20	6
15 min to 2 h	20	10				
3 h	40	20	35	15	40 (12–24 h—Rs. 60)	10 (12–24 h—Rs. 20)
4 h			45	20		
5 h	50	30	50	25		
6 h	60	40	60	30		
7 h	70	50	70	35		
8 h	80	60	80	40		
9 h	90	70	85	45		
10 h	100	80	90	50		
11 h	110	90	100	55		
12 h	120	100	110	60		

Source Compiled from the websites of Kochi Metro, Lulu Shopping Mall and GCDA

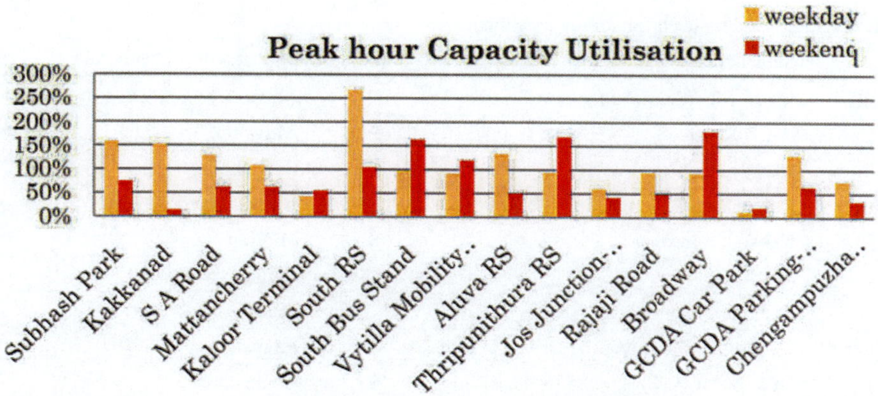

Fig. 15.2 Peak hour parking capacity utilization. *Source* Comprehensive Mobility Plan, 2015

Ernakulam Town, Thripunithura, Ernakulam Junction main entry and Aluva Railway Stations. The profit shared on a 50:50 ratio basis, between 'Kudumbashree' and the Southern Railway. The parking violations are enforced by Kochi city police. There are no heavy parking regulations and policies or parking infrastructure or parking management in Kochi, which has led to underutilization of available infrastructure and overutilization of on-street parking, illegal parking in the city causing traffic congestion, increasing pedestrian accident rates and increasing the travel time for the city dwellers (Fig. 15.2).

Most of the observed areas comprise mixed land use or primarily commercial which generates high parking demand. On-street parking along roads with Right of Way less than 10 m creating traffic chaos in the city core. Encroached parking on footpaths forcing the pedestrians to use carriageway made them more vulnerable to accidents. Parking near junctions reducing the traffic speeds and has increased the delay at junctions. In case of parking near public transit terminals, the railway stations specifically, Ernakulam South Railway Station, Aluva Railway Station and Thripunithura Railway Station showed higher peak hour accumulation and peak hour capacity utilization. This indicates the higher use of park and ride facility at railway stations by commuters on weekdays (Table 15.6; Fig. 15.3).

Parking turn over in the on-street and off-street locations ranges from 2 to 11 vehicles. It shows the higher demand for parking spots and shorter duration of parking. In many spots, the demand for parking spaces is higher than the parking capacity and parking spill over to streets and carriageway causing congestion. A lower parking charge has to be considered as a factor in the highest parking demand in many commercial areas. Parking charges are exclusively collected in selected street stretches public transport interchanges, metro stations and off-street parking spots at major landmarks. No parking charge is levied on major government and private workplaces (Table 15.7).

Table 15.6 Coding used for survey location

A	Subhash Park	O	Mattancherry Dutch Palace
B	Jos Jn-SRV School Junction	P	Aluva (U)
C	Rajaji Road	P′	Aluva (O)
D	Broadway	Q	Pulinchode
E	Kaloor Bus Terminal	R	Ambattukavu
F	Kakkanad Infopark	S	Muttom
G	S A Road	T	Kalamassery
H	Chengumpuzha Park	U	Cusat
I	GCDA Car Park- Marine Drive Ground	V	Edapally
I’	GCDA Car Park-Menaka Building	W	Changapuzha
J	South Railway Station	X	Palarivattom (LHS)
K	South Bus Stand	Y	Palarivattom (RHS)
L	Vytilla Mobility Hub	Z	Kaloor
M	Aluva Railway Station	Z′	MG Road
N	Thripunithura Railway Station		

Source Comprehensive Mobility Plan, 2015

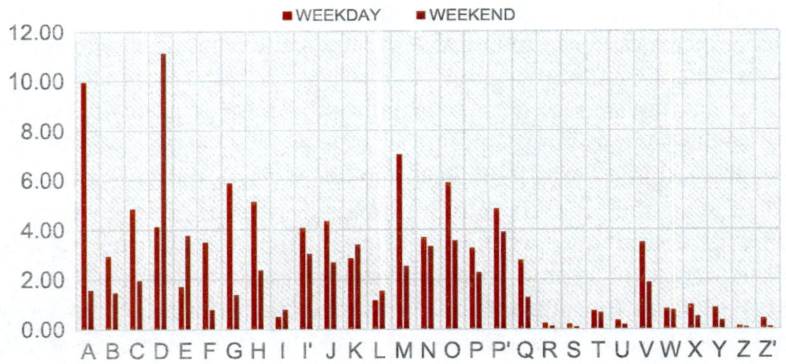

Fig. 15.3 Parking turn over. *Source* Comprehensive Mobility Plan, 2015

15.6.6 Public Transportation in Kochi

Public Transport is the backbone of the Kochi city’s transportation, which includes Metro, city buses, auto-rickshaws and ferry service. Public mass transport facilities share about 51% of all travel (Table 15.8). Bus-based public transportation in the city region is mainly operated by Kerala State Road Transport Corporation, Kerala Urban Road Transport Corporation and private bus. At present, there are 773 city bus routes and 1390 buses running in Greater Kochi Region (CMP 2015). The bus occupancy is obtained as 38.9. 18 km length of Kochi Metro rail corridor has been

Table 15.7 Off-street minimum parking standards

Sl. No	One parking space for every or fraction of
	Group A1—Residential Apartment Houses/Flats
1	(a) 8 units (with each unit up to 100 m^2 of carpet area) (b) 4 units (with each unit above 101 m^2 and up to 150 m^2 of carpet area) (c) 2 units (with each unit above 151 m^2 and up to 200 m^2 of carpet area) (d) Single unit (exceeding 200 m^2 of carpet area)
	Group A2—Lodging and Tourist homes and hostels, Dormitories without any attached eating facility such as a restaurant. Canteen, Cafeteria, mess or dining
	(i) Rooms with attached bath and w.c (a) 8 rooms (with each room up to 12 m^2 carpet area) (b) 5 rooms (with each room above 12 m^2 and up to 20 m^2 carpet area) (c) 3 rooms (with each room above 20 m^2 carpet area) (ii) Rooms without attached bath and w.c (a) 18 rooms (with each room up to 5 m^2 carpet area) (b) 12 rooms (with each room above 5 m^2 and upto 12 m^2 carpet area) (c) 6 rooms (with each room above 12 m^2 carpet area) Note: At the rate of one parking space for every 30 m^2 carpet area of dining space/20 seats of dining accommodation shall be provided in addition to the above, in case of Special Residential. Buildings attached with eating facility
	Group B—Educational (i) High Schools, Higher Secondary Schools, Junior Technical Schools, Industrial Training Institute, etc. (ii) Higher educational institutes
	(i) 300 m^2 of carpet area. (ii) 200 m^2 of carpet area
	Group C—Medical/Hospital: 100 m^2 of carpet area
	Group D—Assembly 25 seats of accommodation Note: (i) In case of wedding halls and community halls, for calculating the carpet area or seating accommodation, for the purpose of off-street parking, the carpet area of either the auditorium or the dining hall, whichever is higher, alone need to be taken (ii) for the purpose of this rule, 1.50 m^2 carpet area shall be considered as one seating accommodation
	Group E—Business/Office Building: 100 m^2 of carpet area
	Group F—Mercantile/Commercial building exceeding 75 m^2 carpet area 100 m^2 of carpet area
	Group G1—Industrial Building exceeding 100 m^2 of carpet area: 200 m^2 of carpet area
	Group G2—Small Industrial exceeding 100 m^2 of carpet area: 200 m^2 of carpet area
	Group H—Storage: 200 m^2 of carpet area

Source Kerala Municipal Building Rules, 1999

open since June 2017. The Kochi Metro is an expanding project with the part still under construction. The first phase of the corridor is set to extend from Aluva to Petta with a distance of 26.6 km (KMRL 2016). At present Kochi Metro carries close to 0.1 million passengers per day. The peak hour frequency of metro is 7 min and buses are 5 min. 67.70% of the households in the Kochi city region have public transport

Table 15.8 Modal split of trips in Kochi

Mode	2014	Mode	2007
Public transport (bus, ferry)	51.10	Public transport	72.32
Car	7.40	Car	4.05
Two-wheeler	13.80	Two-wheeler	5.99
Walk	14.10	Ferry	9.75
Auto-rickshaw	10.40	Auto-rickshaw	4.79
Taxi	1.70	Walk	NA
Cycle	1.10	Cycle	2.51
Train	0.40	Train	0.6
	100.00		100.00

Source Comprehensive Mobility Plan, Kochi

coverage. The government of Kerala has prepared a bill for the creation of the Unified Metropolitan Transit Authority to coordinate the mobility management in the city. A route rationalization process is underway to remove overlapping of public transport buses and metro. Many bus routes will be realigned to work as a feeder to the metro rail service. Kochi metro has introduced close to 50 electric auto-rickshaws and in the process of making arrangements for auto-rickshaws for last mile connectivity. Kochi has a common mobility card for all the public transport public transport modes (Fig. 15.4).

Kochi has a good network of inland waterway systems consisting of backwaters, canals, lagoons and estuaries. Kerala State Water Transport Department (KSWTD) operates the water transport in Kochi. KSWTD operates services from 10 jetties and ferry terminals. The average trip length of ferry passengers is 8.9 km. As per the comprehensive mobility plan prepared for Kochi City Region, 81.36% of the ferry passengers use the ferry services on a daily basis. The achieved speeds of the boats are 5 knots. With dedicated inland water corridors, it could be increased to 12 knots (CMP 2015). There is a potential demand for travel between the island communities and from the island to the mainland. With some of the islands being solely dependent on the ferry services, it is important to improve the ferry services. With a reliable water transport system, it would be possible to increase usage thereby reducing the pressure on the road network (Table 15.9; Fig. 15.5).

15.6.7 Public Transport Accessibility and Parking Cap Zoning

It was observed that over 54.70% of KCR has good access to public transport services and over 67.70% of the households had at least one bus stop within 500 m. 51% of the trips are undertaken by public transport modes. The highest accessibility was observed along the metro corridor and the least values were observed towards the

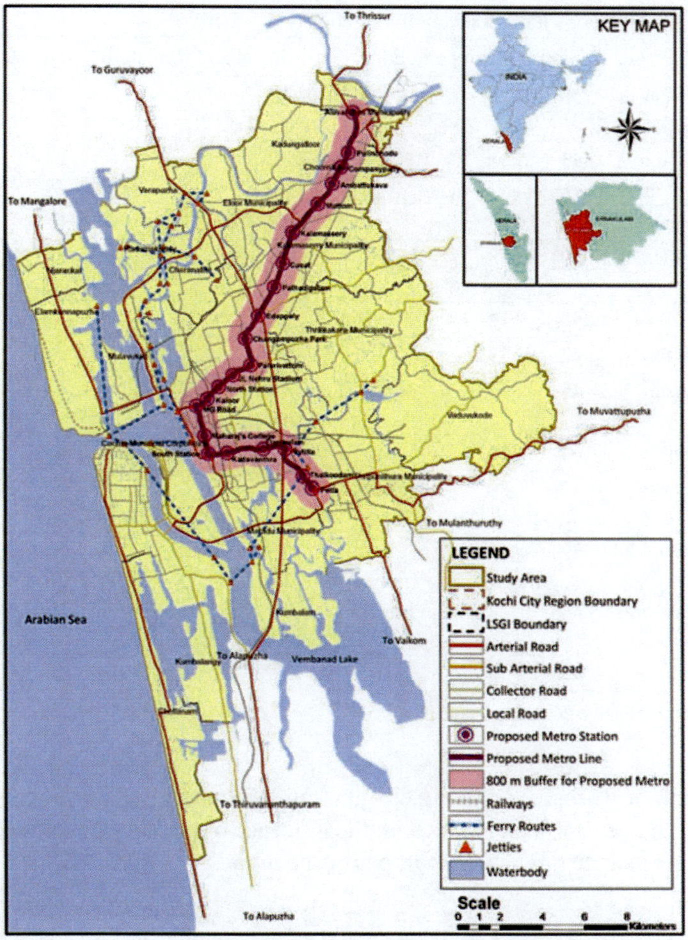

Fig. 15.4 Kochi Metro Rail Corridor. *Source* Comprehensive Mobility Plan, 2015

city peripheries. Figure 15.6a shows the public transport accessibility assessment of the KCR, and it is possible to divide the city into at least four different accessibility zones, A, B, C and D. Figure 15.7 shows the Clustering of Bus Stops and Public Transport Accessibility in the study area. Data from the comprehensive mobility plan employment estimation shows that accessibility zone A and B will have around 80% of the employment in the region. However, Zone A will accommodate more employment in the future due to the proposals of transit-oriented development proposals. Estimations of the employment for various traffic analysis units in each accessibility zone are given in Table 15.10. A survey on the purpose of the trips was revealed that 45% of the trips are made for work. Another 15% of the trips are made by people own their own business enterprises and 7% of the trips are for educational purposes. It means that considering the 21.20% modal split of travel by car and two-wheeler,

Table 15.9 Trip purpose at different cordon locations

Name of the roads	Work	Business	Education	Social and recreation	Tourism	Others
NH 47 Trivandrum Aroor	38	13	10	26	11	2
Kottayam Vaikom Road	43	13	6	30	2	6
Thiruvankulam	37	13	7	34	6	3
Kakkanad Pallikara Road	60	8	7	8	3	14
Vengola road	50	12	4	11	2	21
Vattakattupadi	48	13	4	10	2	23
Aluva Munanr Road	45	11	5	11	4	24
Chellamattom	44	24	6	11	4	12
NH47 Thrissur Road	40	30	5	7	5	13
North Paravoor NH 17	39	16	7	19	7	12
Moothakunna m Road	46	8	15	13	9	8
Average	45	15	7	16	5	13

Source Comprehensive Mobility Plan, 2015

introduction of workplace parking charges can stop modal shift from public transport to private vehicles and in fact with the influence of resource cost-based parking pricing can push the public transport patronage high.

15.6.8 Parking Maximum, Accessibility Sensitive Parking Standards and Workplace Levy

At present Kochi City follows the generic minimum parking requirements provided in the Kerala Municipal Building Regulations (KMBR). Table 15.7 gives the minimum parking norms followed in Kochi. As per the KMBR, Each off-street parking space provided for parking motor cars shall be not less than 15 m^2 of the area (5.5 m × 2.7 m) and for scooters and cycles, the area of each parking space provided shall be not less than 3 m^2 and 1.5 m^2, respectively. It gives a flat parking standard for the entire urban area without considering the availability of transit, land use density, non-motorized transport improvement measures, etc. The current paradigm of parking supply is based on income and vehicle ownership. It is assumed that higher the income people would own more vehicles, hence it is necessary to make regulations for more parking spaces. The core ideal of Transit-Oriented Development recommends a reduction in

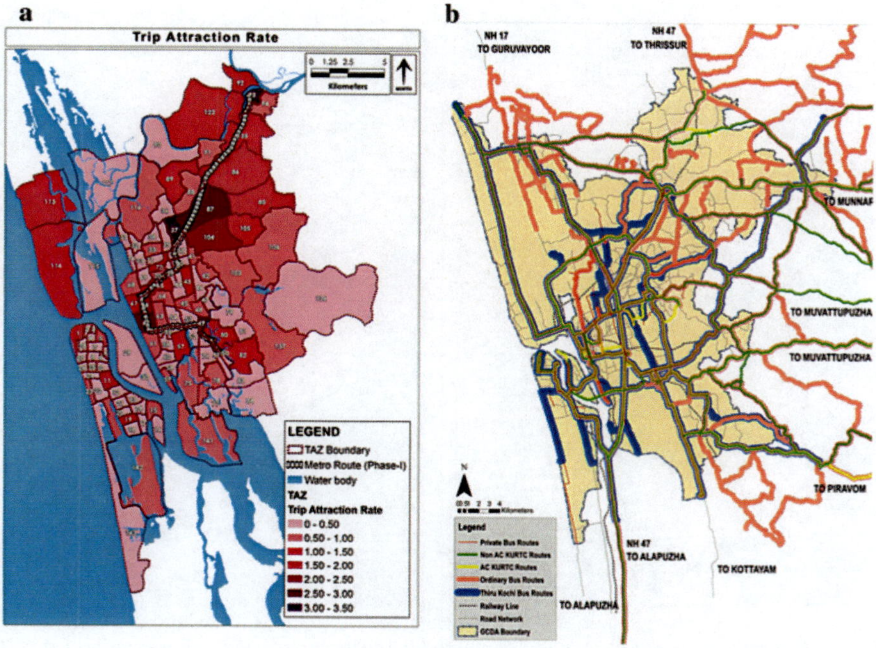

Fig. 15.5 **a**, **b** Trip attraction and Public Transport Routes in Kochi. *Source* **a** Comprehensive Mobility Plan, 2015 **b** Road Transport Office, Kochi

parking supply. Considering the accessibility in Zone A and B, it would be possible to reduce the parking requirement per unit of developed area and parking supply caps can be kept along the transit corridor, around station areas and mobility hubs. Such a reduction in parking requirements can make way for the availability of affordable housing units and valuable land would be used optimally in commercial areas. Study Conducted in Kochi Metro corridors shows that higher the carpet area of the houses lesser is the chance of using the Mass Transit. The study revealed that dwelling units with carpet area size ranging from 60 to 200 m^2 have a higher chance of Metro usage. Metro preference is highest among the respondents with 90–120 m^2 dwelling units. But development control regulations are not modified to encourage construction housing typologies that would be used by mass transit users (Mukundan and Kuriakose 2018). Making such modifications can avoid gradual gentrification of the TOD Zone (Chava et al. 2019). Restrictions on parking supply through parking maximums and supply caps can bring actual transit users closer to the transit.

The concept of workplace levy could be implemented in Kochi. Workplace levy could be implanted in private and government employment locations. Since 67% of the trips are made for work, business and educational purposes, implantation of workplace levy could make a huge difference in inducing a modal shift from private modes of transport to public transport. Information Technology parks, educational institutions and other government offices should be offered with the offered to select

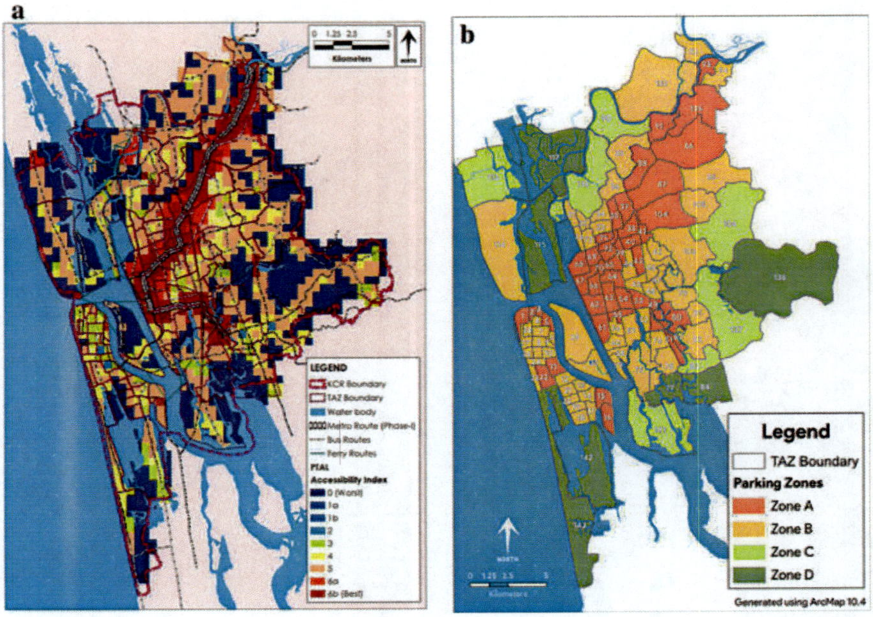

Fig. 15.6 **a**, **b** Public transport accessibility and parking control zones. *Source* Based on the Author's Analysis

free parking or public transport fare with tax-saving benefits. Around 80% of the employment and students are in Zone A and B, a progressive parking charge and restrictive parking supply can be implemented based on the accessibility all over the city.

A reduction from the current minimum requirements can have a remarkable impact in parking supply in high accessibility index zones. Table 15.11 explains the difference between parking supply creation by following the existing parking norms and an assumed reduction of 15% in norms. For the demonstration, a carpet area of 5,000 m^2 is considered under every sub-category of built use. For example, if there a commercial building with 5000 m^2 carpet area is constructed with the existing norms, it is required to provide 50 Equivalent Car Space (ECS) or 750 m^2 area as parking space. But with 15% reduction in norms it ECS could come down to 43 and there will be saving of 115.5 m^2 parking space. This is more than the area of three low-income group housing unit. 15% reduction adopted is for the purpose of demonstrating the benefits, but it requires further studies to arrive at the exact figures for reducing the parking standards.

Based on the recommendation of National Urban Transport Policy, if the parking charges are calculated based on the rent of the land on which the car is parked the parking charges could be as follows:

Rent of a property is calculated by considering 5–6% of the value of the property. As per the KMDR mandatory space for one off-street car parking is 15 m^2 (KMBR

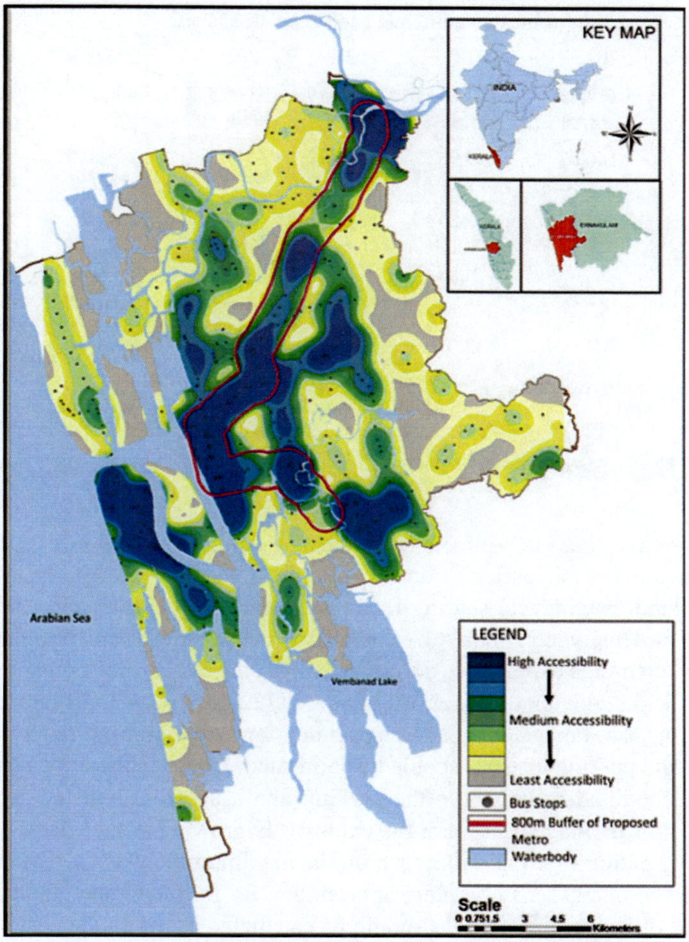

Fig. 15.7 Clustering of bus stops and public transport accessibility. *Source* Based on the Author's Analysis

1999). The average value of 1 m^2 land around the metro stations in Kochi is Rs. 86,483 (One US Dollar is equal to Rs. 70). The total land value for one parking space would be Rs. 1,297,245. So the Rental value at 5% of the land value for one car parking space per month would be Rs. 64,862. At 25 working days a month, and 12 h utilization of parking spot, the total parking spot utilization would be 300 h per month. Then the parking charge per hour works out to be Rs. 64,862/300 = Rs. 216.21, at 100% utilization and at 50% utilization, this will be Rs. 432.42 per hour. But with the current parking charges of Rs. 10 per hour in the city is offering an invisible subsidy to the private vehicle users to commercial areas. All other works, business and educational trips by private vehicles get free parking in their respective places (MoUD 2006).

Table 15.10 Workforce available in different parking control zones

Parking zones	2011	2015	2021	2026	2031	2035
Zone A	114,437 (35.28%)	122,055 (35.68%	123,419 (31.95%)	316,577 (44.73%)	400,265 (47.30%)	414,120 (45.36%)
Zone B	146,637 (45.20%)	149,933 (43.83%)	166,326 (43.06%)	248,871 (35.17%	278,915 (32.96%)	338,068 (37.03%)
Zone C	36,315 (11.19%)	38,484 (11.25%)	52,929 (13.70%)	73,014 (10.32%)	85,708 (10.13%)	76,675 (8.40%)
Zone D	27,020 (8.33%)	31,598 (9.24%)	43,558 (11.28%)	69,218 (9.78%)	81,252 (9.60%)	84,065 (9.21%)
Total	324,409	342,070	386,232	707,680	846,140	912,928

Source Comprehensive Mobility Plan, 2014

15.6.9 How Can India Migrate to a New Parking Paradigm?

Parking standards need to be amended by incorporating parking supply caps, parking maximums and flexible parking standards. Public transport accessibility, availability of mixed land use, bicycle-sharing facilities, walkable streets, residential density, employment density, etc. should be considered in deciding the parking standards. This study attempted to delineate the city only by using public transport accessibility, but other factors like land use density, residential density, walkability improvement measures, etc. The concept of generic minimum standard for the entire city should be abolished and parking districts should be delineated. House registrations procedures should be modified to unbundle the parking and a separate registry needs to be maintained for parking spaces. For the explosively growing private parking market, quantity regulation (putting a tight cap on the maximum number of parking spaces of new developments) may be more appropriate. By following the National Urban Transport Policy, parking charges should be calculated based on the cost of land on which the private vehicle is parked.

There is a need for strong legislative support to implement car parking free development in Central Business Districts, residential areas within the accessible catchment area of public transport nodes, allowing parking fee in private development, fines for parking fee violations, implementation of parking maximums, land value and congestion responsive parking pricing and workplace levies. With the introduction of Goods and Service Tax parking space, managing vendors are paying tax on parking fee. Parking fee and fines are can be a major source of revenue for municipal corporation and proper ring-fencing of this revenue can help the urban local bodies to push sustainable mobility. Considering the bigger role of parking charge in average generalized cost of trip private parking supply should be adequately charged and the concept of free parking at the workplaces should be abolished. Legal tools should ensure the availability of parking space or ownership certificate of parking spaces for vehicle registration. First, it is a straightforward and relatively cheaper method to

Table 15.11 Possible ECS reduction and carpet area saving

Built use	Built use sub-category ECS/Carpet area	No. of units	No. of ECS	Parking area	15% reduction	Area saved	Final ECS
Commercial	>75 m^2 1 ECS/100 m^2 (5000/100)	1	50	750	637.50	112.5	43
Residential	8 units (with each unit up to 100 m^2) (5000/90)	56	7	105	89.25	15.75	6
	4 units (with each unit above 101–150 m^2) (5000/125)	40	10	150	127.50	22.50	9
	2 units (with each unit above 151–200 m^2) (5000/175)	29	14	210	178.50	31.50	12
Lodging/Hotels (i) Rooms with attached bath and w.c	8 rooms (with each room up to 12 m^2 (5000/12)	1	52	780	663.00	117.00	44
	5 rooms (with each room above 12–20 m^2) (5000/15)	1	67	999	849.15	149.85	57
	3 rooms (with each room above 20 m^2) (5000/25)	1	67	1000	850.00	150.00	57
Educational	Schools: 300 m^2 (5000/300)	1	17	250	212.50	37.50	14
	Higher educational institutes: 200 m^2 (5000/200)	1	25	375	318.75	56.25	21

Source Authors Calculations

reduce illegal parking in residential land uses. It can also help to slow down motorization and the strategy is spatially and dynamically efficient because it helps to impose a higher tax for registering a vehicle where/when land is more expensive. Similar strategies are implemented in Japan and successfully reduced the pressure on parking supply and illegal parking in residential land uses (ADB 2011). Parking enforcement is an important facet in using parking as a travel demand management. Parking enforcement responsibilities are is fragmented with traffic police and municipal corporation in India. It reiterates the need for United Metropolitan Transport Authorities (UMTA) for metropolitan areas. There should be a dedicated wing created for parking management within the bigger principle of public transport promotion and travel demand management. UMTAs should be maintaining the database of the entire on-street, off the street and public and private parking supply for effective policy implementation and monitoring.

15.7 Conclusion

Parking should be considered as an economic activity to revive the urban transport in the city and hence a comprehensive parking policy and a parking action plan are necessary to revitalize the existing parking situation in the study area. Many cities in different countries have moved away from generic minimum parking standards to parking maximums and parking supply caps. It appears that the time for India to move towards maximum-based parking supply, parking caps and workplace levy has arrived. It would be advisable to implement parking-based transport demand management strategies along with the investments in public transport. The assessment conducted for Kochi as a sample city reveals that a major share of the city has public transport accessibility and it is possible to delineate the city into different accessibility zones for the purpose of implementing parking supply caps and parking maximums. Half of the trips per day are made by public transport and it is a good scenario to give further fillip to public transport ridership by resource cost-based parking pricing.

References

Asian Development Bank (ADB) (2011) Parking policy in Asian cities. The Asian Development Bank, Manila

Census of India (2011) Rural urban distribution of population. https://censusindia.gov.in/2011-prov-results/paper2/data_files/india/Rural_Urban_2011.pdf. Accessed 10 Oct 13

Chava J, Newman P, Tiwari R (2019) Gentrification in new-build and old-build transit-oriented developments: the case of Bengaluru. Urban Res Pract 12(3):247–263. https://doi.org/10.1080/17535069.2018.1437214

CMC (2010) Draft development plan for Kochi City region 2031. Department of Town and Country Planning Government of Kerala. https://cochinmunicipalcorporation.kerala.gov.in/documents/10157/17825/Vol1_Study%26Analysis.pdf?version=1.0. Accessed 01 Dec 2018

CMP (2015) Comprehensive mobility plan for Greater Kochi Region. Kochi Metro Rail Limited, Kochi

Dale S, Frost M, Ison S, Quddus M, Warren MP (2017) Evaluating the impact of a workplace parking levy on local traffic congestion: the case of Nottingham UK. Transp Policy 59:153–164. https://doi.org/10.1016/j.tranpol.2017.07.015

Evangelinos C, Tscharaktschiew S, Marcucci E, Gatta V (2018) Pricing workplace parking via cash-out: effects on modal choice and implications for transport policy. Transp Res Part A: Policy Pract 113:369–380. https://doi.org/10.1016/j.tra.2018.04.025

Guo Z, Ren S (2013) From minimum to maximum: impact of the London parking reform on residential parking supply from 2004 to 2010? Urban Stud 50:1183–1200. https://doi.org/10.1177/0042098012460735

Joshua EY, Dylan P (2010) Assessing alternative approaches to setting parking requirements. ITE J 80(12):30–34

KMBR (1999) Kerala municipality building rules. Government of Kerala. https://buildingpermit.lsgkerala.gov.in/content/rules/kmbr_rule.pdf. Accessed 01 Dec 2018

KMRL (2016) Transit oriented development action plan for Kochi. Kochi Metro Rail Limited, Kochi

Kodrasky M, Hermann G (2011) Europe's parking U-turn: from accommodation to regulation. Institute for Transportation and Development Policy, ITDP. https://www.itdp.org/wp-content/uploads/2014/07/Europes_Parking_U-Turn_ITDP.pdf. Accessed 01 Dec 2018

Kuriakose PN (2015) Parking policy—a tool for inducing public transport ridership: strategies and lessons from the developed world. Indian J Transp Manag

Li F, Guo Z (2014) Do parking standards matter? Evaluating the London parking reform with a matched-pair approach. Transp Res Part A: Policy Pract 67:352–365. https://doi.org/10.1016/j.tra.2014.08.001

Litman T (2016) Parking management strategies, evaluation and planning, Victoria Transport Policy Institute. https://www.vtpi.org/park_man.pdf. Accessed 12 Oct 18

Marsden G, Frick KT, May AD, Deakin E (2011) How do cities approach policy innovation and policy learning? A study of 30 policies in Northern Europe and North America. Transp Policy 18:501–512. https://doi.org/10.1016/j.tranpol.2010.10.006

McDonnell S, Madar J, Been V (2011) Minimum parking requirements and housing affordability in New York City. Hous Policy Debate 21:45–68. https://doi.org/10.1080/10511482.2011.534386

MoRTH (2017) Road accidents in India—2017. Ministry of Road Transport and Highways, New Delhi. https://www.indiaenvironmentportal.org.in/files/file/road%20accidents%20in%20India%202017.pdf. Accessed 01 Dec 2018

MoHUA (2018) Urban transport metro rail projects. https://mohua.gov.in/cms/Urban-Transport-Metro-Rail-Projects.php. Accessed 12 Oct 18

MoUD (2006) National urban transport policy-2006. https://urbanindia.nic.in/policies/TransportPolicy.pdf. Accessed 04 Oct 13

Mukundan P, Kuriakose NP (2018) Regulating transit oriented development for inducing transit friendly housing: a case study of Kochi. School of Planning and Architecture Bhopal

Murali RM, Kumar PKD (2014) Implications of sea level rise scenarios on land use/land cover classes of the coastal zones of Cochin, India. J Environ Manag (Elsevier Ltd.). https://doi.org/10.1016/j.jenvman.2014.06.010

NBC (2016) National building codes, vol I. Bureau of Indian Standards the National Standards Body of India, New Delhi

Shoup DC (1997) Evaluating the effects of cashing out employer-paid parking: eight case studies. Transp Policy 4:201–216. https://doi.org/10.1016/S0967-070X(97)00019-X

Shoup DC (1999) The trouble with minimum parking requirements. Transp Res Part A: Policy Pract 33:549–574. https://doi.org/10.1016/S0965-8564(99)00007-5

Transport for London (2015) Assessing transport connectivity in London 60. https://content.tfl.gov.uk/connectivity-assessment-guide.pdf. Accessed 03 April 2019

Chapter 16
Assessing the State of Homeless People to Plan Inclusive Smart Regions

Vinita Yadav

Abstract Regions are Smart Regions if they are administered, managed and governed well. The regions having large size cities in its core can become smarter with greater emphasis on 3Es i.e. Equity, Effectiveness and Efficiency. Many questions remain unanswered while developing an understanding of inter-dependence and inter-linkage of cities upon regions. These questions are: Are cities sustainable or depend upon larger region for their sustenance? Would the smart city development proliferate the gain to a larger region leading to holistic development? Which parameters and factors help homeless to be part of smart city in the regional context? Is smart a reality or a misnomer for homeless people? How useful is the geospatial approach for identifying homeless people, their condition and to formulate strategies for their wellbeing in the regional context? The paper tries to answer these questions by assessing the conceptual, theoretical and on ground condition of homeless people using smart-based approach by adopting geospatial technology.

Keywords Effectiveness · Equity · Geospatial · Inter-dependence · Smart regions

Acronyms

CO	Carbon monoxide
CO_2	Carbon dioxide
ISBT	Inter State Bus Terminal
IGSSS	Indo-Global Social Service Society
ITO	Income Tax Office
NO_2	Nitrogen dioxide
OFFER	Organization Functioning for Eytham's
PM	Particulate Matter
ppb	Parts per billion

V. Yadav (✉)
Department of Regional Planning, School of Planning and Architecture, New Delhi, India
e-mail: v.yadav@spa.ac.in

P. Sharma (ed.), *Geospatial Technology and Smart Cities*, The Urban Book Series,
https://doi.org/10.1007/978-3-030-71945-6_16

SO_2	Sulphur dioxide
TERI	The Energy and Resources Institute
US	United States
WAYU	Wind Augmentation Purifying Units

16.1 Introduction

A smart region requires innovative approach to cater to infrastructure, livelihood and livability questions related to homeless people. Smart is a concept, which envisages upon provision, access and availability of the facilities for all but needs to focus on homeless more as a special group. For example, Helsinki (Finland) smart region is a region famous for innovation, digitalism and security whereas Pimpri Chinchwad (Maharashtra, India) is an example of economically viable region. To turn regions into Smart Regions, the focus is mostly on their management and governance whereas it shall be on tackling the real-life issues such as homelessness.

Most of the class I size cities depend upon the extended region. While under smart city mission, focus is on provisioning of infrastructure in such cities. The proportionate amount of money is neither being provided to creatte the facilities for poor people nor to manage the extended region. This would have helped to minimize the shift of marginalized people to such cities for their sustenance and in turn, reduces the labour movements. Thus, it hampers the pro-poor growth and negates the equitable basis of planning for the region. There is a lack of inter-departmental co-ordination amongst institutions involved in the provisioning of services and management of their delivery to the poor people leading to ineffective and inefficient implementation.

In a highly urbanized world, smart, innovative and sustainable form of urban development is possible only with the application of latest geospatial tools for assessing and providing solutions. The developing countries face challenges of poverty and homelessness. It is homeless people who further faces severity of existent mobility and pollution related issues in region. The practical proposition to assess the challenge and find technological solution is the need of hour in 21 century. The usage of geospatial technology helps to assess the situation, which requires urgent attention and requires differential approach to find out the solutions of the problems faced by homeless people. Both fundamental and applied research are required to understand the usage of geospatial technologies for smart solutions for homeless people.

While developing an understanding of homeless people and facilities provided for them in large size city region, geospatial technology helps to find out their location and plan for them in the context of a smart region. Geospatial approach involves the tagging of location using geo-referencing, capturing the location using radar and capturing the exact image of the ground-level situation for ground trothing using drone.

The paper attempts to analyse the inclusivity of region especially for homeless people. The conceptual and practical basis of smart-based approach is used to assess homeless people who are less secure and vulnerable. To achieve inclusivity of smart regions, parameters and factors are identified to reduce homelessness as part of smart movement in the regional context. The smart cities are dependent upon region for their sustenance but their smartness need to be studied in the light of city's inter-dependence upon region for tackling homelessness. The pictorial evidences from different parts of Delhi is used as segment is highly under reported. Usability of geospatial approach to identify homeless people and their condition is assessed and strategies are formulated to plan smart regions with the help of geospatial tool and technology.

16.2 Homelessness and Security

Homeless people are fragile both physically and mentally. The poverty and homelessness are two sides of a coin as mostly poor people are homeless and face physical and social assault. In 2015, 10% of the world's population lived on less than US$1.90 a day, compared to 11% in 2013. In India, 270 million (22%) people out of 1.2 billion lived below the poverty line of $1.25 in 2011–12 (Mehra 2016). India has been the biggest contributor to poverty reduction between 2008 and 2011, with around 140 million or so lifted out of absolute poverty still it has huge number of people below the poverty line (Chakravarty 2014). Among the people below the poverty line, 400 million are children up to 18 years and 60 million children are under the age of 6 in India. Amongst the poor, 4, 49,761 are houseless households/ families, and 17, 73,040 (47%) are homeless people in India (Thiyagarajan et al. 2018). Out of homeless people, 53% are residing in urban areas and 47% in rural areas (Census of India 2011). Amongst homeless people, street children are a significant category and the estimates of street children in India range from 4 to 11 million (Singh 2016). These numbers are only the estimates and different data source enumerate poor and homeless differently (Sattar 2014). Going through the statistics given above, it is salient that poor requires a space in dynamic changing environment.

Smart City Mission launched in 2015 aims at creating 100 smart cities. The mission did talk about turning a city into smart city but with no focus on homeless, migrants or women. In Delhi, shelters for homeless people are 230 in number in the year 2021 but homeless people still prefer to live on the streets (DUSIB 2021). This is either due to their unawareness of such facilities or they are fearful of staying inside due to pathetic sanitary conditions (IGSS 2018). Despite government's effort, homelessness still exists and is further increasing. The homeless poor people sleep largely on the footpath in an unsafe environment. This situation is worrisome especially for women and children (Fig. 16.1). Smart City Mission aims to create 100 smart cities in India but is largely gender neutral (Chaudhry 2019). It has not focused on security of tenure, and access to affordable services, which are primary for homeless people (IGSSS and OFFER 2019).

Fig. 16.1 Homeless person sleeping on a footpath at Connaught place, Delhi. *Source* Picture courtesy author

Amongst homeless, physically challenged are in huge numbers requiring utmost attention. Figure 16.2 reveals that many of them are stationed near religious institutions for survival and meeting their basic requirement. Their wheel chair serves multiple purposes. Other than a mode of transportation, it is also used for entertainment (reading newspaper) or storage or seeking alms or to sleep. For a few, redesigning the wheelchair is the solution to cater to all their requirements. However, for large number of homeless physically challenged persons, it is important that an inventory of their location, basic profile such as age, sex, health status shall be prepared.

For identification of their location, geospatial technologies and drone usage have a much bigger role to play. Geo-referencing requires internal coordinate system of a map or aerial photo image to be related to ground system of geographic coordinates of location. The geo-referencing helps to identify the location of homeless people residing on unsafe spaces. The location is an attribute required for preparing spatial

Fig. 16.2 Physically challenged homeless in Connaught Place, Delhi. *Source* Picture courtesy author

strategies. Further, geo-tagging is required to assess what is the success rate of the intervention to decrease the number of homeless people and to bring them above the poverty line. The three-dimensional analysis of the space, occupied by physically challenged homeless people in Delhi, will help to locate their exact location and also to redesign the public places so as to provide an ease of access and overall improvement in their quality of life. By catering to the special need of such homeless people, the objective of equitable distribution of resources would also be achieved.

16.3 Environment Pollution and Smart Tools

In India, seven out of ten worlds' worst cities in terms of air pollution are located (Economic Times 2019). At a regional scale, air pollution rises due to construction work and burning of stubble. Amongst the world top most polluted cities were Gurugram, Ghaziabad, Faridabad, Noida and Bhiwadi in 2018 but Delhi also had an average air quality index of 113.5 in the same year. According to a study by The Energy and Resources Institute (TERI), 34% of Delhi's Particulate Matter (PM) 2.5 level of air pollution is generated locally in winters (Hindustan Times 2018). Such pollution levels do affect all the residents but homeless people are most affected by air pollution in Delhi.

The pollution creates maximum impact on the homeless people staying on the pavement, along the corridor and behind the main commercial buildings. To tackle the worst impact of air pollution, Delhi government announces ban on construction and excavation annually in the month of October till beginning of November, adopts plying of only Odd and Even vehicle for a period of 15 days, closes the school and sprinkles water using the vehicles whenever PM 2.5 increases beyond permissible level. All such initiatives are designed to cater to the requirements of middle and high-income residents with limited thinking for their effect on poor homeless people.

For homeless people, specific solution, i.e. monitoring the point wise data of pollution levels and micromanagement of pollution at their locations will minimize the ill effects of pollution. The installation of air pollutant absorbent's equipment helps to reduce the harmful impact of pollution. This further requires determining the location best suited for its installation to derive maximum benefit of the healthier environment for the homeless people. The latest techniques such as geospatial technology, high altitude balloon technology with camera and biosensors to monitor the air quality provides solution (Press Trust of India 2019). For micromanagement, installation of device like Wind Augmentation Purifying Units (WAYU) can curtail the pollution levels at the locations where homeless people are residing (Fig. 16.3).

The space is always a constraint in Delhi, which is one of the main factors for homeless people being forced to reside on footpaths or besides the drains. Nearer to such areas, open green space has shrunken. In such a scenario, greenery through vertical garden reduces the pollution levels, enhance the aesthetics of dismal space and improve the overall environment (Fig. 16.4). Currently, vertical gardens have been placed along the highway or near the railway station for curtailing pollution and

Fig. 16.3 Installations of WAYU at ITO Junction. *Source* Picture courtesy author

Fig. 16.4 Vertical Garden *Source* Picture courtesy author clicked on 1 February 2019

beautification. For finding out the ideal location for installation of vertical gardens to improve micro-environment, drone technology needs to be used to pinpoint the location where such homeless people are residing. However, their identification shall not be used by authorities to create atrocious living environment for such people.

By using nanomaterial, detection limit will decrease to parts per billion (ppb) concentration levels. This helps to microlocate the areas occupied by homeless people, which is also worst affected by pollution. The levels of CO_2, CO, NO_2 and SO_2 can be detected with the help of a chip and by using four different sensing elements. This will help planners to gauge the presence of black carbon aerosol produced from burning of biomass, fossil fuels, vehicular emissions and coal-fired thermal power plants in the locations where homeless people resides.

16.4 Public Space and its Usage by Homeless People

Public space is not a static place or object rather it is a struggle to define and create a public space (Dehaene and Cauter 2008). Homeless people use public space either for selling goods or for accommodation or for storage purposes (Fig. 16.5). Footpath is one such public space, which is a pedestrian way but mostly encroached upon either by vendors or homeless dwellers. Street vendors and hawkers offer access to a wide variety of goods and services in public spaces especially to homeless people (Basu and Basu 2016). In Delhi, homeless people are residing mostly on public spaces as '*Rain Basera*' (shelter homes) provided for them are either insufficient in number or nonhabitable due to poor living conditions (Jha 2015). This necessitates the requirement of mapping the public spaces along with its type of usage. The public space usage in terms of their location, area covered, type of usage, etc. shall be mapped three-dimensionally using geo-referencing, geo-tagging, and drone technology. The images from Bhuvan shall be accessed to locate the homeless persons' attribute. The space occupied for accommodation and earning livelihood shall be mentioned separately in the records. Such mapping and record keeping exercise will help to make a comprehensive plan to make public spaces, especially footpaths, usable for homeless persons. The data will help to assess the number of homeless individuals who has to be rehabilitated as well as provided shelter, food and clothing. The formal means of employment for homeless persons will increase their earning capabilities and in turn, lead to affordability.

To conclude, above field-based examples help development professionals to provide practical solutions to the societal issues related to homelessness in general

Fig. 16.5 Public space used as accommodation (storage, sleeping, Parking) in front of Inter State Bus Terminal (ISBT), New Delhi. *Source* Picture courtesy author clicked opposite Inter State Bus Terminal (ISBT) Delhi on 8 April 2019

but in case of Delhi specifically. This is possible with active involvement of stakeholders from public sector, research-based organizations, industries, enterprises and think tanks. The concept of citizen-based region shall be in place for citizen centric planning for homeless people by adopting region wide solutions using geospatial technology. Both human centric and technology intensive service provision are must for the well-being of homeless peoples.

References

Basu M, Basu S (2016) The role of Hawkers in creating encroachment: a case study of Kolkata. Int J Human Soc Sci Stud 3(1):70–80

Census of India (2011) Primary Census Abstract. Registrar General of India, New Delhi

Chakravarty M (2014) The World bank on India's poverty. Live Mint, 13 Oct 2014. Accessed from https://www.livemint.com/Opinion/xrATLLP8ojKEVEQgJV0UxJ/The-World-Bank-on-Indias-poverty.html. Retrieved 17 May 2019

Chaudhry S (2019) Smart city just city-making 'Smart Cities' inclusive, June 3, 2019. Accessed from https://www.ihcglobal.org/2019/06/03/smart-city-just-city-blog-series-2-2-2-2-2/. Accessed on 9 March 2020.

Dehaene M, Cauter LD (2008) Heterotopia and the city: public space in a post civil society. Routledge, Oxon

Economic Times (2019) 7 of the top 10 most polluted cities in the world are in India, 6 March 2019. Accessed from https://economictimes.indiatimes.com/news/politics-and-nation/7-of-the-top-10-most-polluted-cities-in-the-world-are-in-india/articleshow/68264913.cms?from=mdr. Accessed on 17 May 2019.

Hindustan Times (2018) Ban on construction work in Delhi-NCR till Nov 10 to combat pollution, violators to face action, October 28, 2018, 08:44:15. accessed on 1 July 2019 from https://www.google.com/amp/s/m.hindustantimes.com/delhi-news/ban-on-construction=work-in-delhi-ncr-till-nov-10-to-combat-pollution-violators-to-face-action/story-UxlwTUgSqV5duoYpxOieyN_amp.html

Indo Global Social Service Society (IGSSS) (2018) Homelessness Shelters in Delhi: social audit report on problems, status and challenges. Indo-Global Social Service Society, New Delhi

IGSSS (Indo-Global Social Service Society) and OFFER (Organisation Functioning for Eytham's) (2019) Enabling inclusive city for the homeless, Indo-Global Social Service Society (IGSSS), & Organisation Functioning for Eytham's (OFFER) New Delhi

Jha BK (2015) Rain Basera—help the homeless. https://www.mapsofindia.com/my-india/government/rain-basera-help-the-homeless. January 18, 2015

Mehra P (2016) 8% GDP growth helped reduce poverty: UN report. The Hindu, 2 April 2016. Accessed from https://www.thehindu.com/news/national/8-gdp-growth-helped-reduce-poverty-un-report/article6862101.ece. Retrieved on 17 May 2019.

Press Trust of India (2019) High flying baloons, quality air sensors being used to fight airpollution, India Today. accessed from https://www.google.com/amp/s/www.indiatoday.in/amp/science/story/high-flying-baloons-quality-air-sensors-to-fight-air-sensors-to-fight-air-pollution-1542534-2019-06-04. Retrieved on 4 June 2019

Sattar S (2014) Homelessness in India. Shelter—HUDCO Publication. 15:9–15

Singh IP (2016) City Makers. Perfect Press, New Delhi

Thiyagarajan A, Bhattacharya S, Kaushal K (2018) Homelessness: an emerging threat. Int J Healthcare Educ Med Inform 5(2):18–20

Delhi Urban Shelter Impreovement Board (DUSIB) (2021), Occupancy Report, Delhi: DUSIB

Part II
Urban Ecology and Disaster Management

Chapter 17
Fire and Flood Vulnerability, and Implications for Evacuation

Alan T. Murray, Richard L. Church, Jing Xu, Leila Carvalho, Charles Jones, and Dar Roberts

Abstract Vulnerability to fire and floods is ever increasing, placing people and communities at great risk. Climate change, drought, overgrown vegetation, and naturally prevailing weather conditions make coastal regions in southern California a prime example of this susceptibility, particularly at or near the urban-wildland interface. The recent Thomas Fire in December of 2017 and subsequent flooding and mudslides in Montecito in January of 2018 highlight what coastal vulnerability means under the new normal of multifaceted risk. This chapter discusses unique fire and flooding hazards along with local weather conditions that contribute to vulnerability. We then detail geo-spatial technologies to assess, model, and predict risks as well as methods to mediate vulnerabilities. Of particular focus here is the role of evacuation, and how redundant alerting systems are critical.

Keywords Climate change · Vulnerability · Flood · Fire · Urban-wild land interface

A. T. Murray (✉) · R. L. Church · J. Xu · L. Carvalho · C. Jones · D. Roberts
Department of Geography, University of California at Santa Barbara, Santa Barbara, CA 93106, USA
e-mail: amurray@ucsb.edu

R. L. Church
e-mail: rick.church@ucsb.edu

J. Xu
e-mail: jxu@ucsb.edu

L. Carvalho
e-mail: lvdeca@ucsb.edu

C. Jones
e-mail: cjones@ucsb.edu

D. Roberts
e-mail: daroberts@ucsb.edu

P. Sharma (ed.), *Geospatial Technology and Smart Cities*, The Urban Book Series,
https://doi.org/10.1007/978-3-030-71945-6_17

Acronyms

GIS Geographic Information System
MCLP Maximal Covering Location Problem

17.1 Introduction

There is continued and increasing interest in fire and flood impacts on people. Many reasons are behind such interest. One aspect is changing climate conditions across the globe over recent decades. This has altered seasonal weather patterns, bringing spans of drought as well as periods of intense precipitation (Easterling et al. 2000). In vegetated areas drought substantially increases fire vulnerability (McDowell et al. 2015). Drought, and associated loss of vegetation cover, also alters the ability of soils to absorb precipitation, especially during intense events. The result is flooding. Perhaps the most significant issue is what appears to be the intensity of such events and their interaction. Drought combined with fire followed by intense precipitation equals a multifaceted disaster.

Water shortages and vegetation stress are a major byproduct of drought. Moritz et al. (2014) note that fires are a natural process, and therefore inevitable. Clearly drought heightens risks. But of course, drought and fire, alone or in combination, pose further risk when followed by precipitation events. All of these issues combine to impact flora, fauna, soils, aquifers, etc. in different ways. While these impacts are significant in their own right, the involvement of humans presents even more pressing concerns. In many ways fire and flooding represent a range of vulnerabilities, particularly to where humans live, work, interact, and otherwise occupy.

According to Wisner and Adams (2002), vulnerability has to do with the degree to which a population, individual, or organization is unable to anticipate, cope with, resist and recover from the impacts of an incident. Addressing vulnerabilities therefore necessitates capabilities to both characterize and understand the incidents of interest. In this case the focus is on fire and flooding, so there is a need for expertise related to climate, weather, vegetation, wildfire, land use, human behavior, and many other related issues. That is, interdisciplinary science is an essential step toward developing insights and knowledge. Significant progress has indeed been made along these lines, reflected in the work of Carvalho and Jones (2015), Cannon et al (2017), Williams et al. (2018), and Veraverbeke et al. (2018), among others. The ability to bring disparate information and knowledge together in meaningful ways no doubt remains. With greater understanding comes the ability to undertake various mitigation efforts. This may include things like communication, response, and change, as suggested in Table 17.1. Under the heading of communication might be quantification of event likelihood probabilities, service announcements that detail risks, but also various forms of warnings. The response side might include better management of fuels, increasing investment in firefighting resources, enhancing spatial positioning

Table 17.1 Vulnerability mitigation efforts

Communication	Response	Change
• Quantify event likelihood • Public service announcements • Emergency broadcasting • Sirens • Television • Radio • Social media • Internet	• Fuels management • Additional firefighting resources • Strategic positioning of firefighters/equipment • Investment in more sophisticated monitoring technology	• Greenhouse gas reductions • Point source pollution elimination • Enhanced flood control infrastructure • Use of non-flammable building materials • Depopulate the urban-wildland interface • Reduce urban-heat island

of firefighting crews to reduce damages/impacts, prohibiting development in areas where flooding is possible/inevitable, etc. Finally, change may be pursued through efforts focused on climate triggers, such as reducing greenhouse gases, eliminating sources of pollution, decreasing usage of non-renewable resources and the like, but also altering the built environment to be impervious to fire and flood vulnerabilities.

The remainder of the chapter examines the communication aspects of mitigating vulnerabilities associated with fire and floods. The unique situation encountered in Santa Barbara, California provides the context for vulnerability communication. Geospatial technology and aspects of the smart city offer the foundation for undertaking communication, with specific spatial models enabling enhanced capabilities that would complement contemporary approaches. Findings are detailed to support a range of potential decision-making scenarios. This is followed by a discussion in the context of fire, flood and debris flow along with concluding observations.

17.2 Background

While fire is naturally occurring and part of the germination process for many flora, its impact on people and communities is substantial. In the U.S., federal government expenditures alone on wildfire suppression exceeded $3 billion in 2018 according to the National Interagency Fire Center.[1] The U.S. National Park Service indicates that 85% of wildfires are caused by humans.[2] Flooding too is a naturally occurring phenomenon, but as highlighted in a recent National Academies of Sciences, Engineering, and Medicine (2019) report, the impacts and costs are exceptional. The significance of fires and flooding is undeniable. Some areas are more susceptible to fire and/or flooding, depending upon geographic location and associated weather, climatic, and other conditions. Irrespective of causes and conditions, of particular interest, is generally the urban-wildland interface, the areas where communities are

[1] https://www.nifc.gov/.

[2] https://www.nps.gov/articles/wildfire-causes-and-evaluation.htm.

within or next to significant amounts of vegetation (Moritz et al. 2014). The reason for this is fire vulnerability and associated potential for debris flows and flooding following a fire incident, because people and property are at elevated risk along the urban-wildland interface. The Santa Barbara region (including the cities of Goleta, Santa Barbara, Montecito, Summerland, Toro Canyon and Carpinteria) is reflective of a community at or along this interface given its unique geographic position. Nestled between the Santa Ynez Mountain range and the Pacific Ocean, this region is prone to regularly occurring wildfire. Kolden and Henson (2019) discuss the historical significance of fire in this area. Table 17.2 provides a summary of major wildfires in the area over the past decade. What makes this all the more noteworthy is that there is a substantial number of people residing in the region (approximately 150,000), with access into and out of the area effectively limited to U.S. Route 101 and California State Route 154. Beyond this, topography, coastal barriers, dry vegetation, and unique micro-climate conditions in the form of Sundowner winds make fire at the urban-wildland interface particularly challenging to deal with. As if wildfire is not enough, add to this the potential for flooding and debris flows following fires.

These local characteristics combined with real dangers mean that people and property are vulnerable. Geo-spatial and smart city technologies offer great potential for mitigating vulnerability in many ways. Geo-spatial technologies include a range of quantitative approaches that support analysis, policy, and planning involving geographic space. Murray (2010, 2017) discusses the associated approaches in this area, including geographic information system (GIS), remote sensing, spatial statistics, metrics/measures, simulation, spatial optimization, geovisualization, and the like. Perhaps most significant is that such methods are used in combination, often in an integrated fashion, in order to support decision-making processes.

Table 17.2 Santa Barbara area wildfires in past decade, 2009–2019

Name	Dates	Acres	Structures	Deaths
Holiday	July 2018	113	25 destroyed, 3 damaged	0
Thomas	December 2017	281,893	1,063 destroyed, 280 damaged	2
Alamo	July 2017	28,687	1 destroyed, 1 damaged	0
Whittier	July 2017	18,430	46 destroyed, 7 damaged	0
Rey	August 2016	32,606	0	0
Sherpa	June 2016	7,474	1	0
Gibraltar	October 2015	21	0	0
La Brea	August 2009	91,622	1	0
Jesusita	May 2009	8,733	80	0
Tea	November 2009	1,940	210	0

Sources Santa Barbara County Fire (https://www.sbcfire.com/wp-content/uploads/2018/08/Major-Wildfires-in-Santa-Barbara-County-1955-2016.pdf) and The Tribune (https://www.sanluisobispo.com/news/state/california/fires/article214778480.html) (accessed 7/17/19)

Batty (2013) and Kitchin (2014) have written much about the smart city. It is conceived that digital technology in the form of sensors (e.g., Internet of Things, closed circuit cameras, traffic induction loops, radar, etc.) representative of the smart city will connect, protect and enhance our lives, especially through enabling better decision-making. One implication of the smart city is the continuing trend of big data, with digital sources of data coming from sensors, digitizers, scanners, cellular phones, Internet, video, email, social media, GPS, loyalty cards, etc. This necessarily means a variety of data types, such as text, location, path/route, image, video, and sound. Collectively, we already experience difficulties in managing, processing, manipulating, etc. data, so the flood of data continues. Geo-spatial technologies, therefore, become ever more essential as they represent the possibility for creating knowledge from big data.

Much research has focused on the development of advanced geo-spatial technologies for understanding, monitoring, and prediction of wildfire risk using of remote sensing (see Cannon et al. 2017; Williams et al. 2018; Veraverbeke et al. 2018). Significant work too has been undertaken on the development of mathematical models to support evacuation efforts and emergency response (see Cova and Church 1997; Scaparra and Church 2015). Of particular interest in this chapter is technology to support better communication. Li et al. (2015) focused on household-level triggers for evacuation, integrating spatial models with fire spread approaches in order to get better estimates of when and to whom warnings should be issued. Work along these lines has continued in Cova et al. (2017) as well as Li et al. (2019). Of relevance is that emergency notification and communication is about getting information to the public regarding an imminent threat (Kuligowski and Dootson 2018). However, in the case of wildfire and flooding it is also about getting people away from areas of danger, as highlighted in Li et al. (2015), Cova et al. (2017), and Li et al. (2019). Choy et al. (2016) suggest that radio, television, Internet, social media, sirens, etc. all play an important role in communicating emergency warnings.

The above review serves to underline the importance of geo-spatial technologies in various ways. Quantitative methods combined with data must be brought to bear on pressing issues and problems that society faces. How they are combined, integrated, and utilized is critical for effective management and decision-making.

17.3 Fire, Flooding, and Debris Flow

In many ways California reflects modern vulnerabilities brought about by climate change. Past decades have seen significant drought, associated dry vegetation, and increases in wildfire. Noted above was that the Santa Barbara region in particular was highly susceptible to fire and subsequent debris flows. As indicated in Table 17.2, the Thomas fire in December 2017 was major, burning 281,893 acres, destroying/damaging more than 1,300 buildings, causing two deaths and costing some $230 million to fight. Kolden and Henson (2019) provide a lot of the associated context and details of this event. Of note is that the fire started east of Santa

Barbara in Ventura County, nearly 40 miles away. It was driven by prolonged Santa Ana winds. Following the wildfire in January of 2018 was an intense localized rain event near Montecito that resulted in major debris flow and flooding (see Jeffery and Steinberg 2018). The impacts were significant, with 23 deaths, over 150 injured, more than 100 homes destroyed and some 300 homes damaged, and disrupted major infrastructure. Water, power, sewer, and roads were all impacted, with the primary transportation access route to, from, and within the region, U.S. Route 101, closed for nearly 2 weeks.

The situation associated with Montecito was indeed unique in many ways. Much of the property in Montecito was spared damage from the Thomas fire, but surrounding mountains were not. The weather forecasts predicted rain, with expectations of local pockets of heavy rain. On January 6 (9:40 PM) a flash flood watch was issued by the National Weather Service for the area. Accordingly, police, fire, and rescue personnel were on alert throughout the region. On Tuesday January 9, 2018 at 2:32 AM the National Weather Service issued a flash flood warning. What followed was that an hour later, around 3:30 AM (January 9, 2018), there were over 0.50 inches of precipitation from a relatively small single cell in the span of minutes over the mountains above Montecito, far exceeding thresholds for a flash flood event.

A challenge no doubt was the timing of the event, but also its isolated nature. Beyond this, contemporary approaches for communication of the need to evacuate were somewhat limited. News sources (television, radio, and online) are certainly one avenue, but perhaps not effective in the very early morning hours like this. Wireless emergency alerts are another avenue, but are dependent on a number of things, including that those in need of evacuation have a cellular phone, have settings turned on to receive associated government alerts, and are within an audible range of the alert when it goes off. Social media and informal networks of communication too are another avenue, but again are likely ineffective at this time of the morning. The point is that more effective forms of communication are lacking, and yet are critical in situations like these. In particular, as noted in Choy et al. (2016), outdoor warning sirens are a traditional and proven means of communicating emergency notifications. Their usage is commonplace in areas where tornados are likely, as detailed in Current and O'Kelly (1992), Murray et al. (2008a, b), and Mathews and Ellis (2016), for example. Many communities in California do employ a siren system for emergency notification, including San Luis Obispo County, City of Huntington Beach, City of Oakland, etc. The rationale for these systems includes notifications associated with power plant emergencies, dam failure, earthquakes, chemical spills, large fires, terrorist acts, tsunami, or other instances of threat or danger. Interestingly, Santa Barbara does have a few sirens, some on the campus of University of California at Santa Barbara and one at the former Venoco Ellwood Onshore Facility.

Of interest in this chapter is the use of geo-spatial technologies to support the evaluation and planning of emergency communication equipment like sirens for alerting the public of imminent danger, whether it be fire, debris flow, flood, tsunami, chemical spill, etc. Questions regarding how this can be done effectively remain in order to meaningfully address the unique context and situation discussed in this chapter.

17.4 Integrating Geo-spatial Technologies and Data

An integrated framework combining geo-spatial technologies and smart city data is essential for supporting planning, management, and decision-making associated with emergency communication. In what follows, we advance the integration of exploratory spatial data analysis capabilities using GIS, remote sensing, spatial statistics, and spatial optimization. Such a framework is highlighted in Fig. 17.1 for these components in order to addressing emergency communication issues associated with fire and flooding.

GIS, remote sensing, and spatial statistics are essential in many ways, but in particular their use to formalize costs, risks, and relative importance is critical. With this information, geo-spatial model formalization is possible to identify a siren-based system that is both efficient and effective in providing emergency notifications during all hours of the day to people that may be indoors, outdoors, at home, at work, engaging in recreational activities, etc.

In the case of siren services, the planning situation can be conceived of as a spatial optimization problem where the decisions correspond to

- What are the total costs?
- Where will facilities be located?
- Who will be served?

A range of approaches to support this type of decision-making are discussed in Church and Murray (2018). For this particular planning context, an approach that makes the most sense is the continuous maximal covering problem. The mathematical model formulation is the following:

Fig. 17.1 Integrated system of geo-spatial technologies

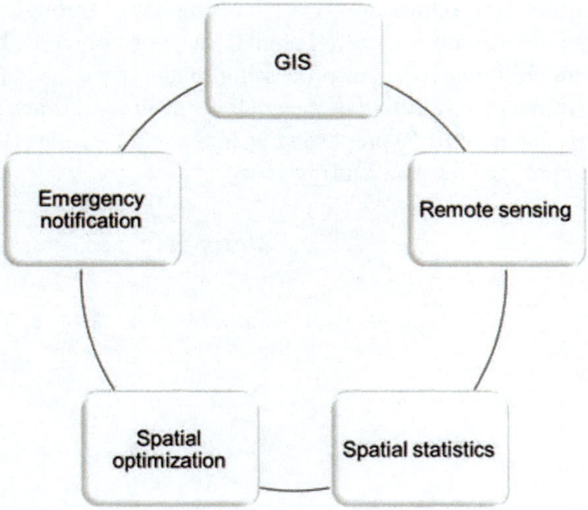

$$Maximize \sum_i \alpha_i$$

$$Subject\ to \iint f_i(\Psi)dR \geq \alpha_i\ \forall i$$

$$|\Psi| = p$$

$$(\phi_j, \lambda_j) \subset R\ \forall j \in \Psi$$

$$\alpha_i \geq 0 \forall i$$

where given a region R, i is the index of demand areas, α_i is the demand distributed across area i, $f_i()$ is a coverage function equating service to the distribution of demand across area i, p is the number of facilities to be sited (level of investment), and Ψ is the set of facilities with the location of each facility j denoted by the coordinate pair (ϕ_j, λ_j). The objective of this model is to maximize the total demand covered, or suitably served. In this case, service is based on being within the audible range of the sited sirens. The constraints of this model track the amount of demand coverage achieved for a given investment level. This is considered a continuous space optimization model since facilities may be located anywhere. Accordingly, the primary decision variables of the model are the location of each siren j, (ϕ_j, λ_j).

Interest in continuous space coverage along these lines has been considerable because the problem has general and broad utility. Academic research continues as well owing to such problems being challenging to structure and difficult to solve optimally (or exactly). Noteworthy approaches for solution include the work of Church (1984), Murray and Tong (2007), Murray et al. (2008a, b), Tong and Murray (2009), Matisziw and Murray (2009), Murray and Wei (2013), Wei and Murray (2015) and Murray (2018), among others.

A common approach for dealing with continuous space is to adopt a discrete space approximation. This is accomplished through the representation of demand as discrete objects, often points, lines, or polygons, but also employing screening, suitability analysis, or processing to identify a discrete set of potential facility sites from which selection is made. Under such conditions, the maximal covering location problem (MCLP) proposed in Church and ReVelle (1974) is an equivalent modeling approach (see also Murray 2016):

$$Maximize \sum_i a_i Y_i$$

$$Subject\ to \sum_{j \in N_i} X_j \geq Y_i\ \forall i$$

$$\sum_j X_j = p$$

$$X_j \in \{0, 1\}\ \forall j$$

$$Y_i \in \{0, 1\}\ \forall i$$

where i is the index of demand objects, j is the index of potential facility sites, a_i is the amount of demand at i, N_i is the set of potential facilities that can cover (or suitably serve) demand i, and p is the number of facilities to be sited. This information is derived in advance, needed as model input in order to determine a solution. Decision variables in the model include X_j, where a value of 1 indicates siting a facility at j and a value of 0 it is not sited, and Y_i, where 1 signifies demand i is covered (or suitably served) and 0 denotes it is not covered. The model then seeks to maximize total demand covered with associated constraints necessary for tracking coverage and specifying the number of facilities to site.

The MCLP is accessible is commercial GIS software (see Murray et al. 2019), where it is solved heuristically. Through the use of integer programming approaches, the MCLP is amenable to exact solution. Commercial optimization software is commonly used for general solution.

17.5 Vulnerability Mitigation

Communication of wildfire risks and danger is fairly well controlled and accounted for in the region, with well-defined evacuation zones and an aggressive process of imposing mandatory evacuation in advance of an event. This has not been the case for post-fire dangers. In the aftermath of the Thomas fire, Santa Barbara County identified extreme risk areas associated with debris flow and flooding.[3] This data was utilized for vulnerability mitigation through the establishment of a system of warning sirens that can provide communication of imminent danger. There are many different types of sirens, with varying performance characteristics. Consistent with other studies, it is assumed that a siren has an audible range of 5,280 feet. The associated costs for a siren along these lines can exceed $50,000 (see Murray 2018). In addition to the extreme risk areas identified by the Santa Barbara County, parcel data was also acquired. Additionally, satellite imagery was obtained, enabling vegetation, land use, slope, and other characteristics to be derived. Such information is an important factor in the determination of risk associated with recent wildfire events.

ArcGIS was used for spatial data processing, manipulation, and analysis. ERDAS IMAGINE was used for remotely sensed data acquisition, processing, and manipulation. GeoDaSpace was utilized for spatial statistical analysis. FICO Xpress, a commercial optimization package, was used to solve all reported problem instances. All processing and computation are done on a desktop personal computer (Intel Xeon E5 CPU, 2.30 GHz with 96 GB RAM).

The recent Thomas fire burn area is shown in Fig. 17.2 (gray shade), along with identified extreme risk areas (red shade) where debris flow and flooding is likely in the event of significant precipitation. Of particular note is the fire perimeter in the mountains above Carpinteria to Montecito. The intervening area from hilltops down to the Pacific Ocean, therefore, represents pockets of extreme risk for debris flow

[3] Data available at https://readysbc.org/.

Fig. 17.2 Thomas fire burn area along with identified debris flow/flooding risk areas

and flooding, and this is precisely what occurred in Montecito. A more detailed view of this situation can be seen in Fig. 17.3. While there were other recent fires in the region (see Table 17.2) that also present extreme risks of debris flow and flooding along the foothills spanning from Santa Barbara to Goleta, the analysis that follows is limited to the Thomas Fire for illustrative purposes. Similarly, the emphasis on emergency communication in the case of debris flow and flooding could be extended to account for wildfire vulnerability as well.

Fig. 17.3 Extreme risk areas

Relative risk estimates can be further elaborated on through a combination of GIS and remote sensing-based analysis combined with spatial statistical modeling. A necessary input for the MCLP is a_i for each area of demand i along with potential facilities sites that can cover each demand area i, N_i. The demand, weight, importance, etc. reflected in a_i is generally a function of many different factors, but also the associated context and geographic setting. Mathematically, specification can take the following form:

$$a_i = g(elevation, slope, vegetation, landuse, people, propertyvalue, \ldots)$$

where $g()$ is a function relating the associated input factors. The precise functional form depends on the important relationships identified. In this case, the County has undertaken analysis to identify the extreme risk areas, as shown in Figs. 17.2 and 17.3. Subsequent processing and evaluation required identification and delineation of incident areas within these risk zones. Intersecting these risk zones are 4,460 parcels along with roads and easements (shown in Fig. 17.3). While parcels are one way to approach demand within the extreme risk areas, the existence of roads and other land use types complicate this as does significant variation in the relative size of parcels. Accordingly, we derived 12,067 demand units within the extreme risk zones, each approximately 120 × 120 feet in size, illustrated in Fig. 17.4. Further processing and analysis identified 74,599 potential siren locations. The MCLP application involved 74,599 decision variables associated with siren placement and 12,067 decision variables associated with demand area coverage, giving 86,666 total decision variables. The number of primary constraints totaled 12,068. Solution time using Xpress was less than 258 s in each case.

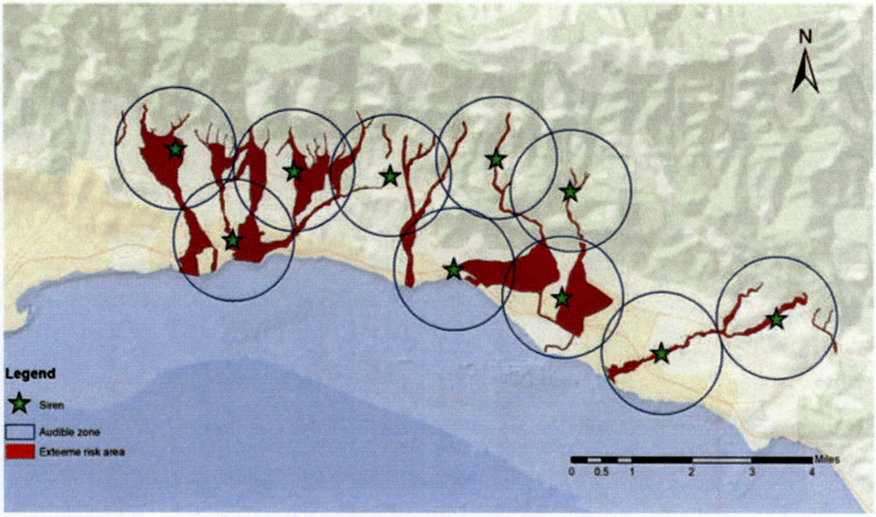

Fig. 17.4 Spatial configuration of sirens ($p = 10$)

Systematic evaluation of different investment scenarios was considered by varying the value of p and resolving the model. One such scenario is shown in Fig. 17.4, where 10 sirens are sited ($p = 10$). This pattern of coverage is interesting in a number of ways, highlighting that one seeks the most coverage possible but doing so under conditions of limited resources. This configuration of sirens is able to provide audible coverage to over 99.50% of the demand in the extreme risk areas.

Of course, other scenarios are possible as well. Along these lines, Fig. 17.5 summarizes the range of possibilities for siting configurations. The investment scenarios begin with one siren and go up to 13 sirens, reflecting consideration of $p \in [1, 13]$. Siting only one siren, for example, would mean that only 25.11% of demand coverage in the risk areas is possible. An addition siren enables demand coverage of 49.92% of the risk areas. Further coverage is possible through investment in more sirens, where 13 sirens are capable of providing complete demand coverage of the risk areas. Further inspection of Fig. 17.5 demonstrates that the marginal gains in coverage relative to investment level are insightful. For example, in the case of six sirens 87.94% of demand coverage is possible, and this progressively increases with additional sirens. However, the marginal increase in coverage steadily decreases. Seven sirens achieve 92.43% coverage, but total coverage provided by eight sirens only increases 3.19%, up to 95.62%. As is typical with other MCLP applications, complete coverage of all demand in the extreme risk areas requires many additional sirens, yet they collectively cover less than 1% of total demand. Specifically, 99.45% coverage is achieved with 10 sirens, but an additional three sirens are necessary to achieve 100% coverage.

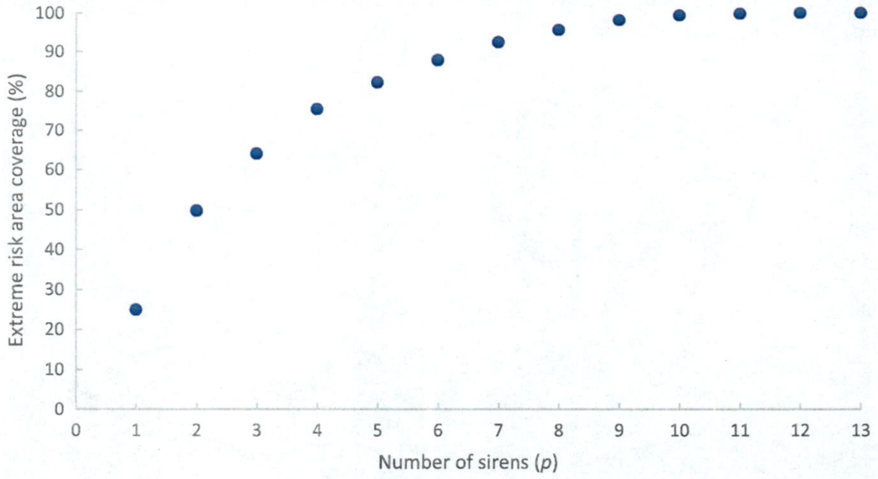

Fig. 17.5 Tradeoff of siren system investment and risk area coverage

17.6 Discussion and Conclusions

There are a number of items worth further discussion associated with emergency communication related to wildfire, debris flow, and flooding. Public agencies must contend with a vast array of considerations in the issuance of timely warnings, alerts, evacuation notices, etc. Public safety is clearly paramount, but certainly false alarms that can lead to public apathy and ignoring notifications altogether are part of an important process to tailor and customize communication to the public. Communication approaches that operate on effective triggers are essential, as are targeted methods of identifying who is at risk and the actions they need to take in order to be safe from impending danger.

The research reported in this chapter arises from deficiencies in emergency notification experienced in the aftermath of the Thomas Fire. There is a role for enhanced and targeted efforts to inform the public of emergency situations. Geo-spatial technologies and supporting spatial information have much potential to contribute to such problems and issues encountered in everyday life.

The detailed geo-spatial technologies are an important step in planning and decision-making for communicating emergencies of various types, but clearly debris flow and flooding notification along the lines discussed in this chapter offers significant system improvement. A number of points can be made regarding the reported approaches and findings. One item to note is that the findings for any level of investment are rather unique. In particular, spatial configuration is likely to change for each investment level. This is because the MCLP is non-hierarchical in the sense that an optimal solution for one level of investment, p, in not linked in any manner to other levels of investment, e.g., $p - 1$ or $p + 1$. Thus, the implication is that an entirely unique configure might be identified for a different level p. In this case, the optimal configuration solution shown in Fig. 17.4 with 10 sirens only has seven sirens in common with the optimal spatial configuration for the case of 9 sirens. Similarly, the optimal configuration for 10 sirens (Fig. 17.4) has no sirens in common with the optimal spatial configuration for the case of 11 sirens. The spatial configuration for 11 sirens is depicted in Fig. 17.6 to illustrate this situation, and clearly shows the variability that is inherent to obtaining the most efficient spatial configuration possible for a given level of investment.

Much of the discussion has revolved around investment scenarios and that different levels of investment can be evaluated and considered. However, this has been approached by focusing only on the number of facilities to site, sirens in this case. The reason for this is that each potential facility site is assumed to have the same basic costs. That is, there is no significant variability in fixed or annual costs. Accordingly, the investment decision becomes how many facilities can be afforded. If there is significant heterogeneity in fixed costs to build a siren at a site or the annual costs to maintain a siren at a site vary, then the budget constraint can be modified to reflect this, as discussed in Church and ReVelle (1974) and Church and Murray (2018). Whether this is necessary and important to consider depends on context and the facilities being considered. In this particular case, the facility is a siren. Often times

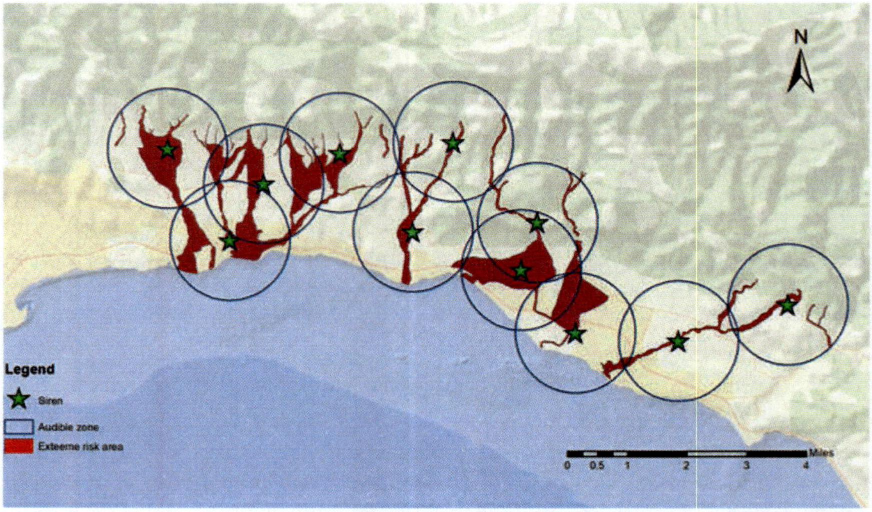

Fig. 17.6 Alternative spatial configuration of sirens ($p = 11$)

a siren is mounted on top of a pole structure, or a building. The fixed costs in such cases may simply be the equipment itself, as the infrastructure is already in place and the structure can be erected within an easement. Again, for situations where this assumption is problematic, then associated data on costs can be acquired and the modeling approach modified accordingly.

A final feature of the modeling reported in this chapter is that the audible zone for a siren was assumed to be regularly shaped, in this case a circle of radius 5,280 feet. Given the nature of siren notification transmission, this was reasonable. However, sound transmission can be impacted by various conditions, so audible zones might be irregularly shaped. Through the use of GIS and remote sensing, it would be easy to consider irregular zones if siren audibility is impacted by local topology, terrain, or other conditions. This is precisely why the notation N_i is utilized in the model formulation because those facilities (sirens in this case) capable of serving a demand area i can be defined in any manner consistent with facility service, and can be regular, irregular, non-contiguous, etc. in shape.

Emergency communication is a challenging task charged to public officials. The Thomas Fire and subsequent debris flow and flooding in Montecito highlight such challenges, but also the potential for supplementing notification systems. Carlson (2018) quotes Cova as saying that text-based alert sent to cellular phones "comes really close to being perfect" for communicating emergency situations. Clearly this is not true for all situations. Traditional approaches like siren systems have an important role for local communication, where contemporary methods may be limited. Geospatial technologies and associated data have much to offer planning and management efforts going forward. Fire, debris flow, and flooding vulnerability can be more effectively communicated through system design along the lines outlined in this chapter.

Acknowledgements This material is based upon work supported by the National Science Foundation under Grant No. 1664173.

References

Batty M (2013) Big data, smart cities and city planning. Dialogues Hum Geogr 3(3):274–279

Cannon F, Carvalho LM, Jones C, Hall T, Gomberg D, Dumas J, Jackson M (2017) WRF simulation of downslope wind events in coastal Santa Barbara County. Atmos Res 191:57–73

Carlson C (2018) As fires claim more victims, California bills to fix alert, warning systems have yet to pass. Ventura County Star Aug. 10, 2018. https://www.vcstar.com/story/news/local/2018/08/10/fires-burn-california-bills-fix-warning-systems-have-yet-pass-legislature-emergency-alert/768143002/. Accessed 19 July 2019

Carvalho LM, Jones C (2015) The monsoons and climate change: observations and modeling. Springer

Choy S, Handmer J, Whittaker J, Shinohara Y, Hatori T, Kohtake N (2016) Application of satellite navigation system for emergency warning and alerting. Comput Environ Urban Syst 58:12–18

Church RL (1984) The planar maximal covering location problem. J Reg Sci 24(2):185–201

Church RL, Murray A (2018) Location covering models: history, applications and advancements. Springer

Church R, ReVelle C (1974) The maximal covering location problem. Pap Reg Sci 32(1):101–118

Cova TJ, Church RL (1997) Modelling community evacuation vulnerability using GIS. Int J Geogr Inf Sci 11(8):763–784

Cova TJ, Dennison PE, Li D, Drews FA, Siebeneck LK, Lindell MK (2017) Warning triggers in environmental hazards: who should be warned to do what and when? Risk Anal 37(4):601–611

Current J, O'Kelly M (1992) Locating emergency warning sirens. Decis Sci 23(1):221–234

Easterling DR, Meehl GA, Parmesan C, Changnon SA, Karl TR, Mearns LO (2000) Climate extremes: observations, modeling, and impacts. Science 289(5487):2068–2074

Jeffery A, Steinberg M (2018) Fire and rain: mudslides now sweep across Santa Barbara County after devastating wildfires. CNBC, January 10, 2018. https://www.cnbc.com/2018/01/10/devastating-mudslides-sweep-across-santa-barbara-county-california.html. Accessed 17 July 2019

Kitchin R (2014) The real-time city? Big data and smart urbanism. GeoJournal 79(1):1–14

Kolden CA, Henson C (2019) A socio-ecological approach to mitigating wildfire vulnerability in the wildland urban interface: a case study from the 2017 Thomas Fire. Fire 2(1):9

Kuligowski E, Dootson P (2018) Emergency notification: warnings and alerts. In: Manzello SL (ed) Encyclopedia of wildfires and wildland-urban interface (WUI) fires. Springer

Li D, Cova TJ, Dennison PE (2015) A household-level approach to staging wildfire evacuation warnings using trigger modeling. Comput Environ Urban Syst 54:56–67

Li D, Cova TJ, Dennison PE (2019) Setting wildfire evacuation triggers by coupling fire and traffic simulation models: a spatiotemporal GIS approach. Fire Technol 55(2):617–642

Matisziw TC, Murray AT (2009) Siting a facility in continuous space to maximize coverage of a region. Socioecon Plann Sci 43(2):131–139

Mathews AJ, Ellis EA (2016) An evaluation of tornado siren coverage in Stillwater, Oklahoma: optimal GIS methods for a spatially explicit interpretation. Appl Geogr 68:28–36

McDowell NG, Coops NC, Beck PS, Chambers JQ, Gangodagamage C, Hicke JA, Huang CY, Kennedy R, Krofcheck DJ, Litvak M, Meddens AJ (2015) Global satellite monitoring of climate-induced vegetation disturbances. Trends Plant Sci 20(2):114–123

Moritz MA, Batllori E, Bradstock RA, Gill AM, Handmer J, Hessburg PF, Leonard J, McCaffrey S, Odion DC, Schoennagel T, Syphard AD (2014) Learning to coexist with wildfire. Nature 515(7525):58

Murray AT (2010) Quantitative geography. J Reg Sci 50(1):143–163

Murray AT (2016) Maximal coverage location problem: impacts, significance, and evolution. Int Reg Sci Rev 39(1):5–27

Murray AT (2017) Regional analytics. Ann Reg Sci 59(1):1–13

Murray AT (2018) Evolving location analytics for service coverage modeling. Geogr Anal 50(3):207–222

Murray AT, Tong D (2007) Coverage optimization in continuous space facility siting. Int J Geogr Inf Sci 21(7):757–776

Murray AT, Wei R (2013) A computational approach for eliminating error in the solution of the location set covering problem. Eur J Oper Res 224(1):52–64

Murray AT, Matisziw TC, Wei H, Tong D (2008a) A geocomputational heuristic for coverage maximization in service facility siting. Trans GIS 12(6):757–773

Murray AT, O'Kelly ME, Church RL (2008b) Regional service coverage modeling. Comput Oper Res 35(2):339–355

Murray AT, Xu J, Wang Z, Church RL (2019) Commercial GIS location analytics: capabilities and performance. Int J Geogr Inf Sci 33(5):1106–1130

National Academies of Sciences, Engineering, and Medicine (2019) Framing the challenge of urban flooding in the United States. National Academies Press

Scaparra MP, Church RL (2015) Location problems under disaster events. In: Laporte G, Nickel S, da Gama F (eds) Location science. Springer, pp 623–642

Tong D, Murray AT (2009) Maximising coverage of spatial demand for service. Pap Reg Sci 88(1):85–97

Veraverbeke S, Dennison P, Gitas I, Hulley G, Kalashnikova O, Katagis T, Kuai L, Meng R, Roberts D, Stavros N (2018) Hyperspectral remote sensing of fire: state-of-the-art and future perspectives. Remote Sens Environ 216:105–121

Wei R, Murray AT (2015) Continuous space maximal coverage: insights, advances and challenges. Comput Oper Res 62:325–336

Williams AP, Gentine P, Moritz MA, Roberts DA, Abatzoglou JT (2018) Effect of reduced summer cloud shading on evaporative demand and wildfire in coastal southern California. Geophys Res Lett 45(11):5653–5662

Wisner B, Adams J (eds) (2002) Environmental health in emergencies and disasters: a practical guide. World Health Organization

Chapter 18
An Information and Communication Technology (ICT)-Driven Disaster Management System: A Case of Firefighting in Mumbai

Vaibhav Kumar, Shyan Kirat Rai, Arnab Jana, and Krithi Ramamritham

Abstract The increasing frequency and intensity of disasters pose a big challenge while building resilient cities. Unplanned urbanization has further complicated the matter. In the era of building technology-driven smart cities, there is a lack of effective Information and Communication Technology (ICT)-based disaster management implementation and strategic planning. To address these concerns, an ICT-based system is developed and tested during the fire drills. Testing of the system was done in sync with the workflow of a proposed framework for the city of Mumbai. The framework consists of an upgraded local disaster body, namely Smart Disaster Management Body (SDMB). The body is built upon the idea of ICT. Framework components and gaps are identified by attending fire drills, reviewing the existing literature, and conducting pre- and post-system implementation controlled group surveys with stakeholders. Results of the stakeholder surveys indicate the solution to be more efficient in disaster management actions. And, it can address the challenges arising in quick response situations and therefore can help in building resilient smart cities.

Keywords Smart cities · Sustainable cities · Disaster response · ICT · Urban sustainability

Acronyms

ASDMA	Assam State Disaster Management Authority
ICBDMS	Cloud-Based Disaster Management System
DM	Disaster Management
DMCR	Disaster Management Control Room

V. Kumar (✉) · S. K. Rai · A. Jana
Centre for Urban Science and Engineering, Indian Institute of Technology Bombay, Mumbai, India

K. Ramamritham
Computer Science and Engineering, Indian Institute of Technology Bombay, Mumbai, India

P. Sharma (ed.), *Geospatial Technology and Smart Cities*, The Urban Book Series,
https://doi.org/10.1007/978-3-030-71945-6_18

DR	Disaster Response
DRR	Disaster Risk Reduction
FICCI	Federation of Indian Chambers of Commerce and Industry
ICT	Information and Communication Technology
MCGM	Municipal Corporation of Greater Mumbai
NDMA	National Disaster Management Authority
QRT	Quick Response Teams
ODC	On Demand Community
SDMB	Smart Disaster Management Body
SDRB	Smart Disaster Response Body
SFAC	Standing Fire Advisory Council
ULB	Urban Local Bodies
WRI	World Risk Index

18.1 Introduction

Disaster management needs to be considered while developing resilient urban spaces. According to the World Risk Report published in 2015, India ranks 77th in the World Risk Index (WRI) on natural disasters (United Nation University 2015). Sharma and Tomar (2010) pointed out the increasing disaster risk for Indian cities due to unplanned growth, deficits in infrastructure, and poor services. To address rapid urbanization and resource constraints, the Government of India (GoI) has taken a significant step by building 100 smart cities across the country. As per the plan managed, urban spaces are created through greenfield development and retrofitting the already developed spaces. The plan also intends to provide a better quality of life to citizens through efficient resource management ("Smart City Mission, GoI" 2015). However, the lack of ambitious goals in planning poses challenges to city managers attempting to build smart cities.

According to Chong et al. (2018), the lack of ICT-driven disaster strategies affects a large proportion of the population in gathering proper information for quick action. This was evident during our field study. It was observed that, due to less citizen-centric ICT interventions, the current system might not support existing and future requirements. This is mainly because ICT has become an essential component of most city activities, including disaster management. Coordination among organizations without a framework of information management will not be able to capitalize on the technical and social benefits of a city (Lee and Lee 2014).

E-Governance and mobile-based systems are going to be the most prominent features of fast-growing Indian cities (Lee and Lee 2014). The capacity to capture, process, and transfer data has made ICT an essential component in service delivery (Karnatak and Kumar 2015). ICT also serves as a tool to bridge the socioeconomic divide (Chourabi et al. 2012). Mhaske and Choudhury (2010) and Allenby (2005) stressed the importance of the Internet and social media in building resilient cities through data acquired from citizens. Wentz (2006) pointed out the role of organizations and their dependence on technology to serve technology-aware citizens, thus enhancing the role of ICT in future cities. ICTs are useful to capture, store, and transmit the data. The data can be analyzed and processed using various models to support decision-making processes and resource mobilization. At times of disaster, this might help to lower the response time in resource mobilization and decision-making.

This research is focused on the development of a holistic solution for efficient disaster management in Indian cities. This study proposes a disaster management system dedicated to meeting the city requirements by combining agencies, authorities, working bodies, and citizens using an ICT platform. It is developed on top of the existing platform, with modifications proposed to achieve better disaster management.

The rest of the study is organized as follows: the second section presents the literature review of studies related to the role of ICT in building resilient cities and ICT's role in service delivery to various socioeconomic classes and coordination among agencies. In the same section, the gaps in Mumbai's current disaster management and fire response systems are discussed. This section also presents a pre-implementation survey with the stakeholders. The objectives of the study are presented in section three. Section four presents the methodology of the study. This section is followed by a discussion of the implementation of the methods in a fire drill scenario. Conclusions and recommendations are presented in section six.

18.2 Background and Motivation

18.2.1 Disaster—A Concern for the Indian Cities

As per the report published in 2016 by the National Disaster Management Authority (NDMA), most of the proposed Indian smart cities lie in disaster-prone areas (NDMA 2016). Apart from floods, landslides, earthquakes, and cyclones, an increase in the frequency of heatwaves is a serious concern. The Federation of Indian Chambers of Commerce and Industry (FICCI) risk report lists major cities that lie in the "highly impacted" category based on the frequency of unprecedented events like fire and building collapses (FICCI 2016). In many instances, the impacts from these events are exacerbated by rapid, unplanned urbanization, overcrowding, lack of inclusive growth, poor planning and maintenance, and lack of adequate infrastructure. With the frequency of disasters expected to increase in the future, the cities lack a detailed road map of disaster management. The paper argues that there is a lack of focus on climate resilience in addition to underfunding for disaster preparedness. Thus, disaster management (DM) planning is especially important in the context of smart cities, where there is attention on building and improving infrastructure which must be able to withstand future disaster impacts (Anu et al. 2017).

18.2.2 Resource Constraints in Urban Disasters

Disaster Response (DR) and preparedness are crucial dimensions in making cities resilient through data-intensive techniques, especially given limited ability to invest sufficient funds in meeting infrastructural needs (Mishra et al. 2017). The India Risk Survey report, published in 2015, found that India has an existing deficiency of around 97.59% fire stations (FICCI 2016). Currently, 1705 fire stations exist as compared to the 70,868 stations that should exist as per Standing Fire Advisory Council (SFAC) norms. The report also presented the data related to the deficiency of 72.75% of fire stations in urban areas. The average police-public ratio in India is 152 police officers per 100,000 population, while the standard set by the United Nations suggested a ratio of 222 officers per 100,000 people (Pal and Ghosh 2014). Mumbai has 34 existing fire stations, and it still faces the deficiency of 66 fire stations to match the international standards (Kumar et al. 2019). With two smart cities planned in Mumbai suburbs, these details pose a great challenge for decision makers, especially when the existing disaster management authority is responsible for DM without any technological up-gradation.

18.2.3 Role of ICT in DRR and Bridging Socioeconomic Gaps in Service Delivery

Mukhopadhyay and Revi (2009) pointed out the complexity of capturing interconnected economic, social, and territorial drivers. Revi (2008) suggested a radical shift in public policy, mobilization, and enterprise from mitigation to adaptation for reducing the vulnerability of future Indian cities. He also called for the inclusion of all economic classes in any policy shift related to Disaster Risk Reduction (DRR). However, this should not suggest abandoning remedial and compensatory approaches, especially given the levels of high risks and the lack of resources to reduce vulnerability (Lavell and Maskrey 2014). Strategic principles for smart cities align the three main dimensions: technology, people, and institutions (Nam and Pardo 2011; Alawadhi et al. 2012). Nam and Pardo (2011) stress the importance of technology as the key to filling the socioeconomic divide in service delivery. India is one of the fastest-growing economies in the world, and the next phase of evolution in the Internet-driven ecosystem has been led by many institutions, who began leveraging the ubiquity of mobile phones to provide services to Indian users across platforms such as e-commerce, payments, and mobile apps (NASSCOM 2016). The report by NASSCOM states that India added 88 million Internet users from 2008 to 2012, and at the end of that period, the total number stood at 137 million. 60% of users connected to the Internet through mobile phones (NASSCOM 2016). The report predicts that India will have around 371 million mobile Internet users by June 2016. 71% of this number will belong to the ever-growing urban areas. These numbers provide strong encouragement for using ICTs as a backbone of DR systems.

ICT applications from various domains play an essential role during disasters in smart cities. Batty (2013) discussed the role of big data in cities and city planning from the perspective of sustainability. Alazawi et al. (2012) developed an Intelligent Cloud-Based Disaster Management System (ICBDMS) for vehicular networks. Arfat et al. (2017) developed a mobile application, a backend cloud-based big data analysis system, and a middleware platform based on cloud and fog technologies for disasters. Aqib et al. (2018) applied deep learning-based techniques for forecasting traffic plans during emergencies in smart cities. Similar studies which use technologies like big data analytics during disaster phases are presented by various researchers (Suma et al. 2017, 2018).

18.2.4 ICT and Building Resilient Cities

The importance of ICT is undeniable during disasters. Many countries have used ICTs as the key entity to strengthen their disaster management strategies. (Mishra et al. 2017; Wentz 2006; Dorasamy et al. 2011) have listed some successful case studies. The Korean government started DR efforts using ICTs in 2004 and, since then, have developed, upgraded, and stabilized the system with new and improved technology backed up by relevant laws and policies, indicating the commitment of the government to such technologies (Washburn et al. 2009). Various DR systems have been implemented in Indian cities. For example, applications like the PEOPLE FINDER blog site were used to share details of the missing people in the aftermath of the 2005 flood and 2006 terrorist train attack in Mumbai (Shankar 2008). During the 2015 Chennai floods, International Business Machine (IBM) officials employed On Demand Community (ODC) to understand the needs of the victims and used the Sahana application to keep track of resources (IBM 2014). The Municipal Corporation of Greater Mumbai (MCGM) has launched a disaster management mobile application which keeps track of the weather forecast, rainfall, and tide information affecting the city during the monsoon period. The Assam State Disaster Management Authority (ASDMA) released a Disaster Ready Assam mobile application to alert citizens during various disasters (EastElegant 2017). However, to the best of our knowledge, no systems dedicated to fire emergencies in India have been developed.

18.2.5 Role of Urban Local Bodies in Disaster Management

The role of government and concerned agencies to deliver required services at the expected time is the key to better disaster management (Alawadhi et al. 2012). According to Kusumasari et al. (2010), the role of local government and the need for disaster management institutions is a significant concern in the disaster discourse. This is because local government plays the most active role during disasters. They also state that in most cases, "agencies and practical resources remain unclear about their role during various disaster phases." Bremer (2003) stressed the role of local bodies during disasters and called for a dedicated working structure of entities based on the disaster type. In the same study, he states the various challenges related to the lack of proper information channels during most disaster events in India. Lee and Lee (2014) emphasized organizational involvement in DM through a common framework of information management, with a focus on regulatory, technical, and support requirements to sustain the system.

18.2.6 Firefighting Process in Mumbai (Study Area): Identifying the Pain Points

Mumbai, the study area, is the capital city of the state of Maharashtra. It is a hub of economic activities and is known as the commercial capital of India. It extends 12 km east to west at its broadest point, and 40 km north to south with an area of 437.71 Km^2 (Mumbai Fire Brigade 2014). Mumbai is the second most densely populated city in the world, with a population of around 20 million (Global Risks Report 2017). With one of the highest urban concentrations on the globe and an ever-growing population requiring a complex system of infrastructure and services, the risk of human-made and natural disasters in Mumbai is a constant threat. Mumbai's vulnerability to disasters, high economic diversity, and a large working population, as well as being the location of two smart cities makes it suitable for our study.

The Disaster Management Control Room (DMCR) is the central agency responsible for disaster management in the MCGM. Every ward has its local DMCR which coordinates with the Urban Local Bodies (ULB) and central DMCR during a disaster. Figure 18.1a shows the participating agencies at the local level in case of an emergency. The gray color boxes represent the active agencies during disaster events, such as fires. These organizations record distinct information from various sources and many channels before taking further action. This increases the response time. The data flows from DMCR toward other involved agencies and vice versa. On many occasions, the information flows from other ULBs like fire stations, police stations, and hospitals or citizens to the central or ward DMCR. Figure 18.1b shows the working model and technology involvement at the Mumbai central DMCR.

In Fig. 18.1, Event 0 indicates a fire event. Due to the lack of ICT components like an auto alarm with location information sender, citizens (shown in Event 1) are the first to report the incident. A quick and well-organized response depends on the quality of information; however, many times, the reported information turns out to be vague, incomplete, or false. The absence of ICT applications, like geo-tagged images or CCTV, that can provide a credible source of information, further adds to woes. The lack of a dedicated single channel for communication also adds to the confusion and the delay in response time, as a citizen can make a call to the local bodies, central DMCR or ward DMCR (Event 2). The participating organization updates its local database. However, due to the non-mandatory update policy or lack of detailed event

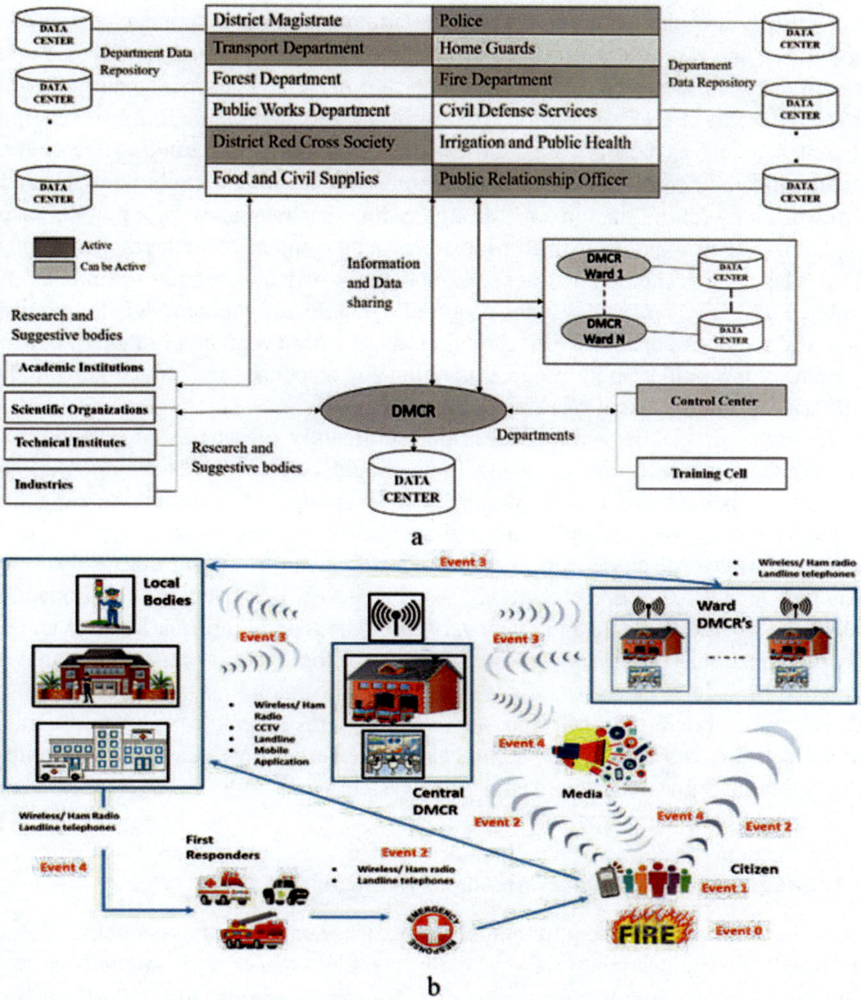

Fig. 18.1 **a** Participating and active entities during a fire incident **b** Working model of involved entities in a fire breakout

information, most of the time, non-participating organizations fail to update their information database.

When any of the involved organizations get the first information from the citizen, they broadcast it using telephone or ham radio to other organizations (Event 3). The broadcasting, preparation to act, and coordinating activities require considerable time, due to the delay in real-time information exchange between the involved actors. The lack of ICT interventions, segregated data centers, and multiple information channels are some of the reasons for the delay in the coordination.

As soon as the information reaches the authorities, first responder vehicles are dispatched in response (Event 4). The vehicle crew is equipped with ham radios to communicate between rescuers and decision makers. The mainstream media is informed by the public relation body of central DMCR about the ongoing proceedings (Event 4). This is the only medium to inform citizens, thus limiting the role of aware citizens. For example, traffic on the possible firefighting vehicle route could be diverted by telling the citizens through radio, alert messages, or a mobile-based application. This would help in en-route traffic management and reduces the response time. The citizens should be aware enough to make way for firefighting vehicles and ambulances. If ICT-based systems acquired dynamic environment details like real-time traffic, victim details, virtual simulations of a fire event, and reliable reporting systems, this would help in resource planning and response time reduction, which is still lacking in the current working system.

Centralization of real-time data is not sufficiently present in current working procedures. Data at various organizations should be available across centralized and common data processing platforms. The actions of participating organizations depend on the availability of credible real-time information. Over the years, researchers have discussed the benefits of centralization of data during disasters. Wang et al. (2017), Zêzere et al. (2014), Pineda (2014) discussed the impact of data centralization on data consistency, resulting in general information flow among the organizations. Organizations can use ICT as a tool for information exchange to improve the silo working situations for better coordination (Alazawi et al. 2014). The absence of dedicated Information Technology (IT) specialists and an interdivisional cell that can support the technical requirements shows the gap in planning. This research argues that the existing DM system needs to be re-engineered, using ICTs to address the current problem.

A pre-implementation survey among the stakeholders

To perceive the gap and requirements, stakeholder surveys were conducted. People from different socioeconomic strata and diverse urban spaces were part of the survey. The responses were collected and analyzed to gain insights. In total, 100 people participated in the survey; 42% of them were females. The age group that participated in the survey was between 17 and 60 years. Out of the total, around 90% of people knew how to read or write in at least one language. The online and offline surveys included socioeconomically disadvantaged people living in slums. About 98% of the total surveyed people owned a mobile phone. 80% of them knew how to operate installed applications on the smartphone. However, out of the remaining 20, 11% used the smartphone for only making calls. This indicates the feasibility of creating a mechanism to communicate at times of a disaster. A lower percentage (46%) of respondents had previously used ICT during an emergency, out of which only 40% of users were satisfied with the functionalities of the mobile application and websites. The percentage is low because Mumbai faces yearly floods and frequent fire events. 95% of respondents stressed that the determining factor for their use of ICT systems is improved functionalities; the lack of good functionality may be the reason for low usage rates. Around 80% of respondents believe ICT can help them in better ways during disasters. The survey results reinforce the idea of citizen-centric topology during disaster management processes. From the survey, it was deduced that different organizations maintain actionable data during disaster events, and they have not been able to use it for the collective benefits. This supports the proposed idea where concerned organizations can use ICT for better information management and decision-making.

18.2.7 Summary of Research Gaps and Motivation

- Little focus on disaster management in smart cities.
- Lack of citizen-centric ICT interventions and functionalities in disaster management.
- The issue of coordination between agencies due to decentralized real-time data and multiple information channels.
- Dearth in research addressing social equality in ICT-based service delivery during disasters.
- High acceptance of mobile-based technology by citizens.
- Better functionalities with timely response remain a key aspect in technology acceptance.

18.3 Objectives

To overcome the discussed challenges, this study proposes a smart disaster management framework, with ICT as its primary component, having the following features:

- Identification of gaps in the existing system using field drills and stakeholder surveys.
- Development of a disaster management system suitable for a smart city environment. The proposed system demonstrates the role of ICT in effective service delivery during disasters like fires.
- To test the system and its components using developed mobile and web applications along with the existing ICTs during a fire drill in Mumbai.
- To study the effectiveness and the response of the system using semi-structured qualitative surveys with stakeholders.

18.4 Research Methodology

The methodologies presented in this study address the feasibility of ICT-based systems for DM and quick response in Indian cities. This study proposes a dedicated DM body at the local level, as recommended by researchers across the world, to address the issues related to the existing situation (Hall et al. 2000; Reuters 2015). This body would use ICT as the central component to bridge the gap between disaster service deliveries across socioeconomic classes and improve coordination among agencies. The first subsection details the approach to identify gaps in the existing system. Driven by the results, the second subsection proposes the new disaster management system which addresses the current challenges. The last subsection presents the implementation process of the developed ICT solution in the fire drill.

18.4.1 Identifying the Gaps

The existing disaster management strategy of Mumbai city was reviewed. Further, the study of related literature, semi-structured surveys with stakeholders, and attending fire drills helped to identify the gaps in existing disaster management strategies. Complete details of the survey and the working pattern of the local disaster authority can be found in Sect. 18.2.6.

18.4.2 Conceptualizing Smart Disaster Response Body (SDRB)—The Proposed Solution

Technological interventions are required in modern city management. This requires trained human resources and specific departments, including IT and interdivisional cell. These departments can help incorporate technical components into the existing organizational structure for efficiency. As pointed out by Wentz (2006) and Greer et al. (1999), the reasons for the slow advancement in the use of modern ICT in organizations are the lack of direct citizen participation in the response system, unavailability of skilled human resource, and fast-paced technological development.

To address the issues related to existing disaster management, a restructured version of DMCR was introduced for robust DM solutions: a Smart Disaster Management Body (SDMB). A generic DR model for different cities with different disaster profiles can be addressed using ICT. ICT is a robust tool and can be developed as per the city requirements. Our solution involves ICT intervention and a common information channel coupled with the administrative components. A significant percentage of citizens in cities are gradually becoming aware of technology (Chourabi et al. 2012). Thus, ICT-based solutions become effective by connecting informed citizens (Nam and Pardo 2011; Lee and Lee 2014). Available ICT tools, like mobile devices and applications, can bring every class of society together on a single information platform. One of the success stories is the m-Indicator app (m-indicator 2017), which is used by more than 10 million people across the socioeconomic class for accessing information related to modes of public transportation like trains, buses, and taxis in Mumbai. The government has planned to provide free Wi-Fi services and low-cost smartphones to citizens, which can further boost the delivery of ICT solutions (MoCIT and GoI 2016). These initiatives would encourage the use of ICT solutions by citizens during disasters. Further, the system and its functionalities were developed based on the response and feedbacks collected in the form of qualitative surveys among agency officials, rescuers, and the citizens (Sect. 18.3). These surveys point out the delays associated with non-centralized real-time information transfer and lack of citizen involvement in the current system.

18.4.3 System Testing: A Case of Fire Drill

This section introduces the methodology applied to develop a mobile- and web-based system, and its implementation in a fire drill as per the practical guidelines of SDMB. A detailed literature survey about various disaster apps and their classification was done by Krishna Murthy et al. (2014) and Tejassvi et al. (2014). These applications lack disaster implementation scenarios and functionalities and thus are less preferred by the citizens. Our system is in-line with the recommendations and suggestions provided by the citizens and city managers during the surveys.

Implementation environment

To develop the system, it was essential to observe the patterns and identify the key areas that need to be addressed during the disaster. An iterative process of gap identification, system development, and testing in the fire drill was performed. Table 18.1 lists the desired and developed functionalities of an ICT system based on the survey (Sect. 18.3) with the officials of DMCR, Maharashtra Fire Services (MFS), and the citizens. Figure 18.2 shows the real-time data exchange and functionality for the developed ICT application.

Two mobile applications were developed, one for the victims and another for the rescuers, to serve rescue operations and relief management. The victim application was installed in a closed group of 25 citizens, and seven rescuers used the rescue application. A central database is used to store, exchange, and analyze the real-time data transfer among victims, first responders, rescuers, and decision makers. The Google-based navigation system was implemented in the applications for routing. The victim, or citizen, application has two modules: victim and volunteer modules. The victim module has the following functionalities:

- Victim can send emergency SMS.
- Victim can send auto SMS about his/her location to the registered numbers.

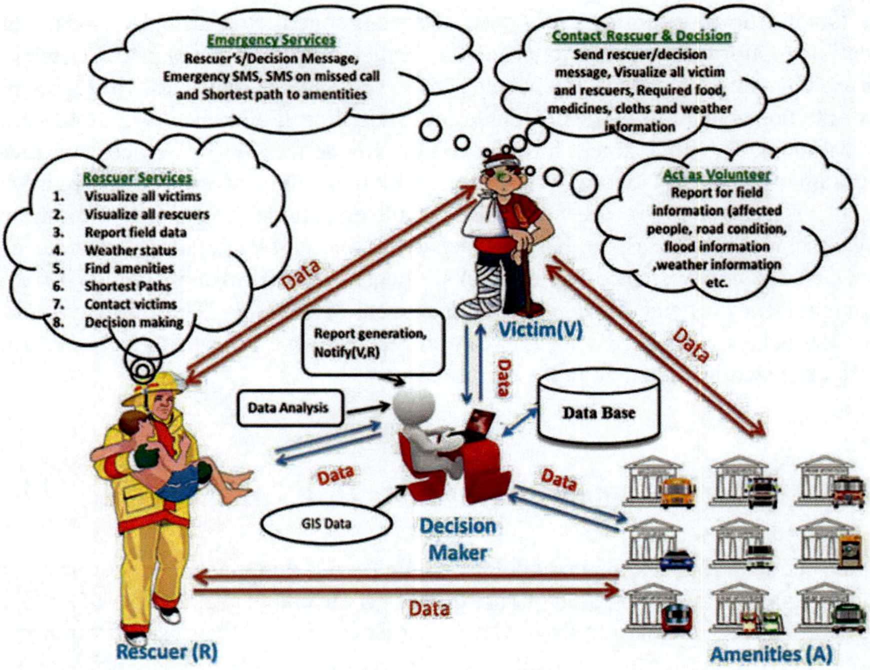

Fig. 18.2 Detailed functionality and data sharing diagram of the actors

Table 18.1 Desired and developed functionalities

Desired ICT-based system	Our developed system
Citizen involvement	The victim and volunteer module
Citizen to the rescuer data transfer	The victim and volunteer real-time information exchange module
Rescuers to control center data transfer	Real-time data exchange between rescuers and control center as a reporting module
Interface to decision makers in the control center	Dashboard to visualize real-time data of victims and rescuers and volunteers for better decision-making
Information exchange between rescuers	Module for victim information exchange and command broadcasting for coordination
Database	Central database at a command center
Participating amenities	Police station, hospital, and fire stations
Google-based navigation facility	Navigation modules for victims, rescuers, volunteers, and control center
Social media alerts	Live Twitter tweets monitoring module in the command center dashboard
Other functionalities	Emergency/distress call, reporting system, disaster alerts, geo-visualization, and post-event data analysis

- Victim can visualize rescuers, broadcast the messages to the nearest rescuers, contact the rescuers, police stations and hospitals, and navigate using the navigation window.
- Victim can transmit his/her emergency requirements to the rescuers.

In the second module, citizens can act as a volunteer and provide supportive services during the disaster. All the information was stored in a centralized data center, which helped the stakeholders become aware of all the proceedings in real time. This resulted in quick and better decision-making. The developed web dashboard for the decision makers in SDMB is used to visualize and analyze real-time data transfer. It also shows the real-time data related to the nearest fire stations and navigation information of the first responder vehicles. The dashboard displays the data exchange between rescuers and victims and data reported by rescuers in the real-time environment. The data set collected can be used in community awareness, training of human resources, and better planning.

18.4.4 Evaluating the Effectiveness

Due to the absence of any other such system in the existing scenario, it was not possible to provide a comparative picture of the system's efficiency. Thus, to test the effectiveness of the proposed system, a semi-structured qualitative survey among the stakeholders was conducted. A post-implementation survey with the same set of citizens (total samples: 100) who were involved in the pre-implementation survey was conducted. Further, the officials of MFB, DMCR, and rescuers were also surveyed for their feedback on the system.

18.5 Results and Discussions

This section presents the results of the application of the developed methodology during a fire drill. The proposed SDMB is discussed in the first subsection. In the second subsection, the case of the implementation of the developed system is presented. The last subsection presents the qualitative results of the structured survey. The survey captures the post-implementation efficiency of the system.

18.5.1 SDMB and Components of the System

The proposed system is shown in Fig. 18.3a. The dotted boxes represent the newly introduced ULBs, and information sharing arrows depict the new information flow structure. SDMB plays a direct role in policy formulation for DM strategies. A single channel for information transfer between the agencies and citizens is introduced to achieve response time reduction. In the proposed system, the information flow is coordinated by SDMB, thus providing hierarchical control over the centralized data. ULBs have a data repository that would be in congruence with the SDMB database. The case study presented in this section demonstrates the implementation of the discussed idea.

The forecasting cell receives disaster-related data that can be used to predict any unusual events to alert the citizens and concerned organizations. The ICT-based systems will be developed and supported by the introduced in-house IT cell. As every city has its dynamics, and problems are better understood and solved by the local agencies, the in-house IT cell has an advantage over outsourced solutions. One of the challenges faced by the current working structure is the decentralization of resources and the difficulty of combining effort in information acquisition and sharing for decision-making. A training cell is introduced in the framework to address this challenge. The human resources from various backgrounds in the inter-divisional cell make it a common platform for solution development. The cell shall work on multiple data formats used across the agencies to provide outcomes. One of the concerns highlighted by decision makers and discussed in the literature is the difficulty of the transition of existing ULBs from a current working structure to a new one. The addition of new bodies and their coordination with the existing bodies holds the key to an implementable solution. To address the concern, the solution retains most of the existing ULBs and their characteristics. The central body SDMB ensures coordination among all the ULBs. A single channel for information transfer in the supervision of SDMB makes coordination easier for the ULBs. The working mechanism of the proposed solution is explained in detail in the next section, along with the developed application and its implementation in a closed environment.

The expected working scenario of SDMB during a fire event

This section presents an expected working scenario of the discussed SDMB and the involved stakeholders during a fire event. Figure 3b shows the events and the information exchange details between stakeholders. As the fire event occurs, the sensors available on the incident space will transmit information about location, floor, resident data, temperature, etc., to SDMB and citizens (Event 0). The information is also broadcast to the users and decision makers using the mobile application, or by a phone call for immediate action (Event 1) and (Event 2). CCTV and geo-tagged images are used to gather information for better response and resource planning, which enhances the credibility of received data. SDMB communicates and passes the data to various local bodies, ward DMCR, disaster volunteers, and Quick Response Teams (QRT) (Event 3). Functionalities like navigation, dynamic update of the event information by local bodies, and citizen input help decision makers, rescuers, and victims in route planning and decision making. Using ICT applications, local institutions like hospitals update the status of beds and doctors. Victims and rescuers exchange real-time information like the current floor, the location of rescuers, and responders, etc., which helps in the efficient rescue process and can save many lives. SDMB also broadcasts the information to the media for making citizens aware of the real-time

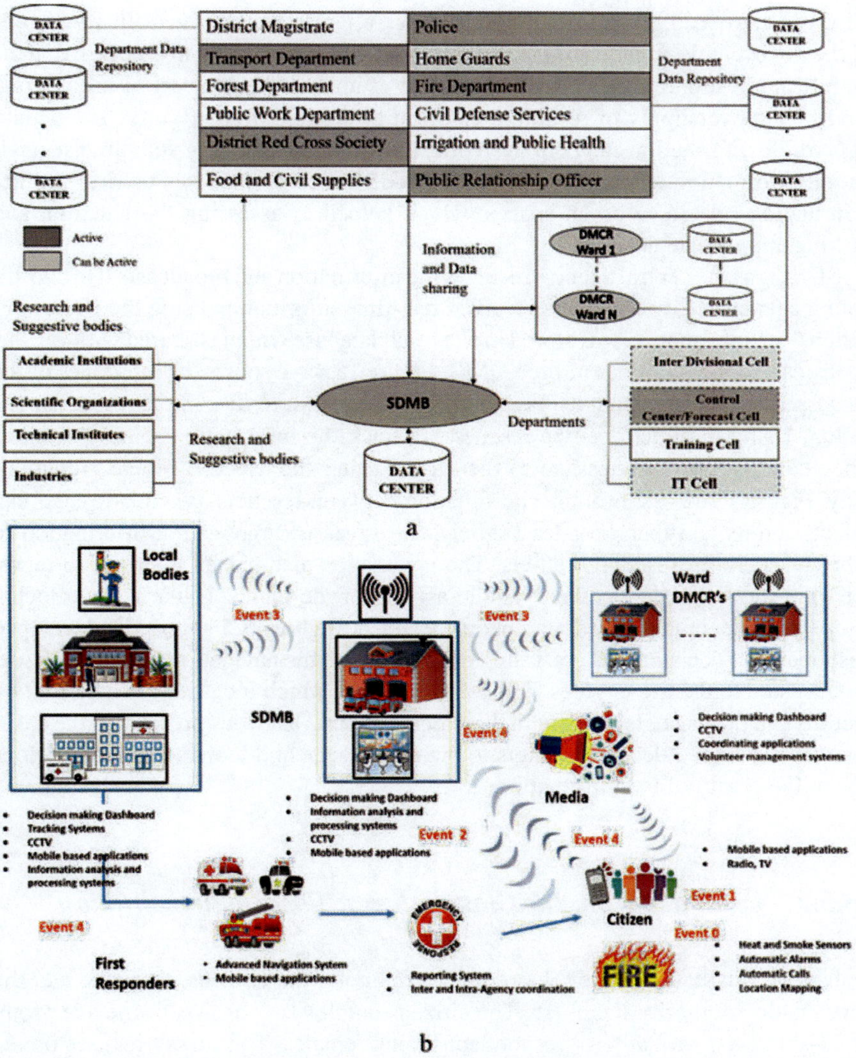

Fig. 18.3 **a** Proposed SDMB **b** Working scenario of the involved entities according to the proposed solution during a fire breakout

situation (Event 4). The reader must note that not all the mentioned technology is implemented in the research. However, some of the techniques as per the presented working scenario are applied in the fire drill discussed in the next section.

System implementation: A case of fire drill
The system was tested in a fire drill in an area of Mumbai city which comprised a mix of medium and high-rise buildings. The area had buildings with commercial, residential, and mixed usage. The fire drill started at 9 a.m. The assigned volunteer

raised an alarm call in the SDMB control room after the start of the drill. The details of the fire event were gathered using geo-tagged images and a form embedded in the mobile application. Proper resource planning is an important aspect of firefighting. Under- and oversupply of resources can lead to life and property loss. The benefit of credible information can help overcome this issue. Firefighting vehicles use navigation information to get the best route. A GIS-based system was used to acquire information related to urban spaces, which helped in assessing the situation and making appropriate decisions.

SDMB, as the central agency, receives the information and broadcasts it to coordinating agencies. City managers received real-time information using the ICT application. The information was stored in a central database which is shared with various agencies on the guidelines of the SDMB. Figure 18.4a–c shows the interfaces of the discussed mobile applications. Figure 18.4d shows the interface of the rescuer application. Using the interface, the rescuer can track the information sent by victims. They can also send a message to the victim using the "send message" functionality. Figure 18.4e–f shows the coordination between rescuers, where a rescuer can select victims and then broadcast crucial messages and navigation information to other rescuers for rescuing victims. The application also has an interface to report the information related to the dynamic aspects of the incident, like wind velocity, temperature, people rescued, and people trapped. Figure 18.5 shows the developed dashboard for decision makers. Using the dashboard, the navigation information was broadcasted to the fire engines. The real-time data, which included the data sent by victims and the rescuers along with their locations, was also viewed through the dashboard. This helped decision makers to plan the rescue and to make quick decisions due to the clarity of the information.

18.5.2 Feedback from the Stakeholders: Post Implementation

A detailed qualitative survey was conducted among the citizens, rescuers, and the participating authority members. The citizen sample size consists of the age group of 17–60 years, with 40% of respondents being females. The survey focuses on the usefulness of functionalities, their relevance to disaster situations, and responder's response. 72% of the responders were married and responsible for reporting on behalf of their family members. To keep the implementation process as close to the real scenario as possible, the message of the optional use of the application during the drills was conveyed to the users. Still, a high percentage (86%) of the users used the installed app during the exercise. It was encouraging to see that about 77% of the people were satisfied with the implementation of the application and mentioned it as a great help during the drill. A total of 68% of users were able to contact the rescuers in less than 2 min, while 22% of the users received a delayed response (over 2 min), and 10% of the users were not able to contact in the threshold time of 5 min, due to resource constraints, i.e., human and equipment. The implementation of the ICT system met the expectations of the users. 72% of the people wanted to use the system

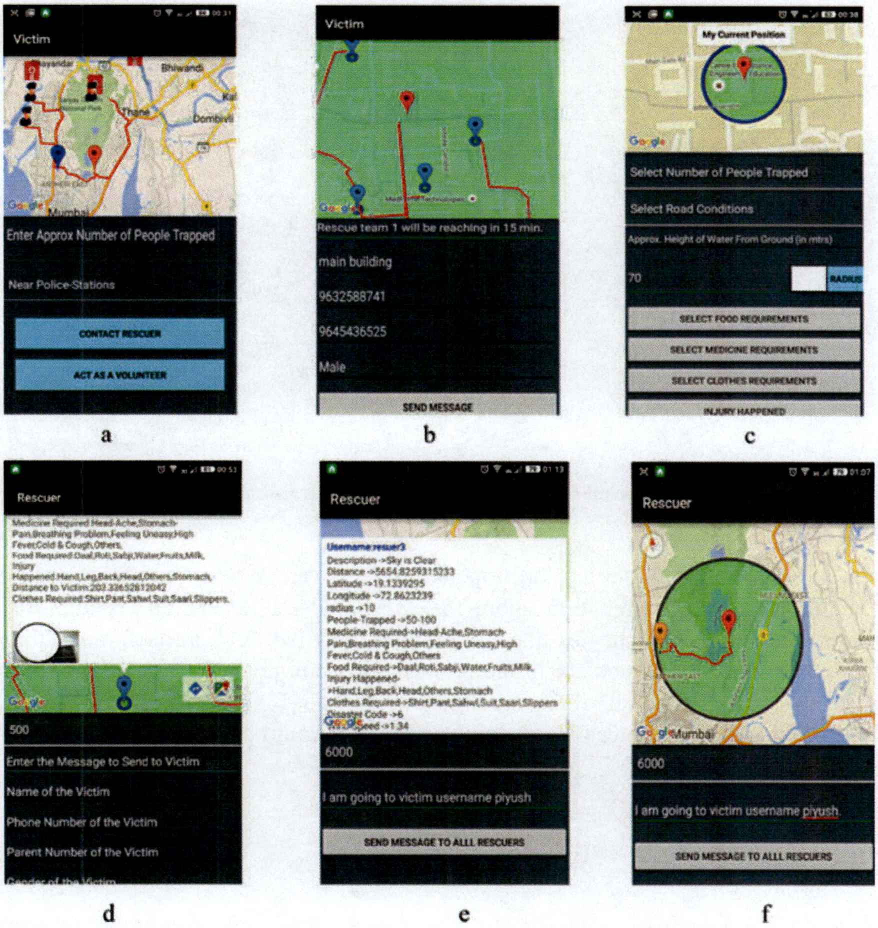

Fig. 18.4 **a** Hospital's, rescuer's, and police station's information with location. **b** Rescuers' status, broadcast message (rescuers and decision makers) and navigation on victims' interface. **c** Act as volunteer interface on victim interface. **d** Rescuer interface for interacting with victims and getting information. **e** Rescuer interface for broadcasting information to other rescuers. **f** Information and navigation broadcast to nearby rescuers

in the future, while 22% out of the remaining 28% of users were not certain about their choice. One of the reasons for the high user satisfaction was the quick response of decision makers and the information transfer using the system. This was evident from the fact that as much as 75% of the users were satisfied with the application functionalities and their implementation. 15% of the remaining users stressed that more features would lead them to use the application in the future. A separate survey was also conducted for the authorities and the rescuers about the usefulness and scope of ICT services in disaster scenarios. The response of the authorities confirmed the successful implementation of the system. They benefited from the response and

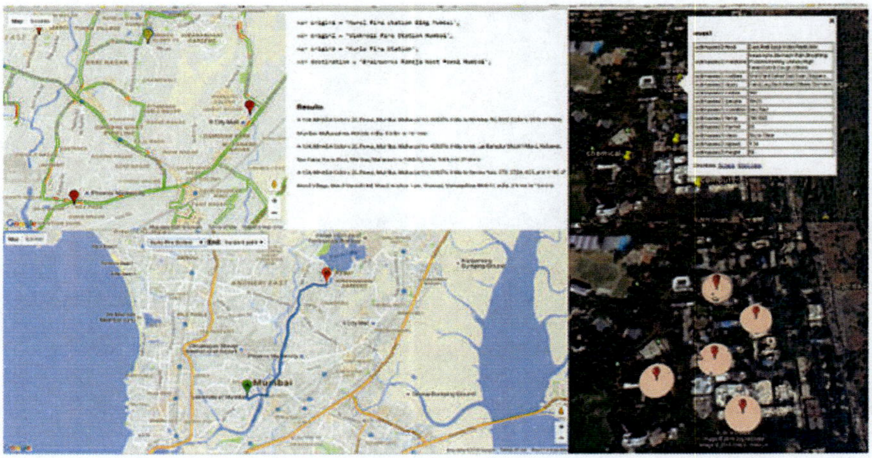

Fig. 18.5 Web-based dashboard for decision makers showing real-time information

rescue modules during the fire drill operations. ICT can be beneficial in managing the much-required coordination among the stakeholders. ICT is going to play a vital role to improve DM strategies in cities by promoting the DR approach rather than the current working approach. In the proposed system, the collection of real-time information makes the job of decision, making it easier for authorities and citizens. This is a critical parameter for developing a resilient smart city.

18.6 Conclusions and Recommendations

Many Indian cities lie in high disaster-prone zones. The study argues that disaster management and its support areas are under-focused and lack the required roadmap for service delivery. Participation in fire drills helped to understand the working process of the current system and related agencies. Based on the drill, surveys with the stakeholders, and existing literature, the need for a dedicated disaster management body with ICT support was identified. A solution is proposed in the form of Smart Disaster Management Body (SDMB). A mobile-based application, along with a web-based system, is developed and implemented in a fire drill under the supervision of city managers to study the feasibility of ICT in disaster management. The working of SDMB was tested in the drill, and post-implementation surveys were conducted to deduce the efficiency of the system. More than 100 citizens and city managers were surveyed to understand their response about ICT and its use in disasters. In the implementation phase of the system, 25 citizens, seven rescuers, and the concerned authorities were involved. Results show around 75% of the citizens were satisfied with the response and functionalities. While 15% out of the remaining 25% suggested more functionalities as the driving force to use such a system in the future, rescuers and authorities also reported better coordination and access to the data.

Any solution must equally benefit every person regardless of socioeconomic conditions. Results show that ICT if accessed by every class of the society, can help in bridging the gap of disaster-based service delivery. It will be early to say that a similar system will work across various cities based on a single implementation scenario, and a larger scale implementation is required to achieve more clarity. However, the results discussed in the study provide way forward strategies that can be used to develop efficient disaster management systems.

References

Alawadhi S, Armando A-N, Chourabi H, Gil-García J, Leung S, Mellouli S, Nam T, Pardo T, Scholl H, Walker S (2012) Building understanding of smart city initiatives. Electron Govern 40–53

Alazawi Z, Abdljabar MB, Altowaijri S, Vegni AM, Mehmood R (2012) ICDMS: an intelligent cloud based disaster management system for vehicular networks. In: Vinel A, Mehmood R, Berbineau M, Garcia CR, Huang C-M, Chilamkurti N (eds) Communication technologies for vehicle. Lecture notes in computer science, pp 40–56. Springer Berlin Heidelberg

Alazawi Z, Alani O, Abdljabar MB, Altowaijri S, Mehmood R (2014) A smart disaster management system for future cities, pp 1–10. ACM Press. https://doi.org/10.1145/2633661.2633670

Allenby B (2005) Toward inherently secure and resilient societies. Science 309(5737):1034–1036. https://doi.org/10.1126/science.1111534

Anu Jogesh U, Chakraborty S (2017) Climate disconnect in india's smart cities mission. Ind Climate Dial (blog). https://indiaclimatedialogue.net/2017/10/11/climate-disconnect-indias-smart-cities-mission/

Aqib M, Mehmood R, Albeshri A, Alzahrani A (2018) Disaster management in smart cities by forecasting traffic plan using deep learning and GPUs. In: Mehmood R, Bhaduri B, Katib I, Chlamtac I (eds) Smart societies, infrastructure, technologies and applications. Lecture notes of the institute for computer sciences, social informatics and telecommunications engineering, pp 139–54. Springer International Publishing

Arfat Y, Aqib M, Mehmood R, Albeshri A, Katib I, Albogami N, Alzahrani A (2017) Enabling smarter societies through mobile big data fogs and clouds. Procedia Comput Sci 109:1128–1133. https://doi.org/10.1016/j.procs.2017.05.439

Batty M (2013) Big data, smart cities and city planning. Dialog Hum Geogr 3(3):274–279. https://doi.org/10.1177/2043820613513390

Bremer R (2003) Policy development in disaster preparedness and management: lessons learned from the January 2001 earthquake in Gujarat, India. Prehosp Disaster Med 18(4):372–384

Chong M, Habib A, Evangelopoulos N, Park HW (2018) Dynamic capabilities of a smart city: an innovative approach to discovering urban problems and solutions. Gov Inf Q 35(4):682–692. https://doi.org/10.1016/j.giq.2018.07.005

Chourabi H, Nam T, Walker S, Gil-Garcia JR, Mellouli S, Nahon K, Pardo TA, Scholl HJ (2012) Understanding smart cities: an integrative framework, pp 2289–2297. IEEE. https://doi.org/10.1109/HICSS.2012.615

Dorasamy M, Raman M, Muthaiyah S, Kaliannan M (2011) Investigating perceived ICT usefulness for disaster readiness: a prelimenary analysis. In: 2011 7th international conference on information technology in Asia (CITA 11), pp 1–7. IEEE

EastElegant (2017) ASDMA—disaster ready assam (version 1.4). Android 4.0.3 and up. EastElegant. https://play.google.com/store/apps/details?id=com.eastelegant.asdma&hl=en

FICCI (2016) India risk survey 2015. https://ficci.in/Sedocument/20328/India-Risk-Survey-2015.pdf

Global Risks Report (2017) S.l.: My sustainable Canada. https://deslibris.ca/ID/10090180

Greer CR, Youngblood SA, Gray DA (1999) Human resource management outsourcing: the make or buy decision. Acad Manag Exec 13(3):85–96

Hall RE, Bowerman B, Braverman J, Taylor J, Todosow H, Von Wimmersperg U (2000) The vision of a smart city. BNL-67902; 04042. Brookhaven National Lab., Upton, NY (US). https://www.osti.gov/scitech/biblio/773961

IBM (2014) IBM—In India, IBM teams respond in flooding disaster. CTB25. https://www.ibm.com/ibm/responsibility/initiatives/stories/IBM-and-Chennai-flood.shtml

Karnatak HC, Kumar V (2015) Performance study of various spatial indexes on 3D Geo-Data in Geo-RDBMS. Geocarto Int 30(6):633–649. https://doi.org/10.1080/10106049.2014.952354

Krishna Murthy YVN, Raju PLN, Srivastav SK, Kumar P, Mitra D, Karnatak H, Saran S et al (2014) Capacity building for collecting primary data through crowdsourcing—an example of disaster affected Uttarakhand state (India). ISPRS Int Arch Photogramm, Remote Sens Spat Inf Sci XL-8(November):1249–1252. https://doi.org/10.5194/isprsarchives-XL-8-1249-2014

Kumar V, Ramamritham K, Jana A (2019) Resource allocation for handling emergencies considering dynamic variations and urban spaces: fire fighting in Mumbai. In: Proceedings of the tenth international conference on information and communication technologies and development—ICTDX '19, pp 1–11, ACM Press, Ahmedabad, India. https://doi.org/10.1145/3287098.3287099

Kusumasari B, Alam Q, Siddiqui K (2010) Resource capability for local government in managing disaster. Disaster Prev Manage: Int J 19(4):438–451. https://doi.org/10.1108/09653561011070367

Lavell A, Maskrey A (2014) The future of disaster risk management. Environ Hazards 13(4):267–280. https://doi.org/10.1080/17477891.2014.935282

Lee J, Lee H (2014) Developing and validating a citizen-centric typology for smart city services. Gov Inf Q 31(June):S93-105. https://doi.org/10.1016/j.giq.2014.01.010

Mhaske SY, Choudhury D (2010) GIS-based soil liquefaction susceptibility map of Mumbai city for earthquake events. J Appl Geophys 70(3):216–225. https://doi.org/10.1016/j.jappgeo.2010.01.001

m-indicator (2017) M-indicator now in a new avatar—The Hindu. https://www.thehindu.com/news/cities/mumbai/m-indicator-now-in-a-new-avatar/article18460041.ece

Mishra AP, Sen A, Kumar A (2017) Exploring potentials and challenges in making smart cities in India: a case study of Allahabad City, Uttar Pradesh. In: Sustainable smart cities in India, pp 123–42. The Urban Book Series. Springer, Cham. https://link.springer.com/study10.1007/978-3-319-47145-7_8.

MoCIT, DeitY, GoI (2016) "Digital India." New Delhi: DeitY. https://www.nio.org/nio/uploads/digital-india-ebook.pdf

Mukhopadhyay P, Revi A (2009) Keeping india's economic engine going: climate change and the urbanisation question. Econ Polit Wkly 59–70

Mumbai Fire Brigade (2014) Fire hazard response & mitigation plan. Municipal Corporation of Greater Mumbai. https://www.mahafireservice.gov.in/Site/PDFs/NewsEvents/FireHazardResponseMitigationPlan/MillenniumCityMitigationPlan/mumbai_mitigation_plan.pdf

Nam T, Pardo TA (2011) Conceptualizing smart city with dimensions of technology, people, and institutions. In: Proceedings of the 12th annual international digital government research conference: digital government innovation in challenging times, pp 282–291. ACM. https://dl.acm.org/citation.cfm?id=2037602

NASSCOM (2016) The future of the internet in India. NASSCOM. https://www.communicationstoday.co.in/images/reports/20160820-nasscom-the-future-of-the-internet-in-india-19082016.pdf

NDMA (2016) Standard operating procedure for disaster management NDMA. Ministry of Home Affairs

Pal I, Ghosh T (2014) Fire incident at AMRI hospital, Kolkata (India): a real time assessment for urban fire. J Bus Manage Soc Sci Res 3:9–13

Pineda MVG (2014) PROJECT YOLO: a disaster response coordination system. https://www.dlsu.edu.ph/conferences/dlsu_research_congress/2014/_pdf/proceedings/SEE-III-019-FT.pdf
Reuters (2015) India's 'smart cities' would be stupid without disaster strategy. https://www.reuters.com/article/us-disaster-risk-indiacities/indias-smart-cities-would-be-stupid-without-disaster-strategy-idUSKBN0LZ0CA20150303
Revi A (2008) Climate change risk: an adaptation and mitigation agenda for Indian cities. Environ Urban 20(1):207–229. https://doi.org/10.1177/0956247808089157
Shankar K (2008) Wind, water, and wi-fi: new trends in community informatics and disaster management. Inf Soc 24(2):116–120. https://doi.org/10.1080/01972240701883963
Sharma D, Tomar S (2010) Mainstreaming climate change adaptation in Indian cities. Environ Urban 22(2):451–465. https://doi.org/10.1177/0956247810377390
Smart City Mission GoI (2015). https://smartcities.gov.in/content/
Suma S, Mehmood R, Albeshri A (2018) Automatic event detection in smart cities using big data analytics. In: Mehmood R, Bhaduri B, Katib I, Chlamtac I (eds) Smart societies, infrastructure, technologies and applications. Lecture notes of the institute for computer sciences, Social Informatics and Telecommunications Engineering. Springer International Publishing, pp 111–22
Sugimiyanto S, Mehmood R, Albugami N, Katib I, Albeshri A (2017) Enabling next generation logistics and planning for smarter societies. In: Procedia computer science, 8th international conference on ambient systems, networks and technologies, ANT-2017 and the 7th international conference on sustainable energy information technology, SEIT 2017, 16–19 May 2017, Madeira, Portugal, vol 109 (January), pp 1122–27. https://doi.org/10.1016/j.procs.2017.05.440
Tejassvi T, Karnatak HC, de BY RA (2014) An android application to support flash flood disaster response management in India. Thesis, Remote Sensi Thesis, Dehradun
United Nation University (2015) World risk report. Bündnis Entwicklung Hilft (Alliance Development Works), and United Nations University—Institute for Environment and Human Security (UNU-EHS)
Wang C, Shiqiang Du, Wen J, Zhang M, Honghuan Gu, Shi Y, Hui Xu (2017) Analyzing explanatory factors of urban pluvial floods in shanghai using geographically weighted regression. Stoch Env Res Risk Assess 31(7):1777–1790. https://doi.org/10.1007/s00477-016-1242-6
Washburn D, Sindhu U, Balaouras S, Dines RA, Hayes N, Nelson LE (2009) Helping CIOs understand 'smart city' initiatives. Growth 17(2):1–17
Wentz L (2006) An ICT primer: information and communication technologies for civil-military coordination in disaster relief and stabilization and reconstruction. National Defense Univ Washington DC Center For Technology and National Security Policy
Zêzere JL, Pereira S, Tavares AO, Bateira C, Trigo RM, Quaresma I, Santos PP, Santos M, Verde J (2014) DISASTER: a GIS database on hydro-geomorphologic disasters in portugal. Nat Hazards 72(2):503–532. https://doi.org/10.1007/s11069-013-1018-y

Chapter 19
Selection of Suitable Site for Biomedical Waste Disposal in Lucknow City, India Using Remote Sensing Data, GIS, and AHP Method

Virendra Kumar, Reetanjali Singh, Ajay Mishra, and Shashank Shekhar Mishra

Abstract Biomedical waste is among the most significant sources, which threaten the global environmental health hazard. The adverse effects on our environment due to biomedical waste can be reduced through the selection of appropriate landfill site. Multi-criteria of site selection depending on consideration of the various factors concerning road/transport network, land use, surface water, geomorphology, geology, habitation/residential areas, soil texture, and groundwater table and the use of different thematic layers assessment method seems unavoidable. The present investigation involves the utility of geospatial data and GIS techniques, which is an important tool for the creation of database for various thematic maps. Analytic Hierarchy Process (AHP) method is an especially useful and structured approach to compare and assigning of weight, relative importance, and decision-making analysis. In the study area IRS-IC/ID satellite's 23.5 m, Landsat 30 m spatial resolution, and ground truth data have been used to couple with the AHP method. The study presents the results based on geospatial data, GIS and AHP method is an endeavor to identify and mark the most suitable site for disposed-off the biomedical waste in Lucknow city, Uttar Pradesh, India.

Keywords Remote sensing data · Biomedical waste · Multi-criterion layers · GIS · AHP

19.1 Introduction

In almost lower middle income group countries of the world in like India, there has been an exponential increase in Biomedical Waste; this is largely due to unprecedented population growth, urbanization, industrialization, and increasing microbiological laboratories and health care facilities. Chitnis et al. (2005), Glenn and Garwal

V. Kumar (✉) · R. Singh
Remote Sensing Applications Centre-Uttar Pradesh, Lucknow, India

A. Mishra · S. S. Mishra
Lucknow University, Lucknow, India

P. Sharma (ed.), *Geospatial Technology and Smart Cities*, The Urban Book Series,
https://doi.org/10.1007/978-3-030-71945-6_19

(1999), Anonymous (1998) suggested that presently in India, about 0.33 million tons of Biomedical Waste (BMW) is being generated per annum and this range may vary from 15 to 35% or 1–2.0 kg per bed per person everyday subjected to the total amount of waste generated. India has reached up to the numbers of 7933 urban local bodies according to India Census population data-2011. Kumar et al. (2013, 2014, 2015), Anugya et al. (2017) advised that in most of the developing countries, the traditional method for management of waste material/ Biomedical Waste is still common through landfilling without applying any specific technology.

In the northern part, Uttar Pradesh is the largest populous state in India. Lucknow, the capital town of this state, is the largest generator of Biomedical Waste. The town produces around 30.0 metric tons of biomedical waste every day. The total population of the city is 28.81 lacs as per India Census-2011, and the entire area of the district is 3244 km^2, and as per Lucknow Development Authority Master Plan Map-2011, the planning area of the town is 525 km^2, and Lucknow Nagar Nigam still has no proper landfill for biomedical waste and all the biomedical waste is being thrown in open low land pits and roadsides without taking any precautions and proper treatment. Collection, transportation and processing of biomedical waste and disposing off it into suitable site is a crucial issue because of the scarcity of land in urban periphery and opposition of local people too, due to its serious threat for detrimental of environment and human health. It is utmost necessary that an integrated system of disposal should be considered in achieving sustainable development. Ojha et al. (2007) advised that such a system typically intensifies on the serviceable element of biomedical waste for handling and assessment of more quantity of geospatial and non-spatial data concerning different aspects for appropriateness of a site. Geospatial data and GIS techniques coupled with the AHP method can play an important role in the analysis and also handling of huge quantum of geospatial data and data collected from various sources in less time for the selection of potential landfill site for biomedical waste management.

19.2 Study Area

Lucknow city has been taken up for study. The spatial extent of the study area lies between 26. 50°′ North latitude to 80. 50°′ East longitude (Fig. 19.1). The capital city of the biggest populated states in India is situated in the central part of the state. This capital city is situated on right and left, both banks of the river Gomati and these both banks of river divide the city into two unequal halves, the southern part is larger than the northern part. The major part of Lucknow is situated in the southern-western direction of the river Gomati. Trans-Gomati is also an important sub-urban area that is relatively more developed than the old part of the city due to the contemporary planning. The city area is around 415 km^2 with 1815 person per km^2 population density (Lucknow city guide, http://www.elucknow.com).

Fig. 19.1 Location map

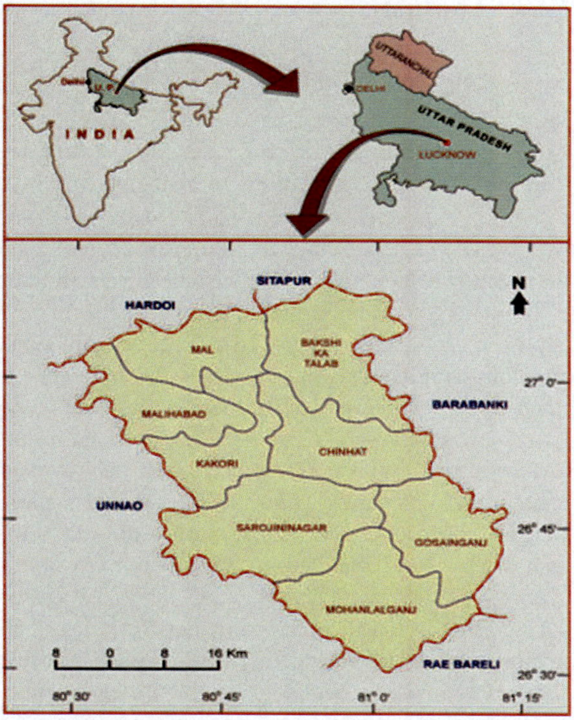

19.3 Materials and Data Used

The datasets used for the study are as follows:

i. SOI Map sheet no. 63B/12, 63B/13, and 63F/1 on 1:50 K for the study area of 1973–1976.
ii. IRS Series satellite optical data of 1C/1D satellites-LISS-III having a spatial resolution of 23.5 m of 10, February, 2001–2002 and Landsat 30 m resolution data of 2014 downloaded from Google.
iii. IKONOS satellite data-2014 downloaded from Google of having 1 m spatial resolution.
iv. Biomedical waste generation quantity data-2012, collected from Lucknow Municipal Corporation.
v. Census of India, population-2011.
vi. Groundwater table data-2010 collected from Govt. Departments.

19.4 Methodology

At first, Survey of India map sheets of 1: 50 K scale and geospatial data corresponding to the study area have been used to obtain the objectives of the study, and various thematic layers were generated. The data assembled from various sources were converted into shapefiles by using Arc-GIS 10.0 version and projection of Universe Transverse Mercator (UTM) zone_44N and World Geodetic System_84 was assigned to each and every criterion layer. Field survey/G.P.S. survey data for the location of landfill sites has also been carried out.

In complex decision-making, the AHP tool is an organized and prominent approach. Which was developed by T. L. Satty in 1980. It has been used for the study area for assigning weightage to objects between one to nine numbers based on their importance and alternatives of the elements were recorded in the ranking of identifying the potential site. In the study, multi-thematic map/data, i.e., land use/cover, soil texture, lithology/geology, groundwater, habitat/residential areas, restricted areas, geomorphology, surface water bodies, and transport network have been used and weightage was assigned with respect to the suitability of the sites. To consider the goal of the study, weightage assigned to each option of the particular theme has been calculated. Typically, the AHP model experiences two disadvantages: firstly, during relating the facts of the objects in terms of their variants, then some cases ranking regularities can occur, however, present investigations do not bear from this constraint, Secondly, which exists the use of irregular weight on a nine-point scale, consequently becomes significant. Occasionally, it becomes an awkward situation for decision makers to differentiate among them. The present paper is also based on 1–9 point scale weightage to different themes about the relative significance of the object based on the consciousness of administrator/decision maker in terms of potential landfill sites. During assigning the numbers/weightage to objects, software takes some weightage irrationally till getting the consistent value.

Eight factors have been used in this study through the AHP tool referring to geomorphology (Fig. 19.2), land use/cover (Fig. 19.3), geology (Fig. 19.4), transport network (Fig. 19.5), soil texture (Fig. 19.6), groundwater level (Fig. 19.7) and water bodies (Fig. 19.8), Habitation (Fig. 19.9). These multi-thematic maps used in the study area contribute a very important role to determine any waste disposal site. A model gives a complete and understandable configuration for ordering a problem of decision-making in terms of achieving the overall goals and for assessing the possible finding used all over the world in a broad array of a series of actions taken in order to achieve a particular end or decision-making activities and operations, which has proved an essential technique to locate and decide the suitable sites for various purposes.

In the AHP method, Saaty (1980) suggested that Consistency Ratio ($\leq$0.1) is an acceptable value. In which 'n' defines the size of the comparison matrix and 'RI' defines the randomly consistent value. If Consistency Ratio remains >0.1, in that case, there are requirements for the revision of the individual opinion until to obtain the Consistency Index (C.I) within the 0.1 or lesser. While the goal is not to totally

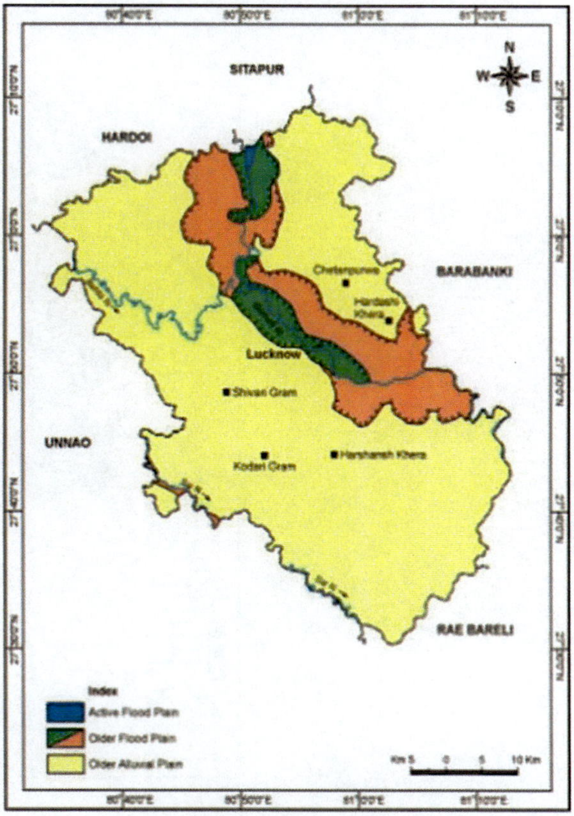

Fig. 19.2 Geomorphology map

eliminate the inconsistency, rather bring it within tolerance limits. The proposed method appears to be mathematically sound, but the pairwise comparison between criteria may differ. In the present study, a single scenario has been considered to find out the best suitable site for a landfill. The overall methodology and summarized results for suitability of selecting final potential sites for landfill have been shown in Table 19.2 and Fig. 19.11.

19.5 Geospatial Data Analysis

IRS series satellite remote sensing data of IRS-1C/1D-LISSS-III, and Landsat, IKONOS satellite imagery has been used to create the various thematic/criterion maps, i.e., land use/cover, transportation, habitation, geomorphology, water bodies, soil texture, geology, and groundwater table.

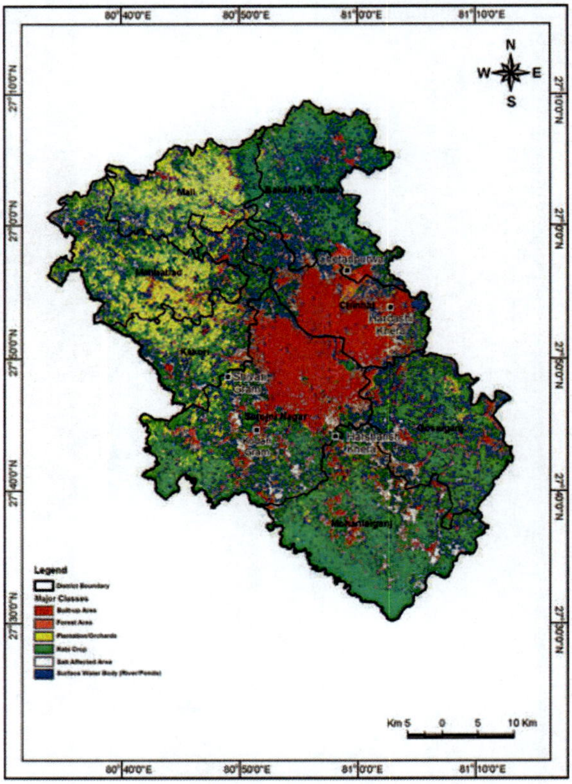

Fig. 19.3 Landuse/Landcover

19.5.1 Geomorphology

The digital database of geomorphology for the study area was prepared using IRS-IC/ID satellite's 23.5 m resolution data and it was updated using IKONOS satellite data. During the course of the study four classes namely flood plain (active), flood plain (older), alluvial (older), and alluvial (younger) are identified and delineated as shown in Fig. 19.2. It is found that all five sites from Site-1 to Site-5 lie in older alluvial, which is suitable for landfill siting.

19.5.2 Habitation

The digital database for settlement categories was interpreted using the IKONOS satellite 1 m resolution image and it has been shown as built-up (Fig. 19.3) in land use/land cover theme. Rahman et al. (2007) suggested that urban built-up area should be at a minimum distance of 1 km and 500 m from isolated houses, respectively, to locate a landfill site. Whereas, 500 m distance is considered suitable in case of towns

Fig. 19.4 Geology map

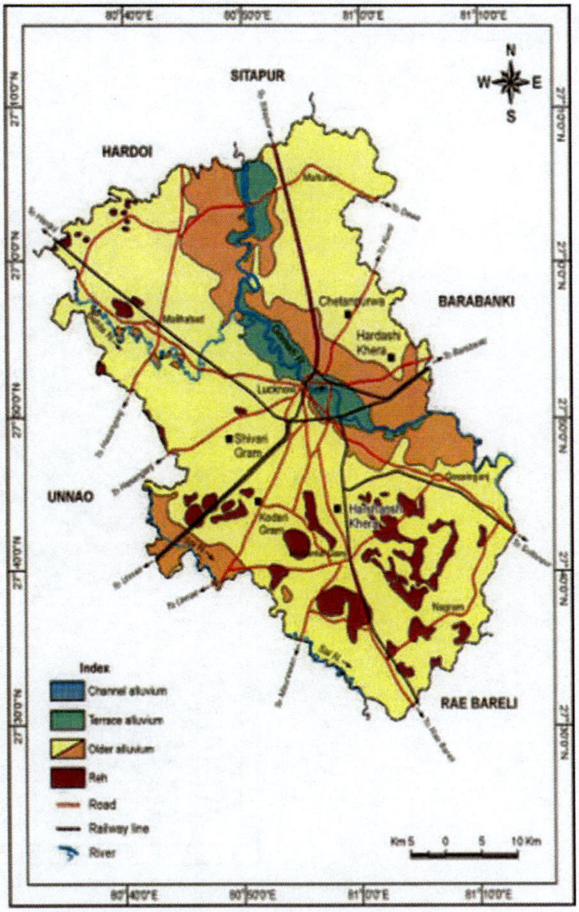

and villages with more than 500 people. Siddiqui (1996) and Jafri (2013) advised that for the collection of waste from ten or more dwelling units, the distance of the new site should not be closer than 250 m.

19.5.3 Geology

The geological information related database for the study area was derived using Geological Survey of India maps. Geology of the area has been grouped into four classes as channel alluvium, terrace alluvium, older alluvium, and reh as shown in Fig. 19.4. It is evident from Fig. 19.4 that all five sites namely from Site-1 to Site-5 are situated in older alluvium. According to different classes of geology older alluvium

Fig. 19.5 Transport-network map

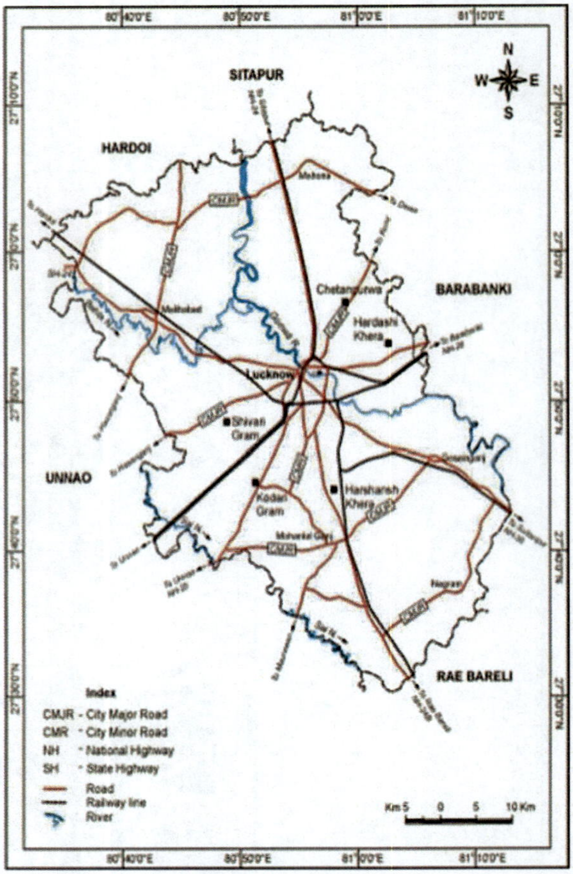

seems to be highly suitable because of deep groundwater prospects and low rate of permeability.

19.5.4 Soil Texture

The digital database for soil texture was prepared using the soil atlas of Uttar Pradesh state. Although soil texture can be classified into various classes, it has been grouped and mapped into seven soil classes such fine loamy calcareous, fine silty calcareous, fine and coarse loamy, coarse sandy silt, fine loamy, calcareously fined, and coarse loamy soil as given in Fig. 19.6. Oweis et al. (1990) suggested that Silt to much fined clay, clay, and mixed soil are very high, high, moderately suitable, respectively. Whereas sandy, clean sand/gravel are unsuitable or having very less suitability for

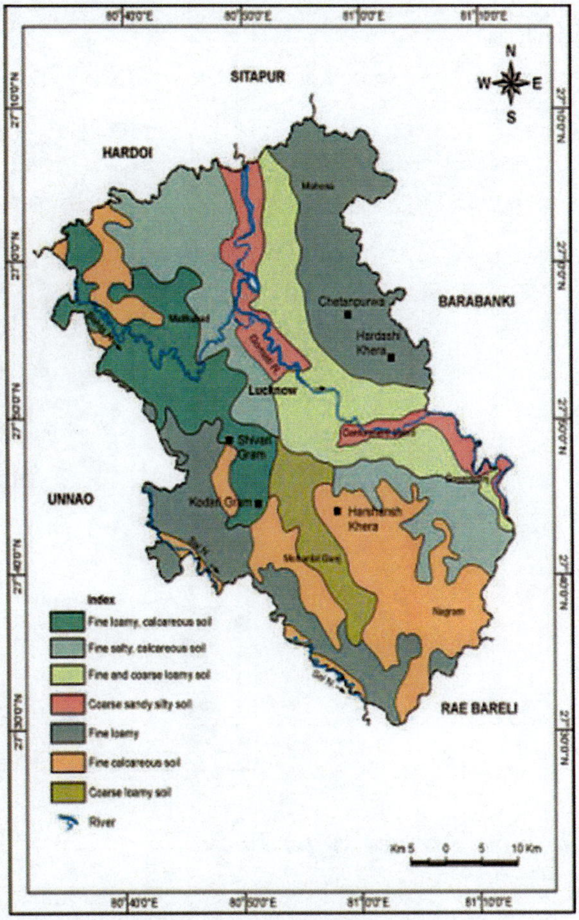

Fig. 19.6 Soil texture map

landfills because fine-grained texture has very less permeability of leachate than coarse-grained soils.

19.5.5 Landuse/Landcover

The digital database for land use/cover theme was created using Indian Remote Sensing series satellite-IRS-1C/1D LISS-III data and it was updated by using IKONOS satellite data. Six major land use/land cover classes have been identified and demarcated in the study area, i.e., cropland, urban (built-up), forest, village/settlement, waste land-salt affected land, and surface water as shown in Fig. 19.3. It is evident from Fig. 19.3 that wasteland-salt affected land/open scrubs are considered the most suitable for landfill.

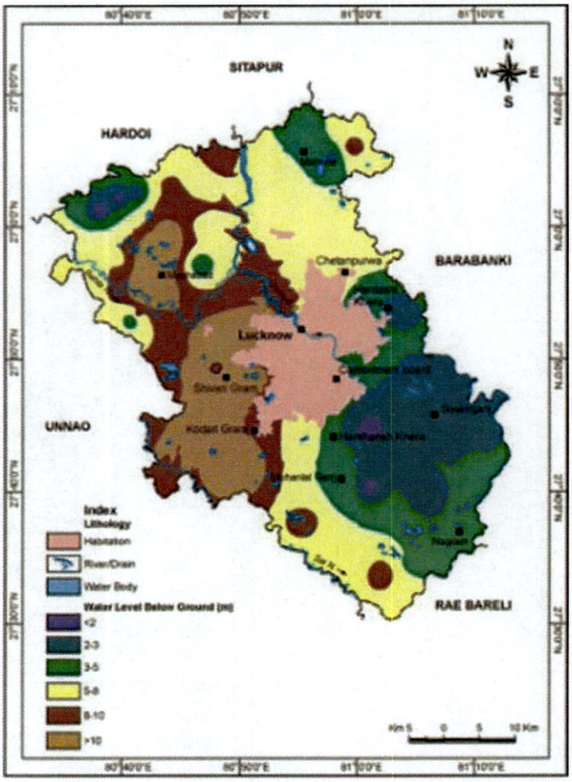

Fig. 19.7 Ground water depth map

19.5.6 Surface Water Body

The information related to surface water bodies was delineated using IRS-1C/ID LISS-III satellite data and was updated using IKONOS satellite data, and first level streams/drainage were delineated as line data on 1: 50 K scale by using Survey of India maps corresponding to area and river/lake/ponds were digitized as a polygon on the same scale as presented in Fig. 19.8. The categories under this theme were grouped into four classes as ox-bow lake, marshy/swampy land, ponds, and river/drainage. Distances of sites concerning each waterbody related to appropriateness from landfill were measured in the software. Rahman et al. (2007) and Thoso (2007) suggested that distance in terms of suitability point of view from river and canal, drain, and from large water bodies should be at least 1000 m, 500 m, and 200, respectively.

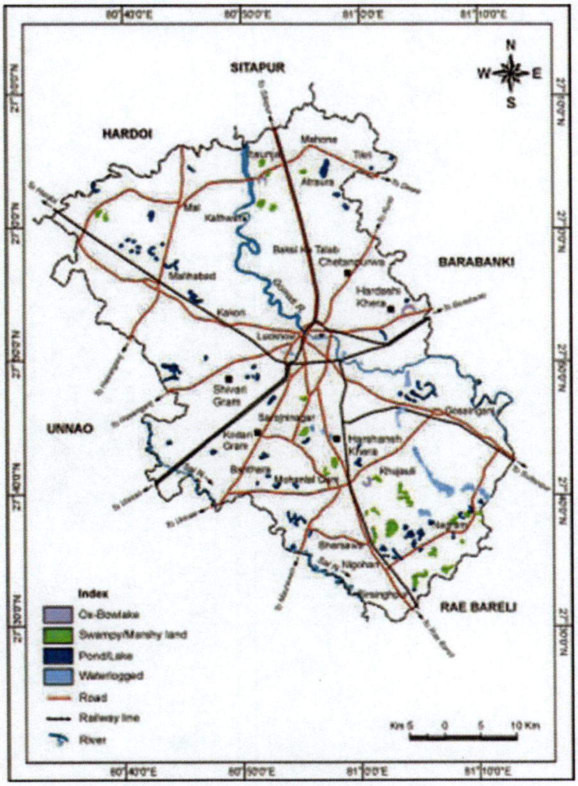

Fig. 19.8 Surface waterbodies map

19.5.7 Transport Network

The digital database for road/transport network was digitized using IKONOS satellite google earth imageries of 1 m resolution. In this theme, information is grouped into five categories as railway line, national/state highway, city major/minor roads as given in Fig. 19.5. Allen (2000), Lin and Kao (1999) has suggested that distance >1000 m from NH, SH, and CMjR should not be proposed for landfill because of the more expensive cost than that of the existing road network. Using Fig. 19.5, distances of all five sites concerning proximity to transportation have been measured and it has been observed that Site-2, Site-4, and Site-5 are near to SH, CMjR, and CMnR, and are found to be suitable, whereas, Site-1 and Site-3 because of not having the adequate distance from NH, SH, CMjR, and CMnR are not suitable.

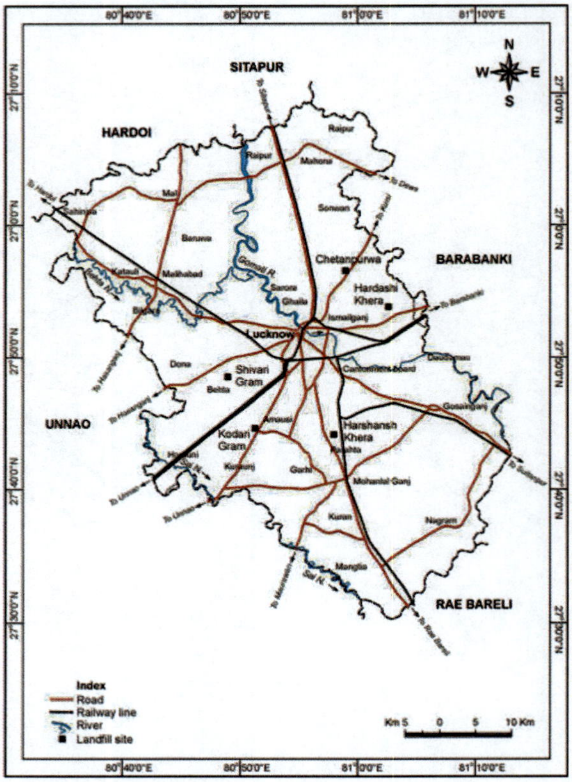

Fig. 19.9 Settlement map

19.5.8 Ground Water Depth

To conserve subsurface potable water, landfill should not be located over high quality groundwater depth areas. Bagchi (1994), and Bolton (1995) suggested that fresh groundwater should be saved with a composite parallel system and observation wells because of penetrating the potential leachate towards the down gradient.

19.6 Weight Calculation

19.6.1 Relative Weights Calculation of Alternatives

Relative weights of each option concerning to different theme of geomorphology, geology, soil texture, land use/cover, transportation, settlement, ground and surface water have been calculated after assigning the weightage between 1–9 based on the importance of each and every object using Analytic Hierarchy Process (AHP) tool for obtaining the values to all five landfill site. Pairwise comparison of each and every criterion layer has been made to decide the potentiality of sites.

Calculation of Relative Weight

The steps for calculation of relative weight are given below:

1. Each column sum calculation of the reciprocal matrix.

$$\mathrm{A} = \begin{pmatrix} a_{ii} \; a_{ij} \; \ldots \; a_{in} \\ a_{ji} \; a_{ji} \; \ldots \; a_{jn} \\ \ldots\ldots\ldots\ldots \\ a_{ni} \; a_{nj} \; \ldots \; a_{nn} \end{pmatrix} \qquad \text{e.g.} \quad \mathrm{A} = \begin{pmatrix} 1\,1\,1\,1 \\ 1\,1\,1\,1 \\ 1\,1\,1\,1 \\ 1\,1\,1\,1 \end{pmatrix}$$

$$Sum \; \sum_{i=1}^{n} a_{ni} \; \sum_{j=1}^{n} a_{nj} \; \ldots \; \sum_{n=i}^{n} a_{nn} \qquad Sum \; 4\,4\,4\,4$$

2. Normalization of the component in each column by dividing the sum of column.

$$\mathrm{A} = \begin{pmatrix} \frac{a_{ii}}{\sum_{i=1}^{n} a_{ni}} & \frac{a_{ij}}{\sum_{j=1}^{n} a_{nj}} & \cdots & \frac{a_{in}}{\sum_{n=i^n} a_{nn}} \\ \frac{a_{ji}}{\sum_{i=1}^{n} a_{ni}} & \frac{a_{ji}}{\sum_{j=1}^{n} a_{nj}} & \cdots & \frac{a_{jn}}{\sum_{n=i}^{n} a_{nn}} \\ \cdots & \cdots & \cdots\cdots \\ \frac{a_{ni}}{\sum_{i=1^n} a_{ni}} & \frac{a_{nj}}{\sum_{j=1^n} a_{nj}} & \cdots & \frac{a_{nn}}{\sum_{n=i^n} a_{nn}} \end{pmatrix} \qquad \text{e.g} \quad \mathrm{A} = \begin{pmatrix} \frac{1}{4} & \frac{1}{4} & \frac{1}{4} & \frac{1}{4} \\ \frac{1}{4} & \frac{1}{4} & \frac{1}{4} & \frac{1}{4} \\ \frac{1}{4} & \frac{1}{4} & \frac{1}{4} & \frac{1}{4} \\ \frac{1}{4} & \frac{1}{4} & \frac{1}{4} & \frac{1}{4} \end{pmatrix}$$

$$Sum \; 1\,1 \ldots 1 \qquad Sum \; 1\,1\,1\,1$$

3. Average values calculation across the row for getting the relative significance.

$$A = \frac{1}{n} \begin{pmatrix} \frac{a_{ii}}{\sum_{i=1}^{n} a_{ni}} & \frac{a_{ij}}{\sum_{j=1}^{n} a_{nj}} & \cdots & \frac{a_{in}}{\sum_{n=i}^{n} a_{nn}} \\ \frac{a_{ji}}{\sum_{i=1}^{n} a_{ni}} & \frac{a_{ji}}{\sum_{j=1}^{n} a_{nj}} & \cdots & \frac{a_{jn}}{\sum_{n=i}^{n} a_{nn}} \\ \cdots & \cdots & \cdots\cdots \\ \frac{a_{ni}}{\sum_{i=1}^{n} a_{ni}} & \frac{a_{nj}}{\sum_{j=1}^{n} a_{nj}} & \cdots & \frac{a_{nn}}{\sum_{n=i}^{n} a_{nn}} \end{pmatrix} = \begin{pmatrix} W1 \\ W2 \\ \ldots \\ W3 \end{pmatrix}$$

$$A = \frac{1}{4}\begin{pmatrix} \frac{1}{4}+\frac{1}{4}+\frac{1}{4}+\frac{1}{4} \\ \frac{1}{4}+\frac{1}{4}+\frac{1}{4}+\frac{1}{4} \\ \frac{1}{4}+\frac{1}{4}+\frac{1}{4}+\frac{1}{4} \\ \frac{1}{4}+\frac{1}{4}+\frac{1}{4}+\frac{1}{4} \end{pmatrix} = \begin{pmatrix} 0.250 \\ 0.250 \\ 0.250 \\ 0.250 \end{pmatrix} \text{Relative Weights}$$

$$\mathit{Sum} \;\; 1$$

Verification of Consistency:

Step 1: Multiplication of values of all columns after one by one of their relative values and for all rows also, and getting a vector of "weighted sum".

$$\text{e.g.} \quad A = \begin{pmatrix} \frac{1}{4} & \frac{1}{4} & \frac{1}{4} & \frac{1}{4} \\ \frac{1}{4} & \frac{1}{4} & \frac{1}{4} & \frac{1}{4} \\ \frac{1}{4} & \frac{1}{4} & \frac{1}{4} & \frac{1}{4} \\ \frac{1}{4} & \frac{1}{4} & \frac{1}{4} & \frac{1}{4} \end{pmatrix}$$

$$= \begin{pmatrix} 0.250 \\ 0.250 \\ 0.250 \\ 0.250 \end{pmatrix} 0.250 \begin{pmatrix} \frac{1}{4} \\ \frac{1}{4} \\ \frac{1}{4} \\ \frac{1}{4} \end{pmatrix} + 0.250 \begin{pmatrix} \frac{1}{4} \\ \frac{1}{4} \\ \frac{1}{4} \\ \frac{1}{4} \end{pmatrix} + 0.250 \begin{pmatrix} \frac{1}{4} \\ \frac{1}{4} \\ \frac{1}{4} \\ \frac{1}{4} \end{pmatrix} + 0.250 \begin{pmatrix} \frac{1}{4} \\ \frac{1}{4} \\ \frac{1}{4} \\ \frac{1}{4} \end{pmatrix}$$

$$= \begin{pmatrix} 0.0625 \\ 0.0625 \\ 0.0625 \\ 0.0625 \end{pmatrix} + \begin{pmatrix} 0.0625 \\ 0.0625 \\ 0.0625 \\ 0.0625 \end{pmatrix} + \begin{pmatrix} 0.0625 \\ 0.0625 \\ 0.0625 \\ 0.0625 \end{pmatrix} + \begin{pmatrix} 0.0625 \\ 0.0625 \\ 0.0625 \\ 0.0625 \end{pmatrix}$$

$$= \begin{pmatrix} 0.250 \\ 0.250 \\ 0.250 \\ 0.250 \end{pmatrix}$$

Step 2: Dividing the components of values by the corresponding relative value and further divided by the corresponding value.

$$\text{e.g.} \quad \begin{pmatrix} 0.250/0.250 \\ 0.250/0.250 \\ 0.250/0.250 \\ 0.250/0.250 \end{pmatrix} = \begin{pmatrix} 1 \\ 1 \\ 1 \\ 1 \end{pmatrix} \quad \begin{pmatrix} 1 \\ 1 \\ 1 \\ 1 \end{pmatrix} / \begin{pmatrix} 0.250 \\ 0.250 \\ 0.250 \\ 0.250 \end{pmatrix} = \begin{pmatrix} 4 \\ 4 \\ 4 \\ 4 \end{pmatrix}$$

Step 3: Compute the mean of the values in step 2 for getting the principal priority value.

$$\lambda_{max} = \frac{4+4+4+4}{4} = 4$$

$(\lambda_{max} \geq n)$ where n defines the size of matrix i.e.4

Step 4: Consistency Index computation.

$$CI = \frac{\lambda_{max} - n}{n-1} = \frac{4-4}{4-1} = 0.00$$

Step 5: Consistency Ratio computation.

$$CR = \frac{CI}{RI} = \frac{0.00}{0.58} = 0.00 \leq 0.10 \text{ (Acceptable)}$$

19.6.2 Maximal Priority Weight, Consistency Index, and Consistency Ratio

Maximal priority weight (λ_{max}), consistent value, and ratio of consistency for each and every alternative concerning each selection criteria have been calculated and Consistencies are obtained during pairwise comparison in terms of each selection criteria that are found to be acceptable. Saaty (1977) suggested that CR = 0.00 indicates that the judgments have not any limit of consistency. Overall computed RWs of alternatives in terms of each selection criteria and RWs of selection criteria with respect to biomedical waste disposal site selection are summarised in Table 19.1, Eastman et al. (1993), advised that overall composite value is calculated through combining the relative values or weight of substitute along with the relative weights of the factors come after by a summation of results to produce a suitability index. Equations given by Eastman are as follows:

$$SI = \sum W_i X_i$$

where $S.I.$ = defines the appropriateness Index of each option

W_i = RW defines the relative weight of individual selection criteria

X_i = RWs of substitutes concerning each criterion.

Table 19.1 Calculation of relative weight for alternatives

Goal	Criteria (Wi)	Alternatives (Xi)	
Biomedical waste disposal site selection (1.000)	Geology (0.035)	S-1	0.200
		S-2	0.200
		S-3	0.200
		S-4	0.200
		S-5	0.200
	Geomorphology (0.081)	S-1	0.200
		S-2	0.200
		S-3	0.200
		S-4	0.200
		S-5	0.200
	Landuse/landcover (0.137)	S-1	0.171
		S-2	0.171
		S-3	0.264
		S-4	0.197
		S-5	0.197
	Habitation (0.090)	S-1	0.083
		S-2	0.109
		S-3	0.241
		S-4	0.152
		S-5	0.415
	Transport network (0.026)	S-1	0.077
		S-2	0.153
		S-3	0.056
		S-4	0.414
		S-5	0.301
	Soil (0.027)	S-1	0.137
		S-2	0.137
		S-3	0.326
		S-4	0.326
		S-5	0.074
	Ground water (0.205)	S-1	0.055
		S-2	0.129
		S-3	0.278
		S-4	0.138
		S-5	0.399

(continued)

Table 19.1 (continued)

Goal	Criteria (Wi)	Alternatives (Xi)	
	Surface water (0.119)	S-1	0.150
		S-2	0.096
		S-3	0.345
		S-4	0.101
		S-5	0.308

S- Defines the identified landfill site as alternative in Table 19.1

19.7 Results and Discussion

In the study area, five landfill sites are identified and selected their ranked according to the suitability for biomedical waste disposal. Analysis of these sites reveals that the suitability/appropriateness of the index varies from 0.116 for site-1 (Harshansh khera) to 0.277 for Site-5 (Kodari gram). Site-2 (Chetanpurwa), Site-3 (Shivari gram), Site-4 (Hardashi khera) sensitivity values are 0.143, 0.232, and 0.231 accordingly. Sites having a higher suitability value are considered more suitable in the study area. One site out of five sites is found to have higher suitability index (S.I.) value as 0.277 namely Site-5 (Kodari gram). Site-3 (Shivari gram) 0.232, Site-4 (Hardashi khera) 0.231 are found to have moderately suitable. Whereas Site-2 (Chetanpurwa) 0.143, and Site-1 (Harshansh khera) 0.116 are found to have lower suitable site due to lower suitable suitability index values. Figures 19.10 and 19.11 shows the overall performance tendency of each site related to each and every criteria of selection.

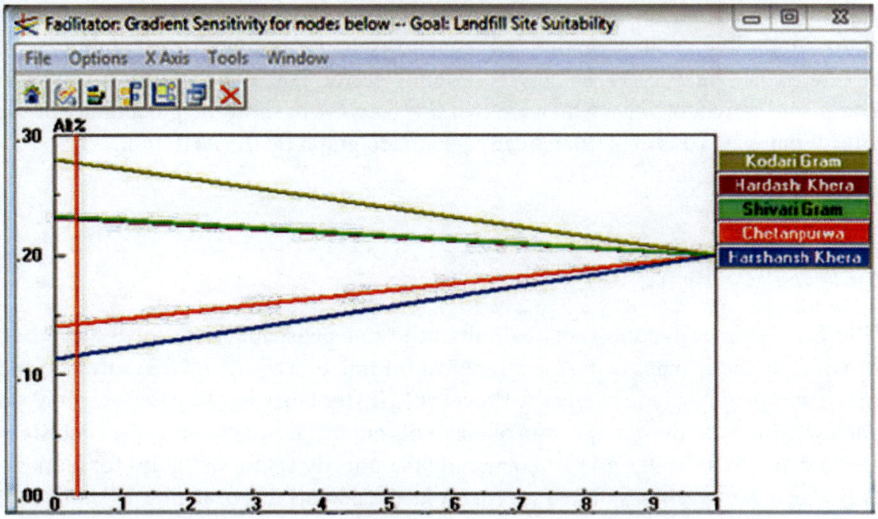

Fig. 19.10 Overall performance sensitivity of sites

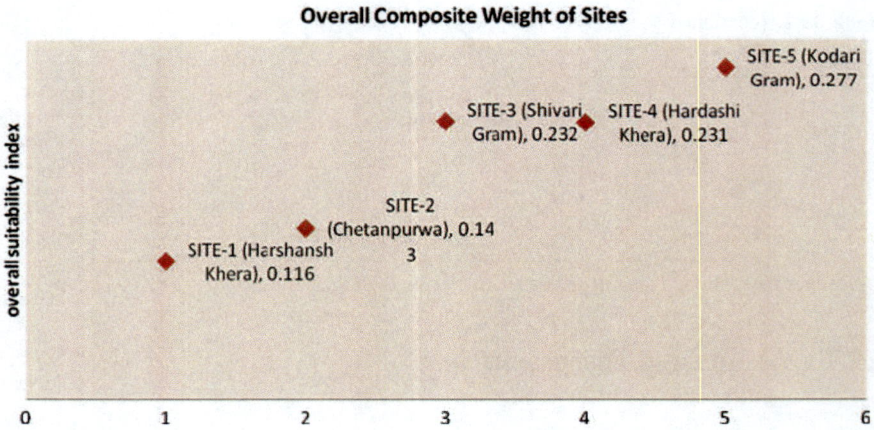

Fig. 19.11 Overall composite weight of the sites

The overall suitability indexes of all alternatives are summarized in Table 19.2, and Fig. 19.11 shows the S.I. values of overall composite weight for each landfill site with respect to each selection criteria, respectively, to their suitability. Site-5 (Kodarigram) is found to be the best site in the area. During field verification, all five sites have been checked in which Site-5 is found the most suitable landfill site concerning all the eight criteria, i.e., geomorphology, land use/cover, geology, surface water bodies, road/transport network, habitation, groundwater, and soil texture used in the analysis.

Weight of Biomedical Waste Disposal Sites

To obtain the overall composite weight or overall suitability index through combining the relative values of substitutes and relative values of the selection factors followed through an aggregation of results to yield a suitability index (Table 19.2). Figures 19.10 and 19.11 shows the overall performance sensitivity and overall composite weight of sites, which are two similar ways to show the potentiality of the sites. Figure 19.10 is an automatically generated graph by the AHP tool.

19.8 Conclusion

The present investigations conclude the utility of geospatial data for the creation of multi-thematic maps using Geographical Information System (GIS) software and applicability of Analytic Hierarchy Process (AHP) tool for assigning the weightage to each selection factors, comparison of elements or objects, and multi-object decision-making analysis (MODMA) and obtaining the priority vector or values for marking and selection of the best suitable site for biomedical waste disposal. Site-1 (Harshansh khera), Site-2 (Chetanpurwa), Site-3 (Shivari gram), Site-4 (Hardashi khera), and

Table 19.2 The overall suitability index of sites (priority vector)

W_i									X_i	$SI = \sum W_i X_i$
	GEOLOGY	GEOM	LU/LC	HAB	TRANSPORT	SOIL	GWT	SWT		
SITE-1	0.200	0.200	0.171	0.109	0.077	0.137	0.055	0.150	0.035	0.116
SITE-2	0.200	0.200	0.171	0.241	0.153	0.137	0.129	0.096	0.081	0.143
SITE-3	0.200	0.200	0.264	0.152	0.056	0.326	0.278	0.345	0.137	0.232
SITE-4	0.200	0.200	0.197	0.415	0.414	0.326	0.138	0.101	0.090	0.231
SITE-5	0.200	0.200	0.083	0.077	0.301	0.074	0.399	0.308	0.206	0.277
									0.127	
									0.205	
									0.119	

Site-5 (Kodari gram) are identified and marked in the study area, respectively. Site-5 (Kodari gram) is found to be the most appropriate site in the study area.

Acknowledgements Authors are obliged to Director, R.S.A.C., Uttar Pradesh, Lucknow for granting the permission to publish this research paper. Authors are also thankful to Dr. Udairaj and Shri V.J. Ganveer for giving us the suggestion to complete this investigation.

References

Allen AR (2002) Attenuation: a cost effective landfill strategy for developing countries. In: Proceedings of 9th congress of the international association for engineering geology and the environment, Durban, South Africa, 16–20 September 2002

Anonymous (1998) Biomedical waste (management and handling) rules. The Gazette of India, Extraordinary, Part II, Section 3(ii), dated 27th July, pp 10–20, 460. Ministry of Environment and Forests, Notification N. S.O.630 (E)

Anugya Kumar V, Jain K (2017) Site suitability evaluation for urban development using remote sensing, GIS and analytic hierarchy process (AHP). Springer Science+ Business Media Singapore 2017. In: Proceedings of international conference on computer vision and image processing. Advances in intelligent systems and computing, vol 460. https://doi.org/10.1007/978-981-10-2107-7_34(pp-377-388)

Bolton N (1995) The handbook of landfill operations, Blue Ridge Services Blue Ridge Solid Waste Consulting

Census of India-Population Data-2011

Chitnis V et al (2005) Biomedical waste in laboratory medicine: audit and management. Ind J Med Microbiol 23(1):6–13

Eastman JR, Jin W, Kyem PAK, Toledano J (1993) Raster procedures for multicriteria/multiobjective decisions. Photogramm Eng Remote Sens 61–5:539–547

Kumar S et al (2013) Selection of a landfill site for solid waste management: an application of AHP and spatial analyst tool. J Indian Soc Remote Sens 41(1):45–56

Kumar V et al (2014) Selection of suitable site for solid waste management in part of Lucknow city, Uttar Pradesh using remote sensing, GIS and AHP method. Int J Eng Res Technol 3(3):1461–1472. ISSN-2278-0181. www.ijert.org

Kumar V et al (2015) Landfill site selection for solid waste disposal in part of Lucknow city using remote sensing, GIS and A.H.P techniques. In: Proceeding of Sardinia symposium 2015, Fifteenth international waste management and landfill symposium, S Margherita di pula Cagliari, Italy, 5–9 October, 2015. CISA Publisher, Italy

Lin H, Kao JJ (1999) Enhanced spatial model for landfill siting analysis. J Environ Eng 125–9:845–851

Glenn McR, Garwal R (1999) Clinical waste in developing countries. An analysis with a case study of India, and a critique of the Basle—TWG guidelines

Ojha SP, Goyal MK, Kumar S (2007) Applying Fuzzy logic and the point count system to select landfill sites

Oweis IS et al (1990) Geotechnology of waste management. Butterworths, London, p 273

Rahman MM et al (2007) Site suitability analysis for solid waste disposal using GIS: a case study on KCC area. J Geo-Environ 6:72–86

Saaty TL (1977) A scaling method for priorities in hierarchical structures. J Math Psychol 15:234–281. https://doi.org/10.1016/0022-2496(77)90033-5

Saaty TL (1980) The analytic hierarchy process. McGraw-Hill

Siddiqui (1996) Landfill siting using geographic information systems: a demonstration. J Environ Eng 122(6):515–523

Siddiqui MJA, Jafri (2013) A review on impact and management of biomedical waste. Period Res II(I)

Singh R (2011) Unpublished PhD thesis on landuse and land suitability classification for safe disposal and waste management of biomedical waste in urban centre—a case study from Lucknow City. Dept. of Geology, University of Lucknow

Thoso M (2007) The construction of a geographic information systems (GIS) model for landfill site selection. Department of Geography, University of the Free State, Bloemfontein, Thesis of M.Tech.

Chapter 20
How Does Tourism Affect Urban Ecological Standards? A Geospatial Analysis of Wetland Transformations in the Coastal Resort Town of Digha, West Bengal, India

Asit Kumar Roy, Suman Mitra, and Debajit Datta

Abstract The growth of tourism often expedites the process of urbanization in terms of expanding built-up areas and other impervious surfaces at the cost of the original natural serenity. The coastal resort town of Digha, situated at the Medinipur coastal plain of West Bengal, India, had been developed after substantial land use conversions and loss of perennial interdunal wetlands. The effects of wetland transformations are more evident in terms of continuous ecosystem degeneration and reduction of subsistence-based livelihood provisions for the local populace. However, this trend of urban expansion was severely contested by the 'smart city' concept which the State Government had envisaged for the sustainable development of Digha Township in recent years. This new concept advocated for a balanced land use planning giving adequate attention to its green infrastructure and urban ecological standards towards developing the cities of the new century. An appropriate assessment of the present patterns of urbanization and consequent environmental transformations thus becomes the prerequisite of any such development endeavour. In this context, the present study aimed to quantify the cumulative stress of expanding buildups on the interdunal wetland ecosystems around the renowned coastal resort town of Digha of this region through a coupling of geospatial technologies with statistical analyses. For this purpose, Normalized Difference Built-up Index and Soil Moisture Index were derived from the multispectral Landsat images of the year 2000 and 2018, respectively, as indicative physical and bio-physical parameters. The linear maximum likelihood regression model was then applied on these derivatives to infer the spatio-temporal relationships between the expansion of buildups and changes in wetland characteristics of this area. Results indicated that the magnitude of wetland encroachment was more severe within the newly developed high-density built-up areas. Moreover, the interdunal wetlands were found to be shrinking more rapidly in 2018 compared to that of the 2000 scenario in direct correspondence with the enhanced growth of built-up zones. Remarkably, a few sites in the rural fringes

A. K. Roy · D. Datta (✉)
Department of Geography, Jadavpur University, Kolkata, West Bengal, India

S. Mitra
Department of Geography, University of Calcutta, Kolkata, West Bengal, India

P. Sharma (ed.), *Geospatial Technology and Smart Cities*, The Urban Book Series,
https://doi.org/10.1007/978-3-030-71945-6_20

were also experiencing aggravated loss of soil moisture contents chiefly due to the establishment of isolated resort compounds and gated housing complexes in spite of being quite far from the core urban zones. Incessantly changing tourist preferences towards secluded lifestyles and demand for serene landscapes as well as lacklustre implementation status of land development regulations were primarily attributed to this sporadic nature of land use conversions in this region. Based on the findings, a few realistically attainable management guidelines have been recommended towards developing a true 'smart city' in terms of both ecological composure and sustenance of tourism initiatives.

Keywords Built-up area · Interdunal wetland · Land use conversion · Remote sensing · Resort tourism · Smart city

Acronyms

DSDA	Digha-Sankarpur Development Authority
ETM	Enhanced Thematic Mapper Plus (ETM+)
GCPs	ground control points (GCPs)
ICT	Information communication technology (ICT)
LULC	land use/land cover
LST	Land Surface Temperature
LUDCP	Land Use and Development Control Plan
MPCS	Multipurpose Cyclone Shelter
NIR	Near Infrared
NDBI	Normalize Difference Built-up Index
NDVI	Normalized Difference Vegetation Index
OLI-TIRS	Thermal Infrared Sensor
RMSE	root mean square error
SMI	Moisture Index
SWIR	shortwave infrared
TIRS	thermal infrared
USGS	United States Geological Services

20.1 Introduction

Urbanization has become one of the most distinct global phenomena of the Anthropocene in order to provide liveable spaces for more numbers of human populace owing to the unprecedented demographic growth seen in the last century (Ruddick 2015; Boone and Fragkias 2012). A series of developmental planning and strategies were adopted across the globe to achieve this coveted concept of liveability (Staffans and Horelli 2014). However, most of these efforts have failed chiefly due to the uncontrolled demographic pressure on natural resources as well as the exploitative relationship of mankind with nature (Staffans and Horelli 2014). Amidst all these, a new approach of urban planning, termed as Smart City, was developed in the early twenty-first century (Caragliu et al. 2011; Nam and Pardo 2011; Staffans and Horelli 2014). Smart City is an intricately designed and systematically planned urban space where information communication technology (ICT) is merged with the internal structure of the city to enhance the quality of life through efficient usage of basic urban facilities and optimization of production, consumption, and waste of natural resources (Nam and Pardo 2011; Dameri 2013; WEF 2016). It has been estimated that by 2050, approximately 6.2 billion of the total world population will be settled in urban spaces, which may create multiple positive opportunities on one hand but may generate adverse socio-economic effects, such as enhanced poverty, unemployment, crime, pollution, and migration (Shapiro 2006; Angel et al. 2012; WEF 2016). To overcome these problems of billions, smart city planning can potentially serve as a universal remedy to achieve a sustainable future for both man and environment (Nam and Pardo 2011; Khansari et al. 2014).

Smart city planning uniquely focuses on the quality and convenience of urban amenities like green energy, smart building, fresh drinking water, waste management, fuel-efficient transportation, and real-time information system (Shapiro 2006). Furthermore, it also promotes the sustainable use of natural resources, which is one of the most challenging objectives of this planning approach (Caragliu et al. 2011; Shapiro 2006). In reality, very few cities have succeeded in maintaining the true spirit of smart city owing to the strong presence of political instability, resource conflict, and lack of environmental awareness (Naphade et al. 2011; Nam and Pardo 2011). The perpetual focus on profit maximization and economic growth often distracts the smart city's ideals and projected goals of environmental sustainability, making a so-called 'smart city' the same as an ordinary one (Dameri 2013).

Among different environmental parameters, soil moisture is considered as an important determinant of plant growth and soil quality (Jackson 1993; Dorigo et al. 2011). In urban areas, topsoils are mostly covered with concrete impervious surfaces, but the remaining interspersed urban and peri-urban lands and water bodies play key roles in urban greening and groundwater recharging. Over the last two decades, urban green spaces came into focus for their contributions towards mitigating heat island effects and conserving aesthetic values, as well as for retaining the surface soil moisture amidst greenhouse effect induced microclimatic changes (Jim 1998; Gill et al. 2007). Notably, significant quantities of moist lands (i.e. interdunal wetland, mudflat, salt marsh, and mangrove) are eventually lost as a direct consequence of rapid built-up expansions encroaching the low-lying coastal areas throughout the world (Li et al. 2010).

The coastal plains of India are densely populated with one-fourth of the country's demographic strength, among which the majority are settled along the low elevation coastal zones frequently encroaching on the surrounding farmlands and wetlands (Prasad et al. 2002; Revi and Singh 2007). In the Indian states of West Bengal and Odisha, numerous coastal wetlands are reclaimed and converted into human settlements riding on tourism and aquaculture-based economy (Mascarenhas 2004; Candela and Figini 2012). Since coastal wetlands are highly valuable for their biodiversity, carbon sequestration, agro-ecosystem potentialities, and recreational services, their conservation and protection have become a major challenge of urban sustainability research (Mitsch and Gosselink 2000). In this context, the coastal resort town of Digha and its surroundings under the Purba Medinipur district of West Bengal were selected as the area of investigation considering the existence of substantial research gaps in tourism-dependent urbanization and consequent environmental degradation in these parts of India. The objective of this study was to quantify the cumulative stress of expanding buildups on the wetland ecosystems of urban and peri-urban areas of Digha over the last two decades through the coupling of geospatial technologies with statistical analyses. Several notable reasons contributed to the decision to choose this area of interest: Firstly, the mass appeal of this long-standing coastal resort in not just West Bengal but throughout entire eastern India and consequent rapid expansion of the Township area in the last 40 years (Chattopadhyay 2000). Secondly, repetitive developmental actions taken by the government to convert this Township area into a more attractive tourist destination with all the facilities of a smart city. Based on the findings, a few realistically attainable management guidelines were also formulated to maintain the urban ecological standards in more sustainable ways.

20.2 Materials and Methods

A hybrid methodology was adopted, consisting of integrated geospatial techniques with intensive ground-based observations, as well as analyses of several published reports of both governmental and non-governmental organizations, as supportive information for validating the findings. Collection of ground control points (GCPs) for the purpose of ground-truthing was conducted through field investigations during 2018.

20.2.1 Delineation of the Study Area

Large cities are yet to develop along the coastline of West Bengal but some medium-sized towns and urban centres such as Digha have flourished in recent times (Chattopadhyay 2000; Mukherjee 2018). Each year, around 2.6 million people visit this place and for accommodating this enormous tourist population, the number of hotels and resorts has also increased very rapidly (DSDA 2008; Pahari 2013). Accordingly, restructuring of the Township area has also been conducted through the transformation of the natural landscapes into cityscapes (Fig. 20.1). Due to the two-fold economic potentials of this area, i.e. saltwater fishing and beach tourism, Digha has become one of the most popular places for investment opportunities in West

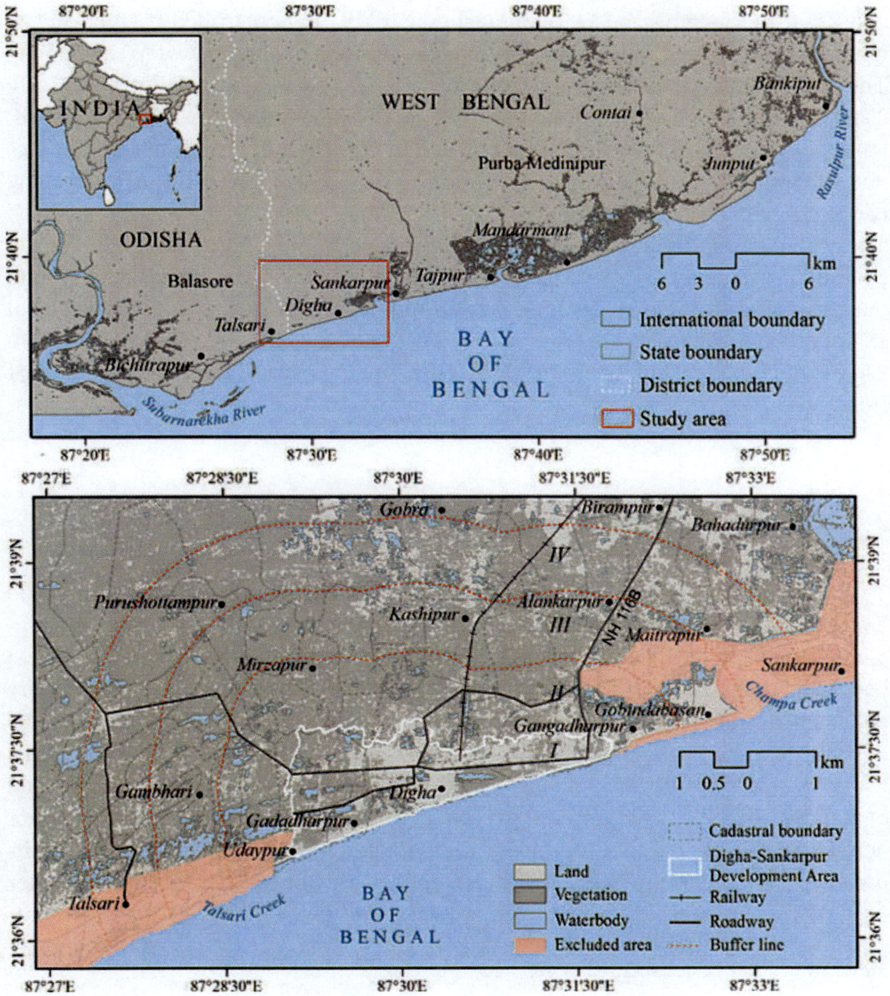

Fig. 20.1 Location map of the study area

Bengal (Mondal and Sen 2017). However, sustainable integration of natural serenity of the coastal environment with tourism was somehow neglected in town planning and infrastructural development for tourism. Concrete embanking of shoreline, reclamation of interdunal wetlands, fragmentation of foreshore dunes, sporadic construction of resorts, etc. are visible signs of unsustainable town planning and mass tourism initiatives which have proven detrimental for the fragile coastal wetlands still remaining here (DSDA 2013).

Digha is under the jurisdiction of Digha Sankarpur Development Authority (DSDA), an autonomous body established in 1990 under the West Bengal Town and Country Planning and Development Act, 1979, and controlled by the Department of Urban Development and Municipal Affairs, Government of West Bengal. The DSDA has approximately 480 ha of the area and the total population was nearly 27,713 in 2001 spread across 42 cadastres of Digha and Sankarpur (DSDA 2008; Pahari 2013). The entire coastal tract of DSDA had been divided into four consecutive zones, viz., core built-up area (Zone I), urban fringe (Zone II), urban shadow (Zone III), and rural area (Zone IV) for this study (Bryant et al. 1982; Antrop 2004). The spatial variability in surface soil moisture with respect to the changing built-up conditions was analysed across these four zones, which had been demarcated through the generation of three buffer lines around the DSDA boundary. The geographical extent of the entire study area is from 87° 27′ 14.66″ E to 87° 33′ 30.13 ″E and from 21° 35′ 47.54 ″N to 21° 39′ 20.07″ N along a 10 km-long coastline with the Bay of Bengal.

20.2.2 Data Sources

Two multispectral Landsat images acquired during the months of November 2000 and November 2018 through Enhanced Thematic Mapper Plus (ETM+) and Operational Land Imager with Thermal Infrared Sensor (OLI-TIRS) respectively, have been used in this study (Table 20.1). These Level-1 multi-band satellite images were collected from the open-source satellite data archive of United States Geological Services (USGS) and orthorectified with UTM projection and WGS84 datum. Specifically, the red near-infrared (NIR), shortwave infrared (SWIR) 1, and thermal band of Landsat 7 and thermal infrared (TIRS) 1 of Landsat 8 were selected for further derivation of physical and bio-physical parameters (Roy and Datta 2018). Additionally, the cadastral boundary was also collected from the data archive of Water

Table 20.1 Satellite dataset used in this study

Satellite	Sensor	Date of acquisition	Path/Row	Bands	Wavelength (μm)	Resolution (m)
Landsat 7	ETM+	8th November 2000	139/045	Red	0.63–0.69	30
				NIR	0.77–0.90	30
				SWIR 1	1.55–1.75	30
				Thermal	10.40–12.50	60* (30)
Landsat 8	OLI-TIRS	18th November 2018	139/045	Red	0.64–0.67	30
				NIR	0.85–0.879	30
				SWIR1	1.57–1.65	30
				Thermal (TIRS1)	10.60–11.19	100[a] (30)

[a]Thermal bands were acquired at 60/100 m resolution and resampled into 30 m for Level-1 data (*Source United States Geological Survey*)

Resource Information System of India. Furthermore, the information regarding the socio-economic scenario of the study area, its tourism and on-going developmental projects, and the boundary of DSDA planning area were collected from the web portal of DSDA (www.dsda.gov.in) (DSDA 2008). The web portal of Smart City Mission, under the Ministry of Housing and Urban Affairs, Government of India (www.smartcities.gov.in), was also used to acquire information related to smart city guidelines and projects.

20.2.3 Data Processing

The Erdas Imagine™ 2014 v14.01 software was used to process the Landsat images. Particularly, geometric corrections were done on these images to maintain the positional accuracy of pixels between two-time frames. Using the collected GCPs, the Landsat OLI-TIRS image of 2018 was georeferenced for using it as the primary reference image for registering the Landsat ETM+ image of 2000. During the image co-registration process between two successive time frames, the root mean square error (RMSE) was kept as 0.02 pixel (Schroeder et al. 2006; Roy and Datta 2018). Two land use/land cover (LULC) maps have also been prepared using those selected multispectral images of consecutive time frames to give a better understanding of the temporal dynamics of LULCs along with their direct impact on the status of derived physical and bio-physical indicators.

20.2.4 Mapping Through Physical and Bio-physical Indicators

Physical indicator-based maps of Normalized Difference Built-up Index (NDBI) and bio-physical indicator-based maps of Soil Moisture Index (SMI) had been prepared for assessing the relationship between soil moisture condition and built-up coverage. NDBI is one of the most widely used indices for identifying the built-up areas as it differentiates between covered and uncovered or paved surfaces. The range of the NDBI values lies between + 1 and –1. Bare or paved areas having higher reflectivity depict higher NDBI values (near + 1) than the moist land or vegetation-covered areas (Yang and He 2010; Angiuli and Trianni 2014). The following equation was used to obtain the index:

$$NDBI = (SWIR - NIR)/(SWIR + NIR) \tag{20.1}$$

In order to achieve higher accuracy level, the bare areas having high reflectivity without any built-up coverage; such as sandy beaches, sand dunes, and dry bare lands, and were excluded from the retrieved data through masking. Those pixels carrying high values can be erroneous as they do not show the actual scenario of urbanization or concretization (Zha et al. 2003).

Calculation of SMI through geospatial technology is a cost-effective method than the in-situ observations for a larger spatio-temporal resolution (Parida et al. 2008). SMI is an empirical parametric relationship between Land Surface Temperature (LST) and Normalized Difference Vegetation Index (NDVI) as these are important parameters to exhibit the dry or wet condition of the soil (Zhang et al. 2014; Zhan et al. 2004; Parida et al. 2008). NDVI alone is not sufficient for exhibiting the status of soil moisture as vegetation remains green even after initial loss of field capacity. Conversely, LST is more sensitive than NDVI towards detecting small changes in soil moisture conditions. Accordingly, a combination of LST and NDVI had been used here to calculate SMI for each pixel through the following equations:

$$SMI = ((LST_{\max} - LST_i)/(LST_{\max} - LST_{\min})) \tag{20.2}$$

where LST_{max} and LST_{min} are the maximum and minimum LSTs for a given NDVI and LST_i is the land surface temperature of a pixel for a given NDVI. LST_{max} and LST_{min} were calculated using following Eqs. (20.3) and (20.4) respectively:

$$LST_{\max} = a_1 * NDVI + b_1 \tag{20.3}$$

$$LST_{\min} = a_2 * NDVI + b_2 \tag{20.4}$$

where a_1, a_2, b_1, and b_2 are the empirical parameters obtained by the linear regression (a presents slope and b presents intercept) defining both warm and cold edges of the data (Parida et al. 2008). SMI value ranges from 0 to 1 where higher values close to 1 represent higher estimated soil moisture levels and values close to 0 represent the lower ones (Potić et al. 2017).

20.2.5 Correlation Analyses

Bivariate linear regressions were performed between NDBI and SMI values derived from the images of 2000 and 2018 to know the degree of stress of expanding buildups on interdunal wetlands (Zhang et al. 2009; Julien and Sobrino 2009). Here, these regression analyses were conducted separately for all four demarcated zones from the town centre towards rural fringes.

The calculated R^2 (coefficient of determination) values of 2000 and 2018 were further compared to each other to analyse their changes over time for a particular zone. Through these comparisons, the changing patterns of soil moisture corresponding with the expanding buildups from the town centre towards the fringe area were also revealed.

20.3 Results

20.3.1 Spatio-Temporal Distribution of Built-Up Areas

The result of NDBI showed a noticeable change in buildups from 2000 to 2018. In this short time frame, built-up areas expanded rapidly around the major tourist attractions of Digha (Appendix 1). From east to west, places like Gobindabasan, Gangadharpur, Digha, Gadadharpur, Udaypur, and Talsari were experiencing substantial stresses from both emerging sectors of coastal tourism and commercial brackish water aquaculture, which accelerated the rate of expansion of buildups (Fig. 20.2a). In the NDBI map of 2000, two prominent patches of buildups were detected with higher values, among which the major patch belonged to the Digha Township area and the minor patch was the tiny settlement of Gobindabasan, located near the left bank of the Champa creek. Except these two, other recreational centres like Udaypur and Talsari had just started to develop at that time and thus did not produce notable reflectance owing chiefly to their lesser built-up coverage.

However, the urban area expanded immensely between 2000 and 2018 (Fig. 20.2b). In reality, Digha Township had two nodes of growth, which were New Digha and Old Digha. Only the Old Digha settlement had prominent urban spectral signatures during 2000 but a new extension of a settlement named New Digha developed in the west later on as a result of locational accessibility and interconnectivity with National Highway 116B and Digha Bypass (Appendix 2). Eventually, these two settlements merged together and created an almost continuous built-up surface (Chattopadhyay 2000; Pahari 2013). In 2018, it was also found that an elongated stretch of interdunal wetland was converted near Udaypur to create recreational sites, which would also be transformed into buildups in the imminent years. Moreover, a large area with high built-up reflectivity was detected near the Gobindabasan settlement in 2018. Due to the economic importance of the local fish market, this settlement patch had grown rapidly after engulfing the associated low-elevation moist lands.

The natural terrain of this coastal tract is undulated due to the presence of sequential swales and troughs of coastal dunes. Subsequently, most of the built-up areas were developed after flattening these dune ridges and filling up the marshy interdunal wetlands. In the last 50 years, many of these wetlands were engulfed and converted into hotels and other paved surfaces (Fig. 20.3). Till 1980, only 10 hotels were developed within the boundary of DSDA, but this figure escalated to 336 in 2018 (DSDA 2013). Though these interdunal wetlands are extremely valuable from both ecological and local livelihood perspectives, the environmental importance of these wetlands are often neglected even now and left unaddressed in the regional developmental plans.

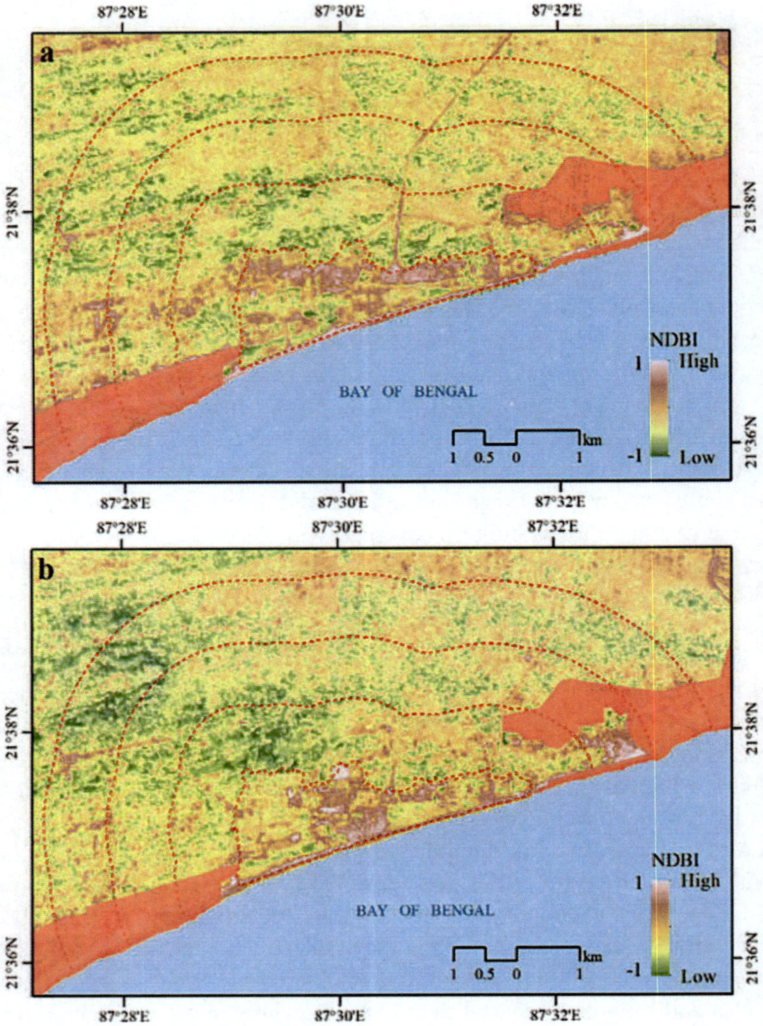

Fig. 20.2 Physical indicator-based mapping of DSDA. NDBI maps of 2000 (**a**) and 2018 (**b**)

20.3.2 *Changing Patterns of Soil Moisture Content*

In 2000, higher moisture content was observed primarily in the inland portions of the study area which are close to Gobra, Birampur, and Bahadurpur of the study area (Fig. 20.4a). The low-lying areas situated between two paleo-coastal dunes had abundant soil moisture content, whereas less amount of moisture content had been detected near the coast because of the presence of course-textured sandy deposits of this area having less moisture retention capacity. Furthermore, the absence of

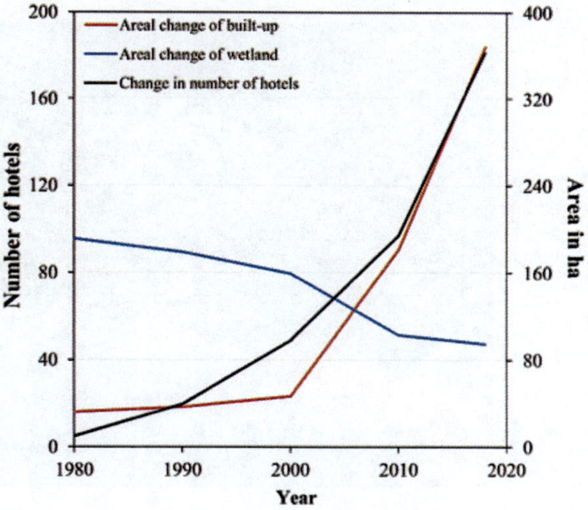

Fig. 20.3 Temporal dynamics of built-up coverage, number of hotels, and wetland areas in DSDA

dense canopy coverage in the coastal vegetated areas and the occurrence of high-velocity winds increased the rate of first stage evaporation from the topsoil, which also acted as the prime determinant of moisture deficit at the beachfront areas. A few big patches of high moisture content (SMI > 0.8) had been identified in and around the concretized areas of Digha Township in 2000 which could be attributed to the networks of interdunal wetlands. During monsoon, interdunal swales were completely inundated with freshwater and subsequently transformed into perennial wetlands.

On the contrary, higher SMI values were detected in the eastern parts than the western parts in 2018 (Fig. 20.4b). Reluctance towards the continuation of subsistence agricultural practices among rural masses of the western part was identified as the major cause of land conversion here thereby resulting in lesser surface moisture contents. Furthermore, alarmingly lower SMI values were also found within the DSDA boundary. In some places, the soil moisture was completely depleted due to the alteration of erstwhile interdunal wetlands into buildups (Fig. 20.5). In addition, the remaining moist lands of the Digha Township area were also suffering from the increasing demand for land for infrastructural development of tourism.

20.3.3 Relationship Between NDBI and SMI

The R^2 values derived from the regression analyses between NDBI and SMI data along the four consecutive zones of the study area for 2000 and 2018 helped in explaining the dependency of moisture content over the buildups for a specific zone (Fig. 20.6). In the core built-up areas, the R^2 value was 0.70 during 2000, which meant that about 70% of the total variation in SMI values (Y) was being explained

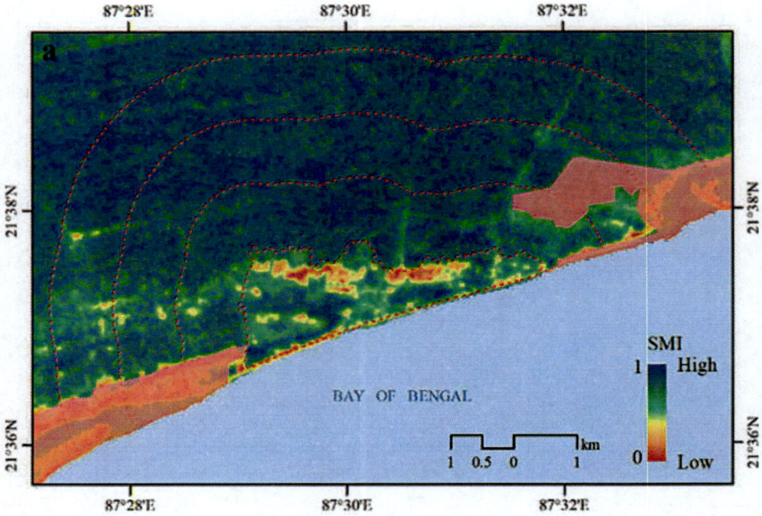

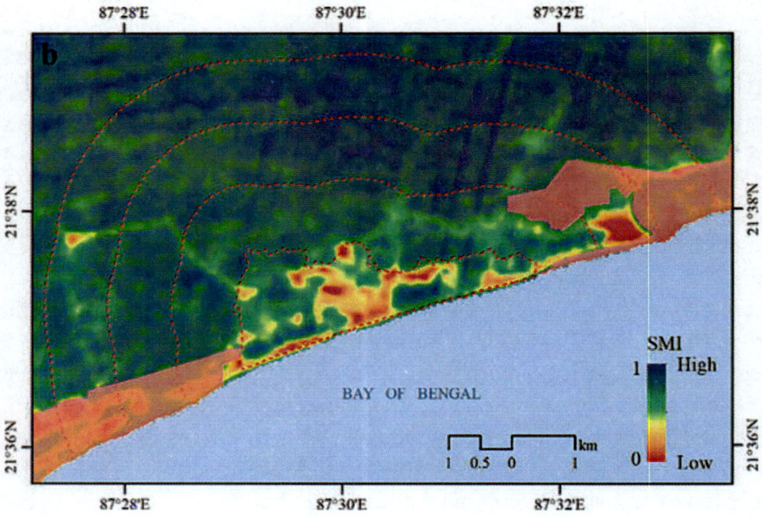

Fig. 20.4 Bio-physical indicator-based mapping of DSDA of 2000 and 2018. SMI maps of 2000 (**a**) and 2018 (**b**)

by NDBI values (*X*) and the scatter plot depicted a very strong negative relationship between buildups and soil moisture content in that particular zone, while the R^2 value was lowered to 0.59 or for that similar zone during 2018 indicating the presence of other intermediate land use conversions. In the urban fringe zone, the R^2 value was 0.51 in 2000 while it was 0.39 in 2018. Here, the relationship was recognized as a moderately strong negative one. In the subsequent zone of urban shadow areas, the R^2 value was 0.49 0.48 during 2000 and 2018, respectively, indicating a stable negative relationship between the variables. In the last zone of rural areas, the R^2

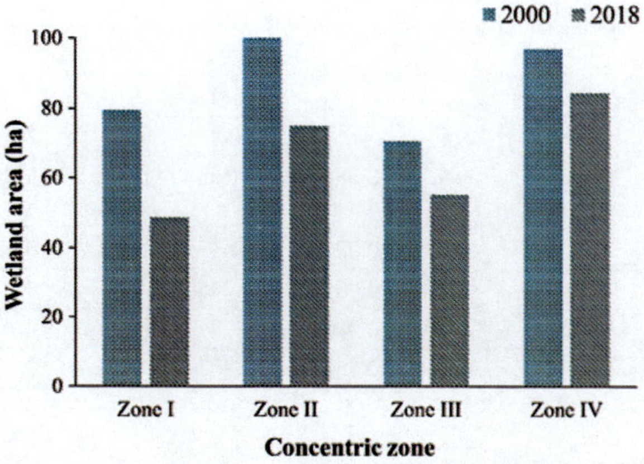

Fig. 20.5 Zone-wise changing pattern of interdunal wetlands from 2000 to 2018

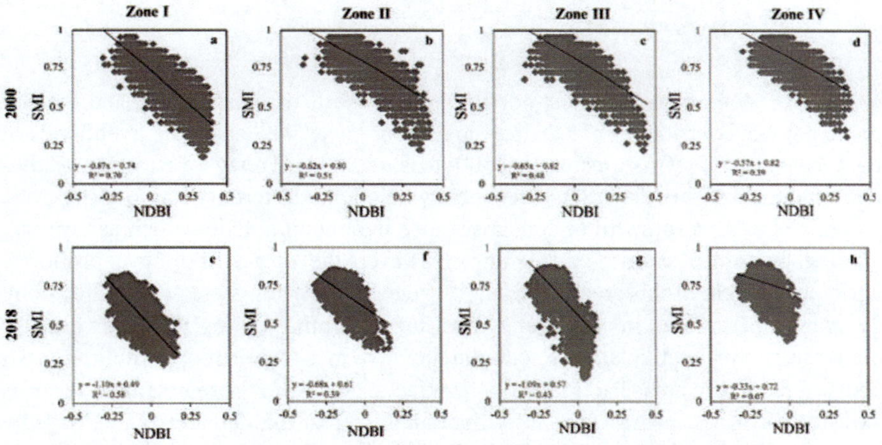

Fig. 20.6 Zone-wise R^2 values between NDBI and SMI estimates for 2000 and 2018

value was found as 0.39 in 2000 representing a very weak negative relationship. In 2018, almost no correlation was observed with a value of 0.07.

A unique pattern was observed among the R^2 values of four consecutive zones. The values were highest in Zone I in both years. Following that, they were decreasing afterwards in Zone II and Zone III and reached their lowest extreme in Zone IV. This decreasing pattern from the town centre to the fringe areas could be explained as the decreasing dependency of *Y* (SMI) on *X* (NDBI) variable towards the rural areas with increased distance from the core built-up areas. In general, the built-up signatures were detected lesser in the rural areas compared to the core built-up areas while moist land signatures showed just the opposite trend (Table 20.2). Thus, a constant negative relationship was established among buildups and soil moisture content, which was stronger in the core but weaker towards the fringes of the study area.

Table 20.2 Comparative summary of NDBI and SMI statistics for consecutive concentric zones of the study area

Zone	NDBI				SMI			
	2000		2018		2000		2018	
	Mean	Standard deviation	Mean	Standard deviation	Mean	Standard deviation	Mean	Standard deviation
I	0.107	0.142	−0.095	0.080	0.649	0.147	0.599	0.116
II	−0.004	0.120	−0.161	0.064	0.786	0.122	0.722	0.076
III	0.002	0.115	−0.159	0.066	0.799	0.127	0.750	0.094
IV	0.003	0.114	−0.160	0.064	0.802	0.122	0.755	0.092

20.4 Discussion

20.4.1 Socio-ecological Drivers of Urban and Peri-Urban Transformations

Digha has been one of the most popular beach resorts of West Bengal and thereby made notable contributions to the regional economy as well as on the livelihoods of the local residents (Chattopadhyay 2000; Mishra 2004). Though the tourism activities conducted within the DSDA area mostly belong to the tertiary economic sectors, its peripheral areas are still under subsistence livelihood practices such as farming, grazing, livestock rearing, and fishing. However, the ever-dwindling unprofitable agricultural yield of this area was found to generate unwillingness among the young generation of farmers in continuing subsistence farming during the fieldwork. By comparing unprofitable farming with the lucrative mass tourism opportunities, residents of the fringe areas had gradually decided to engage in the recreational services akin to that of the nearest beach tourism and cater to the demands of the rapidly expanding tourism industry. In this way, tens of hectares of moist agricultural lands and interdunal wetlands were transformed into impervious surfaces rapidly through the construction of hotels, parking lots, and shopping avenues for tourists without any major conflict.

A series of projects under governmental initiatives had been undertaken to restructure the Township area through integrated developmental planning giving preference to the coastal tourism of Digha (Table 20.3). However, no such projects were initiated in this area where town planning and natural resource management secured equal and adequate attention particularly with respect to the interdunal wetlands. In the past, some asphalt roads were constructed to increase the connectivity among Digha, Udaypur, and Talsari areas, but these constructions were only possible after fragmenting the erstwhile continuous stretch of those wetlands.

Several prominent patterns of LULC transformations were identified within the study area, namely, wetland to built-up conversion; wetland to bare land and then bare land to built-up conversions; and wetland to agricultural land and then agricultural

Table 20.3 Governmental initiatives for 'smart' urban development in DSDA

Project name	Date	Status	Objective
Digha-Sankarpur Integrated Beachfront Area Development Plan	February 2014	Phase 1 is ongoing	i. Beautification of embankments and the adjacent areas of the embankments ii. Development of picnic spots, parking lots, toilets, etc. iii. Vendor rehabilitation iv. Increasing urban amenities such as paving of the unpaved streets, street lighting, water facilities, wastewater treatment plants, electrification, and demarcation of no vehicle zone
Digha-Sankarpur Coastal Zone Management Project	September 2008	Completed	i. Phyto-remediation for the sewerage water treatment ii. Improvement of beach cleanliness and sanitation, establishing skill development centres of local handicrafts for generating livelihood options iii. Improvement of urban amenities such as street lighting and parking facilities
Land Use and Development Control Plan (LUDCP) Digha-Sankarpur planning area	September 2013	Project is ongoing	i. Dispersal of urbanization from the present congested core area of DSDA ii. Increase in urban amenities such as parks, playground, and public conveniences iii. Preservation of the wetlands, tanks, ponds, and water bodies following the existing CRZ rules and regulations iv. Permitting recreational activities in those preserved areas
National Cyclone Risk Mitigation Project Phase-I	May 2015	Project is ongoing	i. Development of Multipurpose Cyclone Shelter (MPCS) using the school buildings

(continued)

Table 20.3 (continued)

Project name	Date	Status	Objective
National Cyclone Risk Mitigation Project Phase-II	May 2015	Project is under progress	i. Conversion of existing overhead electrical distribution network (33 kV) into underground cable network ii. Enhancement of service quality and reliability of the power scenario of the project area and development of necessary management plan to combat disaster risks

land to built-up conversions (Roy and Datta 2018). These one-way conversions of wetlands to buildups were indicating towards a completely modified and humanized coastal environment without any remaining signs of its original serenity.

20.4.2 *Impacts of Expanding Impervious Surface on Soil Moisture Contents*

Marshy dune swales play a crucial role in replenishing groundwater aquifers during the wet season. Furthermore, they serve variegated ecosystem services to mankind and help in maintaining the ecological balance. The peat content stored in these moist lands has great potential in the sequestration of atmospheric carbon. Additionally, provision of water purification facility, coastal flood control, and providing habitable space to aquatic organisms are the foremost ecosystem services of the interdunal wetlands (Mitsch and Gosselink 2000). However, in the form of unplanned built-up development for the sake of growing mass tourism, moist lands of this area were reclaimed drastically and converted into multi-storey concrete structures (Fig. 20.7). Here, the amount of surface soil moisture had been found to be increasing towards the town fringe and rural areas compared to the decreasing coverage of the impervious surface. Accordingly, the moisture deficiency in soil was proving to be detrimental regarding the maintenance of a sustainable cityscape in the Digha Township.

20.4.3 *Guidelines Towards Soil Moisture Conservation*

During this study, it was found that all of the developmental projects had focused on reshaping the Township area of Digha and strengthening the civic amenities to provide a better tourism experience rather than focusing on the conservation of

Fig. 20.7 Evidences of expanding buildups and vulnerability of interdunal wetlands; Google Earth image comparison shows the rapid expansion of built-up coverage at New Digha from 2011 to 2018 (**a**, **b**); still existing patches of interdunal wetland (**c**); encroachment of wetlands for hotel construction (**d**); deterioration of wetland health caused through waste disposal by tourists (**e**); evidence of stage-wise engulfment of wetland areas (**f**)

natural environmental serenity. To some extent, during the planning of this coastal resort town, either the approach of wetland conservation was completely ignored in the planning framework or it was incorporated without assessing their applicability in reality. From the findings, it could be clearly inferred that the entire region was shifting towards becoming an unsustainable Township with a highly degraded environment as the biased urbanization processed had proven to be a menace for soil moisture conservation of the entire study area. Depending on the field experiences gained during the course of the study, few achievable guidelines were formulated towards developing a true 'smart city' in this area by maintaining perfect harmony between balanced wetland ecology and sustainable tourism initiatives. These suggestions are as follows:

i. Sustainable ecotourism initiatives should be introduced to showcase the pristine environment of interdunal wetlands within the study area and promote the wetland-based tourism laterally along with the prevailing beach tourism.

ii. Local authority should encourage sustainable agroecosystem practices (i.e. paddy cum fish culture, fresh and brackish water aquaculture, and farming of aquatic plants such as *Nymphaea alba* and *Trapa natans*) in these remaining interdunal wetlands, which can help to satisfy the increasing demand of food within the emerging Township area.

iii. During the construction of beach link roads, instead of fragmenting those wetlands by constructing continuous roadways, low elevated concrete bridges, wooden bridges, and wooden walkways should be constructed, which will help to keep intact the natural condition of these wetlands.

iv. The local administrative body and executive planning authority should engage in regular monitoring, which will effectively reduce the unscientific encroachment of these wetlands.

v. The government should understand the ecological sensitivity of these interdunal wetlands in order to initiate an awareness campaign for visitors and participatory management initiatives for local inhabitants.

20.5 Conclusions

Application of a hybrid methodology by integrating the physical and bio-physical indicator-based maps produced through a geospatial database with a detailed field-based observation ensured that this study had attended its primary objective of a comprehensive understanding of wetland exploitation triggered along the proposed 'smart' Township of Digha. This study depicted that not only the development of a smart city, but any kind of liveable or sustainable urban development is not possible when its environmental standards have not been properly taken care of. As the study pointed out, the major victims of unplanned and haphazard urban sprawls can have highly detrimental effects on the ecology of the surrounding areas in general and biodiversity in particular. A smart city thrives on the assurance of proper administrative control and synchronized implementation of jurisdictional powers of different governing agencies. Unfortunately, this kind of ideal situation is not generally observed in most of the proposed smart cities of India, the Digha Township being not an exception. Implementation of Township planning in a slipshod manner, ineffective measures of environmental protection, and lack of coordination between project planners and urban users had been identified as the pertinent reasons for ecological degradation in the study area. Abysmal growth of sea-beach-based mass tourism activities had clearly moved way beyond the carrying capacity of this hazard susceptible coastal region and making the lives and livelihood of thousands of local inhabitants more vulnerable. In such a scenario, the achievement of the smart city ideals remains a distant dream for the Digha Township. However, immediate implementation of the proposed management guidelines along with long-term strategic

Supplementary Table 20.4 LULC statistics of the study area from 2000 to 2018

LULC class	Year				Change (2000–2018)	
	2000		2018			
	Area (ha)	Area (%)	Area (ha)	Area (%)	ha	%
Agricultural land	104.55	21.56	36.54	7.54	−68.01	−65.05
Bare land	78.05	16.10	32.13	6.63	−45.93	−58.84
Built-up area	23.32	4.81	184.18	37.99	160.86	689.75
Casuarina plantation	60.72	12.52	85.59	17.65	24.88	40.97
Non-agricultural land	138.63	28.59	99.02	20.42	−39.62	−28.58
Wetland	79.58	16.41	47.40	9.78	−32.18	−40.44
Total	484.86	100.00	484.86	100.00	–	–

planning and adoption of ecologically resilient urban development measures could only save this already fragile coastal stretch from utter environmental degradation and might pave the path for sustainable urban development in a smarter and more efficient way.

Acknowledgements This study received financial supports extended by the University Grants Commission, India, to the first author (Fellowship Reference No. 3261/NET-JUNE 2015) and DST-SERB (File No. ECR/2017/003380) to the corresponding author.

Appendix 1

See Supplementary Table 20.4

Appendix 2

See Fig. 20.8

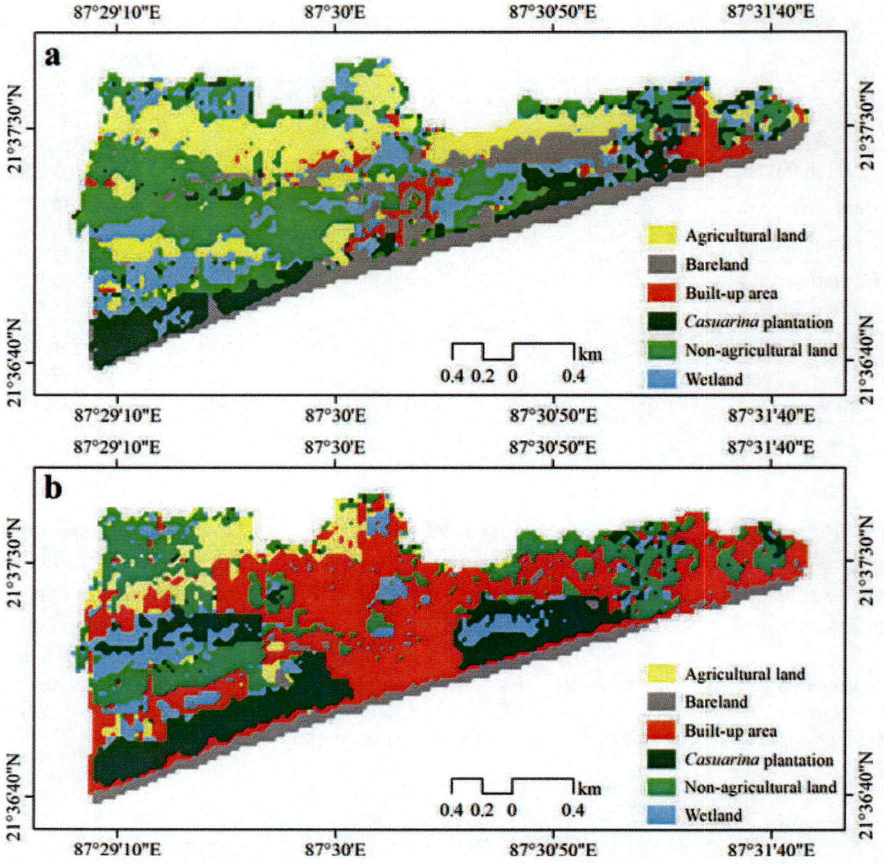

Fig. 20.8 LULC Map of the study area

References

Angel S, Blei AM, Civco DL, Parent J (2012) Atlas of urban expansion. Lincoln Institute of Land Policy, Massachusetts

Angiuli E, Trianni G (2014) Urban mapping in Landsat images based on normalized difference spectral vector. IEEE Geosci Remote Sens Lett 11(3):661–665

Antrop M (2004) Landscape change and the urbanization process in Europe. Landsc Urban Plan 67(1–4):9–26

Boone CG, Fragkias M (2012) Urbanization and sustainability: linking urban ecology, environmental justice and global environmental change. Springer, Berlin

Bryant CR, Russwurm LJ, McLellan AG (1982) The city's countryside: land and its management in the rural-urban fringe. Longman, UK

Candela G, Figini P (2012) The economics of tourism destinations. Springer, Berlin

Caragliu A, Del Bo C, Nijkamp P (2011) Smart cities in Europe. J Urban Technol 18(2):65–82

Chattopadhyay K (2000) Economic impact of tourism development: a linear programming approach. Ind Econ J 47(3):105–109

Dameri RP (2013) Searching for smart city definition: a comprehensive proposal. Int J Comput Technol 11(5):2544–2551

Dorigo WA, Wagner W, Hohensinn R, Hahn S, Paulik C, Xaver A, Gruber A, Drusch M, Mecklenburg S, van Oevelen P, Robock A, Jackson T (2011) The international soil moisture network: a data hosting facility for global in situ soil moisture measurements. Hydrol Earth Syst Sci 15(5):1675–1698

DSDA (2008) Statistics on DSDA. https://www.dsda.gov.in/index.php/home/dsda_statistics. Accessed 25 August 2018

DSDA (2013) Digha-Shankarpur integrated beachfront area development plan https://www.iczmpwb.org/main/pdf/pea/Digha%20Shankarpur%20Integrated%20Beachfront%20Area%20Development%20Plan.pdf. Accessed 25 August 2018

Gill SE, Handley JF, Ennos AR, Pauleit S (2007) Adapting cities for climate change: the role of the green infrastructure. Built Environ 33(1):115–133

Jackson TJ (1993) III. Measuring surface soil moisture using passive microwave remote sensing. Hydrol Process 7(2):139–152

Jim CY (1998) Urban soil characteristics and limitations for landscape planting in Hong Kong. Landsc Urban Plan 40(4):235–249

Julien Y, Sobrino JA (2009) The Yearly Land Cover Dynamics (YLCD) method: an analysis of global vegetation from NDVI and LST parameters. Remote Sens Environ 113(2):329–334

Khansari N, Mostashari A, Mansouri M (2014) Impacting sustainable behavior and planning in smart city. Int J Sustain Land Use Urban Plan 1(2):46–61

Li Y, Zhu X, Sun X, Wang F (2010) Landscape effects of environmental impact on bay-area wetlands under rapid urban expansion and development policy: a case study of Lianyungang. China. Landsc Urban Plan 94(3–4):218–227

Mascarenhas A (2004) Oceanographic validity of buffer zones for the east coast of India: a hydrometeorological perspective. Curr Sci 86(3):399–406

Mishra S (2004) Tourism management: the socio-economic and ecological perspective. Universities Press, Hyderabad

Mitsch WJ, Gosselink JG (2000) The value of wetlands: importance of scale and landscape setting. Ecol Econ 35(1):25–33

Mondal S, Sen J (2017) A framework of predictive analysis of tourist inflow in the beaches of West Bengal: a study of Digha-Mandarmoni beach. In: Mandal J, Dutta P, Mukhopadhyay S (eds) international conference on computational intelligence, communications and business analytics. Springer, Singapore, pp 161–176

Mukherjee S (2018) Journeys into the past historical and heritage tourism in Bengal. New Academic Publishers, New Delhi

Nam T, Pardo TA (2011) Conceptualizing smart city with dimensions of technology, people, and institutions. In: Proceedings of the 12th annual international digital government research conference: digital government innovation in challenging times, pp 282–291. ACM Publications, New York

Naphade M, Banavar G, Harrison C, Paraszczak J, Morris R (2011) Smarter cities and their innovation challenges. Computer 44(6):32–39

Pahari DP (2013) Coastal resorts of West Bengal: an environmental appraisal. PhD thesis, University of Burdwan, India

Parida BR, Collado WB, Borah R, Hazarika MK, Samarakoon L (2008) Detecting drought-prone areas of rice agriculture using a MODIS-derived soil moisture index. GISci Remote Sens 45(1):109–129

Potić I, Bugarski M, Matić-Varenica J (2017) Soil moisture determination using remote sensing data for the property protection and increase of agriculture production. In 2017 annual world bank conference on land and poverty. The World Bank, Washington DC

Prasad SN, Ramchandra TV, Ahalya N, Sengupta T, Kumar A, Tiwari AK, Vijayan VS, Vijayan L (2002) Conservation of wetlands of India-a review. Trop Ecol 43(1):173–186

Revi A, Singh AK (2007) Cyclone and storm surge, pedestrian evacuation and emergency response in India. In: Waldau-Drexler N, Gattermann P, Knoflacher H, Schreckenberg M (eds) Pedestrian and evacuation dynamics 2005. Springer, Berlin, pp 119–130

Roy AK, Datta D (2018) Analyzing the effects of afforestation on estuarine environment of river Subarnarekha, India using geospatial technologies and participatory appraisals. Environ Monit Assess 190(11):1–16

Ruddick S (2015) Situating the Anthropocene: planetary urbanization and the anthropological machine. Urban Geogr 36(8):1113–1130

Schroeder TA, Cohen WB, Song C, Canty MJ, Yang Z (2006) Radiometric correction of multi-temporal Landsat data for characterization of early successional forest patterns in western Oregon. Remote Sens Environ 103(1):16–26

Shapiro JM (2006) Smart cities: quality of life, productivity, and the growth effects of human capital. Rev Econ Stat 88(2):324–335

Staffans A, Horelli L (2014) Expanded urban planning as a vehicle for understanding and shaping smart, liveable cities. J Commun Inf 10(3):1–28

World Economic Forum (2016) Inspiring future cities & urban services shaping the future of urban development & services initiative. https://www3.weforum.org/docs/WEF_Urban-Services.pdf. Accessed 20 August 2018

Yang ZX, He XF (2010) Automatic extraction of urban land-use information from remote sensing images based on improved NDBI method. J Hohai Univ Nat Sci 38(2):181–184

Zha Y, Gao J, Ni S (2003) Use of normalized difference built-up index in automatically mapping urban areas from TM imagery. Int J Remote Sens 24(3):583–594

Zhan Z, Qin Q, Wang X (2004) The application of LST/NDVI index for monitoring land surface moisture in semiarid area. Proceedings 2004 IEEE international geoscience and remote sensing symposium. IEEE, USA, pp 1551–1554

Zhang F, Zhang LW, Jing-Jing SHI, Huang JF (2014) Soil moisture monitoring based on land surface temperature-vegetation index space derived from MODIS data. Pedosphere 24(4):450–460

Zhang Y, Odeh IO, Han C (2009) Bi-temporal characterization of land surface temperature in relation to impervious surface area, NDVI and NDBI, using a sub-pixel image analysis. Int J Appl Earth Obs Geoinf 11(4):256–264

Chapter 21
Urban Housing in Itanagar: Mountain Geomorphology and Hazard Vulnerability Vis-a-Vis Smart City Framework

S. K. Patnaik

Abstract Itanagar, Arunachal Pradesh, is a township established in 1978 to function as the administrative centre and capital of the State. Since then, it has been a hub for administration and a residential place for employees. It has also attracted businessmen and families to settle in the capital for various purposes. Now, there is a stream of migration from rural areas to the town to avail opportunities in the form of livelihood, health care, education, connectivity and better amenities. This has resulted in choking the urban space and facilities. At this juncture of visible stress, the town has been included in the list of Smart Cities of India with the hope to reduce the pressure on land and resources and for ease of living. The town is located on the Siwaliks of the Himalaya Mountain, where terrain conditions are very fragile and susceptible to degradation and hazard. As there is no rule or guideline for site selection and housing types, people are using whatever place is available to them for housing and commercial activities, thereby making themselves vulnerable to hazards associated with mountains. A GIS-based study has been carried out by evaluating and integrating geomorphological parameters for the town area. Subsequently, categories of houses and buildings have been cross-tabulated in estimating proportions of each category versus vulnerability. The analysis shows that about twenty-two per cent of houses have been constructed in vulnerable areas to landslide, landslip, channel shift and channel bank erosion. Government residential units and office buildings constructed during 1980–90s have better terrain condition scores and these are not vulnerable to hazard. Privately owned houses of higher income and wealth groups have their houses built on safe areas. The recent migrants and lower income groups have crowded vulnerable areas. Due to scarcity as well as the high cost of land, in many cases setbacks are not maintained leading to the collapse of houses on the upper elevation. Building bylaws of 2010 and approval of commercial building by authority are not rigidly followed leading to violation of master plan and safety norms. The present requirement for Itanagar is to create a framework and guidelines that factor into hazard and vulnerability issues on a priority basis before development activities like housing, infrastructure, drainage, etc. take place for the Smart City.

S. K. Patnaik (✉)
Department of Geography, Rajiv Gandhi University, Doimukh, Itanagar, Arunachal Pardesh, India
e-mail: santanu.patnaik@rgu.ac.in

P. Sharma (ed.), *Geospatial Technology and Smart Cities*, The Urban Book Series,
https://doi.org/10.1007/978-3-030-71945-6_21

Keywords Himalaya terrain · Degradation · Housing vulnerability · Infrastructure · And smart city

Acronyms

CDP	City Development Plan
NEFA	North Eastern Frontier Agency
TWI	Topographic Wetness Index

21.1 Introduction

Arunachal Pradesh, covering an area of 83,743 km^2, is located in the extreme north-eastern part of India. Physiographically, it forms a part of the mighty Himalayas. The state is highly dissected with equally mighty rivers, viz., Lohit, Dibang, Siang, Kameng and Subansiri. The rich flora and fauna diversity has promoted and equally sustained diverse tribes, culture and their traditional institutions. It is home to different early migrants from Bhutan, Tibet, Burma, Yunan and recently from Myanmar and Bangladesh. As of 2011, population of the state was 1,383,727, with 68.79% of the scheduled tribe population. At present there are more than 125 tribes, however, only 15 of them have a population of more than 5000.

The first census of the state was conducted in 1961, and the population of the state was 336,558 with a population density of only 4 per km^2. The growth of the population was rapid after independence due to the establishment of healthcare facilities and consistent food supply. During the 1961–2011 period, the population grew 4.11 times (Fig. 21.1) with the growth of the non-indigenous population outpacing the growth of the indigenous Scheduled Tribe population during 1971–1981–1991. The

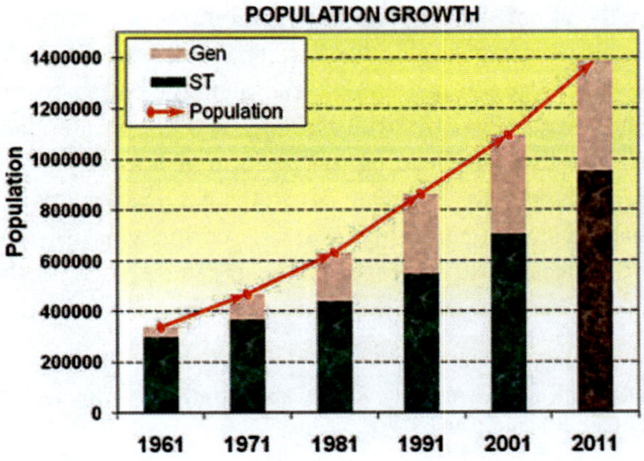

Fig. 21.1 Population Growth during 1961–2011

latest population figure for Arunachal Pradesh as of the 2011 census is 1,383,727 persons with 713,912 males and 669,815 females. It amounts to only 0.11% of India's population. It had a high decadal growth of 26.06% between 2001 and 2011 with still a low population density of only 16.5 per km^2.

The urbanization process took off at a very late stage in Arunachal Pradesh with four subdivisional headquarters being declared as Census Towns in the year 1971. In the state, only 3.04% lived in the urban area in 1971 and that increased from 20.75% in 2001 to 22.94% in 2011. Till 2001, there were no towns in four out of sixteen districts. Since 2011, all districts have either Census or Notified Towns with Hawai Town of Anjaw District having a population of 982. The state capital, Itanagar, was established in 1978 and was designated as a town in the 1981 census with a population of 6406. Since then, the population has been increasing at a higher rate due to in-migration for search of job or profession in the trade and service sectors in this administrative town. The population of the town in 1991 was 16,545; in 2001, it was 35,022; and in 2011, it rose to 59,490. The projected population for Itanagar by 2025 is 117,707.

One of the push factors for urban growth in Arunachal Pradesh is the higher inter-state immigration than the intra-state rural–urban migration (Planning Commission 2005). During the decades 1981–1991 and 1991–2001, percentages of in-migration components were 77.4% and 77.1% of the total growth of the population of the town. The towns of Arunachal Pradesh have a unique urban administration, where the State Government is directly responsible for urban development and the municipality has minimal role unlike other parts of India. These are Department of Public Health Engineering, Power, Public Works Department and Directorate of Urban Development & Housing. The latter was established in 1996 and acts as the nodal agency for the development of urban areas of the State. The responsibility of this directorate is to prepare master plan, action plan, building design for government establishments, traffic and transportation plan and a host of services.

This rapid growth of population since the 1980s and lack of a proper urban management action plan have created a vacuum of urban management. Another dimension of this unplanned growth is due to a combination of lack of a cadastral map, lack of land ownership title and lack of revenue record. Land in Arunachal Pradesh is owned traditionally by the community as 'Community land' without any individual ownership. Progressively with an increase in population in urban areas, individuals are now obtaining land ownership documents from the Office of Deputy Commissioner with a purpose to lease it to migrants in urban areas for business or institutional setups or to have a legal claim over the parcel of land, be it plain, river valley area, hill slopes, hilltops or even the forest area. Due to administrative regulations during North Eastern Frontier Agency (NEFA), non-natives (Arunachal Tribes) cannot purchase land and settle permanently.

The migrants arriving in the town from the countryside basically look for a job opportunity or hang on after studies to get a job in the government sector. The main reasons for this are (i) natives are traditionally from agrarian community and have agriculture and hunting as the primary occupation and (ii) lack of industry and professional service sectors to absorb these unskilled graduates. This imbalance in demand and supply of job scenario has resulted in high population growth vis-à-vis urbanization in the state especially in the capital town Itanagar. The visible change in the urban landscape is the rapid construction of residential and commercial buildings and huts to accommodate the growing population and their need.

21.2 Area of Study

The study area (Fig. 21.2) is located in the northeastern part of Arunachal Pradesh, India, with the town centre at 27° 05′ 54″ N and 93° 37′ 19″ E covering an area of about 30 km^2. Geologically, Itanagar is a part of the product of plate movements, and it is a product of incipiently lithified fluvial sediments of unclassified late Neogene Siwaliks. Tectonic activities along Himalayan Frontal Thrust and Main Boundary Thrust are associated with the Himalayan Thrust system and associated transverse normal and strike-slip faults. Tectonically, it is in Earthquake Zone V (Thakur 2004). The northeast Himalayas are considered to be a high seismically active region with 1897 and 1950 earthquakes. This has made the Itanagar area highly landslide prone (Singh et al. 2008; Devi et al. 2011), and it is evident from frequent and large-scale landslides every year. Geomorphologically, the terrain is highly dissected and the rugged hills are constituted of unconsolidated sediments. Pachin River, that pass

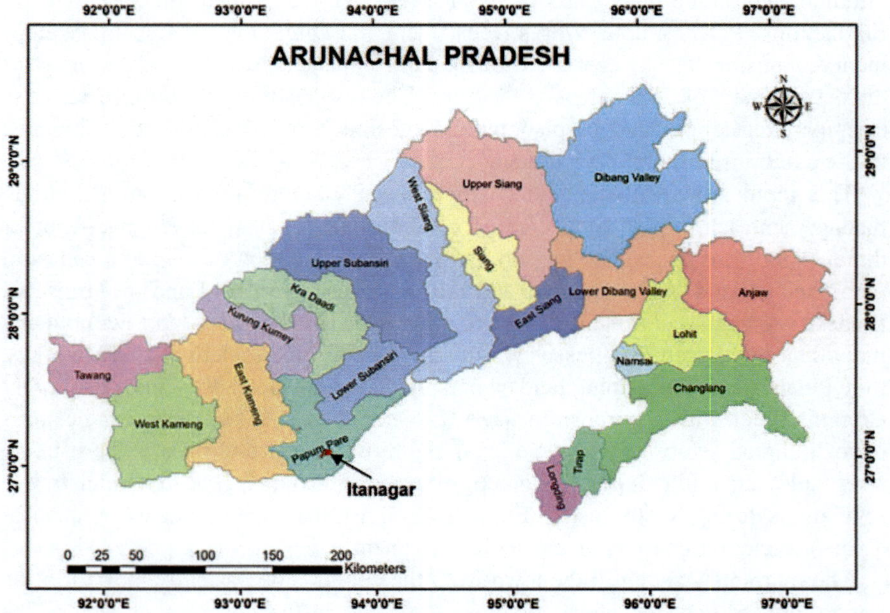

Fig. 21.2 Location of Itanagar in Arunachal Pradesh, India

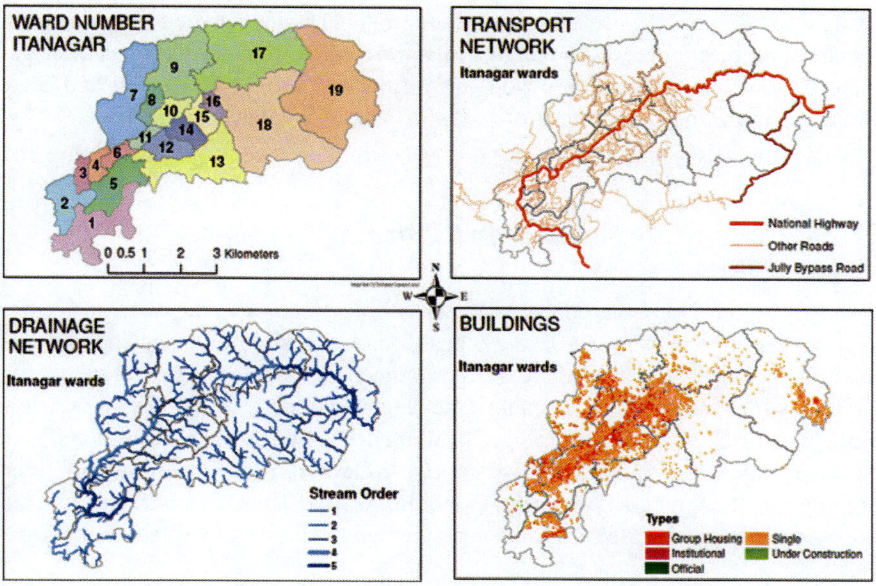

Fig. 21.3 Basic information about Itanagar Smart City

through this area as well as its tributaries drain this area; from west to east direction and finally outflows into Dikrong River, having a fairly large catchment.

The town had no limiting boundary as there was no municipality jurisdiction till 2007. In 2009, Arunachal Pradesh Municipality Act was notified. In 2010, Rule for Ward Delimitation in Arunachal Pradesh was notified (Fig. 21.3). In 2011, Arunachal Pradesh Reservation of Urban Land for Housing of Urban Poor Rules was notified. Recently, The Arunachal Pradesh (Land Settlement and Records) (Amendment) Bill, 2018 was introduced to allow lease of land to others for a period of 30 years which is renewable for another 30 years.

Itanagar Smart City Development Corporation Limited is a registered Public Limited Company of State Government of Arunachal Pradesh. It was incorporated on 10 September 2018 at Registrar of Companies, Shillong. Its authorized share capital is Rs. 100,000 and its paid-up capital is Rs. 100,000. The aim of this Company is related to Real estate activities with own or leased property, which includes buying, selling, renting and operating of self-owned or leased real estate such as apartment buildings and dwellings, non-residential buildings, developing and subdividing real estate into lots, etc. Also included are the development and sale of land, operating apartment hotels and residential mobile home sites, etc. Itanagar Smart City Development Corporation Limited's corporate registered address is Directorate of Town Planning and Urban Local Bodies, Govt. of A.P.

Itanagar Smart City Development Corporation Limited invited e-tender from reputed multidisciplinary consultancy firms who have experience in Project Management Consultant services for various infrastructure projects of Central and State Government for the appointment of PMC for Smart City Itanagar.

21.2.1 Smart City Features and Proposal

Smart City is an amorphous concept with ten core elements covering water and power supply, transport, governance, safety, health and education and housing. Typical features included for comprehensive development of Smart Cities include mixed land use with open space and housing. One major key feature is to make the area less vulnerable to disasters. Accordingly, Government of Arunachal Pradesh submitted a 92-page proposal document to Ministry of Urban Development, Government of India, highlighting the Strength, Weakness, Opportunities and Threats and the check-listed information required. Major threats perceived are a fragile ecosystem and natural hazards.

Government of Arunachal Pradesh in its City Development Plan (CDP) has focused on Urban Infrastructure and Governance and Basic Services to the Urban Poor. The main objectives are targeted towards Uncontrolled Expansion, Under-Developed Infrastructure, Unorganized Public Services, Future Land requirement, Water Supply, Urban Transportation, Solid Waste Management, Sewerage System, Urban Drainage, Prevention of Landslides and Rehabilitation of soil erosion, etc. whereas, the Directorate of Town Planning, Government of Arunachal Pradesh, has prepared a Final Draft for Arunachal Pradesh Building Byelaws—2008 covering

1. Development code pertaining to residential and non-residential premises.
2. General building requirements.
3. Conservation of heritage sites including heritage buildings, heritage precincts and natural feature areas.
4. Additional provisions in development control regulations for safety in natural hazard-prone areas.
5. Additional provisions in building byelaws for structural safety in natural hazard-prone areas.
6. Integrated township.
7. Multi-story buildings and group housing schemes'/apartments' additional requirements.

The CDP prepared for the capital town aimed at achieving urban development and the Building Byelaws prepared by Directorate of Town Planning aimed at urban management for Arunachal Pradesh are good in the context of normal terrain conditions. As the town Itanagar is located on the eastern Himalayas, the geomorphological conditions have a strong bearing on land use and vulnerability of urban infrastructure. Negligence to conform to it will have serious repercussions in future. As commented in the Symposium Discussion on Physical Problems of the Urban Environment that urban geographers have neglected the physical facts of situation and site (Brown et al. 1976) still holds true in the present context. This is not because they have not understood the importance of physical parameters, but because they are not updated with the latest technology and products to understand in a more précise context.

21.3 Urban Geomorphology and Vulnerability

Urban geomorphological studies (Cooke 1976) have provided detailed insight into the various aspects of resource evaluation for urbanization and hazard perception through the prism of geomorphological knowledge. They also help a diverse array of management agencies and individuals responsible for the planning and management of urban environments. Since the seminal work of Cooke, innumerable papers have been published in the domain of urban geomorphology, geotechnical studies, use of remote sensing and GIS in urban mapping and management, and vulnerability assessment of settlements. The present study aims to move a step ahead to understand urbanization and urban reconstruction and management by integrating geomorphology, Remote Sensing and GIS techniques at the individual house level.

The town Itanagar, located in a mountainous terrain coupled with unplanned growth and endeavour by the Government to plan for city's development, has been taken as a pilot study to establish linkage between geomorphological parameters with situational context of each individual house for better reconstruction and management.

21.3.1 Structured Approach

The terrain type, its geomorphometric and hydromorphometric characteristics, defines the nature and strength of land for its use. Mountain terrain owing to its high runoff generating characteristics has severe implications on its use. The geological and structural characteristics of mountains are another set of factors that govern the stability conditions of land. Geology provides a platform, Climate provides energy for the processes to operate and Time provides the requisite duration of the sculpturing processes that carve out a landform. All these conform to the trio of W. M. Davis or a continuum of the never-ending process of T. J. Hack. Landforms are reflections of a balancing act between structure and processes. While characterizing landforms, it is of utmost importance to understand and calibrate the terrain parameters and their functions. Some of the early modelling of landforms that have been experimented are of Speight (1990), Gares et al. (1994), Patnaik (1993, 1995), Mejía-Navarro et al (1994), Gupta and Ahmad (1999), Smyth and Royle (2000), Klingseisen et al. (2004, 2008) and Metternicht et al. (2005). Recent advances in quantitative analysis of landforms have added many valuable methodologies in understanding the topography and physiographic conditions as well as their impacts on infrastructure and settlements

(Drăguţ and Blaschke 2006; Tagil and Jenness 2008; Schilirò et al. 2016; Patnaik and Gogoi 2019).

To characterize terrain for urban vulnerability, topographic parameters like altitude, its deviation from regional trend, slope, aspect, curvature, distance from stream, topographic wetness index and topographic position have been incorporated in the model. Soil in the Siwalik region of the Himalayas is uniformly formed out of unconsolidated aggregates with high permeability. The movement of surface and subsoil water is dependent on topographic parameters in the Siwaliks. Another important parameter is earthquake vulnerability, which is a regional phenomenon. The whole of northeastern India is under Earthquake Zone V. Therefore, these latter two parameters have not been factored into local-scale analysis.

A model of integration of terrain parameters (Fig. 21.4) has been developed and tested to confirm it with the ground realities during the monsoon months of 2019. The Digital Elevation Model at 12.5 m resolution of ALOS PALSAR (Advanced Land Observing Satellite Phased Array type L-band Synthetic Aperture Radar) has been used in analysing terrain parameters.

Altitude

Altitude or elevations of land owe their origin to mountain building or epeiorogenic processes. Altitude is linked to potential energy for geomorphological processes, viz., surface runoff, erosion and sediment transport. It is also linked to temperature variation at the local scale at the rate of 0.65° per 100 m that may control various micro-scale geomorphological processes. Itanagar is located in Siwaliks with an altitude varying from 102 to 588 m. The north and south-central parts of the township have higher elevation. The latter locality is the main residential, commercial and

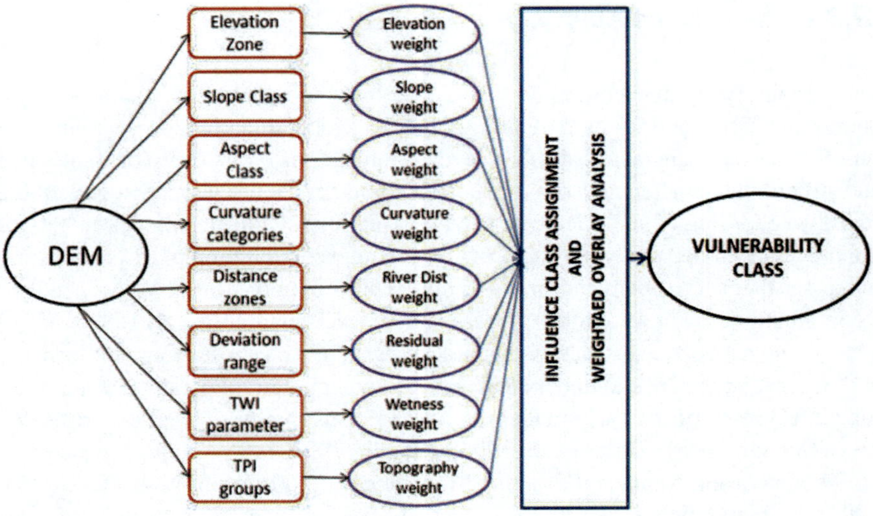

Fig. 21.4 Simplified model flowchart

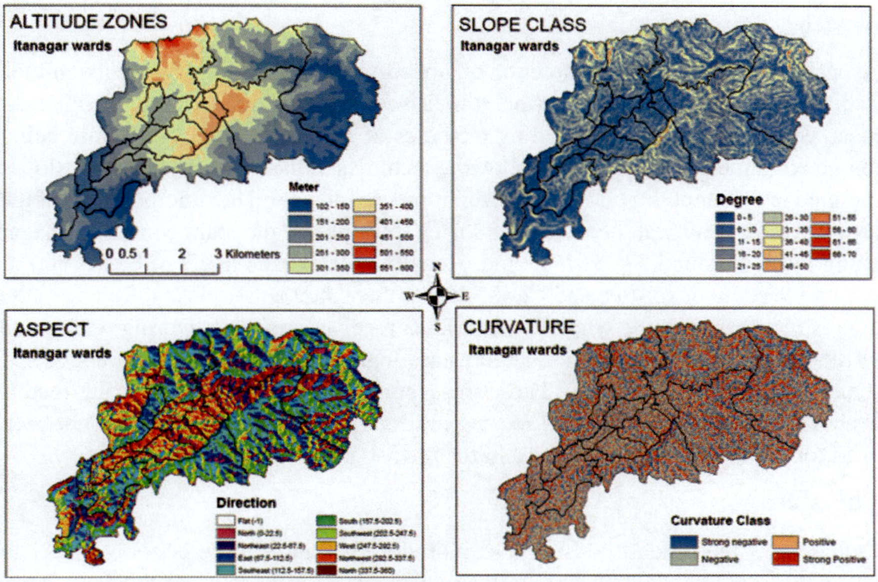

Fig. 21.5 Altitude, slope, aspect and curvature parameters of Itanagar Municipality

administrative hub as well as the CBD of the Itanagar Capital Complex area. For utilizing altitude as a governing parameter, the range of altitude is classified into ten classes at a fixed interval of 50 m. Buildings have been constructed heavily on altitude ranging from 150 to 500 m (Fig. 21.5).

Slope

Slope is the first derivative of altitude. The functional importance of slope is the effectiveness of gravity. Slopes have their origin due to tectonics, depositional and erosional processes and human activities. It has a greater significance in allowing the Hortonian flow of water as well as allowing the subsurface flow of water. As the slopes become steeper, surface runoff thickness and velocity increase and pose a threat to the stability of slopes. Itanagar being located on heavily dissected topography has steeper slopes, and the whole area is vulnerable to mass wasting processes like landslides. Another major risk on the steeper slopes is slope wash. It removes the topsoil along impervious layers of roads, pavements and other built-up area and damages the infrastructure. Slope has been categorized into six categories with 5–10 and 10–15° strongly inclined slope classes covering an area of 21.8 and 20.7% and 15–30° steep slopes area covering 1263 ha, i.e. about 41% of Itanagar area (Fig. 21.5). Most of the buildings are constructed on slopes between 0 and 15° slopes. Newer areas of expansion of Itanagar Smart City have buildings on steeper slopes, thereby increasing vulnerability to hazards related to slope failure.

Aspect

Aspect is the azimuth or the direction of slope. Aspect is governed by the magnitude of dissection in a particular direction. It is also controlled by tectonics. In the Itanagar area, aspect is mostly controlled by tectonics of Himalaya orogeny. Terrain being dissected immensely by rivulets following tectonics, it has smaller patches of slopes inclined in different directions. It has an immense impact on the microclimate of the area as well as soil water content and lower atmospheric moisture content. Itanagar being located around 27° 5″ N latitude, maximum elevation angle of Sun is 84.86° during the summer solstice and 42.48° during the winter solstice. As the slope in the area is less than 84°, the whole Itanagar area receives sunshine. During winter due to the low elevation of the Sun, sunrays are highly inclined and slopes above 43° remain under shade (Fig. 21.5). These areas remain damp and moist due to ineffective evaporation of soil moisture and settling of dew. However, such areas have not been used for the construction of houses in the township.

Curvature

Curvature is the second derivative of altitude. Plan and profile curvature refer to changes in a gradient along contour lines and along the hillslopes, respectively. It has a tremendous impact on hillslope hydrology. Hillslope geomorphological processes include movement of soil, regolith and soil water. This is dependant not only on rainfall but also on the curvature of the terrain (Fig. 21.5). A convex plan form with strong positive curvature will act as an area of dispersal of rainwater thereby thinning the depth of overland flow. It reduces the erosivity of surface runoff. During the post-monsoon period, due to reduction of rainfall, these spur-like features remain relatively dry, thereby are more stable areas. On the other hand, concave plan forms of strong negative curvature are cusp-like features that helps in the accumulation of water. This leads to a higher rate of weathering and higher volumetric weight of soil, thereby making the slopes susceptible to landslide, mudslides, etc.

Trend Surface Residuals

Geomorphological processes act differentially over a space not only on the basis of potential energy due to altitude and slope for the effective velocity of surface runoff, but also on the basis of achievable entropy. Trend Surface Analysis provides an ideal surface close to that entropy level but maintains the general topographic trend. Deviations from the modelled trend surface are the precursors for areas of active and accelerated erosion and areas of depositions. Areas of positive deviations are potentially erosion-prone and hazardous. This carries a lot of credence in the unconsolidated lithology of Siwaliks of the Itanagar area where ridgelines and areas closer to it having higher deviations up to 175 m are vulnerable to soil creep, landslide, rill erosion and gullying. On the other side, areas having negative deviations up to − 181 m are prone to urban flooding, accumulation of debris and talus (Fig. 21.6).

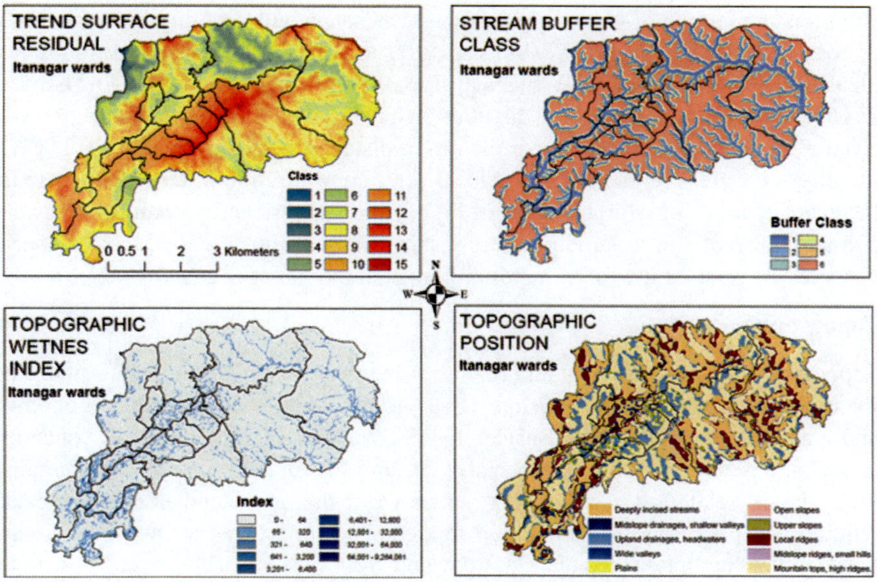

Fig. 21.6 Trend Surface Analysis (TSA) residual, stream buffer, topographic wetness index (TWI) and topographic position (TPI)

Stream Buffer

The northeastern part of India is one of the highest rainfall-receiving regions of the world. Itanagar receives an annual rainfall of 3390 mm. Monsoon season means monthly rainfall from May to September is 453, 640, 610, 505, 492 mm, respectively. Rivulets and streams are pathways for the concentrated flow of surface runoff. Areas closer to channels carry a lot of water and sediment during the monsoon. As the built-up area increases, especially closer to the channel, the vulnerability of high depth surface runoff and flooding increases. Total runoff exceeds the capacity of roadside drains and culverts leading to overflow of stormwater and sewage water on roads and habitations. The presence of water in the subgrade reduces the bearing power of road and load dispersion capacity. Loss of subgrade support leads to the failure of the road pavement under traffic loads. Therefore, Distance from the river has been taken as a major geomorphological parameter in assessing hazard vulnerability. Buffer distance is assigned as per Strahlers' Stream Order with multiples of 10 m, viz., fifth order has a buffer of 50 m (Fig. 21.6).

Topographic Wetness Index (TWI)

The presence of water in soil has many implications. From an engineering point of view for urban landscape, soil is important for deciding the depth of foundation. Drier soil is stronger due to higher cohesion and higher load-bearing capacity than wet soil of the same texture and structure. Distance from water divides and pathways for gravity flow of water in soil are geomorphological parameters that govern the soil moisture regime of an area. In an urban landscape, vulnerability increases with increased TWI. Slopes on the Siwaliks are transport limited slopes. While interfluves areas have a low surface runoff, seepage slopes allow water to infiltrate. An analysis has been carried out to assess the distribution of wetness indices of the Itanagar area. Even if there is a strong spatial correlation of distance from a river with TWI, TWI highlights the areas or pockets where local-scale concentration water will be there in the event of rain. It highlights areas of higher soil wetness and probability of swell and shrinking of soil volume in alternating seasons. In many cases, this has been observed by relating it with tilting of buildings in the Itanagar area (Fig. 21.6).

Topographic Position

Topographic characteristics of an urban area in many ways determine its suitability for its establishment and expansion. It has aesthetic, cultural, climatic, economic and environmental relevance. Besides these governing factors, there are concerns about minimizing earthworks especially cut and fill for all types of construction; safety issues related to waterlogging, storm water drainage, landslide and setback requirements; especially in mountain topography. Itanagar as a township has no level land; its mountain slope facets, ridge and crest lines, open slopes and midslope ridges provide space for the construction of buildings, roads, drainage and sewage networks. As it has been repeatedly stated that Itanagar is located on Siwaliks, it has deeply incised channels that restrict available land for the construction of urban infrastructure, facilities and amenities. These streams flanked by steep slopes on loose soils restrict the connectivity of two sides of streams. The Itanagar area has been analysed for topographic positions to identify areas of vulnerability and safety (Fig. 21.6).

21.3.2 Existing Infrastructure

Naharlagun is the place where the foundation stone was laid by then President of India, late V. V. Giri, on 20 April 1974, for the capital of Arunachal Pradesh. To overcome the constraint of space for expansion in the then seat of administration at Naharlagun, nearby localities of a small village Chimpu were selected for establishing the permanent Capital of Arunachal Pradesh and named it as Itanagar in 1978. Temporary Secretariat building, Line Department buildings, college and schools were established and became an administrative town. It has now grown into a well-developed capital township. The earlier envisaged extent of expansion is saturated with residential, government, institutional and commercial buildings.

The town was governed by the State Government through Urban Development and Housing Department. Itanagar Municipal Council was inaugurated on 14 August 2013. The Itanagar Municipal Council supervises the administration of Itanagar. The Municipal Council works for the development of the capital city as a whole and takes care of the basic necessities of the people. It has a master plan for a total of 30 wards covering Itanagar, Naharlagun, Nirjuli and Banderdewa.

National Highway 415 extends from Banderdewa to Gohpur via Nirjuli, Naharlagun, Itanagar. The railway line stretches up to Naharlagun. Foundation Stone of Greenfield Airport at Holongi close to Itanagar was laid in February 2019. Many more Academic institutions, Central sponsored projects and institutions are being established at a fast pace. All these have increased demand for residential and commercial buildings to cater to the growing demand.

In 2017, a detailed survey was conducted by Government of Arunachal Pradesh as a prerequisite for developing Itanagar into a smart city. A total of 11,651 buildings was mapped, out of which 10,166 buildings are within the demarcated 19 ward boundaries comprising Itanagar. Buildings have been categorized under four categories: Group buildings, Single buildings, Office buildings, and Institutional buildings and buildings under construction. As per the Annual Development Agenda 2018–19 for Department of Urban Development and Housing, Government of Arunachal Pradesh had allocated funds for residential buildings, approach roads, parking place, drain, protection wall, storm water drain, etc. to augment as well as bridge the gap of a shortfall.

As per survey, buildings in Itanagar are clustered and unplanned in their layout. Wards Nos. like 7 and 14 have the highest number of buildings, i.e. above 700. Ward No. 2 has the least number of buildings, i.e. 34. This excludes huts as dwelling units. Ward-wise building concentration is high in 11 and 16 where per hectare there are more than 20 buildings. These two are part of the area of early settling of the township. When the buildings are clustered around some favourable patches, building density goes up thereby putting a lot of constraints on urban amenities and facilities. It also adds to vulnerability and reduced safety standards. An exercise has been carried out to find out the number of buildings per hectare (Figs. 21.7).

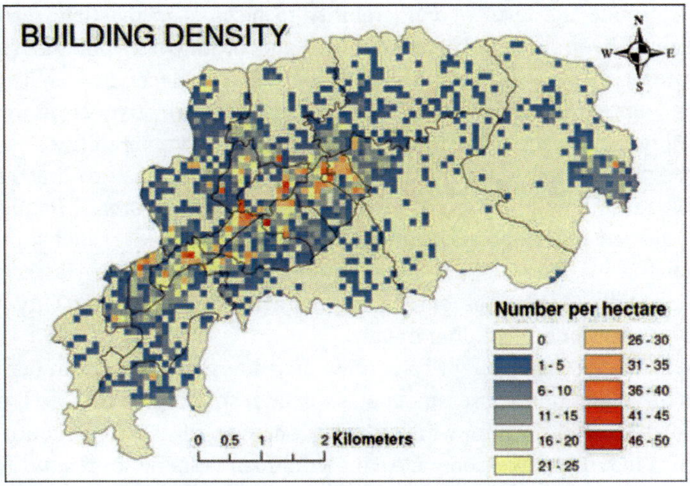

Fig. 21.7 Building density per hectare

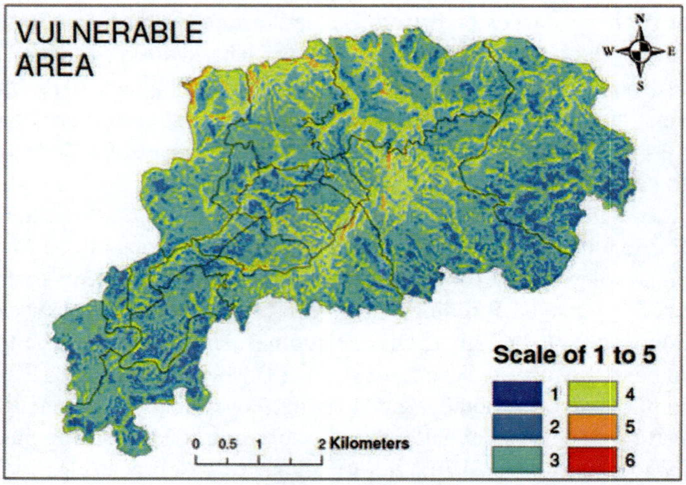

Fig. 21.8 Vulnerable areas of the city

21.3.3 Vulnerability

Vulnerability is assessed as per perceivable factors of threats. As discussed earlier, the most threatening hazard is an earthquake and it is uniform across the Itanagar Smart City. Extreme rainfall is also a perceivable hazard, and it is expected to be similar within an areal extent of 3089 ha of contiguous land. Vulnerability to the hazard varies on the basis of varied land and hydrological parameters. Altitude, slope, aspect, curvature, topographic undulations, distance from river, TWI and topographic position determine the level of vulnerability to hazards in the Itanagar area. All parameters have been classified in certain classes according to an individual range and suitable criteria, then each class is assigned a scale value depending on its influence in inducing vulnerability (Fig. 21.8). Every parameter has relatively different control in aggravating the impact of the hazard, thus reinforcing vulnerability.

During this monsoon season (July 2019), several places had to brace the brunt of excessive rainfall and its impact. Field visits to different parts of Itanagar Smart City were carried out to correlate modelled vulnerability level and ground truth (Photographs 1–14). Some of the ground realities and vulnerability assessment are presented through field photographs. This is also a precursor for Smart City Authority to implement appropriate building bylaws.

The spatial extent of different categories of vulnerability has been derived with six classes of vulnerability assessment at a spatial resolution of 12.5 m. It conforms to an area of 12.5 m by 12.5 m, which is sufficient even for the construction of a one BHK simplex in Earthquake Zone V. An integration of vulnerable areas with building locations reveals the number of buildings in different vulnerable zones (Tables 21.1, 21.2).

Photographs 1–6 Ground realities and vulnerability

21.4 Conclusion

Since the inception of the Itanagar township, land became a precious commodity in this forested inaccessible part of Himalayan State. It was the Government who started constructing office and residential buildings for its officers and staffs. A support system of commercial activities flourished to cater to the needs of employees. Houses were constructed with Assam type design, which is not only cost-effective due to the use of timber, bamboo and tin roofing, but also has advantages against earthquake. Vulnerability factors associated with this building typology are too low with very low chances of suffering collapse resulting in human fatalities. However,

Photographs 7–14 Vulnerable area and damages to building and infrastructure, Itanagar

Table 21.1 Classes, scale value and weightage scheme to assess vulnerability

Parameter	No. of classes	Scale value range	Layer weight
DEM	10	1–9	10
Slope	6	1–9	20
Aspect	9	1–5	5
Curvature	5	1–6	10
TSA residual	15	1–9	10
TWI	5	1–9	20
Stream buffer	6	1–7	15
TPI	10	1–9	10

such building types are replaced by modern bricks, RCC building for safety against burglary, multi-story capabilities and aesthetics.

Smart city framework added another dimension to cityscape development as the expectations are high for modern and ultra-modern facilities and lifestyle. At present, there are large-scale remodelling and renovations going on in the smart city, Itanagar. When older buildings are redesigned or renovated or new buildings are constructed, the built-up area and the super area are stretched up to the plot area, leaving no space for setbacks. Building norms are framed by the authority, but those are not implemented as it has become practically impossible here in the smart city. Architectural design approval or soil testing are not taken seriously, thereby violation of master plan and safety norms are normal. This adds to the vulnerability of buildings to various hazards. Building density has gone up tremendously in the central part especially in Wards 6, 11, 12, 14, 15 and 16 where there exist 40 to 50 buildings per hectare. This is due to a lack of proper land revenue record, encroachment, wilful violations of law and guidelines. This problem is exacerbated as the Government has no control over land as traditionally natives are the owners of land, forest, river course and everything else. Another dimension of population pressure on limited land of the smart city is the influx of population from all parts of the State in search of education, job, health care, livelihood and many more other reasons.

This pressure of higher demand for land has pushed the cost of land to an exorbitantly higher level. It has resulted in unplanned growth, construction of houses close to rivers, steeper slopes and marshy lands, thereby increasing the number of buildings in higher vulnerability categories. As found from the analysis, there are about 21% of total buildings under a high level of threat, and there are 64 buildings under a higher level of vulnerability (Fig. 21.9 and Table 21.2). This also poses a hindrance in provisioning on road, drain and sewage, and emergency services like fire tender, ambulance and evacuation.

Future plan for the growth of Itanagar Smart City must factor these parameters contributing to the assessment of vulnerability. Many least vulnerable areas (category 1) are devoid of any buildings as these are government-owned vacant spaces and parks. Many areas under categories 1 and 2 of vulnerability level are not having

Table 21.2 Ward number-wise buildings under vulnerability classes

Ward no.	Area (hectare)	Total no. of households	Total Population (in thousands)	Number of buildings	Buildings per hectare	No. of Buildings vulnerability class 1	No. of Buildings vulnerability class 2	No. of Buildings vulnerability class 3	No. of Buildings vulnerability class 4	No. of Buildings vulnerability class 5	No. of Buildings vulnerability class 6
1	200.1180	1005	3035	433	2.16	3	114	258	56	2	
2	110.5900	550	3073	34	0.31		1	29	4		
3	36.5034	760	3073	249	6.82		50	158	37	4	
4	37.4730	1731	3457	561	14.97		155	305	99	2	
5	137.1980	1535	3115	618	4.50		111	481	26		
6	33.3939	1480	3176	602	18.03		119	418	65		
7	259.2000	388	3402	780	3.01		97	448	227	8	
8	45.2321	642	3286	471	10.41		76	319	76		
9	228.2520	672	3057	686	3.01		26	459	195	6	
10	55.7254	571	3499	597	10.71		48	326	210	13	
11	25.3760	900	3039	554	21.83		54	363	136	1	
12	70.4772	999	3542	582	8.26		65	380	137		
13	247.0980	1739	3554	372	1.51		76	222	72	2	
14	57.3988	1442	3531	717	12.49		71	438	206	2	
15	55.0938	1131	3690	653	11.85		74	447	129	3	
16	29.6385	2072	3658	621	20.95		18	439	157	7	
17	401.7200	991	3298	486	1.21		57	314	110	5	
18	530.1210	1883	3568	517	0.98		7	293	209	8	
19	528.5700	1540	3298	633	1.20		197	354	81	1	
Sum/mean	3089.1791	22,031	63,351	10,166	3.29	3	1416	6451	2232	64	0

(continued)

Table 21.2 (continued)

Ward no.	Area (hectare)	Total no. of households	Total Population (in thousands)	Number of buildings	Buildings per hectare	No. of Buildings vulnerability class 1	No. of Buildings vulnerability class 2	No. of Buildings vulnerability class 3	No. of Buildings vulnerability class 4	No. of Buildings vulnerability class 5	No. of Buildings vulnerability class 6
					Percentage- >	0.0295	13.9288	63.4566	21.9555	0.6295	0.0000

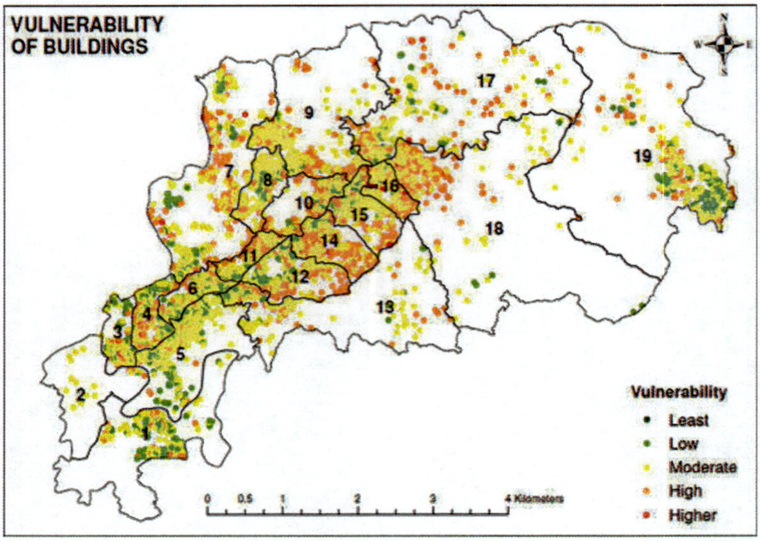

Fig. 21.9 Vulnerability of buildings, Smart City, Itanagar

any connecting roads and are surrounded by difficult terrain. These areas must be surveyed first for vulnerability followed by execution of appropriate provisioning of road, drain and sewage, electricity and water supply system. Only after these essential requirements for a smart city are complete, approval for construction of houses and buildings may be accorded by the Itanagar Smart City Development Corporation Limited.

References

Brown EH, Atkinson BW, Wolf PO, Rodda JC, Collins MP, Penning-Rowsell EC, Lee DO, Brunsden D, Hollis GE, Francis M, Chandler TJ, Cooke RU, Douglas I (1976) Physical Problems of the Urban environment: a symposium: discussion. Geograph J 142(1):72–80

Cooke RU (1976) Urban geomorphology. Geogr J 142(1):59–65

Devi RM, Bhakuni SS, Bora PK (2011) Neotectonic study along mountain front of northeast Himalaya, Arunachal Pradesh, India. Environ Earth Sci 63(4):751–762

Drăguţ L, Blaschke T (2006) Automated classification of landform elements using object-based image analysis. Geomorphology 81(3–4):330–344

Gares PA, Sherman DJ, Nordstrom KF (1994) Geomorphology and natural hazards. In: Geomorphology and natural hazards. Elsevier, pp 1–18

Gupta A, Ahmad R (1999) Geomorphology and the urban tropics: building an interface between research and usage. Geomorphology 31(1–4):133–149

Klingseisen B, Metternicht G, Paulus G (2008) Geomorphometric landscape analysis using a semi-automated GIS-approach. Environ Model Softw 23(1):109–121

Klingseisen B, Warren G, Metternicht G (2004) Landform-GIS based generation of topographic attributes for landform classification in Australia. NA

Mejía-Navarro M, Wohl EE, Oaks SD (1994) Geological hazards, vulnerability, and risk assessment using GIS: model for Glenwood Springs, Colorado. In: Geomorphology and natural hazards. Elsevier, pp 331–354

Metternicht G, Klingseisen B, Paulus G (2005) A semi-automated approach for GIS based generation of topographic attributes for landform classification. In: XXII international cartographic conference (ICC2005). The International Cartographic Association (ICA-ACI).

Patnaik SK, Gogoi P (2019) Geospatial analysis and modelling in characterization of topography and integration of vegetation cover in Panna District. In: Chetry N (ed) A glimpse of geospatial technologies and applications. EBH Publishers (India), pp 120–129

Patnaik SK (1993) Hydrogeomorphologic characterization of Sankh head water catchment. Ph.D. Thesis, CSRD, JNU

Patnaik SK (1995) Effect of spatial resolution in trend surface analysis. In: Jog SR (ed) Indian geomorphology: geomorphology and resource management, vol 1, pp147–58

Planning Commission, Govt. of India (2005) Arunachal Pradesh development report, p 323

Schilirò L, Montrasio L, Mugnozza GS (2016) Prediction of shallow landslide occurrence: validation of a physically-based approach through a real case study. Sci Total Environ 569:134–144

Singh Y, Singh T, Kaushal PD (2008) GIS based landslide inventory of Itanagar-the capital of Arunachal Pradesh. Indian Landslide 1(2):19–26

Smyth CG, Royle SA (2000) Urban landslide hazards: incidence and causative factors in Niterói, Rio de Janeiro State, Brazil. Appl Geogr 20(2):95–118

Speight JG (1990) Landform. Australian soil and land survey field handbook, pp 9–57

Tagil S, Jenness J (2008) GIS-based automated landform classification and topographic, landcover and geologic attributes of landforms around the Yazoren Polje, Turkey. J Appl Sci 8(6):910–921

Thakur VC (2004) Active tectonics of Himalayan frontal thrust and seismic hazard to Ganga Plain. Curr Sci 86(11):1554–1560

Chapter 22
Hydrogeological Studies of Urban–Rural Interface in the Northwest Part of Pune Metropolis, India

Bhavana N. Umrikar

Abstract Rapid and haphazard urbanization is posing a threat to geo-environment in general and water resources in particular along the fringe areas of metropolitan cities in the world. This problem has reached an alarming stage in developing countries like India. Keeping this fact in view, thematic layers of contours, drainage network, geology and Land use/Land cover were prepared in the GIS environment for two sub-watersheds of River Mula in the NW part of Pune, namely Ramnadi and Pirangut. The superimposed picture of these thematic layers was obtained and interpreted. The overlay of contour- and Land use/Land cover layers in the Ramnadi watershed revealed that 44%, 10% and 2% area occurring, respectively, below 660 m altitude, between 660 and 720 m altitude and above 720 m has been urbanized and thus, is responsible for flash floods, reduction in groundwater recharge vis-a-vis minimizing the groundwater storage capacity. In comparison with this, the Pirangut watershed depicts less impact of urbanization on water resources. Moreover, the demand for groundwater is increasing in the Pirangut watershed, and the city development plan (CDP) has declared Pirangut village as a D zone, meaning secured for industrial setup. Hence, before the implementation of the development plan, it would be appropriate to assure the conservation of water and soil resources in both the watersheds for sustainable development.

Keywords Hydrogeology · Urbanization · Basaltic aquifer · Morphometry · Urban–rural watersheds

Acronyms

EIA	Environmental Impact Analysis
PMC	Pune Municipal Corporation
RS-GIS	Remote Sensing and Geographical Information System

B. N. Umrikar (✉)
Department of Geology, Savitribai Phule Pune University, Pune, India
e-mail: bnumrikar@unipune.ernet.in

P. Sharma (ed.), *Geospatial Technology and Smart Cities*, The Urban Book Series,
https://doi.org/10.1007/978-3-030-71945-6_22

22.1 Introduction

The combination of increasing population and haphazard urban growth has resulted in a radical reshaping of both urban and rural landscapes in countries around the world. In the process of urbanization, human-induced changes on land are posing a serious threat to the geo-environment in general and water resources in particular along the fringe areas of most cities in the world. This problem has reached an alarming stage in developing countries but fortunately, there is a realization that development is a complex process which demands researchers, practitioners, developers and policy makers to share current research results and implementation strategies, and to identify knowledge gaps regarding land use changes. Awareness about Environmental Impact Analysis (EIA) is therefore on the rise. The first and the most important aspect of EIA is to collect baseline data and use it not only for future comparison but also for offering necessary inputs right from the planning to implementation stages. The need to collect baseline data in the fringe villages is thus evident. Such a data would help policy makers, planners and implementing authorities to take timely measures for sustainable development. RS-GIS techniques are popularly used to detect these land use changes and generate a sound database for identifying urbanization problems, evaluating their impact and providing locale-specific solutions (Chamley 2003; Singh 2006; Subba and Prathap 2004; Maggirwar and Umrikar 2009; Desai et al. 2009; Chaudhari et al. 2018; Kadam et al. 2018; Lad et al. 2018).

Before 1997, Pune Municipal Corporation (PMC) had jurisdiction of 146 km^2. It was expanded to 244 km^2 after the inclusion of several villages in the year 1997. In the proposed development plan (PMC 2001), about 11 villages would come under the jurisdiction of PMC increasing its area up to 368 km^2. But these villages are undergoing haphazard urbanization even before their inclusion in PMC. The Ramnadi watershed developed in the NW part of Pune city is one such area that offers a textbook example of the ill-effects of urbanization (Umrikar and Iyer 2009), but fortunately the Pirangut watershed is still retaining its natural setup.

The ill-effects of human-induced changes by way of an increasing network of roads, reduction in agricultural land, metamorphosis of hill slopes and stream network, etc. are evident in such areas. On this background, the author felt a need to generate a database that can be utilized in deciding the strategies for expanding Pune city. Hence, an attempt has been made to give RS-GIS-based geo inputs in the context of the Ramnadi and Pirangut watersheds, with a belief that the assessment of geo-environmental parameters can be best carried out within a natural entity of a watershed, since the development plans evolved within a watershed encompass the whole gamut of land, water and human population.

22.2 Study Area

Ramnadi, a tributary of river Mula, originates at Warpewadi near Bhukum and continues its travel through Bhugaon, Bavdhan, Pashan, Someshwarwadi, Aundh and finally meets river Mula near Baner. The longest stretch of the Pirangut stream lies in the SW part of the watershed originating at Magalwadi and continues its travel through Botarwari, Uraode, Bhilarwadi, Ubhewadi, Kasar Amboli, Shelarwadi and finally meets river Mula near Rawatwadi (Fig. 22.1). The areal extent of the Ramnadi watershed is 50.35 km^2 and has an elevation ranging from 550 to 800 m above mean sea level. The areal extent of the Pirangut watershed is 43.619 km^2 and has an elevation ranging from 550 to 980 m above mean sea level. The study area is included in the Survey of India toposheet nos. 47 F/10, 47 F/11, 47 F/14 and 47 F/15. It lies between 18° 25′ 05″ and 18° 34′ 00″ North latitudes and between 73° 36′ 00″ and 73° 49′ 15″ East longitudes. The area is approachable by National Highway No. 4 and is well connected with a network of metalled roads. It has undergone rapid urbanization in the last two decades, particularly around Bavdhan, Pashan, Aundh and Baner, Pirangut villages.

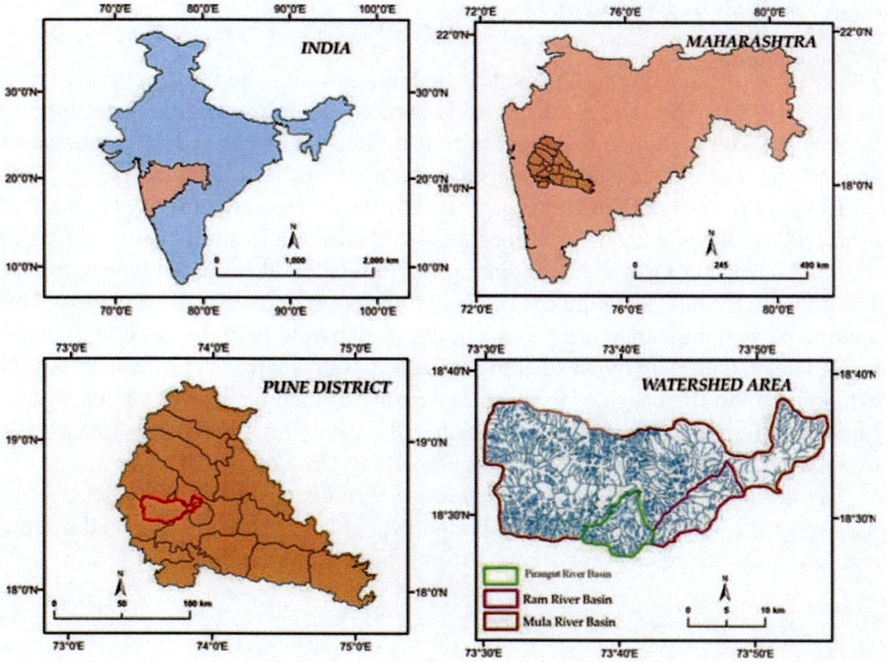

Fig. 22.1 Location map of the study area

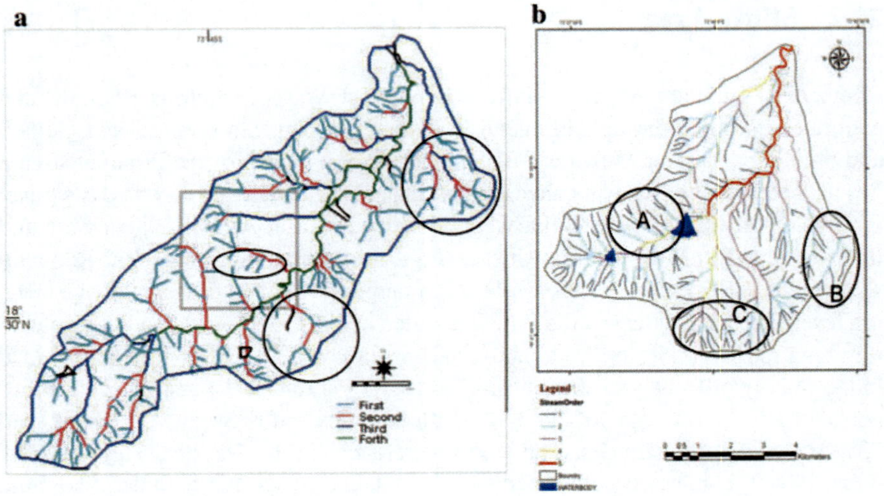

Fig. 22.2 **a** and **b** Drainage map of Ramnadi and Pirangut watersheds

22.3 Physical Framework

The climate of Pune city has been classified into monsoon and tropical-rainy types by IMD (1982). The rainfall of the study area occurs in the months from June to September. The area also experiences return monsoon showers in the months of October and November. The average temperature varies from 15 to 35 °C.

The physiography of the area is governed by the horizontally disposed of basaltic flows. These flows extend for a considerable distance and show features such as spheroidal weathering, columnar jointing, cavities of zeolites and red bole horizons. The flows on the hill slopes are covered by a thin veneer of residual soils (up to 0.4 m) underlain by poor to moderately weathered/jointed basalt; foothills by colluvium (up to 2 m) again underlain by weathered or jointed basalt whereas gently rolling terrain extending up to the banks of third-order streams and the main river stretch by soil (up to 1 m) underlain by alluvium/weathered basalt. Outcrops of fresh basalt are intermittently seen all over the area (Agashe 1990; Umrikar 2007).

The drainage pattern is mainly dendritic to sub-dendritic, but parallel to sub-parallel type and rectilinear type involving lower order streams are seen in the upper reaches of the watershed (Fig. 22.2a, b).

22.4 The Data

The toposheets of Survey of India were scanned and geo-referenced, and used as a base map to create thematic layers of contour and drainage. The thematic layer of the contour is further used for generating the Digital Elevation Model

(Fig. 22.3a, b). The layer of geology was prepared after taking appropriately designed traverses along the road- and stream-cuttings, well sections and across both the watersheds (Fig. 22.4a, b).

The thematic layer of Land use/Land cover was prepared with the help of Google earth images (ETM and SRTM). Categories suggested by Anderson et al. (1976) and Deren et al. (2000) for different Land use/Land cover classes were used (Fig. 22.5a, b).

Morphometric parameters were computed using Horton's method (1945) and data is included in Table 22.1. The bifurcation ratio and total stream length were computed in GIS, and the total length of lost streams was calculated by the superimposition of drainage layer on Land use/Land cover map (Fig. 22.6a, b, Tables 22.2 and 22.3). The ground checks were performed to confirm the status of lost streams during both pre- and post-monsoon periods. The drainage density maps were prepared for both the watersheds (Fig. 22.7a, b). The dug well inventory was carried out in May 2016. The well inventory data of a few wells (Fig. 22.8a, b) was compared with the earlier data

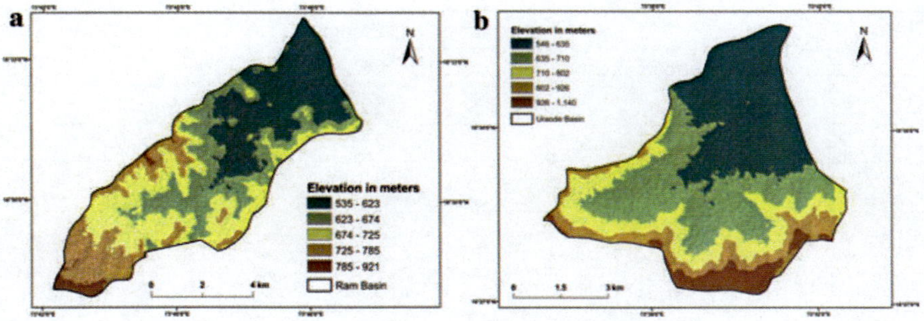

Fig. 22.3 **a** and **b** Digital elevation model of Ramnadi and Pirangut watersheds

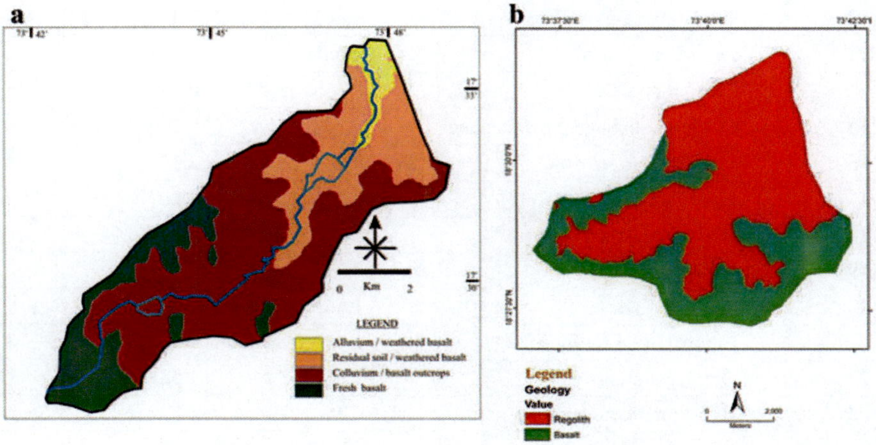

Fig. 22.4 **a** and **b** Geology map of Ramnadi and Pirangut watersheds

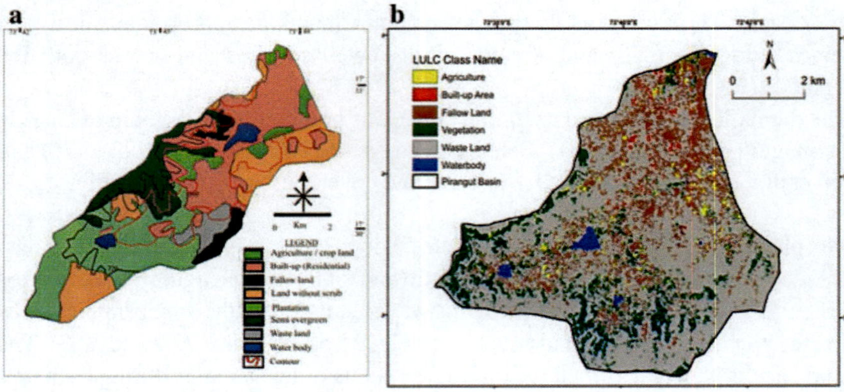

Fig. 22.5 **a** and **b** Land use/Land cover classification of Ramnadi and Pirangut watersheds

Table 22.1 Morphometric parameters of Ramnadi and Pirangut watersheds

Sr. no.	Watershed parameter	Ramnadi	Pirangut
1	Shape	Elongated	Elongated/Pear
2	Highest stream order	04	05
3	Total area km^2	50.35	43.61
4	Main stream length km	11.00	9.70
5	Watershed perimeter km	35.00	29.30
6	Basin/Valley length	14.45	9.50
7	Total number of streams	192	230
8	Stream frequency	3.84	5.34
9	Drainage density	3.80	3.48
10	Overland flow	0.169	0.083
11	Form factor	0.244	0.482
12	Total stream length km	190.32	150.26
13	Lemniscates factor	4.081	2.127
14	Elongation ratio	0.556	0.784
15	Constant of channel maintenance	0.263	0.269
16	Shape factor	4.083	2.12
17	Circulatory ratio	0.701	18.69
18	Drainage texture	5.54	7.847
19	Fitness ratio	0.314	0.331
20	Sinuosity index	1.36	1.35
21	Compactness coefficient	0.233	1.488
22	Drainage intensity	1.563	1.58
23	Hypsometric integral	0.45	0.32

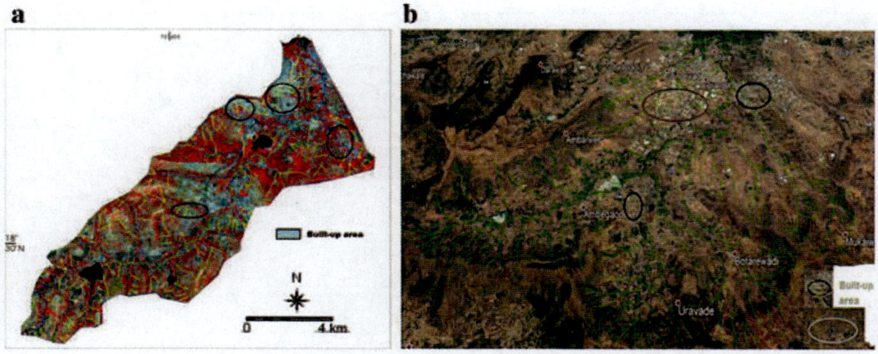

Fig. 22.6 **a** and **b** Drainage lost due to urbanization in Ramnadi and Pirangut watersheds

Table 22.2 Bifurcation ratio and stream length data of the Ramnadi watershed

Order of streams	Total no. of streams A	Bifurcation ratio (Rb)	Total stream length (km) B	Average length per stream (km) B/A	Total no. of lost drainage	Total length of lost drainage (km)
1	147	3.97 5.28 7.00	99.80	0.68	57	38.19
2	37		43.00	1.16	23	26.68
3	07		27.00	3.85	3	11.55
4	01		20.52	20.52	–	–
	Total = 192	Avg. = 5.41	Total = 190.32			

Table 22.3 Bifurcation ratio and stream length data of the Pirangut watershed

Order of streams	Total no. of streams A	Bifurcation ratio (Rb)	Total stream length (km) B	Average length per stream (km) B/A	Total no. of lost drainage	Total length of lost drainage (km)
1	165	3.30 4.54 3.66 3.00	88.17	0.5344	17	9.0848
2	50		35.10	0.7021	05	3.5105
3	11		12.12	1.1018	–	–
4	03		07.68	2.5614	–	–
5	01		07.29	7.2990		
	Total = 230	Avg. = 4.11	Total = 150.26			

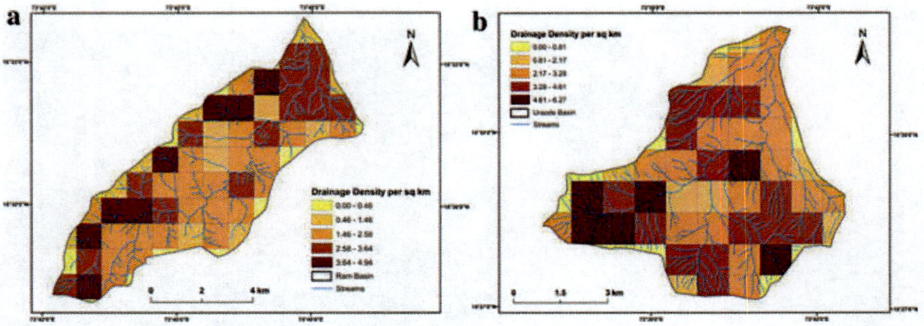

Fig. 22.7 **a** and **b** Drainage density maps of Ramnadi and Pirangut watersheds

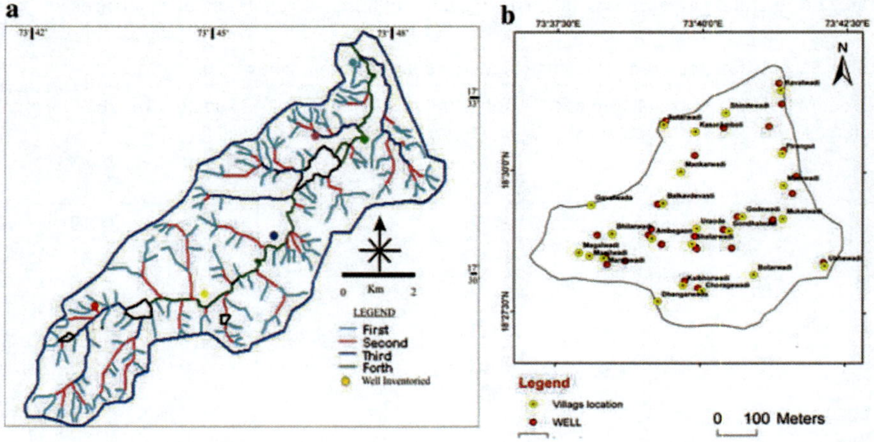

Fig. 22.8 **a** and **b** Well inventory location maps of Ramnadi and Pirangut watersheds

Table 22.4 Comparison of well inventory data from the Ramnadi watershed

Location	Depth of well (m) A	Water levels in summer (m)		Available water column (m)		Difference in available water column (m)
		1986 B	2016 C	1986	2016	
Bhukum	5.90	3.10	3.50	2.80	2.40	0.40
Bhugaon	5.50	3.30	3.60	2.20	1.90	0.30
Bavdhan	6.30	2.80	4.30	3.50	2.00	1.50
Pashan	6.50	3.10	4.70	3.40	1.80	1.60
Baner	5.80	3.20	3.90	2.60	1.90	0.70

Table 22.5 Comparison of well inventory data from the Pirangut watershed

Location	Depth of well (m) A	Water levels in summer (m)		Available water column (m)		Difference in available water column (m)
		B 1986	C 2016	1986	2016	
Ambegaon	7.00	5.10	5.30	1.90	1.70	0.20
Pirangut	7.80	6.30	6.80	1.50	1.00	0.50
Uraode	6.25	5.10	5.40	1.15	0.85	0.30
Kasar Amboli	7.00	5.50	5.70	1.50	1.30	0.20
Ravatwadi	7.70	6.50	6.70	1.20	1.00	0.20
Marnewadi	7.50	5.30	5.60	2.20	1.90	0.30

Table 22.6 Urbanized area % covered by different contour levels in the Ramnadi watershed

Sr. no.	Contour (m)	No. of pixels	Pixel area	Area (sq m)	Area (km^2)	Area urbanized (km^2)	% area urbanized	Depleted groundwater level (m)
1	<660	53,646	400	21,458,500	21.45	9.45	44	1.5–1.6
2	660–720	40,762	400	16,305,100	18.60	1.86	10	1.0–0.7
3	>720	31,350	400	12,540,000	10.55	0.31	2	0.5–0.3

Table 22.7 Urbanized area % covered by different contour levels in the Pirangut watershed

Sr. no.	Contour (m)	No. of pixels	Pixel area (acre)	Area (sq m)	Area (km^2)	Area urbanized (km^2)	% area urbanized	Depleted groundwater level (m)
1	<660	52,755	506	21100244.51	21.10	2.15	5	0.3–0.5
2	660–720	20,214	1997	8,085,887	8.08	0.58	1.3	0.2–0.3
3	>720	35,656	3524	14,259,817	14.25	0.31	0.71	0.1–0.2

(GSDA 1986; Saxena 1986) and incorporated in Tables 22.4 and 22.5. Tables 22.6 and 22.7 include the data of different classes of Land use/Land cover of the study area.

22.5 Discussion

Urban area is defined as an area with at least one million people, in which there is a zone of metropolitan areas with intervening counties never far from a city. This definition holds good for western- but certainly not for eastern countries, which face the problems of population explosion and inadequate finance. It is, therefore, inappropriate to use the western models to tackle the problems arising from urbanization in eastern countries. The need to generate a sound database including relevant geo-inputs to suit different terrains has thus been felt. The database generated through Remote Sensing—Geographical Information System (RS-GIS) techniques for Ramnadi and Pirangut periurban watersheds located in the NW part of Pune city is viewed on this background.

It is evident from the overlay of thematic layers of contour- and Land use/Land cover (Tables 22.6 and 22.7) that 44%, 10% and 2% area of the Ramnadi watershed and 5%, 1.3% and 0.71% occurring, respectively, below 660 m altitude, between 660 and 720 m altitude and above 720 m altitude has undergone metamorphosis converting pervious land surface into impervious one. It is also evident from the overlay of thematic layers of drainage and Land use/Land cover of the Ramnadi watershed (Table 22.2) that out of 147 first and 37 second order streams, 57 and 23 streams, respectively, have been disappeared in the process of leveling of land for construction purposes such as buildings and transportation routes, etc. It amounts to a loss of 65 km out of 190 km of the total length of streams. Such anthropogenic changes over a large area would obviously shift the equilibrium conditions causing flash floods and reducing the storativity of aquifers. This fact is clearly evident from the comparison of well inventory data of 1986 (GSDA 1986; Saxena 1986) and 2016 (Tables 22.4 and 22.5). The data reveals that though the groundwater levels in dug wells located along Ramnadi reach up to the surface during rainy season, they depletes fast due to loss of pervious surface and thereby storage capacity. The groundwater levels in dug wells located beyond the PMC limit have remained more or less the same whereas those located within the PMC limit show depletion in the range of 0.70 m to 1.60 m from 1986 to 2015. The depletion of groundwater levels in the Pirangut watershed is between 0.2 and 0.5 m indicating it is less affected as the area has not yet been covered by haphazard construction. The morphometric data (Tables 22.2 and 22.3) reveals that bifurcation ratios for first- and second-order streams are less than or around 5 indicating erosional control over their development on uniform basaltic lithology. The bifurcation ratio for third- and fourth-order streams in the Ramnadi watershed, however, is 7.0 indicating structural control over their development (Strahler 1957).

It can thus be seen that haphazard development has led to the deterioration of the hydrogeological regime in the Ramnadi watershed compared with the Pirangut watershed. However, most of the land in this area has already been purchased by developers, educational trusts, industrial houses, etc. This fact demands adopting appropriate measures for the conservation of precious natural resources before urbanization. Encouraging roof-top harvesting, maintaining pervious surface by bringing land under gardens and horticulture plots, and cutting trenches to serve as infiltration galleries are some measures suggested for maintaining the groundwater table. The establishment of sewage treatment plants has to be proposed by keeping 10 m tracks along the banks of third-order streams and mainstream course, also the measures for preventing domestic effluent to enter into the streams are few remedies for minimizing both surface and groundwater pollution. But most important long-term measures that should be considered during the planning stage are (1) to declare fringe villages beyond the PMC limit as a 'restricted zone' for controlling haphazard development and (2) perspective of watershed development covering geo-environmental parameters in totality should be involved for adopting measures to prevent not only flash floods, pollution and depletion of groundwater levels but also other probable hazards such as erosion and landslides.

22.6 Conclusion

Geo-inputs in the form of thematic layers of contour, drainage density, geology and Land use/Land cover have been utilized to study the impact of urbanization on the groundwater regime of Ramnadi and Pirangut periurban watersheds. It has been inferred that 54% of the area covered by 1–7 m thick soil/weathered rock/colluvium/alluvium lying below 720 m has lost its pervious status from the Ramnadi watershed whereas out of total 190 km length of streams, 65 km length has been lost in the process of leveling of land for various purposes. These anthropogenic changes have shifted the equilibrium conditions causing flash floods and reducing infiltration. The groundwater level in urbanized areas has been depleted in the range of 0.70 to 1.60 m from the Ramnadi watershed as compared to 0.20 to 0.30 m depletion in the Pirangut watershed.

It is thus evident that haphazard development has led to the deterioration of the hydrogeological regime in the Ramnadi watershed. Only 25% of the total area of Ramnadi and about 7% area of Pirangut watersheds have come under urbanization so far. Most of the remaining area falls outside the jurisdiction of PMC and therefore retained as agricultural land, land without scrub and semi-ever green status. However, these lands have already been purchased by developers, educational trusts, industrial houses, etc. and thus emphasize the need to take appropriate measures before urbanization for maintaining the pervious nature of the land, increasing groundwater recharge and preventing pollution. There is a need to declare fringe villages along the PMC limit as a 'restricted zone' for controlling haphazard development and to consider geo-environmental parameters of watersheds in totality for adopting measures for minimizing the impact of anthropogenic changes.

References

Agashe RM (1990) Scope of artificial groundwater recharge in Deccan Trap areas in Maharashtra-an overview. In: Proceedings of all India seminar on modern techniques of rainwater harvesting, water conservation and artificial recharge for drinking water, afforestation, horticulture and agriculture, Pune, G.S.D.A., pp 121–129

Anderson JR, Hardy EE, Roach JT, Witmer RE (1976) A land use and land cover classification system for use with remote sensor data. Geol Survey Professional Paper 964, U. S. Government Printing Office, Washington

Chamley H (2003) Geosciences, environment and man. Elsevier Publ, Amsterdam, 450 pp

Chaudhari RV, Lal D, Dutta S, Umrikar BN, Halder S (2018) Weighted overlay analysis for delineation of ground water potential zone: a case study of Pirangut River Basin. Int J Remote Sens Geosci (IJRSG) 7(1):1–7. ISSN No: 2319-3484

Deren LI, Kaichang DI, Deyi LI (2000) Land use classification of remote sensing image with GIS data based on spatial data mining techniques. In: International archives of remote sensing, vol XXXIII, part B3. Amsterdam, pp 238–245

Desai CG, Patil MB, Umrikar BN (2009) Application of remote sensing and geographic information system to study land use/land cover changes: a case study of Pune Metropolis. J Adv Comput Res 1(II):10–13. ISSN: 0975-3273

Groundwater Surveys and Development Agency (1986) Information of hydrogeological studies and groundwater development schemes implemented by GSDA. Technical report, 56 pp

Horton RE (1945) Erosional development of streams and drainage basins, hydrogeological approach to qualitative morphology. Geol Soc Am Bull 26:275–370

Indian Meteorology Department (1982) Climate of Maharashtra, 144 pp

Kadam AK, Umrikar BN, Sankhua RN (2018) Assessment of soil loss using revised universal soil loss equation RUSLE: a remote sensing and GIS approach. Remote Sens Land 2(1):65–75. https://doi.org/10.21523/gcj1.18020105

Lad S, Ayachit R, Kadam A, Umrikar BN (2018) Groundwater vulnerability assessment using DRASTIC model: a comparative analysis of conventional, AHP, fuzzy logic and frequency ratio method. Model Earth Syst Environ. https://doi.org/10.1007/s40808-018-0545-7

Maggirwar BC, Umrikar BN (2009) Possibility of artificial recharge in overdeveloped miniwatersheds: RS-GIS approach. e-Journal, Earth Sci India 2(II):101–110. ISSN No. 0974-8350 (https://www.earthscienceindia.info)

Pune Municipal Corporation (2001) Proposed development plan for newly included villages 2001–2021, 88 pp

Saxena IN (1986) Geomorphology and hydrogeology of the Upper Mula River basin, Dist Pune, Maharashtra. Unpubl PhD thesis, Univ of Pune, 168 pp

Singh RB (2006) Sustainable urban development. Concert Publishing Co Publ, Delhi, 431 pp

Strahler AK (1957) Quantitative analysis of watershed geomorphology. Trans Am Geophy Union 38:913–920

Subba RN, Prathap RR (2004) Geoenvironmental appraisal in a developing urban area. J Environ Geol 47:20–29

Umrikar BN (2007) Geo-environmental study of urban/rural interface in the NW part of Pune Metropolis, Maharashtra. Unpubl project report, Univ of Pune, 34 pp

Umrikar BN, Iyer U (2009) Impact analysis of urbanization on surface and ground waters using RS-GIS technique: a case study of Pune Municipal Corporation. J Urban India XXVII(1):45–56

Chapter 23
Groundwater Analytics for Measuring Quality and Quantity

Mukta Sharma

Abstract Groundwater constitutes the key segment of freshwater assets of the world. With increasing urbanization across the globe, quality and quantity of groundwater are the issue of major concern. Water resources are highly dynamic and undergo significant changes with the changing land-use patterns. Rapid industrialization and population growth have strongly accelerated land-use changes globally and have a severe impact on groundwater resources. Accurate monitoring and management of groundwater resources is therefore required for the sustainable development of cities. Mapping groundwater resources is significant for the development and management of same. Estimation of Potential Groundwater zones, Assessment of Ground Water Quality, and Ground Water Vulnerability assessment are the key concerns for Groundwater management. Geospatial technologies have proven to be very useful in the development of optimal approaches for groundwater development, assessment, and management. Studies of various researchers have revealed that interrelated factors of lithology, lineaments or geologic structures and terrain features help in identifying the most promising sites for groundwater exploration. Remote Sensing studies facilitate delineating various geological, topographic, and structural features and further integration in the GIS environment to demarcate potential groundwater zones. Similarly, groundwater models created with sufficient groundwater quality data converted into GIS database give an accurate assessment of the spatial distribution of groundwater quality and changes with time and change in land-use. Information about vulnerability to contamination of groundwater can aid in selecting suitable sites for certain activities and thus minimizing adverse effects on groundwater. GIS-based vulnerability maps based on the subjective rating of hydrogeological factors are commonly used to evaluate groundwater vulnerability and can be effectively used for decision-making and proper groundwater planning. Therefore, an integrated approach including data-driven out from Remote Sensing, ground truth data, and hydrogeological GIS database and groundwater modeling using statistical techniques offers immense potential for groundwater mapping and management.

M. Sharma (✉)
IKG Punjab Technical University, Jalandhar, Punjab, India

P. Sharma (ed.), *Geospatial Technology and Smart Cities*, The Urban Book Series,
https://doi.org/10.1007/978-3-030-71945-6_23

Keywords Groundwater management · Remote sensing and GIS · Groundwater prospecting · Groundwater quality · Vulnerability assessment

Acronyms

AHP	Analytic Hierarchy Process
AVI	Aquifer Vulnerability Index
GRACE	Gravity Recovery and Climate Experiment
GOES	Geostationary Operational Environmental Satellite
GMS	Geostationary Meteorological Satellite
INSAT	Indian National Satellite System
IRS	Indian Remote Sensing
MSS	Multi-Spectral Satellite
NOAA	National Oceanic and Atmospheric
WIOA	Weighed Index Overlay Analysis
WQI	Water Quality Index

23.1 Introduction: Groundwater Scenario of Smart Cities

The speed with which the urban population is increasing, the pressure on resources is increasing simultaneously. Optimum and judicious utilization of the available resources to cater to the growing demands of the society is the need of the hour. Our rural areas too can serve to channelize the surplus resources. Groundwater which is the prime water resource all over the world is under the focus. The United Nations International Decade for Action (2005–2015) of 'Water for Life' embodies groundwater conservation, manageable through remediation and groundwater engineering (Margane 2003). Groundwater constitutes the major portions of the available potable water, but has limited extent and volume; blatant mismanagement in water utilization and wastage is alarming. Thus, the potable water scenario needs an immediate and far-reaching strategy. Government at all levels needs to put their acts together.

The GRACE (Gravity Recovery and Climate Experiment), a joint venture by NASA and the University of California, has confirmed that the biggest aquifers across the world are getting depleted at a very fast pace and it is becoming increasingly tough to replenish them. There is extremely high surface water stress in north-western India as well, where the rate at which groundwater is being withdrawn is very high as compared to other parts of the globe; making the scenario worse. Statistics reveal 54% of 4,000 measured groundwater wells are experiencing decline, which indicates a very alarming situation and calls for appropriate groundwater management(Uprety and Salman 2011) through uniform approach of integrated geospatial technologies, computing and modelling structures.

Water is vital for sustenance of life and to make our smart cities sustainable, we need adequate amount of water supply with high water quality index (Becker 2006; Rainey et al. 2003; Rodriguez et al. 2006). With one-fourth of the world's river basins running dry before reaching the oceans, addressing issues related to water are significant as it can immensely affect the economy of the country and the living conditions of the inhabitants of the peri-urban and urban societies (Molden et al. 2007). The major challenges for the smart cities would be the depletion of the water tables, wastage of water during supply, issues of economic valuation and allocation for domestic as well as agricultural purposes. The smart cities have to be water sustainable and for this various avenues need to be explored.

The spatial distribution of potential groundwater zones, aquifer delineation and temporal dynamics of groundwater can effectively help in its management. The objective is to address various groundwater issues such as projection and aquifer replenishment, water quality index, vulnerability and modelling with the help of geospatial technologies.

23.2 Ground Water Management Enabled by Remote Sensing and GIS

Groundwater, a renewable source and therefore, if located, exploited and managed carefully, can sustain forever (Thenkabail 2015). Estimation of Potential Groundwater zones, Quality of Groundwater, Aquifer Recharge and Groundwater Vulnerability assessment are key concerns. The deterioration of the watershed is attributed to uncontrolled, unplanned and unscientific land-use aided by human interventions. Watershed management is simply watershed 'protection'. The wise and judicious use of the land and water resources facilitates adequate and sustainable production and reduced hazards to the natural resources.

Remote sensing and GIS are the leading tools in the field of hydrogeological science and help in the forecasting, assessment, monitoring and conservation of groundwater resources. GIS know-how provides viable alternatives for efficient management of large and complex databases. The groundwater behaviour is geological diversity dependent, hydro-chemical conditions and climate logical variations also have a significant role to play. The proper spotting and mapping of all these features, the availability of sufficient data of high quality determines the reliability and validity of the sustainable groundwater models. Proper documentation for effective planning and management, geospatial is undoubtedly nonpareil in natural resource management, computational statistics and data mining from inter-operative maps in a geophysical environment. The desideratum-based software provides a seamless interface with imagery and background base maps of groundwater data. The spatial distribution of different groundwater prospect classes based on geomorphology and other associated parameters has been made easy with remote sensing.

For the generation of bottom-line information for groundwater exploration, remotely sensed data is validated with field survey.

Many satellites are used for observing the atmospheric features of earth. These can be broadly put in two groups, earth resource satellite and environmental satellite. Earth resource satellites capture information of the same area relatively infrequently (days) with relatively high resolution, e.g. Landsat (Land Satellite, USA), JERS and ADEOS (Japan), IRS (Indian Remote Sensing, India), OKEAN (meaning Ocean, Russia), RADARSAT (Canada), SPOT (Satellite Positioning and Tracking, France), ERS (European Remote Sensing, ESA and Canada), CBERS (China Brazil Earth Resources Satellite, China). Environmental satellites capture the information of same area relatively frequently (hours) with relatively low resolution, e.g. NOAA (National Oceanic and Atmospheric), INSAT (Indian National Satellite System, India), GOES (Geostationary Operational Environmental Satellite, USA), GMS (Geostationary Meteorological Satellite, Japan) (Jensen 2000). A transformation in the way we look at advanced analytical applications and sustainable approach to groundwater remedial management is visible globally.

23.3 Groundwater (Projection) and Aquifer Replenishment

Identifying and calculating usable aquifer reserves and adopting measures to restore depleting aquifers is a vital step in the groundwater forecasting and management of smart cities (Foster 1998; Haridas et al. 1998; Lattman 1958; Meijerink 1996; Tam et al. 2004; Tripathi et al. 2017; Valstar et al. 2004). The presence of lineaments in the region which are straight linear elements seen on the earth's surface is important in determining the occurrence of groundwater. Presence of lineaments on the surface is indicative of conduits of groundwater flow in fractured aquifers and are targeted for locating production wells (Rosema and Fiselier 1990; Tiwari et al. 2016). Faults, fractures, shear zones, dykes, veins, stratigraphic contacts and bedding planes are the geological features which indicate subsurface faults and fractures and influence the presence of groundwater since they act as reservoirs, canals, conductors, etc. In fact, the presence of dykes in the region controls the regional flow of subsurface water (Foster 1998; Lattman 1958; Meijerink 1996; Tam et al. 2004; Valstar et al. 2004). Thus, lineament density in the region largely determines the groundwater potential of a place as it indicates the permeable zone. Regions with higher lineament density have high groundwater potential.

Several studies have been carried out on delineating groundwater potential zone using geospatial techniques and artificial recharge of Aquifers (Brunner et al. 2007; Gogu et al. 2001; Gumbricht et al. 2005; Hammouri et al. 2012; Jawad and Yahya 2013; Mayilvaganan et al. 2011; Mishra et al. 2010; Nagarajan and Singh 2009; Rao et al. 2009; Slater et al. 2006) (Fig. 23.1).

The thematic maps portray spatial variations and interrelationships of geographical distribution that focuses on a specific theme or subject area. Thematic layers of lithology, geomorphology, slope, lineament, land-use and drainage density of the

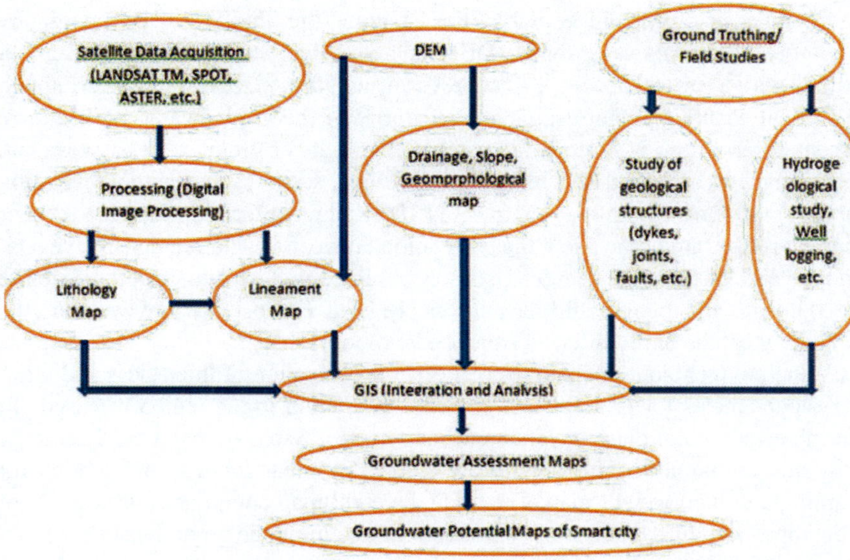

Fig. 23.1 Flowchart showing estimation of groundwater potential of smart city using remote sensing and GIS techniques

area of study are integrated and their weights are computed using Weighed Index Overlay Analysis (WIOA), Analytic Hierarchy Process (AHP), fuzzy logic technique, etc. Maps of Potential groundwater zones and artificial recharge sites are then prepared by calculating the computed weights. The results can be validated with field observations at observation wells as well as water level depth data collected from various authenticated government and non-government bodies. These thematic maps help in more accurate modelling of the real-world behaviour and scenarios, predictive analysis and decision-making. Maps generated from weighed overlay using various statistical techniques are found to be performing very well in forecasting the groundwater surface.

The availability of high-quality data is very vital in generating dependable and authentic groundwater analysis. GIS tools help in logical and coherent arrangement of available data for hydrogeological studies. Integration of GIS technology and groundwater simulation models provides great potential for groundwater exploration. The hydrogeological data stored in the database can be used for various numerical models by following time and spatial queries.

The information for the exploration of terrain characteristic comprising geomorphology, structure, geology, land-use land cover, etc. is created using remote sensing and GIS techniques; it is further used for assessment of the groundwater resources. The thematic maps such as geology, lineaments and land-use can be produced by digitally magnifying colour composites and panchromatic images such as IRS, LANDSAT TM, ASTER and SPOT. For mapping, the topographic parameters such as the slope, surface curvature, drainage systems and landforms, DEMs are used. A large number of DEMs are available. For example, the Shuttle Radar Topography Mission (SRTM) data set provides DEM approachable between 60 °N and 56 °S Latitude with a vertical accuracy of at best 5 m and 90-m pixel resolution. Although the vertical accuracy is not useful for a majority of the groundwater applications, the spatial resolution is sufficient to cater to the needs of groundwater assessment, monitoring, etc. LIDAR data with sufficient vertical accuracy and spatial resolution is apt for groundwater exploration. In case LIDAR or radar techniques, due to cost or accuracy issues, are not availed, the correlation between vegetation density, vegetation type and land surface characteristics as observed in the Multi-Spectral Satellite (MSS) images and topographic elevation can be used. For instance, the wetlands, the topography can be easily inferred from the land-cover maps.

Evaluation techniques of multiple criteria for evaluation of inter-class and inter-map dependencies for groundwater resource evaluation in the region are used. To determine individual class weights and map scores, Satty'a Analytical Hierarchy Technique can be used. The weights are applied in linear summation equation for obtaining a unified weight map. The unified weight map comprises due weights of all the input variables that can be regrouped for obtaining the groundwater potential zone map for the sites selected for artificial recharge structures and the type of structure recommended. The GIS tools are useful in refining the stream densities and lineament intervals and for the identification of most promising sites based on the available groundwater well. These serve as very dependable tools for timely and cost-effective identification and scaling down of target areas for groundwater exploration and facilitate further investigations.

In the course of fieldwork, the fracture patterns and spacing in different rock types are measured and comparison is made with the lineaments. The topographic and hydrogeological settings of the springs and wells such as the well logs, pumping tests, water table depth in dry and wet season as well as location of wells are studied. The integration and analysis of all these thematic layers along with the hydrogeological data is done using GIS, yielding groundwater potential map. Thus, estimation of the groundwater recharge is done on the basis of water level fluctuations in the wells or boreholes. The landform maps helpful in the groundwater potential assessment are created with the DEM data. The presence of large lineaments, joints and corresponding structural features such as dykes increase the probability of high-yielding wells and springs. The highly weathered rocks or young alluvial deposits have high permeability and are significant for water supply in crystalline bedrock. Regional patterns related to lithologies, landforms and drainage systems mark the spatial distribution of groundwater potential zones. With the integration of remote sensing, GIS and traditional field work or ground truthing, models (such as recharge rate, overdrafting rate, etc.) are prepared which are useful in the assessment, management and groundwater development of the smart cities.

Certain relevant entities, viz. water flux, head, transmissivities, etc. cannot possibly be observed directly from remote sensing. Under such circumstances, the observable quantities are linked to the input data by the hydrological models. This can be done by using the remotely sensed data for creating the spatially distributed input parameter sets for the model and constraining of the models at the time of calibration by spatially distributed data that has been derived from remote sensing. It is helpful in improving the models conceptually as well as quantitatively. The application of remote sensing data to groundwater models requires fieldwork, necessary for obtaining sufficient data.

The non-availability of data in the semi-arid and arid regions hampers the assessment, modelling and management of the groundwater resources. Sparse point measurements which are available and the groundwater modelling require spatial and temporal distributions of input and calibration data and in their absence the models fail to serve the purpose as they are underdetermined and uncertain. The potential of remote sensing for the improvement of models despite being immense is untapped largely.

23.4 Ground Water Quality Index

The water quality parameters such as chlorophyll, suspended particles/sediments (turbidity) and temperature can be easily monitored with the help of remote sensing techniques (Christodoulidou et al. 2012; Elumalai et al. 2017; Kaur et al. 2017; Mukundan et al. 2011) using high resolution optical and thermal sensors on satellites, aircrafts or boats. For assessing the waterways, integrated management of remotely sensed data, GPS and GIS serve as valuable tools, which can be used as database for future comparisons and can be used for the management of natural resources.

To assess the physicochemical parameters and heavy metal concentrations of water resources and to check the effectiveness of water treatment and supply by the concerned authorities, water quality parameters are made by Bureau of Indian Standards (BIS 2012). WQI common with many other index systems, relate a group of physicochemical water quality parameters to a common scale and combine them into a single number in accordance with the chosen method of computation. Similarly, heavy metal concentration is assessed with the help of heavy metal pollution index method. The heavy metals in surface and groundwater are analyzed with the help of multivariate statistical tools. The human exposure risk or hazard quotient is calculated to assess whether the concentration of heavy metals, such as aluminium, lead, nickel, boron, cadmium, iron, manganese, zinc, silver and copper are within the permissible limits for drinking at all the locations in the smart cities. The major pollutants are human induced such as the abuse of insecticides and fertilizers, industries, mining and related activities, geogenic and anthropogenic activities. Polluted regions, possibly in the vicinity of the industrial operations and landfill sites, show high concentration of the heavy metals. Understanding of geo-statistics and GIS is a promising tool for the effective analysis of spatial and temporal groundwater quality data.

Use of geospatial tools like the GIS for in-depth mapping of the water quality parameters plays a vital role in policy or decision making. GIS is advantageous over manual cartographic analysis in handling attribute data with spatial features as the software helps in editing, manipulating, analyzing and displaying data in graphical as well as text form. All the spatial and attribute data that is groundwater

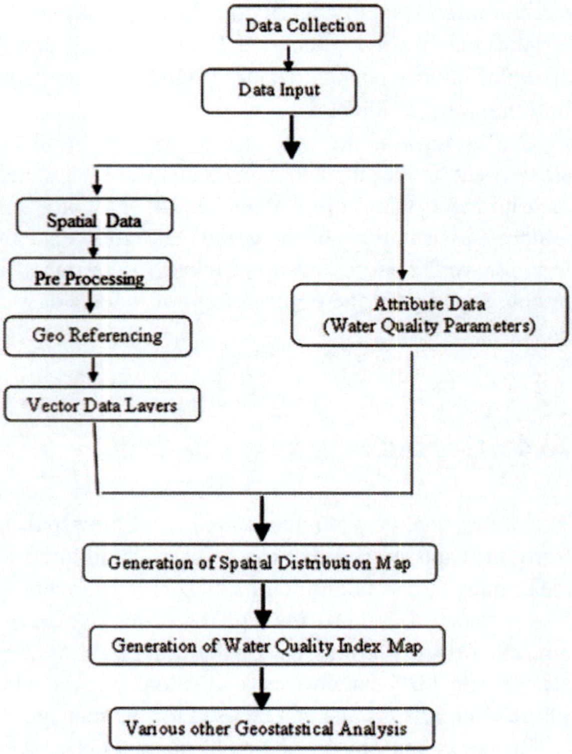

Fig. 23.2 Flowchart showing estimation of groundwater quality index

quality parameters are input in GIS and thematic data layers for all the water quality parameters are generated (Fig. 23.2). Various statistical techniques can be utilized for creating groundwater quality maps.

Case Study: Ludhiana (Punjab)

With the economy of the state of Punjab (India) being predominantly an agrarian one and more than 80% of land under agriculture, the entire state is immensely dependent on groundwater to meet the agricultural, industrial and domestic needs (Ahada and Suthar 2018). Nevertheless, the anthropogenic activities have led to the degradation in the quality of water in many regions of Punjab at an alarming rate (Hundal et al. 2009). For example, the groundwater quality in the Malwa region of Punjab has been rendered unsuitable for human consumption due to infiltration of chemicals such as Arsenic and Uranium in the shallow and deep aquifers (Brown et al. 1970, 1972; Goher et al. 2014; Horton 1965; Jasmin and Mallikarjuna 2014; Singh et al. 2011, 2019; Yin et al. 2013).

Water Quality Index and spatial variation of groundwater of Ludhiana city, Punjab, has been assessed with geospatial techniques in the present study. Groundwater samples from 99 locations (Fig. 23.3) have been collected using grid-based sampling procedure during March 2018. Samples were analyzed in the lab for determining

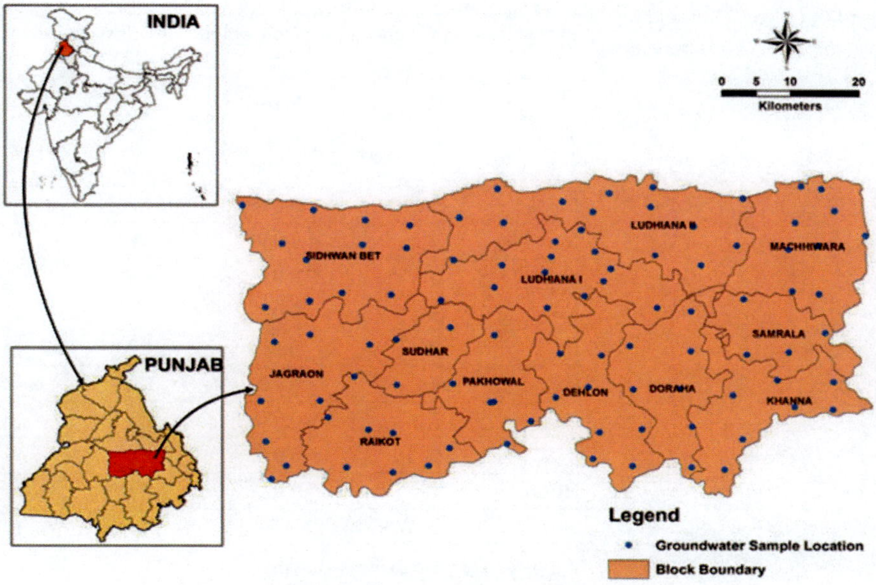

Fig. 23.3 Study area with sampling locations (Singh et al. 2019)

various parameters viz. pH, Total Hardness (TH), Total Dissolved Solids (TDS), Magnesium (Mg^{2+}), Calcium (Ca^{2+}), Potassium (K^{+}), Sodium (Na^{+}), Chloride (Cl^{-}), Fluoride (F^{-}), Nitrate (NO_3^{-}), Bicarbonate (HCO_3^{-}) and Sulphate (SO_4^{2-}).

Thematic data layers of all these parameters, viz., pH, TH, TDS, Ca^{2+}, Mg^{2+}, F^{-}, Na^{+}, K^{+}, NO_3^{-}, $HCO_{3,}$ SO_4^{2-} have been generated in ArcGIS for groundwater quality monitoring using Inverse Distance Weighted (IDW) interpolation technique (Singh et al. 2019). In IDW, on the basis of measured values, the nearby missing values are calculated. The prediction location values are more influenced by the measured values that are more close to them.

WQI Assessment

A Water Quality Index (WQI) depicts quality index of water and helps in taking appropriate measures to address the concerns. WQI outlines the combined influence of different water quality parameters. Horton introduced the concept of water quality index of water of the natural water bodies which was further developed by Rown et al. and many other researchers (Brown et al. 1970, 1972; Goher et al. 2014; Horton 1965; Jasmin and Mallikarjuna 2014; Singh et al. 2011).

Computation of WQI involves three steps. Firstly, all the parameters are assigned weight (W_i) according to its relative importance in drinking water. These weights as suggested by BIS water standards 2003, are presented in Table 23.1. The maximum weight of 5 has been assigned to the parameters which are more significant in water quality assessment and insignificant ones have relatively lower values.

Table 23.1 Weightage of parameters vis-a-vis standards

Parameter[a]	Weight age (W_i)	Unit weight (w_i)	BIS standards (S_i)
pH	5	0.125	6.5–8.5
TDS	5	0.125	500
TH	4	0.100	200
Ca^{2+}	3	0.075	75
Mg^{2+}	2	0.050	30
Na^+	3	0.075	200
K^+	3	0.075	–
F^-	5	0.125	1.0
Cl^-	4	0.100	250
NO_3^-	3	0.075	45
SO_4^-	2	0.050	200
HCO_3^-	1	0.025	500
	$\sum W_i = 40$	**$\sum w_i = 1$**	

[a]All parameters are expressed in mg/l, except pH

Secondly, Unit weight (W_i) for each parameter is calculated by using the following equation

$$w_i = \frac{K}{S_i} \tag{23.1}$$

where
K = proportionality constant $= \frac{1}{\sum_{i=1}^{n} S_i}$
S_i = recommended standard value of ith parameter.
Calculated (W_i) values of each parameter are given in Table 23.1.

In the third step, a quality rating scale (Q_i) for each parameter is assigned by dividing its concentration in each water sample by its respective standard according to the guidelines laid down in the BIS 10500 (2003).

$$Q_i = 100 * \frac{(V_i - V_o)}{(S_i - V_o)} \tag{23.2}$$

where
v_i = estimated concentration of ith parameter in the analyzed water.
v_o = ideal value of this parameter in pure water.
$v_o = 0$ (except for pH where $v_o = 7.0$).
WQI is then calculated as per the following equation

$$WQI = \frac{\sum_i^n W_i Q_i}{\sum_i^n W_i} \tag{23.3}$$

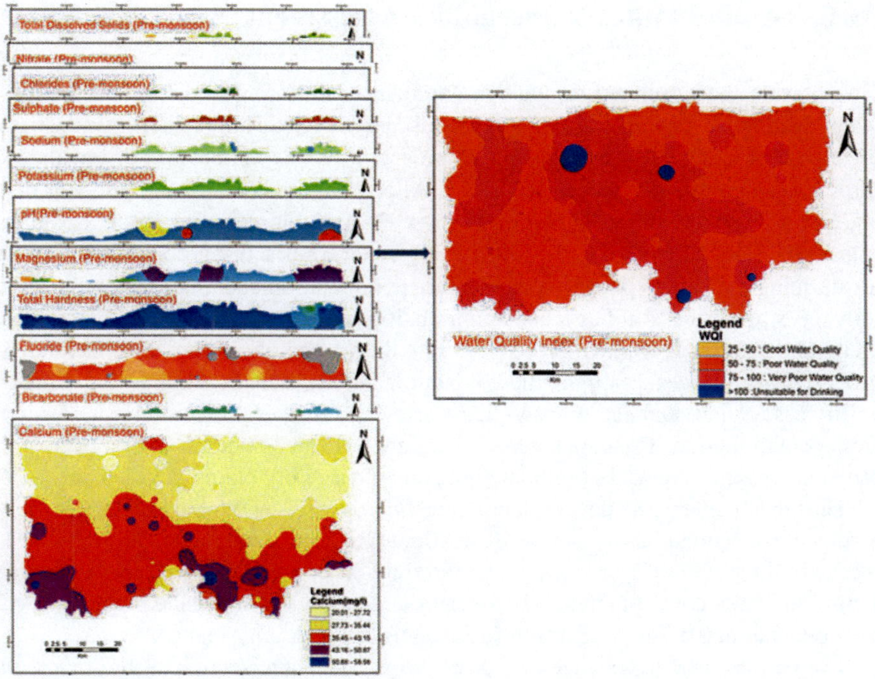

Fig. 23.4 Water quality index map of Ludhiana, Punjab (Singh et al. 2019)

Water Quality values are obtained in GIS environment after performing the above mention calculations; ranged from 49.9 to 150.13. These values are classified into five categories that is Excellent water quality (0–25), Good water quality (25–50), Poor Water Quality (50–75), Very Poor Quality (75–100) and Unsuitable for drinking (Above 100) [38] (Fig. 23.4).

The map clearly indicates that 58.6% of the study area falls under the category of poor water quality, whereas water of 40.4% of the study area is unsuitable for drinking purposes. The groundwater in the entire region has been found to be very hard. In addition, parameters like total dissolved solids, fluoride, magnesium and nitrate exceed the permissible limit as recommended by the BIS. Thus, the results obtained by the geospatial techniques give clear picture of groundwater quality scenario in the Ludhiana region and the technique can be helpful in developing proper strategies for controlling and managing the quality of water so that it caters to the need of smart cities.

23.5 Ground Water Vulnerability Assessment

Groundwater gets polluted through contaminants infiltrating into soil and groundwater from waste disposals in dump yards, discharge of toxic effluents from industries or release of chemicals through agriculture activities like application of fertilizers, insecticides, pesticides, weedicides, etc. (Aller et al. 1987; Connell and Van den Daele 2003; Foster 1987; Palmer and Lewis 1998; Schlosser et al. 2002; Secunda et al. 1998; Van Stempvoort et al. 1992). Wastewater volume that is contaminated is at an alarming rise due to phenomenal increase in population and industrial waste. Not only this wastewater is unfit for consumption, it can mix with other water sources and contaminate them as well. Upon seepage into the ground, it contaminates the underground water sources. Mapping these contaminants can help in groundwater vulnerability assessment and can manage this wastewater further protecting groundwater from contamination. The data layers and parameters can be unified and modified for potential groundwater pollution mapping and vulnerability classification using GIS.

The advancement and development in technology has facilitated planning, monitoring and decision making. As per the National Research Council, 1993 guidelines, there are three methods for comparative evaluation of areas related to the potential for groundwater contamination, viz., overlay and index, statistical and process-based methods that make use of deterministic models based on physical processes.

The overlay and index methods take into account intersection of the maps on regional basis and qualitative interpretation of data by indexing the parameters and assigning suitable weights or numerical indices for each attribute with physical and climatic attributes. The DRASTIC system, developed by the United States Environmental Protection Agency, falls under the index category and helps in the evaluation of groundwater vulnerability index for a variety of land areas. It may not be the absolute measure of vulnerability, but it definitely serves as one of the criteria facilitating decision making (Aller et al. 1987; Secunda et al. 1998; Yin et al. 2013). GOD method developed by Europe (Foster 1987) is equivalent to DRASTIC takes into account three parameters, i.e. overall aquifer class, groundwater occurrence and depth to water table. The Aquifer Vulnerability Index (AVI) method (Van Stempvoort et al. 1992) developed in Canada takes into account the vertical hydraulic gradient and depth and hydraulic conductivity of each sedimentary layer above the groundwater level. According to the UK vulnerability system (Schlosser et al. 2002), nature of the aquifer, presence or absence of drift and soil type are the three important components of groundwater vulnerability.

The deterministic methods for assessing the vulnerability of groundwater require good geochemical and hydro-geologic database coverage. As a result, mathematical models are used for simulation of complex phenomena of flow and contaminant transport in the subsurface.

The statistical methods handle a large number of extensive databases and help in identifying priority pollutant for monitoring soil and groundwater remediation projects. The monitoring data coupled with the organic carbon partition coefficient and half-life directly helps in calculating the vulnerability to groundwater.

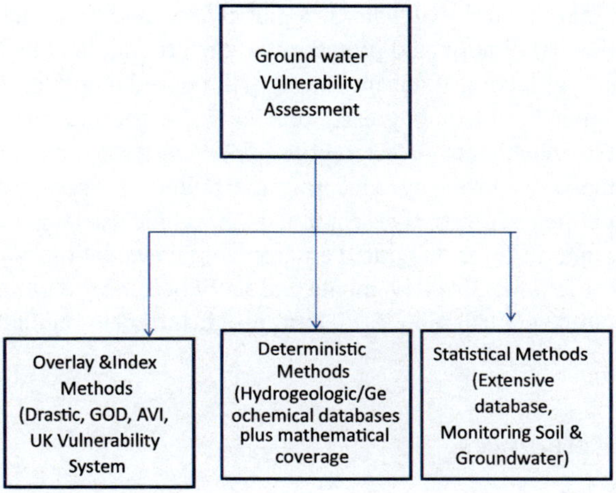

Fig. 23.5 Flowchart showing groundwater vulnerability assessment

For the estimation of expected risk to groundwater contamination on regional scale, various elements of the index methods can be combined with other information such as LULC and contaminant loading (Fig. 23.5).

23.6 Summary

The quality and quantity of groundwater resources are the issues of major concern across the globe. The cause and extent of groundwater pollution can be determined using geospatial technologies. The hydrogeological GIS database and groundwater modelling offer immense potential for groundwater analytics of the smart cities. The advanced database supported by GIS facilitates data verification and validation. Remote sensing and GIS are also essential for process-based models, hydraulic head maps, maps of statistical data and pumping rate allocations which are needed for allowing the view of the aquifer behaviour and stress factors involved. The connection between the aquifer depth, lithology, groundwater hydro-chemical and land-use is easily made with the help of recorded data and statistical procedures implemented by the GIS software. Not to forget, the aquifer vulnerability studies can be performed using the existing spatial database. Nevertheless, the importance of public awareness regarding judicious use of groundwater resources and saving them from contamination is equally important. It becomes imperative that the technical and administrative team works in collaboration to make people aware. To ensure the city accomplish the goal of being a smart city. Social awareness of the judicious use of natural resources, analyzing the water requirement versus availability and promoting rainwater harvesting is an important aspect in the smart cities (Chowdhury et al. 2003; Connell and Van den Daele 2003; Isa et al. 2014; Ramachandra 2006; Tait et al.

2004). Scarcity of water is a global issue and necessitates more vigilant approach in the exploitation and protection of the groundwater resources.

All technical advancements will be useless, unless the protection of the groundwater resources is greatly related to the consciousness of the local authorities and the inhabitants in the region. These measures are required for the protection of these resources by educating the public at large towards this burning issue. The public administrative sector and the technical sector should collaborate for preparing guidelines for integrated environmental evaluation.

Technical achievements can be only effective with the local authorities and the public contributing. Still, currently insufficient regulations should also be revised.

References

Ahada CPS, Suthar S (2018) Assessing groundwater hydrochemistry of Malwa Punjab, India. Arab J Geosci 11(2):17

Aller L, Bennett T, Lehr JH, Petty RJ, Hackett G (1987) DRASTIC: a standardized system for evaluating groundwater pollution potential using hydrogeologic settings. US EPA, Robert S. Kerr Environmental Research Laboratory, Ada, OK. EPA/600/287/035, Ada, OK. https://nepis.epa.gov/Exe/ZyNET.exe/20007KU4.txt?ZyActionD=ZyDocument&Client=EPA&Index=1986Thru1990&Docs=&Query=&Time=&EndTime=&SearchMethod=1&TocRestrict=n&Toc=&TocEntry=&QField=&QFieldYear=&QFieldMonth=&QFieldDay=&UseQField=&IntQFieldOp=0&ExtQFieldOp=

B I S (2012) Indian standard drinking water–specification (second revision). In: Bureau of Indian Standards (BIS), New Delhi

Becker MW (2006) Potential for satellite remote sensing of ground water. Ground Water 44(2):306–318. https://doi.org/10.1111/j.1745-6584.2005.00123.x

Brown RM, Mccleiland NJ, Deiniger RA, O'Connor MF (1972) Water quality index-crossing the physical barrier (Jenkis SH, ed). Proc Intl Conf Water Poll Res Jerusalem 6:787–797

Brown R, McClelland N, Deninger R, Tozer R (1970) A water quality index: do we dare? Water Sewage Works. A Water Quality Index—Crashing the Psychological Barrier. In Indicators of Environmental Quality 117:339–343

Brunner P, Franssen H-JH, Kgotlhang L, Bauer-Gottwein P, Kinzelbach W (2007) How can remote sensing contribute in groundwater modeling? Hydrogeol J 15(1):5–18

Chowdhury A, Jha MK, Machiwal D (2003) Application of remote sensing and GIS In. In: Ground water pollution: proceedings of the international conference on water and environment (WE-2003), December 15–18, 2003, Bhopal, India, vol 3, p 39

Christodoulidou M, Charalambous C, Aletrari M, Kanari PN, Petronda A, Ward NI (2012) Arsenic concentrations in groundwaters of Cyprus. J Hydrol 468–469:94–100

Connell LD, Van den Daele G (2003) A quantitative approach to aquifer vulnerability mapping. J Hydrol 276(1–4):71–88

Elumalai V, Brindha K, Lakshmanan E (2017) Human exposure risk assessment due to heavy metals in groundwater by pollution index and multivariate statistical methods: a case study from South Africa. Water 9(4):234

Foster SSD (1987) Fundamental concepts in aquifer vulnerability, pollution risk and protection strategy. In: Duijvenbooden W, Waegeningh HG (eds), Vulnerability of soil and groundwater to pollutants. TNO Committee on Hydrological Research. The Hague, Proc Info, vol 38, pp 69–86

Foster SSD (1998) Groundwater recharge and pollution vulnerability of British aquifers: a critical overview. Geolog Soc London, Special Publications 130(1):7–22

Gogu R, Carabin G, Hallet V, Peters V, Dassargues A (2001) GIS-based hydrogeological databases and groundwater modelling. Hydrogeol J 9(6):555–569

Goher ME, Hassan AM, Abdel-Moniem IA, Fahmy AH, El-sayed SM (2014) Evaluation of surface water quality and heavy metal indices of Ismailia Canal, Nile River, Egypt. Egypt J Aquat Res 40(3):225–233

Gumbricht T, McCarthy TS, Bauer P (2005) The micro-topography of the wetlands of the Okavango Delta, Botswana. Earth Surf Process Landforms J Brit Geomorphol Res Group 30(1):27–39

Hammouri N, El-Naqa A, Barakat M (2012) An integrated approach to groundwater exploration using remote sensing and geographic information system. J Water Resour Prot 04(09):717–724. https://doi.org/10.4236/jwarp.2012.49081

Haridas VR, Aravindan S, Girish G (1998) Remote sensing and its applications for groundwater favourable area identification. Quart J GARC 6(6):18–22

Horton RK (1965) An index number system for rating water quality. J Water Pollut Control Fed 37(3):300–306

Hundal HS, Singh K, Singh D (2009) Arsenic content in ground and canal waters of Punjab, North-West India. Environ Monit Assess 154(1–4):393

Isa NM, Aris AZ, Lim WY, Sulaiman WNAW, Praveena SM (2014) Evaluation of heavy metal contamination in groundwater samples from Kapas Island, Terengganu, Malaysia. Arab J Geosci 7(3):1087–1100

Jasmin I, Mallikarjuna P (2014) Physicochemical quality evaluation of groundwater and development of drinking water quality index for Araniar River Basin, Tamil Nadu, India. Environ Monitor Assess 186(2):935–948

Jawad TA-B, Yahya YA-J (2013) Application of GIS and remote sensing to groundwater exploration in Al-Wala Basin in Jordan. J Water Res Protect 5(10):962–971

Jensen JR (2000) Remote sensing of the environment an earth resource perspective Prentice Hall. Pearson Education India

Kaur T, Bhardwaj R, Arora S (2017) Assessment of groundwater quality for drinking and irrigation purposes using hydrochemical studies in Malwa region, southwestern part of Punjab, India. Appl Water Sci 7(6):3301–3316

Lattman LH (1958) Technique of mapping geologic fracture traces and lineaments on aerial photographs. Photogramm Eng 24(4):568–576

Margane A (2003) Guideline for preparation of groundwater vulnerability maps and risk assessment for the susceptibility of groundwater resources to pollution

Mayilvaganan MK, Mohana P, Naidu KB (2011) Delineating groundwater potential zones in Thurinjapuram watershed using geospatial techniques. Indian J Sci Technol 4(11):1470–1476

Meijerink AMJ (1996) Remote sensing applications to hydrology: groundwater. Hydrol Sci J 41(4):549–561

Mishra RC, Chandrasekhar B, Naik RD (2010) Remote sensing and GIS for groundwater mapping and identification of artificial recharge sites. In: She QH Shui-Long (eds) Geoenvironmental engineering and geotechnics: progress in modeling and applications, pp 216–223

Molden D, Frenken K, Barker R, de Fraiture C, Mati B, Svendsen M, Sadoff C, Finlayson CM (2007) Trends in water and agricultural development in water for food, water for life: a comprehensive assessment of water management in agriculture (Molden D, ed)

Mukundan R, Radcliffe DE, Ritchie JC (2011) Channel stability and sediment source assessment in streams draining a Piedmont watershed in Georgia, USA. Hydrol Process 25(8):1243–1253

Nagarajan M, Singh S (2009) Assessment of groundwater potential zones using GIS technique. J Indian Soc Remote Sens 37(1):69–77

Palmer RC, Lewis MA (1998) Assessment of groundwater vulnerability in England and Wales. Geolog Soc London, Special Publications 130(1):191–198

Rainey MP, Tyler AN, Gilvear DJ, Bryant RG, McDonald P (2003) Mapping intertidal estuarine sediment grain size distributions through airborne remote sensing. Remote Sens Environ 86(4):480–490. https://doi.org/10.1016/S0034-4257(03)00126-3

Ramachandra TV (2006) Soil and groundwater pollution from agricultural activities. The Energy and Resources Institute (TERI)

Rao PJ, Harikrishna P, Srivastav SK, Satyanarayana PVV, Rao B, Deva V (2009) Selection of groundwater potential zones in and around Madhurawada Dome, Visakhapatnam District-A GIS approach. J Indian Geophys Union 13(4):191–200

Rodriguez E, Morris CS, Belz JE, Chapin EC, Martin JM, Daffer W, Hensley S (2006) An assessment of the SRTM topographic products. Photogramm Eng Remote Sens 72(3):249–260

Rosema A, Fiselier JL (1990) Meteosat-based evapotranspiration and thermal inertia mapping for monitoring transgression in the Lake Chad region and Niger Delta. Int J Remote Sens 11(5):741–752

Schlosser SA, McCray JE, Murray KE, Austin B (2002) A subregional-scale method to assess aquifer vulnerability to pesticides. Groundwater 40(4):361–367

Secunda S, Collin ML, Melloul AJ (1998) Groundwater vulnerability assessment using a composite model combining DRASTIC with extensive agricultural land use in Israel's Sharon region. J Environ Manage 54(1):39–57

Singh CK, Shashtri S, Mukherjee S (2011) Integrating multivariate statistical analysis with GIS for geochemical assessment of groundwater quality in Shiwaliks of Punjab, India. Environ Earth Sci 62(7):1387–1405

Singh DD, Sahoo M, Sharma M, John S (2019) Geospatial analysis of groundwater quality in Ludhiana, Punjab (India). J Geo Environ Earth Sci Inter 20(3):1–12

Slater JA, Garvey G, Johnston C, Haase J, Heady B, Kroenung G, Little J (2006) The SRTM data "finishing" process and products. Photogramm Eng Remote Sens 72(3):237–247

Tait NG, Lerner DN, Smith JWN, Leharne SA (2004) Prioritisation of abstraction boreholes at risk from chlorinated solvent contamination on the UK Permo-Triassic Sandstone aquifer using a GIS. Sci Total Environ 319(1–3):77–98

Tam VT, De Smedt F, Batelaan O, Dassargues A (2004) Study on the relationship between lineaments and borehole specific capacity in a fractured and karstified limestone area in Vietnam. Hydrogeol J 12(6):662–673

Thenkabail PS (2015) Remote sensing handbook: remote sensing of water resources, disasters, and urban studies. In: Remote sensing of water resources, disasters, and urban studies. CRC Press. https://doi.org/10.1201/b19321

Tiwari AK, De Maio M, Singh PK, Singh AK (2016) Hydrogeochemical characterization and groundwater quality assessment in a coal mining area, India. Arab J Geosci 9(3):177

Tripathi R, Shyju K, Jasim HR (2017) Evaluation of ground water potential of Nallatangaal Odai using remote sensing and GIS techniques. Int J Appl Pure Sci Agri 3(7):72–80

Uprety K, Salman SMA (2011) Aspects juridiques du partage et de la gestion des eaux transfrontalières en Asie du Sud: Prévention des conflits et promotion de la coopération. Hydrol Sci J 56(4):641–661. https://doi.org/10.1080/02626667.2011.576252

Valstar JR, McLaughlin DB, te Stroet CBM, van Geer FC (2004) A representer-based inverse method for groundwater flow and transport applications. Water Resour Res 40(5):1–12

Van Stempvoort D, Ewert L, Wassenaar L (1992) AVI: a method for groundwater protection mapping in the prairie provinces of Canada. Prairie Provinces Water Board

Yin L, Zhang E, Wang X, Wenninger J, Dong J, Guo L, Huang J (2013) A GIS-based DRASTIC model for assessing groundwater vulnerability in the Ordos Plateau, China. Environ Earth Sci 69(1):171–185

Chapter 24
Status of Groundwater Water Quality in Bhilwara District of Rajasthan: A Geospatial Approach

Neha Pandey, Chilka Sharma, and M. P. Punia

Abstract Water on the earth is in abundance but its distribution is very much uneven on the land surface. Only 2% of the total water is available for use. Due to its distribution and quality, scarcity of portable water will be the major challenge at global level as most of the water available in surface reservoirs and groundwater are affected by various kind of contaminations of various sources. The situation is more aggravated in arid and semi-arid areas. Rajasthan state of India is located in arid & semi-climatic region with poor water quality. Same is the situation in the Bhilwara district located in the central part of the Rajasthan state where availability of water resource is very poor because of quality, quantity, and distribution issues. At the same time demand for potable water is increasing day by day for irrigation, industrial & domestic purposes. The present study is focused on spatial variability of groundwater quality for the Bhilwara district of Rajasthan, India using geospatial techniques. Four important water quality parameters that is Total dissolved solids, Chloride, Nitrate, and Fluoride (TDS, Cl, NO_3, and F) has been taken into consideration for assessment of water quality. Data on these parameters have been collected and classified with the standard parameter values as suggested by the BIS standards (ISI 10,500:2012). After data normalization appropriate weights have been given according to the contribution of individual parameter in water quality and a ground Water Quality Index (WQI) is generated. The scale of WQI is categorized into (1) Very Good, (2) Good, (3) Average, and (4) Poor. The analysis indicates that good water quality is associated with high water level, more thickness of alluvium, deep bedrock, more water-saturated strata, good groundwater recharge areas, nearness from the river, etc. The results are verified in the field at appropriate locations supported by interviews of local farmers. Status of water quality shows that the 24.65 and 20.18% area of district cover by the "Very

N. Pandey (✉) · M. P. Punia
Department of Remote Sensing, Birla Institute of Scientific Research (B.I.S.R), Jaipur, Rajasthan, India
e-mail: pandeyneha2014@gmail.com

M. P. Punia
e-mail: punia.rsd@gmail.com

C. Sharma
School of Earth Sciences, Banasthali Vidyapith, Niwai, Rajasthan, India
e-mail: chilkasharma@gmail.com

P. Sharma (ed.), *Geospatial Technology and Smart Cities*, The Urban Book Series,
https://doi.org/10.1007/978-3-030-71945-6_24

Good" and "Good" quality of water and 33.72% area show the "Average" quality of water while the 21.45% of area is covered by the "Poor" quality of water.

Keywords Groundwater quality · GIS · ISI standards · Ordinary kriging · Water Quality Index (WQI)

Acronyms

BOD	Biochemical Oxygen Demand
CGWB	Central Ground Water Board
COD	Chemical Oxygen Demand
CCME WQI	Canadian Council of Ministers of the Environment Water Quality Index
DO	Dissolved Oxygen
EC	Electrical Conductivity
GPS	Global Positioning System
IDW	Interpolated Distance Weightage
NSFWQI	National Sanitation Foundation Water Quality Index
NRSC	National Remote Sensing Centre
OWQI	Oregon Water Quality Index
TH	Total Hardness
TDS	Total Dissolved Solids
WQI	Water Quality Index
WHO	World Health Organization

24.1 Introduction

Water is the primary need of human beings to live on the earth. It is valuable for agriculture production, wildlife, industrial production and maintaining ecosystem, etc. In the future, water scarcity will be one of the biggest challenges in the world. In general, water bodies near population areas are polluted with organic and hazardous pollutants. Rajasthan has arid/semi-arid climate with hot summers. There is big issue of water security across the world for various purposes specifically in arid and semi-arid regions. The rising population and increase in urbanization have lead tremendous pressure on water resources of state. The problem is very acute in maximum area of Rajasthan state where the groundwater resources are being steadily depleted. Many districts of Rajasthan are greatly affected due to nitrate contamination of groundwater.

Numerous dental and skeletal diseases are reported due to increased level of nitrate and fluoride concentration in drinking water nearly in the entire state of Rajasthan (Munoth et al. 2015). The water used is being stored in the water bodies like pond, reservoir, etc. which were specifically designed for the purpose. With the increase of population, the demand for water has increased considerably and in order to fulfill the water requirements, exploitation of groundwater has also increased which drastically lowered the groundwater table in most of the state. The government of Rajasthan has launched many state-level projects and developed alternatives water supply schemes for stabilizing the depleted water condition and providing water to different sectors as per their required demand. But still increased water demand and other secondary parameters like seepage, evaporation, sedimentation, etc. lead to lowering water level of reservoirs. Since the state has higher fluoride and nitrate level in ground water, this becomes a necessity to analyze and study the water quality condition in concerned areas. Many districts in Rajasthan have industrial setup which act as a catalyst and further degrade and deplete the groundwater quality.

Bhilwara district is known as textile city of India. It is famous for textile & minerals industries which are located in the central part of Rajasthan falling in semi-arid zone. It is famous for textile and minerals industries with industrial growth water demand is also increasing for drinking, industrial, and other purposes. According to the Government report (Central Ground Water Report 2013), Meja Dam is the main water provider for drinking purposes in Bhilwara district. Entire blocks of Bhilwara district are facing the problem of water scarcity. Groundwater draft is very high and water level is depleting day by day. As water level goes deeper, water quality is deforioraling. Mining and industries have adverse effect on both surface and groundwater quality. Water quality has been assessed on the basis of concentration of major parameters that is TDS, Cl, F, and NO_3.

Bhilwara district is also facing the shortage of drinking water from last decade because the mining and industries have adverse effect on both ground and surface water. As per the rate of population and industrial growth, the drinking water demand of Bhilwara district is increased but the available water resources, both ground and surface water are not adequate enough to fulfill the demand of drinking, irrigation, and industry due to the regular scarcity of water and the district is also facing the problem of industrial growth due to the water crisis.

In this study, the geo-statistical analyst tool in ArcGIS10.4 has been applied for analyzing spatial distribution of water quality parameters.

Prospects related to underground water can be efficiently studied with the emerging application of remote sensing and GIS techniques (Hussain et al. 2013). With the help of GIS, spatial analysis of large volume of data and its integration and correlation can be done effectively (Shankar et al. 2010). Kriging method of interpolation is used to analyze the spatial distribution of water quality data of the district, which gives quality of water at any geographical locations of the study area using the Water Quality Index (WQI) . WQI is a beneficial and unique rating system to depict the overall groundwater quality status in a single term that is useful for the choosing of appropriate and correct treatment technique to meet the concerned issues related to groundwater.

24.2 Literature Context

In this study, we have gone through the large number of previous works from various national journals, books, and published reports and we found that Bhilwara district experiences downfall in both quality and quantity of underground water.

In recent years contamination of groundwater has become major concern in the study area. There are number of researches have been done related to water quality of Bhilwara district (Meena et al. 2016; Munoth et al. 2015). Hydro-chemical characteristics of underground water of Jahazpur tehsil of Bhilwara district were determined with the specific parameters like pH, TDS, NO_3, etc. using water quality index (Meena et al. 2016).

Grographic Information System (GIS) -based groundwater quality study have been carried out through the collected water sample data (Calcium, Sodium, Potassium, Magnesium, bicarbonates, chlorides, and sulfates) for the sub-basin of Paravanar River, Tamil Nadu. Its spatial distribution of water quality parameters is done through the ArcGIS and categorization of different classes which is based on ISI and WHO's standards (Shankar et al. 2010). Assessment of temporal and spatial behavior of groundwater quality for the Lalitpur district of UP state has been done through the Interpolated Distance Weightage (IDW) interpolation method using various water quality parameters like TDS, pH, TH, Cl, F, and iron. Spatial distribution of all these parameters is done through ArcGIS 9.3 software. Different water quality indices like Weighted Arithmetic Water Quality Index Method and National Sanitation Foundation Water Quality Index (NSFWQI) employed various physico-chemical and biological parameters for examining the water quality (Schoups 2006a, b). On the basis of RMSE, VAR, COR, FB, FA2, it is concluded that the geo-statistical methods best fitted for the rainfall data (Pandey et al. 2016). WQI is used to determine the status of groundwater in Karaikal area of Tamil Nadu and Pondicherry areas (Sirajudeen et al. 2014).

24.3 Study Area

The Bhilwara district covers an area of 10,455 km^2. Geographically it is located in the central part of Rajasthan State. It is situated state between 25° 01′ and 25° 58′ north latitudes and 74° 01′ and 75° 28′ East longitude covered by Survey of India Toposheet No. 45 K and 45O.

Administratively, Bhilwara district belongs to Ajmer division and it is divided into four subdivisions, such as Bhilwara, Gulabpura, Mandalgarh, Shahpura, and 11 blocks (Bhilwara, Sahara, Raipur, Mandal, Banera, Shahpura, Jahazpur, Mandalgarh, Kotari, Asind, and Hurda). Location map of study area on IRS P6 LISS-IV Satellite image is presented in (Fig. 24.1).

Bhilwara is situated on an elevated plateau, and the several places district is intersected by Aravalli ranges. The northern part of district is covered by the open

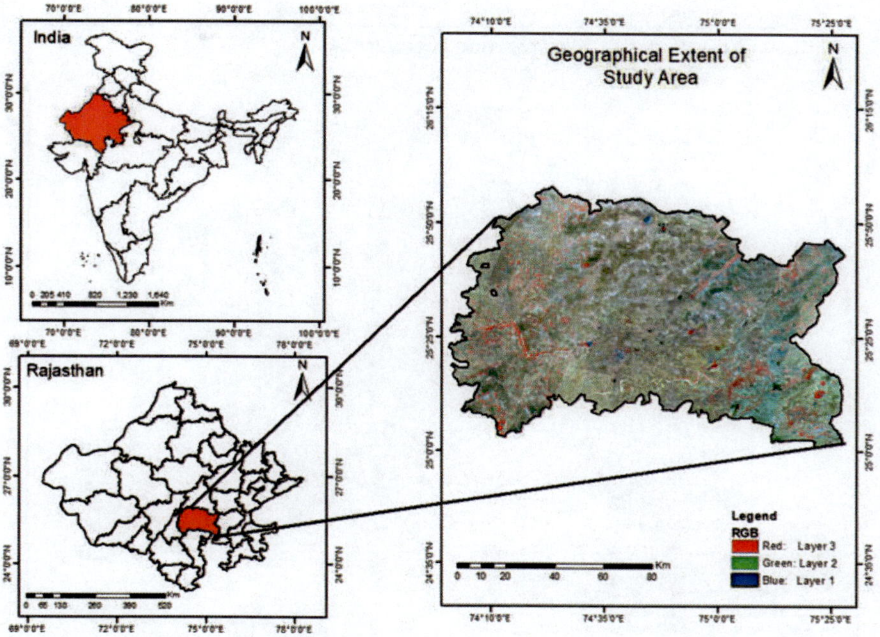

Fig. 24.1 Location of study area with satellite image

planes and the southeast part consists of few hillocks and undulating plains. Majority area of the district is covered through gently slope with an exception in west & northwest part of the district where the slope is high. Soil of the district varies from sandy loam to heavy loams. Different types of soil are found in the Bhilwara district like loam pebbly and stony soil, loam soil, Clay loam soil and black soil, etc. Weathered gneiss forms upper part of the bedrock in central part. Weathered gneiss with schist composition occupies most of the northern part under thin cover of alluvium soil. Bhilwara district is rectangular in shape except for its western portion which is comparatively broader than the eastern one. The major river of the Bhilwara district, Banas flows from northeast to east direction along with its tributaries namely Bedach, Kothari and Khari. Rainfall is the major source of water which is used for agriculture, domestic, and industrial purposes.

24.4 Data Requirements

In the research, high resolution satellite image (LISS-IV) is used for the detail information of study area which was procured from National Remote Sensing Centre (NRSC), Hyderabad, Water quality data (TDS, Cl, NO_3 & F) is collected from Central Ground Water Board (CGWB), Rajasthan which is used to assessment of water quality mapping of the study area.

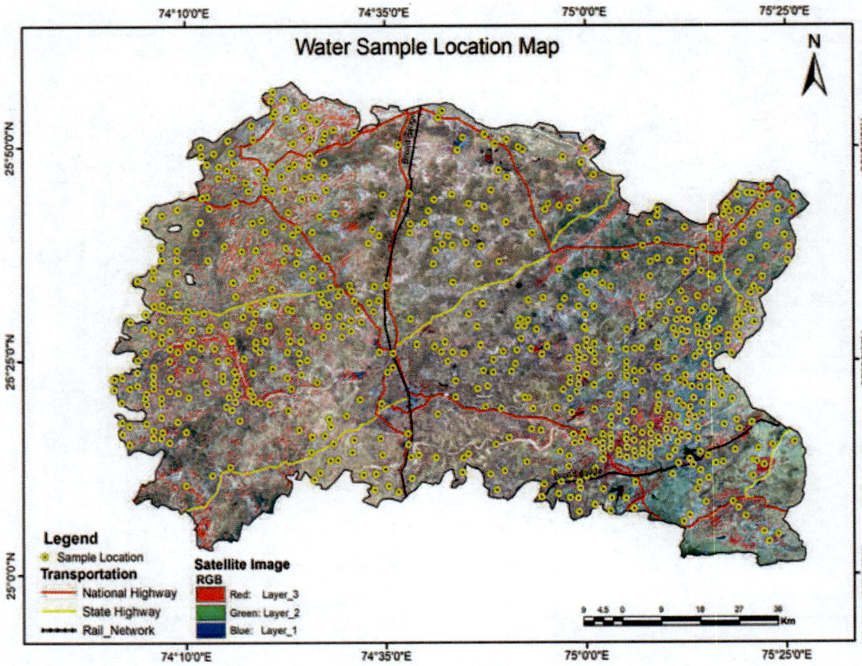

Fig. 24.2 Water sample location map of Bhilwara district

Figure 24.2 represents water sample location of the concerned area which is converted from tabular data to shapefile using ArcGIS 10.4. Figure 24.3 represented the field location map which is created through field survey using Global Positioning System (GPS). Road and railway networks are also presented with water sample location map in Fig. 24.2.

24.5 Materials and Method

Methodology adopted for this study is given below in the form of a flow chart which is presented in Fig. 24.4. The methodology is divided into three sections. First section includes the data collection and organization. Second section "Remote Sensing & GIS analysis" includes the thematic mapping and chemical parameters of water such as TDS, Cl, F and NO_3 as point data used to generated statistical layer through appropriate interpolation techniques. Third section elaborates the status of water quality of the study area.

Data Collection and Preparation

Boundary of the study area has prepared through survey of India topsheet of 1:50,000 scale in ArcGIS 10.4. High Spatial resolution satellite data LISS-IV (5.8 m spatial

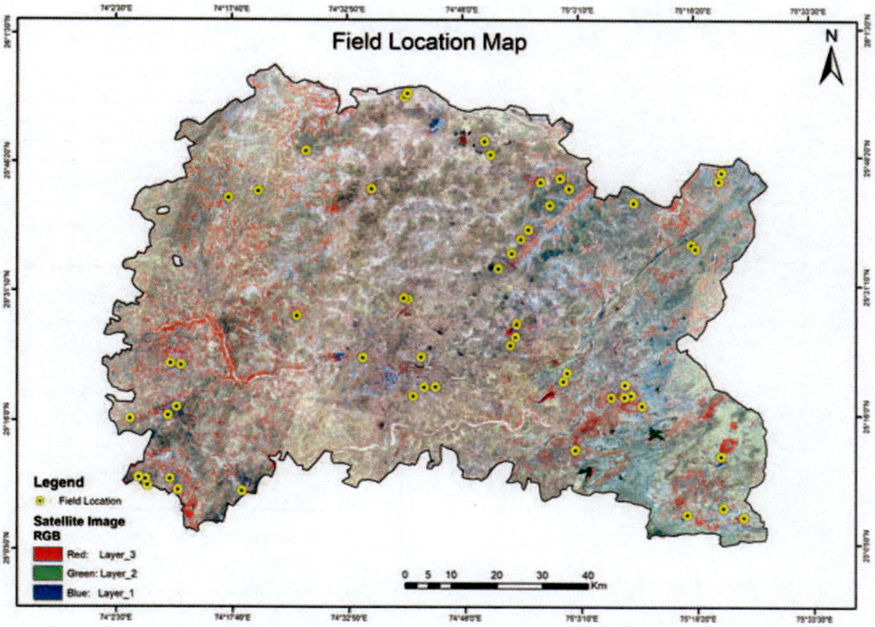

Fig. 24.3 Field location of Bhilwara district

resolution) which is used to prepare the layout map of concern area for 2014 year. Water Quality data is collected from water resource department, Rajasthan. Various samples data were collected in order to cover the different latitude & longitude of the concerned area.

Remote Sensing and GIS Analysis

Spatial distribution of TDS, Chloride, Fluoride, and Nitrate is generated using Ordinary Kriging interpolation method andthe layout map of TDS, Cl, F, and NO_3 is prepared in ArcGIS 10.4. The interpolation techniques were implemented in ESRI's ArcGIS software using Spatial Analyst Tool. This tool offers various interpolation techniques for generating surface grids from point data.

Normalization of Raster Data: In this study, raster data is normalized by using the ArcGIS 10.4 through raster calculator.

$$(X - X_s)/(X + X_s) \quad (24.1)$$

Here,

X is parameter sample value and

Xs is Desired value of a parameter X as per the Indian standard or WHO

Normalized raster surface value lies between −1 and 1. "Normalized raster data" is used to provide fixed upper and lower limits for contamination level of water quality parameters.

Fig. 24.4 Methodology for WQI

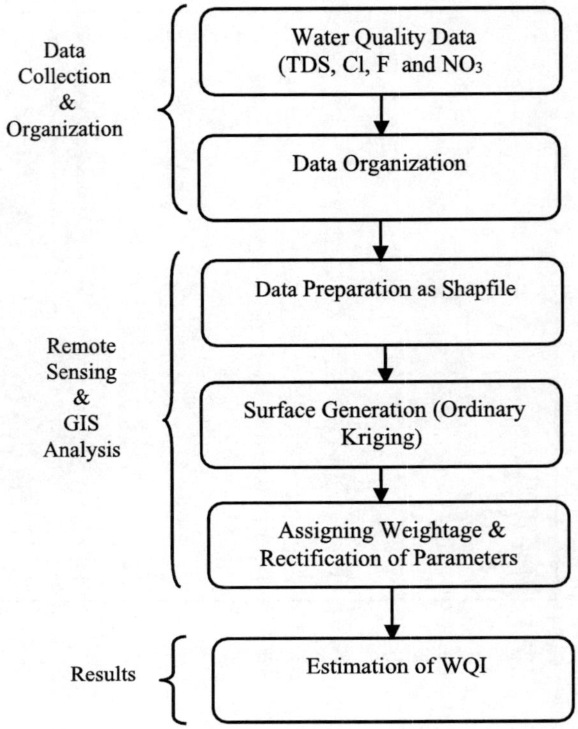

Calculation of Water Quality Index (WQI) :

In the first step, relative weight (Wi) i calculated by using the following formula:

$$W_i = \frac{w_i}{\sum w_i}$$

Here,

W_i is the mean value of ith parameter.

In the second step, Quality Rating Scale for ith parameter (Q_i) is calculated using the following formula:

$$Q_i = \frac{(V_i - V_0) \times 100}{(S_i - V_0)}$$

Here,

V_i is estimated concentration of ith parameter in the analyzed water.

V_O is the ideal value of this parameter in pure water

S_i is recommended standard value of ith parameter

In the third step, WQI is calculated using the following formula:

Table 24.1 ISI standard of water quality parameters

S. no	Parameters	Desirable (mg/l)	Permissible (mg/l)
1	TDS	500	2000
2	Chloride	250	1000
3	Fluoride	1.05	1.5
4	Nitrate	45	100

$$WQI = \frac{\sum Q_i W_i}{\sum W_i}$$

The observed range of WQI is 940.79 to 146.06 in Bhilwara District of Rajasthan.

WQI has been used by many researchers to determine the groundwater quality (Bazargan-Lari et al. 2009; Kumar et al. 2005; Johnson 2008; Michelle et al. 2006; Noori et al. 2014).

In the study, the range of groundwater quality parameter is calculated through the Indian Standard Institution ISI Standards. According to Indian Standard Institution desirable limit of TDS, Cl, F, NO_3 is 500 (mg/L), 250 (mg/L), 1.05 (mg/L) and 45 (mg/L) respectively while permissible limit of TDS, Cl, F, NO_3 is 2000 (mg/L), 1000 (mg/L), 1.5 (mg/L), 100 (mg/L), respectively, which is presented in Table24.1.

The concentration of TDS, Chloride, Fluoride, and Nitrate in the concern area varies from range of 28.84 to 2024.58 (mg/L), 46.67 to 770.84 (mg/L), 0.38 to 3.12 (mg/L) and 6.01 to 87.87 (mg/L), respectively, and presented in Fig. 24.3a–d.

In the analysis, we break the range of water quality standard which is easily shown in Table 24.2. For TDS We take low value 28.84 whose contamination is less than desirable (500) and high value 2024.58 for greater than permissible limit (2000+) contamination value and medium value is taken between 28.84 and 2024.58. For the Cl, we take low value 46.67 whose contamination is less than 250 and high value 770.84 is near to the permissible limit (1000+) contamination value. For the F, we take low value 0.38 whose contamination is less than desirable (1.05) and high value 3.12 for greater than permissible limit (1.5+) contamination value and for the NO_3 we take low value 6.01 whose contamination is less than 45 and high value 87.87 is near to permissible limit (100+) contamination value, presented in Table 24.2.

In the study, the WQI is calculated through weighted values of individual parameter of groundwater quality. Water Quality Index (WQI) values lie between 4 and 15. In which 4 values show the minimum contamination in water or having good water

Table 24.2 Range of TDS, Cl, F, and NO_3

Level	TDS (mg/l)	Cl (mg/l)	F (mg/l)	NO_3(mg/l)	Status of water
Low = 1	<500	<250	<1	<45	Very good
Medium = 2	500–2000	250–1000	1–1.5	45–100	Average
High = 3	2000 +	1000 +	1.5 +	100 +	Poor

Table 24.3 Weights assigned to water quality parameters

Water quality weightage [1, 4 & 6]			
Parameters	Low	Medium	High
TDS	1	4	6
Chloride	1	2	3
Fluoride	1	2	3
Nitrate	1	2	3
Total	4 [very good]	10 [average]	15 [poor]

quality and 15 values represent the maximum contamination in water that shows the poor condition of water which is presented in Table 24.3.

24.6 Results and Discussion

In the study, the assessment of groundwater quality in the Bhilwara district using the four water quality parameter such as Total Dissolved Solids (TDS) , Chloride (Cl), Fluoride (F), and Nitrate (NO_3) and the weightage is given according to the major contribution of individual parameters. TDS have one of higher contamination value among the other parameters of water quality so we provide high weightage to these parameters which is shown in Table 24.3. Figure 24.5(a)–(d) shows the distribution of water quality in the study area.

In this study, the WQI is calculated through the raster calculator in ArcGIS 10.4. The WQI value 940.79 is maximum and the value 146.06 is minimum in the concerned area. The computed WQI shows that 24.65 and 20.18% of water falls in the "Very Good" and "good" water quality while 33.72% WQI shows the average water quality. On the other hand, 21.45% of groundwater comes in the "poor" category that indicates the groundwater is not suitable for the direct consumption presented in Table 24.4.

Figure 24.6 shows that south-eastern part of Bhilwara district is very good where all 4 parameters are within permissible limits because groundwater aquifer in this area is sandstone and limestone. Unlike to other parts of the state calcium carbonate beds are not available in this area. North-western and south-western parts have average groundwater quality on account of high discharge and high level of carbonates and bicarbonates in the upper strata. Large areas in northern and south-western parts have poor groundwater quality due to availability of calcium carbonate noodles near the surface and availability of high concentration fluoride, carbonates, and bicarbonates. The values of groundwater quality parameters are found high above permissible limits in some locations of the study area. The analysis shows that lower the value and higher will be the water quality & vice versa.

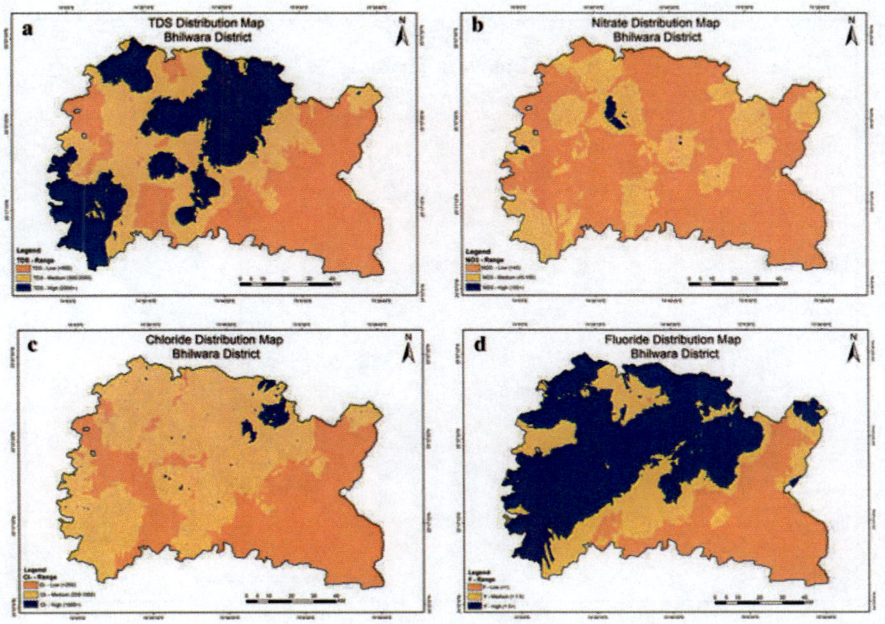

Fig. 24.5 **a** Spatial distribution of TDS. **b** Spatial distribution of nitrate. **c** Spatial distribution of chloride. **d** Spatial distribution of fluoride

Table 24.4 Status of water quality

S. no.	WQI	Level	Area (%)
1	Very good	4 & 5	24.65
2	Good	6 & 7	20.18
3	Average	8 to 10	33.72
4	Poor	11 to 15	21.45

24.7 Conclusion

Quality of groundwater is defined as very good, where all four parameters (TDS, Cl, F, and NO3) are under desirable limits. Quality is considered as average where all these parameters are found between desirable and permissible limits. When all these parameters are going beyond the permissible limits that water is considered to be poor for all purposes. The result shows that 24.65% area of eastern part of the district has "Very Good" quality of water which comes under the desirable limit of ISI standards. While only 21.45% area of the district has "Poor' condition of water. It means this much area beyond the permissible limit of ISI standards. The major part of the area having "Average" quality of water is 33.72% of total study area. Figure 24.6 shows that the southeast of the district has most favorable water quality while northeast & northwest part of district have average & poor quality of water.

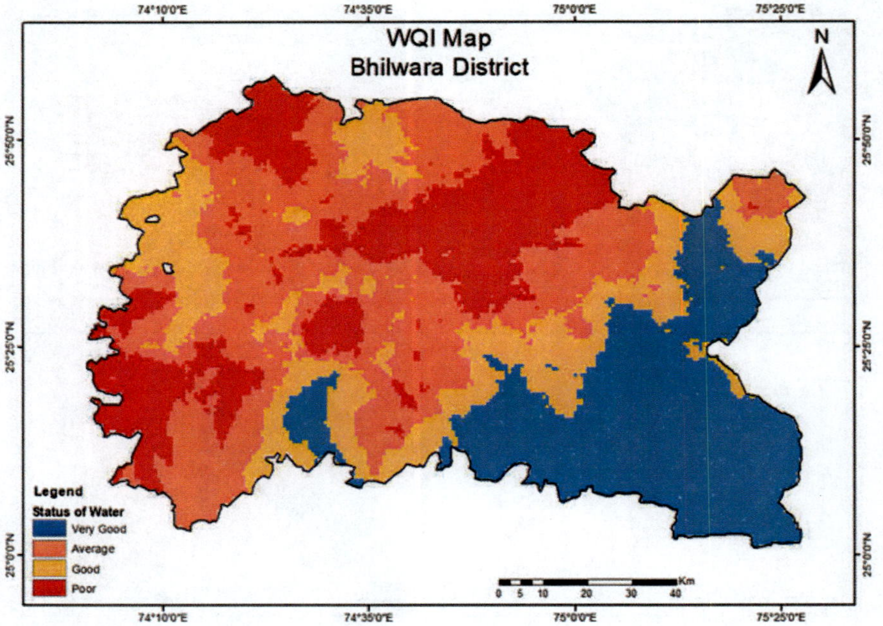

Fig. 24.6 WQI map

Acknowledgements The authors are thankful to Birla Institute of Scientific Research (BISR) Jaipur for providing the resources that enabled us to carry out this study. The author also acknowledge the help provided by School of Earth Sciences Banasthali Vidyapith.

References

Al-Adamat R (2017) Modelling surface water susceptibility to pollution using GIS. J Geogr Inf Syst 9(03):293

Bazargan-Lari MR et al (2009) A conflict-resolution model for the use of surface and groundwater resources that considers water-quality issues: a case study. Environ Manage 43(3):470

Bredehoeft JD, Young RA (1983) Conjunctive use of groundwater and surface water for irrigation agriculture: risk aversion. Water Resour Res 19(5):1111–1121

Chang N-B, Hong Y (2012) Multiscale hydrologic remote sensing: perspectives and applications. CRC Press, Taylor and Francis. ISBN No.: 1439877459

Chapagain AK et al (2006) The water footprint of cotton consumption: an assessment of the impact of worldwide consumption of cotton products on the water resources in the cotton producing countries. Ecol Econ 60(1):186–203

Chaudhary BS et al. (1996) Applications of remote sensing and geographic information systems in ground water (Investigations in Sohna Block, Gurgaon District (India). Int Arch Photogr Remote Sens 31:18–23

Chen Y et al. (2014) GIS and remote sensing in hydrology, water resources and environment. International Association of Hydrological Sciences Press, ISBN No.: 1901502724

Claude R et al. (2005) Remote sensing in northern hydrology: measuring environmental change. American Geophysical Union Press. ISBN No.: 0875904289
Drever JI (2005) Surface and ground water, weathering, and soils: treatise on geochemistry, vol 5. Elsevier
Elamassi KS (2012) Assessment of groundwater quality using multivariate and spatial analyses in Gaza governorate-Palestine. The Islamic University-Gaza, p 106
Engman ET, Gurney RJ (1991) Remote sensing in hydrology. Springer Press. ISBN No.: 9401066701
Fares A (2016) Emerging issues in groundwater resources. Springer Press. ISBN No.: 3319320068
Government of India Ministry of Water Resources Central Ground Water Board (2013) Groundwater Scenrio Bhilwara District, Meja Dam, Bhilwara, Western Region Jaipur
Government of India Ministry of Water Resources, River Development & Ganga Rejuvenation Central Ground Water Board (2017) Central Ground Water Board Western Region Jaipur
Ground Water Department (2013) Rajasthan "Hydrogeological Atlas of Rajasthan Bhilwara District
Gumindoga W et al (2018) Effect of landcover/land-use changes on water availability in and around Ruti Dam in Nyazvidzi catchment Zimbabwe. Water SA 44(1):136–145
Hussain J et al (2013) Fluoride contamination in groundwater of central Rajasthan, India and its toxicity in rural habitants. Toxicol Environ Chem 95(6):1048–1055
Indian Standard for Drinking Water-Specification (1991) IS 10500
Ji W (2007) Wetland and water resource modeling and assessment: a watershed perspective. CRC Press, Taylor and Francis. ISBN No.: 1420064142
Johnson LE (2008) Geographic information systems in water resources engineering. CRC Press, Taylor and Francis
Kaur R, Singh RV (2011) Assessment for different groundwater quality parameters for irrigation purposes in bikaner city, Rajasthan. J Appl Sci Environ Sanitation 6(3).
Khanna RK et al (2008) Solar still an appropriate technology for potable water need of remote villages of desert state of India Rajasthan. Desalination 220(1–3):645–653
Krishan G et al. (2016) Assessment of water quality index (WQI) of groundwater in Rajkot district, Gujarat, India. Earth Sci Clim Change 7(3)
Kumar R et al (2005) Water resources of India. Curr Sci 89(5):794–811
Al-Mamun et al. (2009) Improvement of existing water quality index in Selangor, Malaysia
Lumb A et al (2011) A review of genesis and evolution of water quality index (WQI) and some future directions. Water Qual Expo Health 3(1):11–24
Meena PL et al (2016) Determination of hydro-chemical characteristics of ground water for assessment of quality for drinking and other domestic purposes in North East Bhilwara, Rajasthan, India. Eur J Adv Eng Technol 3(9):20–27
Michelle K et al. (2006) Exploring water resources: GIS investigation for the earth sciences. ArcGIS Edition, Brooks Cole Press. ISBN No.: 0495115126
Mishra P et al. (2014) Assessment of groundwater quality of Lalitpur District, Uttar Pradesh using GIS mapping. IOSR J Environ Sci Toxicol Food Technol (IOSR-JESTFT) e-ISSN: 2319–2402, pp ISSN: 2319–2399.vol 8, Issue 12 Ver. IV
Mujumdar PP et al. (2001) Floods in a changing climate: hydrologic modeling. Cambridge University Press. ISBN No.: 1107033276
Munoth P et al (2015) Fluoride and nitrate groundwater contamination in Rajasthan. Hydro international, India, p 20
Musy A et al. (2014) Hydrology: a science for engineers. CRC Press, Taylor and Francis. ISBN No.: 1466590599
Nikam NG et al (2015) Optimal operation of multipurpose reservoir for irrigation planning with conjunctive use of surface and groundwater. J Water Resour Prot 7(8):636
Noori MJ et al (2014) Spatial estimation of rainfall distribution and its classification in Duhok governorate using GIS. J Water Resour Prot 6(2):75
O'Mara G (1988) Efficiency in irrigation: the conjunctive use of surface and groundwater resources. The World Bank

Pandey N et al. (2016) Analysis of spatial interpolation techniques for rainfall data using various methods: a case study of Bisalpur Catchment Area. In: NCACE-2016 conference proceedings, international journal of engineering research & technology (IJERT)

Peranginangin N et al (2004) Water accounting for conjunctive groundwater/surface water management: case of the Singkarak-Ombilin River basin, Indonesia. J Hydrol 292(1):1–22

Prashant K et al. (2016) Geospatial technology for water resources application. CRC Press, Taylor and Francis. ISSN No.: 9781498719681

Ramesh H et al. (2012) Conjunctive use of surface water and groundwater for sustainable water management. INTECH Open Access Publisher

Rathore MS et al (2005) Groundwater exploration and augmentation efforts in Rajasthan: A review. Institute of Development Studies, Jaipur

Rwanga SS, Ndambuki JM (2017) Approach to quantify groundwater recharge using gis based water balance model: a review

Schoups G et al. (2006) Reliable conjunctive use rules for sustainable irrigated agriculture and reservoir spill control. Water Resour Res 42(12)

Schoups G et al. (2006b) Reliable conjunctive use rules for sustainable irrigated agriculture and reservoir spill control. Water Res Res 42(12)

Schultz GA, Engman ET (2000) Remote sensing in hydrology and water management. Springer Press. ISBN No.: 3642640362

Sehgal KK et al (1960) Rajasthan district gazetteers Bhilwara. Government of Rajasthan, Jaipur, p 1975

Sekar I, Randhir TO (2007) Spatial assessment of conjunctive water harvesting potential in watershed systems. J Hydrol 334(1):39–52

Shah KA, Joshi GS (2017) Evaluation of water quality index for River Sabarmati, Gujarat, India. Appl Water Sci 7(3):1349–1358

Shankar K et al (2010) GIS based groundwater quality mapping in Paravanar river sub-basin, Tamil Nadu, India. Int J Geomatics Geosci 1(3):282

Sharma PK et al (2015) Characterization of groundwater quality of Tonk District, Rajasthan, India using factor analysis. Int J Environ Sci 6:4–2015

Sharma S, Chhipa RC (2016) Seasonal variations of ground water quality and its agglomerates by water quality index. Global J Environ Sci Manage 2(1):79–86

Shishodia A. (1994). Ph.D Thesis, economic and Environmental aspects of conjunctive water use in the Mahi Right Bank Command area, Gujarat.

Siddique-E-Akbor AHM et al (2014) Satellite precipitation data–driven hydrological modeling for water resources management in the Ganges, Brahmaputra, and Meghna Basins. Earth Interact 18(17):1–25

Sirajudeen J, Vahith AR (2014) Applications of water quality index for groundwater quality assessment on Tamil Nadu and Pondicherry, India. J Environ Res Develop 8(3):443

Subramanya K (2013) Engineering hydrology. Published by McGraw Hill Education (India) Private LimitedP-24. ISBN (13): 978-1-25902997-4

Subramoniam SR et al (2011) Estimating variograms of soil salinity properties by Kriging, remote sensing and GIS techniques in Indira Gandhi Nahar Pariyojana irrigation command of Rajasthan, India. Ann Arid Zone 50(2):93–97

Sukumar S, Sankar K (2011) Statistical study on pre & post monsoon variation of groundwater level in Theni district, Tamil Nadu, India. Int J Environ Sci 1(5):798

Thenkabail PS (2015) Remote sensing of water resources, disasters, and urban studies. CRC Press, Taylor and Francis. ISBN No.: 1482217910

Tyagi NK et al. (1995) Modelling conjunctive use of water resources: hydraulic and economic optimization

Tyagi S et al (2013) Water quality assessment in terms of water quality index. Am J Water Resour 1(3):34–38

Van Dijk A, Bos MG (2001) GIS and remote sensing techniques in land and water management. Springer Press. ISBN No.: 940106492X

Waikar ML, Nilawar AP (2014) Identification of groundwater potential zone using remote sensing and GIS technique. Int J Innov Res Sci Eng Technol 3(5):12163–12174
Wang M et al (2017) Comparison of spatial interpolation and regression analysis models for an estimation of monthly near surface air temperature in China. Remote Sensing 9(12):1278
Woessner WW (2000) Stream and fluvial plain ground water interactions: rescaling hydrogeologic thought. Ground Water 38(3):423–429
Yadav AK et al (2010) Water quality index assessment of groundwater in Todarai Singh Tehsil of Rajasthan State, India-A Greener Approach. J Chem 7(S1):S428–S432
Yang CS et al. (2004) Twelve different interpolation methods: A case study of Surfer 8.0. In: Proceedings of the XXth ISPRS congress, vol 35, pp 778–785

Links

https://www.indiawaterportal.org/
file:///C:/Users/Staff/Desktop/Bhilwara/Reports/GW%20senario_Bhilwara.pdf
https://plan.rajasthan.gov.in/statistics
https://bhuvan.nrsc.gov.in/bhuvan_links.php
https://imd.gov.in/Welcome%20To%20IMD/Welcome.php
https://cgwb.gov.in/District_Profile/Rajasthan/Bhilwara.pdf
https://www.censusindia.gov.in/2011census/dchb/0824_PART_B_DCHB%20_BHILWARA.pdf
https://www.nrsc.gov.in/Careers
https://www.crcpress.com/Groundwater-Quality-Sustainability/Maloszewski-Witczak-Malina/p/book/9780415698412

Chapter 25
Green Infrastructure as a Tool for Improving Livability of Area Based Development Projects Under Smart City Mission

Rama U. Pandey, Tapas Mitra, Mrunmayi Wadwekar, Jyotika Nigam, and Kriti Trivedi

Abstract Smart cities concept for sustainable urban development became prevalent in India with the introduction of the 'smart city mission' by the Government in 2015 to make cities more livable and sustainable at the same time. In light of ambitious missions such as 'Smart cities', many cities are aiming to improve livability to provide a better quality of life and day-to-day conveniences to the citizens. A methodological approach for consideration of 'benefits offered by green infrastructure' is required in spatial planning so as to better integrate nature-based solutions into the decision-making process. Many projects have been initiated under various categories. Some of the small-scale projects have already been implemented whereas large-scale projects such as 'Area Based Development' (ABD) are in the process of getting implemented. This paper is part of an ongoing project on 'Climate Informed Environmental Planning for smart cities'. The primary focus of the project is to integrate water bodies and green spaces in spatial planning to address the concerns related to climate change. An attempt has been made to understand the vulnerability of the city to climate change impacts and propose an alternative spatial planning approach using 'green infrastructure' as a nature-based solution for the ABD area under Smart City Mission. ArcGIS software was used extensively for both analyses of existing conditions of the site with respect to the broader context of the city as well as preparation of thematic maps. The primary survey was conducted to verify the ground conditions and computation of the carbon sequestration potential of the site. An interconnected network of green spaces and water bodies is proposed whereas built-up spaces as per the requirement of the mission are integrated into the plan as infill. The methodology adopted helps in addressing: first, the mitigation and adaptation to the impacts of climate change; second, elaborates the consideration of green

R. U. Pandey (✉) · T. Mitra · M. Wadwekar · K. Trivedi
School of Planning and Architecture, Bhopal, India
e-mail: ramapandey@spabhopal.ac.in

T. Mitra
e-mail: tapasmitra@spabhopal.ac.in

J. Nigam
Urban Design Consultant, Opus Design Studio, Bhopal, Madhya Pradesh, India

P. Sharma (ed.), *Geospatial Technology and Smart Cities*, The Urban Book Series,
https://doi.org/10.1007/978-3-030-71945-6_25

infrastructure as a tool for sustainable water management. It would help in not only addressing the mitigation and adaptation to impacts of climate change but would also improve the livability of upcoming development projects.

Keywords Smart cities · Green infrastructure · Area-based development · Improving livability · Adaptation

Acronyms

ABD Area-Based Development
BMC Bhopal Municipal Corporation
ICT Information & Communication Technologies
SDG Sustainable Development Goals
SUDS Sustainable Urban Drainage Systems
UHI Urban Heat Island

25.1 Introduction

The cities across the globe are being selected as the 'habitat choice' by people for obvious reasons. As a result, cities also become the hub for physical, social, economic, political, and cultural activities and thereby putting various resources under stress (White 2010). This increasing urbanisation is posing a threat not only to natural resources but also exacerbating the risks arising out of climate change. The Government of India has been trying to address these challenges witnessed by Indian cities through introducing various 'Urban Development' and 'Infrastructure Upliftment' missions. The aim of these missions is on 'sustainable economic development' and 'high quality of life' that can contribute significantly towards making up of cities livable and at the same time sustainable. The 'smart city mission' is one of the missions initiated by the Government in 2015 wherein many cities are aiming to provide a better quality of life and day-to-day conveniences to the citizens. The 'Mission Statement & Guidelines' of the Government of India for smart cities (Ministry of Urban Development 2015) apart from layers of smartness such as the use of Information & Communication Technologies (ICT) and good governance focuses on 'sustainable development of cities' in making of a smart city. The green infrastructure approach based on the concept of 'ecosystem services' offered by green spaces and water bodies can be adopted as a tool for achieving sustainability in the city development process. Cities are more and more shifting towards the instrumentalisation of nature for human benefit (Olofsdotter et al. 2013) and green infrastructure is one such approach based on concept of 'ecosystem services' (Millenium Ecosystem Assessment 2005) offered by green spaces and water bodies.

Green Infrastructure approach is not only supportive of ecological processes but also contributes significantly to the quality of life in any urban area (EPA 2017). Which in turn affects the general health and well-being of the city.

Seven out of 98 identified cities to be developed as smart cities by GoI are from Madhya Pradesh. All these cities offer immense potential for growth having a city population ranging between two million to one-fifth a million. The focus of the M.P. Government would be to identify economic potentials in these cities and propose developmental activities along with necessary infrastructure. A methodological approach for consideration of 'benefits offered by green infrastructure' is therefore required in spatial planning to address urban stressors and to integrate nature-based solutions into the decision-making process. Bhopal, the state capital of Madhya Pradesh and one of the selected cities under smart city mission having an area of approximately 400 Acres of high-value land within the city assigned for redevelopment known as Area-Based Development (ABD) is selected for the study. This study explores the possibilities of integrating water bodies and green spaces in spatial planning so as to conserve the ecologically sensitive areas and to address the concerns related to climate change. An alternate spatial planning approach using 'green infrastructure' as a nature-based solution has been proposed in the identified land for the ABD area under Smart City Mission for the city of Bhopal.

25.1.1 Green Infrastructure and Livability in Spatial Planning

Green infrastructure is defined as an interconnected network of protected land and water that supports native species, maintains natural ecological processes, prevents flooding, sustains air and water resources, and contributes to the health and quality of life of local communities (Plus Integreat Knowledge 2012). The components like corridors (both green and blue) and spaces (place or setting) that could be assembled so as to contribute to ecosystem health and human well-being forms green infrastructure (Austin 2014). In the context of climate change, the multi-functionality of green infrastructure has been increasingly taken into account these days. The green infrastructure enables preservation, restoration or creation of facilities that utilise natural processes to recycle stormwater, conserve energy and purify air in a way that encourages connectivity, supports the development and is environmentally and economically sustainable (Marić et al. 2015).

Pacione (1990) suggested livability to be a quality that is not an attribute inherent in the environment but is behaviour-related function of the interaction between environmental characteristics and personal characteristics. Veenhoven (1996) equated "livability' with 'habitability' and 'quality of life' in the nation. Livability is about the human requirement for social amenity, health, and well-being and includes both individual and community well-being as recorded by Newman (1999). The links between community issues and local environmental quality offer the potential for livability to move towards a more integrated understanding of the relationship between the physical environment and their underlying social context (Lyndhurst 2004).

The Earth Summit in 1992 brought focus to the idea of sustainability. The recent emphasis on sustainability with clearly defined Sustainable Development Goals (SDG's) in 2015 has increased the attention given to the quality of life. The sustainability concepts that involve the interrelationship of economic, social and environmental systems functioning in a community is about whether these systems are providing a healthy, productive and meaningful life for all community residents, present and future (Sustainable Seattle 2004). Lyndhurst (2004) suggests that at the micro-level, livability could potentially contribute towards sustainable development through the way in which local environmental problems are dealt with. The upcoming development projects under various Missions of GoI provide an opportunity to adopt the ecological approach that integrates the green infrastructure provisions in spatial planning so that livability contributes to achieving sustainability. The Green Infrastructure research of EPA (US Environmental Protection Agency) also highlights that introducing green infrastructure to supplement the existing grey infrastructure can promote urban livability and add to communities' bottom line (EPA 2017). The ecosystem services and products provided to humans through green infrastructure will only benefits human society if the ecosystem is maintained and will contribute to ecosystem health (Austin 2014).

The consequences of climate change can be resolved by making suitable changes in the developmental approach. Applying measures of green infrastructure can help in mitigating climate change impacts such as urban flooding and water scarcity (Christopher Coutts 2015; Per Bolund 1999). Green infrastructure approach gives several methods to establish green spaces in an urban setting (Jennifer Wolch 2014). Also, establishing infrastructure that improve the infiltration capacity of the area would improve the underground reservoir as well as decrease the pressure on the sewerage network (Martijin Kuller 2019). Rain being the main source of groundwater recharge, it is necessary to have the surface which can percolate water inside. Increasing impermeable areas in cities contribute to the surface runoff of rainwater. It is not that only paved surface contributes to runoff, but other types of surfaces such as rooftops and compacted soils also contribute to runoff (Hao Chen 2017). There is no surface that is known for cent percent percolation. It is also very important to look at the soil type, because it is one of the major parameters which affect runoff. Therefore, laying down guidelines and measures that encourage groundwater infiltration can be an effective tool.

25.2 Methodology

The study focuses on highlighting the importance of green infrastructure in making a 'redevelopment/new development' projects livable and sustainable. The city of Bhopal, capital of one of the states, Madhya Pradesh in India having little less than two million population and selected for developing as smart city under GoI 'Smart

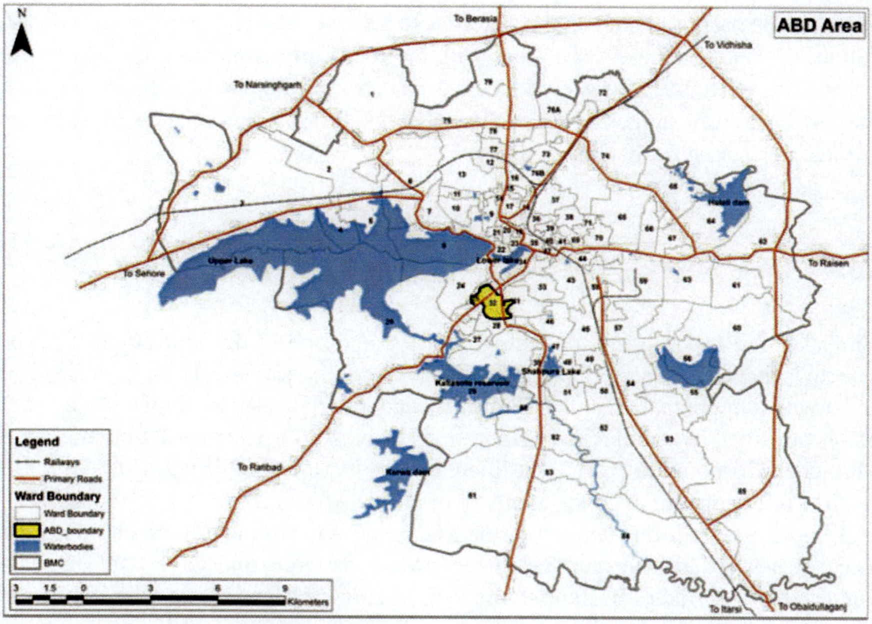

Fig. 25.1 Location of ABD area in Bhopal city

City Mission', has been selected for the study. The Area-Based Development site of 400 Acres as highlighted in Fig. 25.1, it is located centrally within the Bhopal city is the only ABD area in Madhya Pradesh state that focuses on the complete redevelopment of the high-value land. The city is quite cosmopolitan in its character and encompasses features that can be seen in a number of cities across India.

The quantum of built space and distribution of building use in the ABD site is based on design guidelines laid down by the Smart City Cell, Bhopal, further locations are identified in a way to provide optimum accessibility. For the ABD area, an alternative approach for spatial planning has been adopted which unlike the conventional planning processes gives due regard to the existing ecological processes and aims at conserving the local ecosystem, local flora and fauna throughout the design process.

The zoning has been done in such a manner that minimum ecological damage is done and also the natural balance is maintained, for example, conserving of trees within the ABD area on the basis of carbon sequestration within the area, interconnected network of water bodies and drainage channels conservation. Green infrastructure methods are thus adopted parallel to the physical infrastructural methods during the development process for the multidimensional enhancement of quality of life in the city. Surface water management is proposed through Sustainable Urban Drainage System (SUDS) and Rainwater harvesting mechanisms within the area. These mechanisms act as a solution against water quality, quantity and amenity aspects. The proposed plan, therefore, aims to achieve the sustainability and conservation of existing blue-green spaces with the focus on maintaining the ecosystem balance for the future. These guidelines for the area development or development of the city have been suggested for inclusion in the development statutes to make the city more sustainable, healthy and livable.

25.3 Case Study

Bhopal is the capital city of Madhya Pradesh and one of the first 20 cities to be selected under the smart city mission. The city is one of the 15 largest cities of India with a municipal area of 463 km^2 divided into 85 wards as shown in Fig. 25.2 and is governed by a municipal corporation (BMC 2017). The population in Bhopal Municipal Corporation (BMC) has increased from 10,80,802 in 1991 to 17,98,218 in 2011 (Government of India, Ministry of Home Affairs 2018).

Bhopal district forms the part of the Malwa plateau with undulating topography. The city lies in a hilly terrain that slopes towards the north and southeast. There are numerous water bodies in Bhopal, the major being the Upper Lake with the spread of 3100 ha, Lower Lake, Shahpura Lake and prominent dams like Kerwa reservoir, Halali reservoir and Kaliasote reservoir. The city also has lineaments indicating secondary porosity that helps in infiltrating the water into the ground.

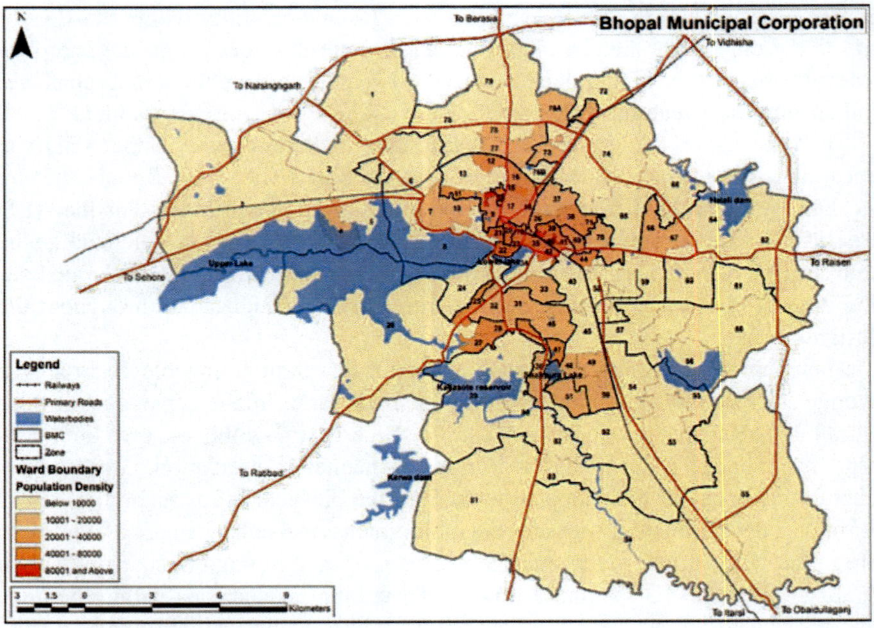

Fig. 25.2 Ward map of Bhopal with zones

In terms of hydrology, there are two major watersheds, part of the Yamuna basin which are further divided into 35 sub-watersheds. The major part of the city drains into the water bodies such as Upper Lake, Kaliasote reservoir, Halali reservoir and Kaliasote River. The watershed analysis was done using a rational method to identify the amount of runoff in cubic meters per second taking into consideration the peak hour rainfall intensity. Runoff evaluation indicates suitable areas having high runoff which can be developed using a green infrastructure approach, as it facilitates groundwater recharge and allow stormwater to soak into the ground as it would in a natural setting. This management approach has the potential to restore groundwater recharge rates and save money in groundwater resources when applying the practices to new development and redevelopment (Li Liu 2018; NDMA 2010; MoHUA 2019).

25.3.1 Vulnerability to Climate Change Impacts

The study area has a moderate climate with temperatures ranging from 10 to 40 °C. It receives an annual rainfall of 1123 mm. The climate in recent decades is witnessing variation due to climate change. Bhopal City has the unique characteristic of having a huge water body in the centre of the city affecting the heating pattern of the city. Therefore, in contrast to the Urban Heat Island (UHI) effect typically seen in urban areas, the temperature along the core of the city is cooler than the outskirts.

The city was analysed for variation in meteorological conditions to understand the climate change impacts in the ABD area. The Mann–Kendall non-parametric trend analysis for all the months from 1980 to 2016 predicts increasing temperatures for the city in near future. Bhopal has seen a significant change in its meteorological conditions over the years witnessing a severe increase in temperature in summer and a sharp drop in winter. The temperatures have been peaking more than average. This has led to hot and humid days across the city.

However, rainfall has become uncertain with a decrease in the overall rainfall throughout the year but an increase in the intensity of rainfall over a shorter number of days, thus creating stress on the city's stormwater and drainage infrastructure. This is a major cause of flash floods and infrastructure damage. The change in intensity and frequency of rainfall has led to localised flooding events across the city as witnessed in 2009, 2011 and 2016. The rainfall is also varied in its arrival pattern. The arrival of the monsoon is delayed consistently in the last decade.

It is evident that being centrally located within the Bhopal City the selected ABD area is vulnerable towards the impacts of climate change in the form of increasing average temperature and variations in intensity and frequency of rainfall. Therefore, Green infrastructure strategies are required as a measure for mitigation against climate change.

25.3.2 *Alternative Spatial Planning Approach Using Green Infrastructure*

In recent years, "ecosystem services" and "ecological networks" have emerged as key concepts in planning and development of urban areas. These services and relationships are being linked together in both academic literature and policy practice in what is termed as the Green Infrastructure approach. There is increasing recognition of the beneficial relationship between access to green space and improved public health and well-being of the city. The Green Infrastructure approach is not only supportive of ecological processes but also contributes significantly to the quality of life in any urban area. An alternative approach for planning is demonstrated with an emphasis on Green Infrastructure over other essential urban infrastructure. A conceptual framework recognising the importance of ecological processes for the development of the ABD area in Bhopal is discussed below.

The Proposed Plan takes into account the topography of the area and its ecosystem. Sustainable Urban Drainage System (SUDS) and retaining existing flora and fauna are major tools in achieving urban development goals. Aspects of the high land value of the ABD area and Transit Orient Development (TOD) ideas are significant aspects considered in the overall development agenda.

The ABD site is predominantly residential in nature with 46% percent residential share, 11% mixed-use development, about 7% commercial, 6.5% educational, 11% Open Space/Parks/Sports and 2% utilities. The area has two major arterial roads on site, i.e. Bhadbhada Road and New Market Road and two other sub-arterial roads touching the site on the southern and eastern end, i.e. Main Road 2 and Main Road 1, respectively.

The site has a low population density (0–140 pph). The urban morphology is loose with 76% unbuilt space. There are 68 different types of trees in the ABD area which are more or less spread all over the site with a comparatively higher concentration in the central and north-eastern part as documented through a primary survey (May 2018). The trees on site vary from a smaller girth of 0.1 m to as huge as 4.2 m. The site also has two major tributaries towards the lower lake and Shahpura Lake and one which further streams down towards Kerwa Dam.

The SUDS scheme will cover three main aspects of surface water management; water quantity, quality and amenity. The aims for sustainable development for housing, commercial, institutional areas including the provision of SUDS, are based on phase-wise strategic construction and water retention and management techniques on site. SUDS includes stormwater design features such as previous paving, soak ways, swales, infiltration trenches, filter strips, sand filters, bio-retention filters/areas and green roofs. It also engages in water harvesting systems, infiltration basins, detention basins, ponds and stormwater wetlands, silt removal devices, pipes and conduits, and subterranean storage systems. The incorporation of SUDS aims to balance environmental, social and economic requirements for site development to provide a sustainable, healthy, pleasurable environment for the community as well as adequate management of surface water drainage by attenuating excess stormwater flow to reduce the risk of flooding.

In the proposed ABD area, SUDS provides an integrated approach to surface water design problems in integrated approach using the techniques mentioned. The proposed master plan is designed with 30% ground coverage which enables us to collect rainwater through rainwater harvesting from buildings, through surface pond collections and groundwater recharge through the permeable surface. Also, with 70% of the site unbuilt, a significant volume of water can be stored within the site for non-potable as well as potable uses (after treatment). The surface water bodies cover about 113,781.58 m^2 within the site. The map in Fig. 25.3 shows how natural drainage channels have been retained on site and areas around them have been marked as construction-free zones. These zones form the primary location for surface ponds and underground water sumps to be interconnected for integrated water collection as demonstrated in Fig. 25.4. During heavy rains, the trickling effect will fill the ponds at different levels and flow into sumps and water treatment plants. Any excess water

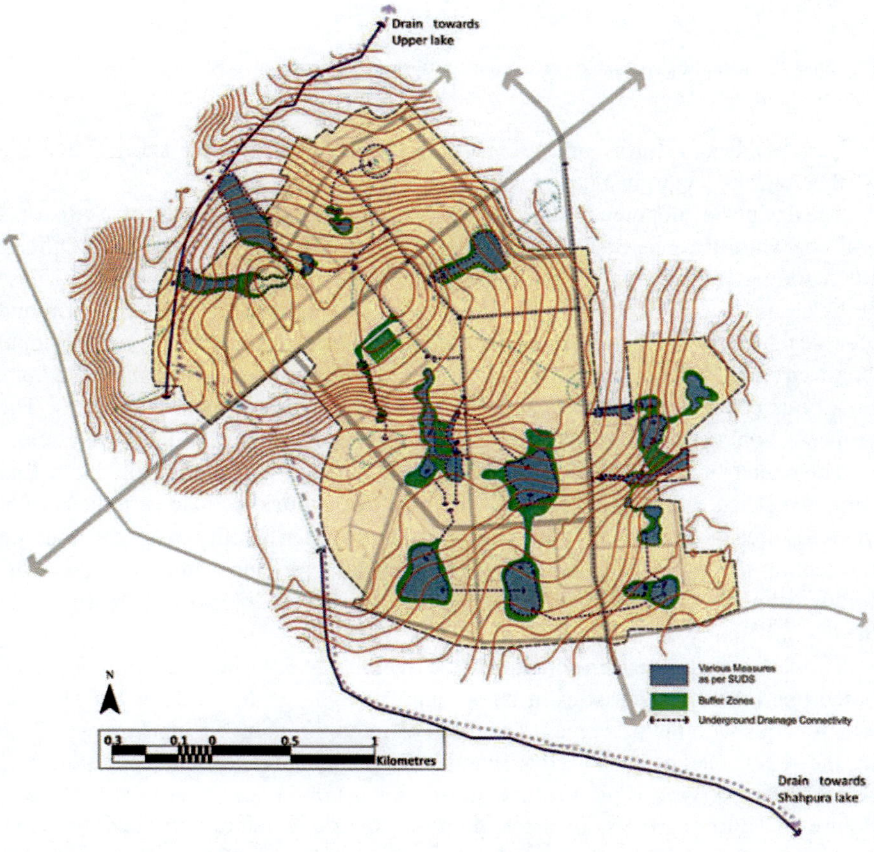

Fig. 25.3 Map showing the adoption of sustainable drainage systems in the proposed ABD

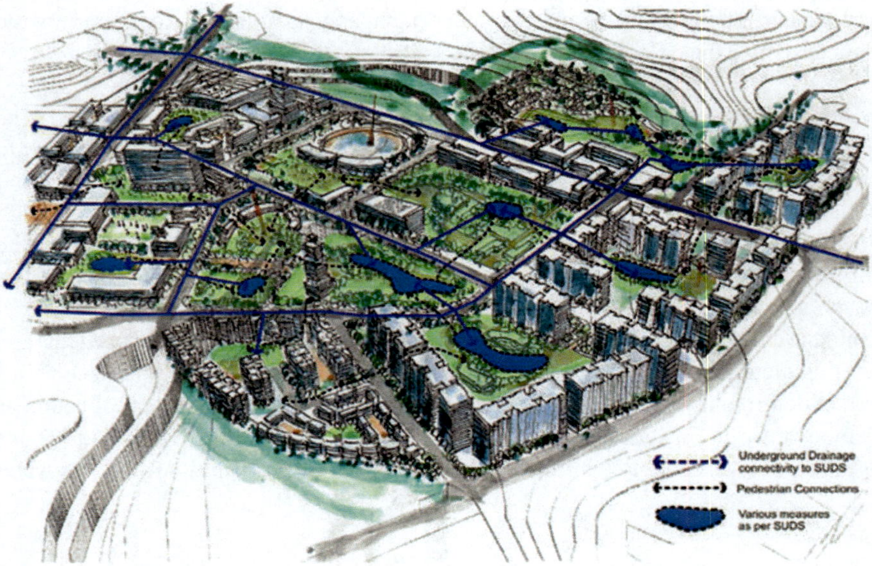

Fig. 25.4 Concept of sustainable urban drainage system for proposed ABD

in these ponds and sumps will discharge into the tributaries to the Lower Lake and Shahpura Lake, creating a larger water network.

The design development process is based upon the key development of six districts and character areas as depicted in Fig. 25.5: Residential, Commercial/IT/Offices, Recreational Hub, Cultural Hub, Knowledge Hub and Administrative Hub. These Hubs are proposed as city-level facilities. The assumptions for the built form and area designation were done as per the demand calculations given by the Bhopal smart city cell. The zoning provides the necessary access to and from these hubs to various centres/areas of the city as well as preserves the local ecosystem. The proposed zoning, therefore, causes minimum damage to the prevailing ecosystem.

The sketch in Fig. 25.6 shows the overall scheme of development indicating various districts and proposed zones. The public amenities are placed in the core of the ABD area which not only gives the advantage of accessibility to the surrounding residential, commercial, hospitality sectors but also provide a huge space for the ecological services which can then be combined with landscapes and contribute to the overall quality of the space.

The carbon sequestration potential of the trees in the ABD area has been calculated based upon the characteristics of trees such as height, girth, green and dry weight which varies with the tree types. The data collected through Primary Survey (2018) for the ABD area has been used to identify the potential for carbon sequestration to be 15,109.64 kg/year as per the number of trees belonging to various types of trees. Figure 25.7 shows the canopy areas of various trees documented for the ABD area. The sequestration levels of each type of tree have been calculated, as mentioned in Annexure 1.

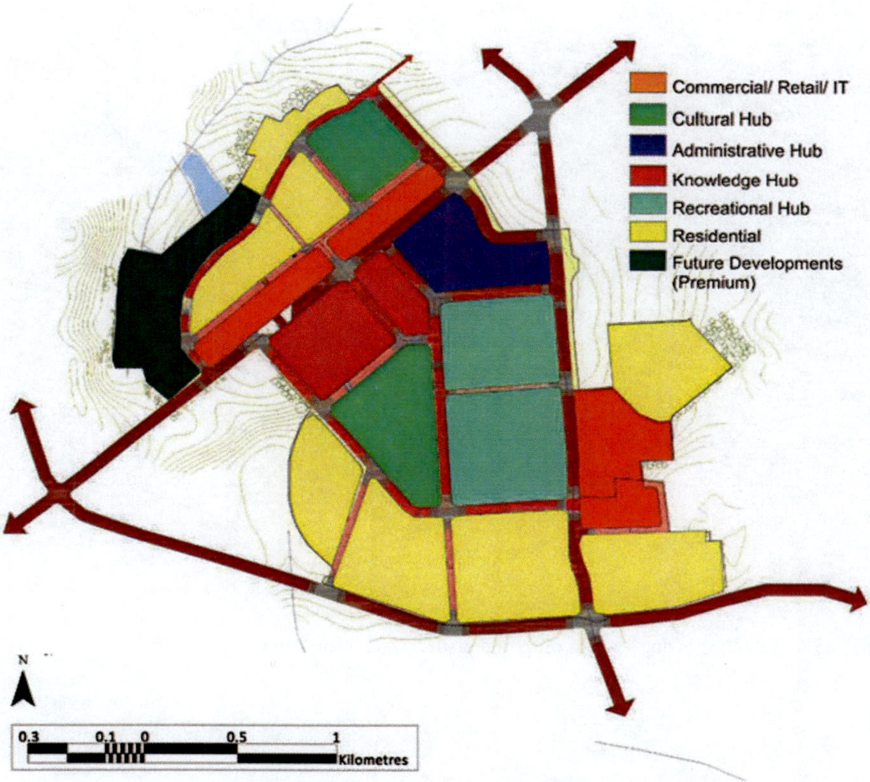

Fig. 25.5 Structure plan showing major roads, their hierarchy and zonal division across the proposed site for ABD

While conceptualising the master plan special emphasis has been laid on protecting the trees present on the site. The plan is proposed in such a way that it retains a lot of trees and also provides ample opportunities for planting new trees. The map below (Fig. 25.8) gives an insight into the status of trees; existing, to be retained, cut or pruned and the ones to be removed and transplanted/replanted.

Based upon the characteristics and locations of trees, suitable areas have been identified for the development process such that the ecological balance is not disturbed. Trees contributing to higher sequestration levels are conserved, the calculations also suggest the additional requirement of trees which should be proposed after the development takes place. It can be observed that in the central part of the ABD area most of the trees are retained to preserve the existing local ecosystem and maintain the environmental balance. Hence, such considerations are duly taken into account in the design process for the targeted area.

The proposed plan shown in Fig. 25.9 complements rapid urban growth taking into account the impact of development on the environment and factors of sustainability. In this alternative approach to spatial planning, spatial design and Green

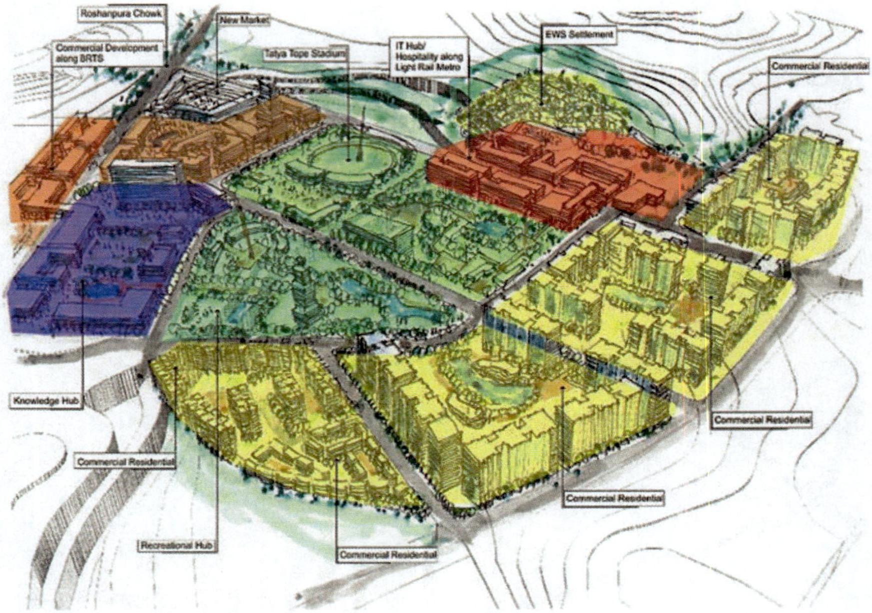

Fig. 25.6 Sketch showing various proposed districts and character zones

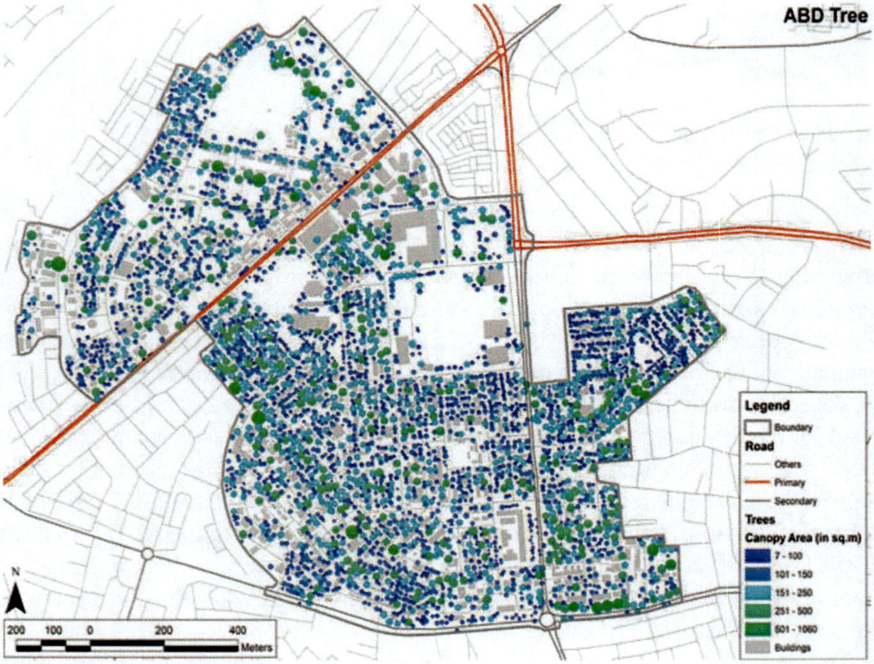

Fig. 25.7 Canopy areas of trees in ABD

Fig. 25.8 Map showing the status of trees in the ABD

Infrastructure are considered in an integrated strategic manner to achieve a sustainable environmental planning agenda. 'Sustainable Urban Drainage System' (SUDS) is a tool adopted to take advantage of the terrain and develop a system of interconnected water reservoirs at the city scale. While this system will help reduce the load on our lakes and other water resources, it will also enhance the groundwater level and in the long run help rejuvenate the dying drains and rivers around Bhopal City. Also, the proposal is sensitive towards the flora and fauna on site and successfully saves uprooting of more than 2000 trees in the ABD area.

This kind of developmental project is incremental in nature and it will be carried out in three phases. The proposed phases are elaborated in Table 25.1. The planning/design programme was conceived keeping in mind that discreet project phasing was necessary to be able to accommodate modifications in the programme and cash flow. The present design move is based on high-rise built forms to release significant unbuilt land to facilitate water movement and create bigger public realms. It is possible to modify the strategy by building low-rise in certain sectors and reduce the overall built volumes, retaining the overall goal of sustainability and ecologically responsive features (Fig. 25.10, 25.11 and 25.12).

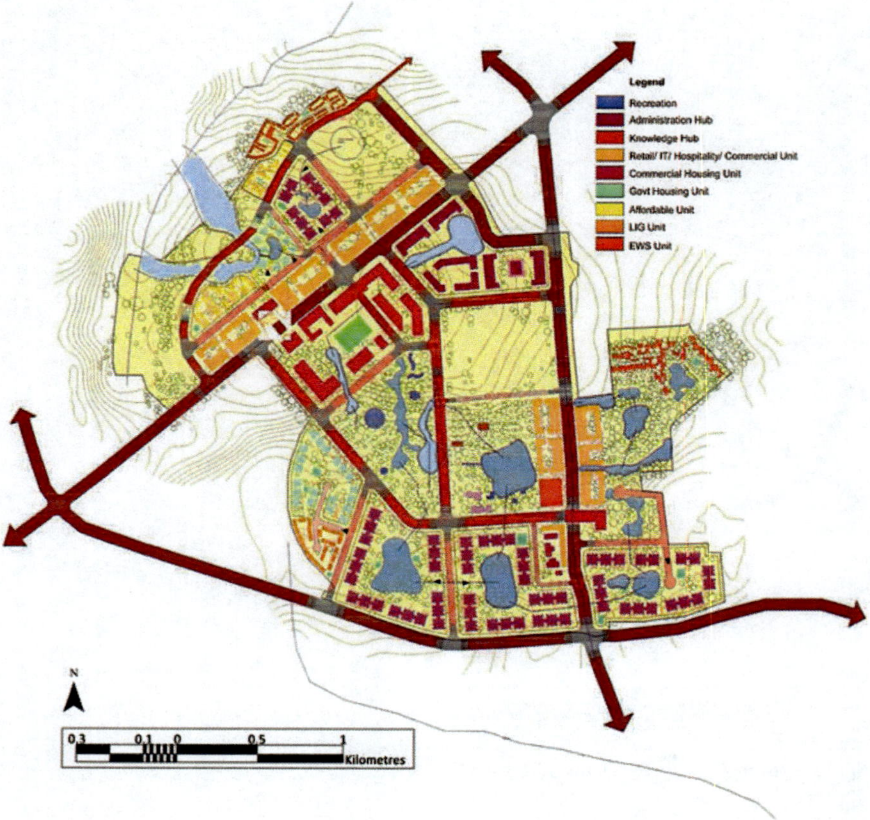

Fig. 25.9 Proposed plan for ABD area

25.4 Discussion

This study has explored the various nature-based spatial planning approaches to minimise the climate change impacts for the city of Bhopal. The potential of green spaces and water bodies was assessed to mitigate not only the climate change impacts but also to enhance the microclimate of the identified pockets. An approach for the identified 'Area Based Development' site has been proposed to demonstrate the use of green infrastructure components while planning for the cities.

The study emphasises on the adoption of 'Sustainable Urban Drainage System (SUDS)' in which the optimum use of city green spaces is done in addition to the advantage of natural slopes. A system of interconnected water reservoirs at the neighbourhood scale has been demonstrated in the proposed ABD area plan which will be used to tap in the city's abundant rainfall through water harvesting. While this system will help reduce the load on our lakes and other water resources, it will also enhance the groundwater level and in the long run help rejuvenate the dying drains

Table 25.1 Proposed phases for development of ABD area

See Fig. 25.10	Phase 1 In Phase 1, the transit nodes will be developed and the transit corridor from Platinum plaza to Roshanpura Chowk will be developed. Major footprint containing commercial, hospitality, office building use and commercial residential zone will be constructed. Revenue for Phase 2 will be generated from Phase 1. Parallel to this, the recreational and knowledge hub will be developed as shown in Fig. 25.10 wherein red colour indicates existing buildings
See Fig. 25.11	Phase 2 In Phase 2, the transit corridor from Jawahar chowk to Roshanpura chowk will be developed. Major footprints containing commercial, IT, Hospitals and Clinics, Retails and Office building use will be constructed along with the development of the commercial residential zone and government housing. The EWS will also be settled in some parts to support the demand of human source. Parallel to this the core district of knowledge hub and the administrative hub will be developed as shown in Fig. 25.11
See Fig. 25.12	Phase 3 In Phase 3, the development will get completed along with plantation, landscaping and integrated sustainable urban development system as shown in Fig. 25.12

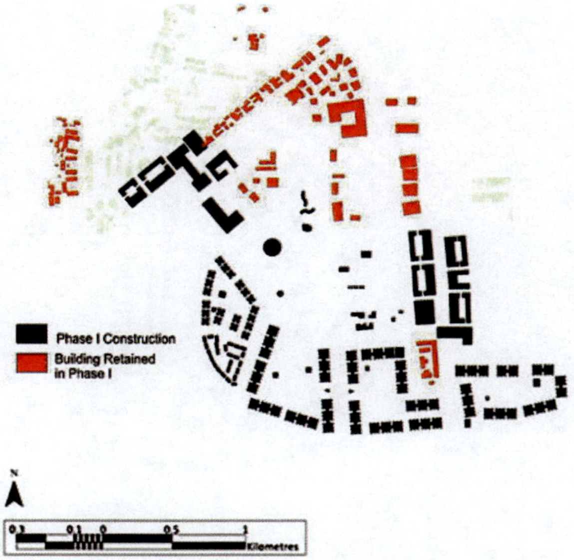

Fig. 25.10 Phase 1

and rivers around Bhopal city. Also, the proposal is sensitive towards the trees that already exist in the proposed site and successfully saves the uprooting of more than 2000 trees.

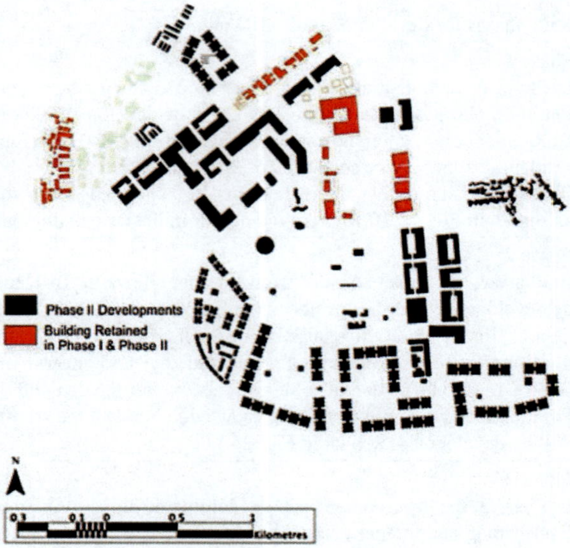

Fig. 25.11 Phase 2

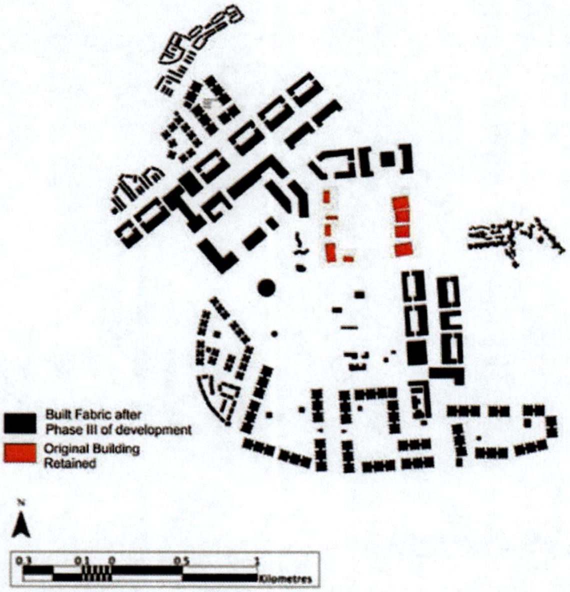

Fig. 25.12 Phase 3

In the proposed plan, nature-based approach is adopted to address the surface water design problems giving equal importance to quality, quantity and amenity aspects to enhance the livability of the area. This study is therefore an attempt to understand and analyse the impact of spatial development, and to propose such strategies which have a minimal adverse effect on the environment and have the capacity to combat the impact of climate change. Climate Informed Spatial Planning to be adopted for redevelopment or new development sites should achieve the following seven milestones:

Mapping of Drainage Pattern of the City: The entire city drainage pattern and water system including the drains as part of the catchment of any water body within the city requires mapping. The network of all rivulets/drains draining the catchment are to be analysed with respect to the proposed site area for the development activities. This will help in protecting the drain which will further carry the runoff during heavy rains to the connecting water body.

Analysing Ecological Conditions of the Site within the City: The ecological conditions of the site for any upcoming redevelopment and new developments to be analysed with respect to the basics of ecology, i.e. biotic and abiotic components to assess the possible impact due to the developmental process. Understanding of the site with respect to local people and their livelihood dependency within and around the site. This includes understanding all the informal activities happening within the site for better integration and for providing new sustainable livelihood opportunities. A survey of the local people through conducting interviews to understand their livelihood pattern and how due to the changing climatic conditions they are getting affected or continuing the work with some coping mechanism.

Water Harvesting Potential: Exploring Water harvesting options through water-gathering and water-spreading slopes and groundwater recharge potential. Conserve water-gathering slopes with a soft landscape. Surface water storage bodies accordingly can be proposed at these points to collect the maximum of rainfall during the monsoon period. In this manner, the rainwater harvesting potential of the site can be fully utilised. The excess water should be channelised appropriately using 'Sustainable Urban drainage System' concepts so that water remains within the site and thus help in slowly diverting the water to the river through aquifers. The possibilities of integrating rainwater harvesting to rejuvenate the natural drains are to be worked out. The proposed surface water bodies (ponds and underground water sumps) are to be interlinked either through surface channels or underground channels. This will enable the excess water within the site rather than runoff from the site under consideration.

Identifying Developable Land: The water spreading slopes are to be identified for built spaces whereas the major drainage channels should be identified for no construction as it may obstruct the natural flow of water. Also, the existing vegetation cover to the extent possible is to be kept untouched. Any other features of heritage importance (built structures, sacred groves, etc.) are to be conserved.

Built Form: Deciding on built form as per the topography of the site and to cater to the market demand. The location of the site plays an important role in deciding on the density and built form of the site. The optimum use of developable land within the site depending on the market value of land is to be done. Inclusive, safe, resilient and sustainable approach as per SDG 11 should be the priority in proposing the type of dwelling units.

Linking Ecology and Aesthetics: History is testament that being ocular is important for preservation and conservation. We don't see oceans and we dump plastics and toxic materials into them. We turn our back to rivers they turn into nallahs; we don't see the underground drains and we have flooded our cities. So, at this point in time, it is important that we see more of our ecology and make the visual soothing for a daily commute. People should see the trees, birds, ponds and lakes and that will probably make them aware and concerned about the health of the prevailing ecology. So, it is important to link ecology and aesthetics. And not in a separate ecological zone but in our day-to-day lives and then we can say we are ready to embrace it.

Carbon Neutrality: All development should cater for carbon neutrality. The estimated emissions both at the construction phase as well as operational phase are to be calculated. Accordingly, green spaces and types of trees are to be proposed/retained are to be planned within the site. The aim is to have provision for surplus carbon sequestration so that the emissions and pollution from transport on the major roads in the vicinity can also be taken care of by the proposed vegetation.

25.5 Conclusion

An alternative approach for development has been demonstrated for the ABD area identified under the smart city mission. In addition to the conventional mixed-used compact development adopted from the smart city proposal, strategies to minimise the ecological damage to the current scenario have also been adopted. Nature-based planning approaches and use of green infrastructure components while planning for the cities will address the surface water design problems, reduce disaster risk and enhance the livability of the areas. The study therefore explores and suggests the combination of spatial planning approaches and ecological considerations to conserve the vulnerable natural resources and minimise the climate change impacts for the city through the adoption of green infrastructure.

In projects of urban retrofit and implant such as this, it is worthwhile to consider the phasing of development, which in this case, has been considered an essential part of the intervention. The core advantage of such an approach helps to make informed decisions at later stages in consultation with diverse stakeholders and acknowledge the dynamic nature of such an intervention initiative. Within the development timeline of the project, new interventions can be dovetailed in response to certain developments at the city level keeping in mind a comprehensive growth synergy that is sustainable and symbiotic to be able to avoid repetition of infrastructure inserts already proposed. Also, an embedded energy audit mechanism should be part of the development proposal which would account for the carbon footprint in the process of development and during the phases of occupancy proposed. This is an essential ingredient of the project which serves as a model for such developments in an already existing urban fabric. Aspects of retaining certain artefacts on site and responding creatively in the design process are another aspect of significance the present project focuses on. The overall goal of the project is to achieve a holistic design intervention in terms of environmental, social and cultural urban environment.

The study can be taken forward as a reference document to suggest specific methods of inclusion of the strategies for sustainability and resilience in master plans and Development Control Regulations. The study is limited to urban areas but can be developed for rural areas addressing the climate change impacts at the district level.

Acknowledgements This study was part of a consultancy project titled "Climate Informed Environmental Planning for Smart Cities of Madhya Pradesh" supported by the Center for Climate Change & Sustainable Development (CCCSD), Bhopal, Department of Environment, Government of Madhya Pradesh. The authors also acknowledge the support extended by the School of Planning and Architecture Bhopal for conducting the study.

Annexure 1

Carbon sequestration potential in ABD area of Bhopal.

S. No	Tree type	Girth (m)	Diameter (m)	Height (m)	Green weight (kg/m2)	Green weight including roots	Dry weight	Carbon content	Carbon sequestered (kg/year)	No. of trees	Total carbon sequestered
1	Amaltas	4.5	1.433121	10	3.08	3.70	2.68	1.34	4.91	23	113.01
2	Apta	2	0.636943	5	0.30	0.37	0.26	0.13	0.49	70	33.97
3	Ashoka	2	0.636943	12	0.73	0.88	0.64	0.32	1.16	77	89.68
4	Babool	2.5	0.796178	20	1.90	2.28	1.65	0.83	3.03	66	200.17
5	Bamboo	2.5	0.796178	20	1.90	2.28	1.65	0.83	3.03	39	118.28
6	Banana	1	0.318471	3.5	0.05	0.06	0.05	0.02	0.08	4	0.34
7	Banyan	6	1.910828	30	16.43	19.72	14.29	7.15	26.20	26	681.31
8	Bombax Cebia	2.5	0.796178	30	2.85	3.42	2.48	1.24	4.55	40	181.97
9	Bougainvillea	0.3	0.095541	5	0.01	0.01	0.01	0.00	0.01	36	0.39
10	Eucalyptus	2	0.636943	25	1.52	1.83	1.32	0.66	2.43	220	533.79
11	Guava	2	0.636943	6	0.37	0.44	0.32	0.16	0.58	86	50.08
12	Gulmohar	4.5	1.433121	12	3.70	4.44	3.22	1.61	5.90	94	554.22
13	Imli	1	0.318471	20	0.30	0.37	0.26	0.13	0.49	103	49.98
14	Jamun	3.6	1.146497	25	4.93	5.92	4.29	2.14	7.86	80	628.90
15	Kala Siris	2	0.636943	25	1.52	1.83	1.32	0.66	2.43	209	507.10
16	Keekar	2.5	0.796178	20	1.90	2.28	1.65	0.83	3.03	18	54.59
17	Maharukh	2	0.636943	15	0.91	1.10	0.79	0.40	1.46	193	280.97
18	Mahua	0.8	0.254777	20	0.19	0.23	0.17	0.08	0.31	13	4.04
19	Mango	4	1.273885	25	6.09	7.30	5.29	2.65	9.71	769	7463.38
20	Moringa	1	0.318471	12	0.18	0.22	0.16	0.08	0.29	2	0.58

(continued)

(continued)

S. No	Tree type	Girth (m)	Diameter (m)	Height (m)	Green weight (kg/m2)	Green weight including roots	Dry weight	Carbon content	Carbon sequestered (kg/year)	No. of trees	Total carbon sequestered
21	Neem	2	0.636943	15	0.91	1.10	0.79	0.40	1.46	566	823.98
22	Palm	2	0.636943	5	0.30	0.37	0.26	0.13	0.49	7	3.40
23	Palm V	2.5	0.796178	10	0.95	1.14	0.83	0.41	1.52	162	245.67
24	Peepal	3	0.955414	30	4.11	4.93	3.57	1.79	6.55	165	1080.93
25	Pluberia Alba	0.3	0.095541	5	0.01	0.01	0.01	0.00	0.01	17	0.19
26	Pseudo Ashoka	2	0.636943	15	0.91	1.10	0.79	0.40	1.46	227	330.47
27	Saptaparni	3	0.955414	25	3.42	4.11	2.98	1.49	5.46	182	993.58
28	Son Pakhad	2	0.636943	15	0.91	1.10	0.79	0.40	1.46	57	82.98
29	Yellow Orleander	0.4	0.127389	2.5	0.01	0.01	0.01	0.00	0.01	174	1.69
											15109.64

Source How to calculate the amount of CO_2 sequestered in a tree per year (BROWARD—www.broward.org)

References

Austin G (2014) Green infrastructure for landscape planning: integrating human and natural systems. Routledge, London and Newyork

BMC (2017) About Bhopal. Retrieved from Bhopal Municipal Corporation. https://www.bhopalmunicipal.com/city-information/about-bhopal.html

Christopher Coutts MH (2015) Green infrastructure, ecosystem services and human health. Int J Environ Res Public Health

EPA (2017) Green infrastructure research. www.epa.gov. https://www.epa.gov/green-infrastructure/green-infrastructure-research. Accessed 12 Sept 2019

Government of India, Ministry of Home Affairs (2018) 2011 census data. Retrieved from Census of India 2011. https://censusindia.gov.in/

Hao Chen DZ (2017) An integrated assessment of urban flooding mitigation strategies for robust decision making. Environ Model Softw 143–155

Jennifer Wolch JB (2014) Urban green space, public health, and environmental justice: the challenge of making cities 'just green enough'. Landsc Urban Plan 234–244

Li Liu MB (2018) Green infrastructure for sustainable urban water management: practices of five forerunner cities. Cities

Lyndhurst B (2004) Livability & sustainable development: bad habits & hard choices. Retrieved from final report for the UK Office of the Deputy Prime Minister, London. https://webarchive.nationalarchives.gov.uk/20120919191556. https://www.communities.gov.uk/documents/corporate/pdf/142424.pdf

Marić I, Crnčević T, Cvejić J (2015) Green infrastructure planning for cooling urban communities: overview of the contemporary approaches with special reference to serbian experiences. SPATIUM, pp 55–61

Martijin Kuller PM (2019) A planning-support tool for spatial suitability assessment of green urban stormwater infrastructure. Sci Total Environ 856–868

Millenium Ecosystem Assessment (2005) Ecosystems and human wellbeing: current state and trends. Island Press. https://www.millenniumassessment.org/documents/document.765.aspx.pdf. Accessed 14 Mar 2018

MoHUA (2019) Guidelines for urban water conservation. Retrieved from Jal Shakti Abhiyan. https://ejalshakti.gov.in/JSA/JSA/Help.aspx

NDMA (2010) Management of urban flooding—national disaster management guidelines. National Disaster Management Authority, New Delhi

Newman P (1999) Sustainability and cities: extending the metabolism model. Landsc Urban Plan 33:219–226

Olofsdotter B, Björnberg K, Chang H-W, Kain J-H, Linn E, Scurrell B (2013) Competing for urban land. www.mistraurbanfutures.org. https://www.mistraurbanfutures.org/sites/mistraurbanfutures.org/files/competing_for_urban_land_sr3_130517_0.pdf

Pacione M (1990) Urban liveability: a review. Urban Geogr 11:1–30

Per Bolund SH (1999) Ecosystem services in urban areas. Ecol Econ

Plus Integreat Knowledge (2012) Planning committee members learning and development program Leeds & Bradford Series module 2: green and blue infrastructure. www.more.lgyh.gov.uk. https://www.more.lgyh.gov.uk/wp-content/uploads/2012/03/Green-BlueInfrastructure-Planning-Series-Handout-.pdf. Accessed 11 Apr 2014

Sustainable Seattle (2004) Indicators of sustainable community 1998, a status report on long-term cultural, economic, and environmental health for Seattle. Seattle, Washington

Veenhoven R (1996) Happy life-expectancy—a comprehensive measure of quality-of-life in nations. Soc Indic Res 39:1–58

White I (2010) Water and the city: risk, resilience and planning for a sustainable future. Routledge, New York

Chapter 26
Evaluating Decadal Change in Green Cover of Dehradun City

Ashish Mani, Dharmendra Kumar, and Deepak Kumar

Abstract This study emphasizes the use of remote sensing parameters NDVI and LST to identify decadal change of three different years 1999, 2009, and 2019 in green cover in Dehradun city. Data analysis was done by using satellite imagery Landsat 8 OLI and Landsat 5 TM. The relationship between NDVI and LST are inversely proportional. The result shows the continuous decrease in green cover with an increase in temperature over the time of Dehradun city. From this study, we came to know that the area under NDVI (Very High) value is decreasing from 10% in the year 1999 to 3.7% in the year 2009 and further 0.6% in the year 2019 and in contrast, the area under LST range(>35 °C) is increasing from 1.68% in the year 1999 to 16.55% in the year 2009 and finally reaches to 39.75% in the year 2019. This results in environmental damage of Dehradun city over the time period. This study helps the decision makers of different organizations who want to do urban planning, climate change analysis, and forest fire risk management study in reducing sustainable damage.

Keywords Decadal change · Dehradun City · GIS · Remote sensing · LST · NDVI

Acronyms

AMSL	Average mean sea level
LSE	Land Surface Emissivity
LST	Land Surface Temperature
NDVI	Normalized Difference Vegetation Index
TI	Temperature Index

A. Mani · D. Kumar (✉)
Amity Institute of Geoinformatics & Remote Sensing (AIGIRS), Amity University Uttar Pradesh, Noida 201313, Uttar Pradesh, India

D. Kumar
Indian Institute of Remote Sensing (IIRS), ISRO, Department of Space, Government of India, 4 Kalidas Road, Dehradun, Uttarakhand, India

P. Sharma (ed.), *Geospatial Technology and Smart Cities*, The Urban Book Series,
https://doi.org/10.1007/978-3-030-71945-6_26

26.1 Introduction

Climate change is one of the most vital issues among researchers, scientists, planners, and politicians across the globe. Of all of the climatic factors, temperature plays a prime role in detecting climatic change introduced approximately through urbanization and industrialization. Artificially triggered weather change and global warming arising from anthropogenic-pushed emissions of greenhouse gases and land-use and land cover change have emerged as one of the maximum crucial environmental troubles among researchers in the last two decades (Nkwunonwo 2013). The traditional methods for detecting climate change to manually collect data and make predictions and decisions on the basis of hard copies of aerial maps were tedious and expansive. Further, it also did not ensure accuracy in the tasks performed and predictions made (Blaschke et al. 2019). The chances of errors were high. New techniques like GIS and Remote Sensing are accurate and time-consuming. And also, the prediction was generally correct (Zhang et al. 2012).

Data utilized in GIS usage and remote sensing are spatial statistics (records represented in map shape with a digital form or pixel form), that is satellite imagery (Jensen et al. 2014). The satellite imagery is obtained based on the differences in electromagnetic energy reflectance of each item on this planet floor (Narayanan 2013). Different reflectance energies through the sensor may be recorded and described as exclusive objects displayed in an image. The investigation centered on the outcomes of vegetation (NDVI) and Land Surface Temperature (LST) to become aware of environmental changes (A-Du et al. 2006).

The Normalized Difference Vegetation Index (NDVI) is a graphical indicator that is used to evaluate and analyze remote sensing measurements usually from space and assesses whether the selected consists of live green vegetation or not. It can also be useful for the estimation of vegetation productivity and rainfall in semi-arid areas (Chen et al. 2006). Mathematically stating, the NDVI of an area containing dense vegetation will be a positive value ranges between (0.3–0.8 nm). Whereas, clouds and snowfields will indicate a negative number on the respective index (Rashed and Jürgens 2010). Other targets that are visible from the Earth are free-standing water (oceans, lakes, and rivers) that depict very low positive or slightly negative values and soil that also represent low positive values (World Bank, Washington DC 2005). This algorithm is simple and can also compress data so that the effect of other factors can be tested on it. NDVI is a graphical indicator and using it for quantitative assessments will restrict the actual usefulness of the index (Alemu et al. 2014). Numerous users of NDVI have analyzed large vegetations that are derived by correlating NDVI values and ground measurements (Jiang et al. 2008). This however creates issues as satellite sensors measure larger areas than field measurements. Similarly, the reflectance measurements should be relative to the respective area and should be acquired simultaneously which becomes a difficult task. Also, misregistration and misinterpretation of the spectral rays lead to errors and deviated and inaccurate results (Weng et al. 2004). However, despite these drawbacks, NDVI is an important vegetation monitoring tool.

Land Surface Temperature (LST) is a crucial parameter related to the energy present at the surface and water balance at local as well as global levels. Simply, it is a measure of how hot or cold the Earth's surface would feel to the touch. LST is developed from space data using various methods. LST is important to many aspects of geosciences. Satellite LST measurements deliver distinctive statistical sources for local and regional coverage. However, LST is not a worldwide indicator for soil moisture. LST and Land Surface Emissivity (LSE) collectively determine the longwave radiation in land surface radiation budgets. These are the main input factors in climatic, hydrological, ecological as well as biogeochemical models. LSE is basically a physical and composite characteristic of the surface (Li et al. 2013). It is used for soil moisture, climate, terrestrial and geological studies, bedrock mapping, resource exploration, and biogeochemistry (Irfan et al. 2020). LST is usually derived from TIR remote sensing and also from passive microwave remote sensing. LST derived by IR thermal sensors results in a better spatial resolution under a clear sky state. Whereas, passive microwave sensors provide coarser spatial resolution and less accuracy in LST (Khandelwal et al. 2017). LST affects the separation of energy between ground and vegetation and regulates the surface temperature also. Many water- and climate-related applications use LST to determine surface temperature. The present study is used to identify the green cover loss over a time period of twenty years with three different years of data (1999, 2009, and 2019).

Tampubolon et al. expressed in their paper "Identification of Environmental Changes with NDVI and LST Parameters in Pakpak Bharat Regency by using Remote Sensing Technique" that the relationship between NDVI and LST is inversely proportional; if the NDVI value is high then the LST value goes down and vice versa (Zhuo et al. 2018). They also emphasize that the NDVI index from 2013 to 2015 proves the decrease in the value of NDVI which represents a decrease in vegetation (Baniya et al. 2018). Temperature changes assessed by the Temperature Index (TI) from 2013 to 2015 prove an increase in the value of (TI) that states the increase in temperature (Yuan and Bauer 2007). Another example of NDVI and LST relation was found in Gorgani et al. paper "The Relationship between NDVI and LST in the urban area of Mashhad, Iran". The present paper studies the correlation between the Normalized Difference of Vegetation Index (NDVI) and Land Surface Temperature for the urban area of Mashhad, using a satellite image from Landsat 5 TM and Landsat 7 ETM+ (Gaohong et al. 2015). The major objective of the current work includes assimilating the relevant spatial data relevant to the study area, developing a NDVI map with respect to LST map of different time period using geospatial sciences, and estimatingthe correlation between the NDVI and LST result. The present study was used to identify the green cover loss over a time period of twenty years with three different years of data (1999, 2009, and 2019). Here the relationship between NDVI and LST is identified and then the correlation between both has been performed. The vegetation loss has been estimated through area change.

26.2 Study Area

The current study area Dehradun city comes under Dehradun district of state Uttarakhand, India and lies between 30° 12′ and 30° 25′ north latitudes and 78° 08′ and 78° 55′ east longitudes (Fig. 26.1). The Average Mean Sea Level (AMSL) is between 540 and 1048 m with a total area of about 195.23 km^2.

Dehradun lies at the foothills of the Himalayas. It is surrounded by river Song on the east, river Tons on the west, *"Queen of the Hills"* Mussoorie on the north, and Rajaji national park in the south. In the summer months, the temperature stages between 44 and 23 °C. The winter months are generally cooler with the maximum and minimum temperatures touching 23.4 °C and 5.2 °C, respectively. Dehradun experiences heavy to mild showers from June to mid-August. Most of the yearly rainfall (approximately 2000 mm) is received during the months from June to September; July and August being the rainiest months of the season. The Dehradun city has emerged as a crucial business, academic, and cultural destination in north India after turning into the capital of newly carved out Uttarakhand state since the year 2000. Moreover, Dehradun city is the trading center for the entire hill vicinity of the Uttarakhand state. The city is also widely known for its salubrious climate, natural splendor, places for visitor's attraction, schools, and Institute of national and international

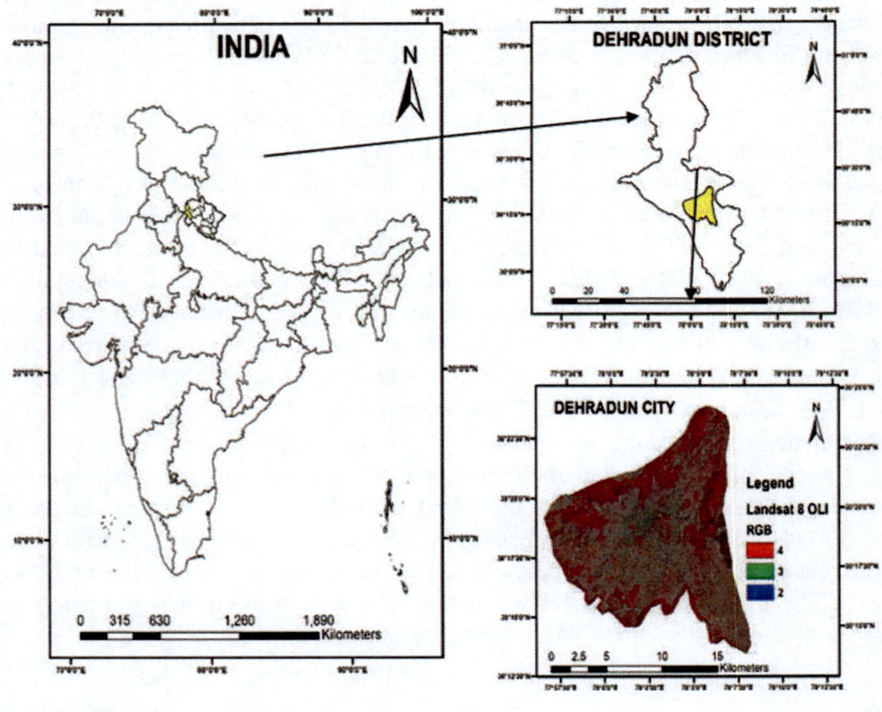

Fig. 26.1 Study area of Dehradun City

standard. The geomorphological types of this area include Alluvium, Piedmont Fan Deposits, Denudational and Structural Hills, and Residual Hills. The Soil surface texture is sandy, loamy, and clayey. The main crops are wheat, sugarcane, and rice. As Litchi is the dominant fruit in most of the region in summers, some other fruits are grapes in high hills, mangoes, litchis, guava, cherries, and pears in warmer and temperate regions.

26.3 Methodology

Satellite imagery of Landsat 8 OLI for the year 2019 of date (28/4/2019) and Landsat 5 TM for the years 1999 and 2009 of date (5/4/1999 and 16/4/2009) has been downloaded from the source USGS earth explorer (https://earthexplorer.usgs.gov/). For image processing and preparing maps, ArcGIS Software (10.3.1) has been used.

26.3.1 NDVI

Normalized Differential Vegetation Index (NDVI) measures vegetation with the difference between near-infrared (where vegetation reflects strongly) and red (where vegetation absorbs).

Formula for calculating NDVI:

NDVI = (NIR – Red)/(NIR + Red).

The value of NDVI ranges between −1 to +1. For example, negative values or near zero show the region of water and clouds and positive values show dense vegetation. For the present study, NDVI values of three different years data have been calculated using map algebra tool in ArcGIS software. For Landsat 8 OLI, band 5 = NIR and band 4 = Red and for Landsat 5 TM, band 4 = NIR and band 3 = Red. The map of three different years has been prepared in ArcGIS.

26.3.2 LST

Land Surface Temperature (LST) is the radiative skin temperature of the land surface. It is a measure of how hot or cold the Earth's surface would feel to the touch. For the present study, LST values of three different years data have been calculated using map algebra tool in ArcGIS software. The thermal band from each satellite has been used. For Landsat 8 OLI, the thermal band is band 10 and band 11 and for Landsat 5 TM, the thermal band is band 6. The map of three different years has been prepared in ArcGIS.

26.3.2.1 Steps for Calculating LST for Landsat 8 OLI

- **Convert Digital Numbers (DN) to Top of Atmospheric (TOA) Spectral Radiance (L)**
  $TOA_{(L)} = ML * Q_{cal} + AL$
  where

  $TOA_{(L)}$ = Spectral Radiance
  ML = Radiance_Multi_Band (found in MTL file which will be downloaded at the time of data download)
  Q_{cal} = Band 10 or Band 11
  AL = Radiance_Add_Band (found in MTL file which will be downloaded at the time of data download)

- **TOA to Brightness Temperature conversion (BT)**
  $BT = (K_2/(\ln (K_1/L) + 1)) - 273.15$
  where

  BT = Brightness Temperature conversion
  K_1 = Band-specific thermal conversion constant from the metadata or MTL file (K1_CONSTANT_BAND).
  K_2 = Band-specific thermal conversion constant from the metadata or MTL file (K2_CONSTANT_BAND).
  L = Spectral Radiance

- **Calculate the NDVI**
  NDVI = (Band 5 – Band 4)/(Band 5 + Band 4)
  where

  NDVI= Normalized Differential Vegetation Index
  Band 5= NIR and Band 4= RED

- **Calculate the proportion of vegetation (P_v)**
  P_v= Square ((NDVI – $NDVI_{min}$)/($NDVI_{max}$– $NDVI_{min}$))
- **Calculate Emissivity (ε)**
  $\varepsilon = 0.004 * Pv + 0.986$
- **Calculate the Land Surface Temperature (LST)**
  $LST = (BT/(1 + (Q_{cal} * BT/1.4388) * \ln (\varepsilon)))$

26.3.2.2 Steps for Calculating LST for Landsat 5 TM

For calculating LST of Landsat 5 TM, only the first step is different from the Landsat 8 OLI which is given below.

- **Convert Digital Numbers (DN) to Spectral Radiance L_λ**
  $L_\lambda = L_{min} + (L_{max} - L_{min}) * DN/255$
  where

L_λ = Spectral radiance,
L_{MIN} = Spectral radiance of DN value 1
L_{MAX} = Spectral radiance of DN value 255
DN = Digital Number

Finally, after all the values of NDVI and LST have been calculated, the Maps has been prepared for the years 1999, 2009, and 2019.

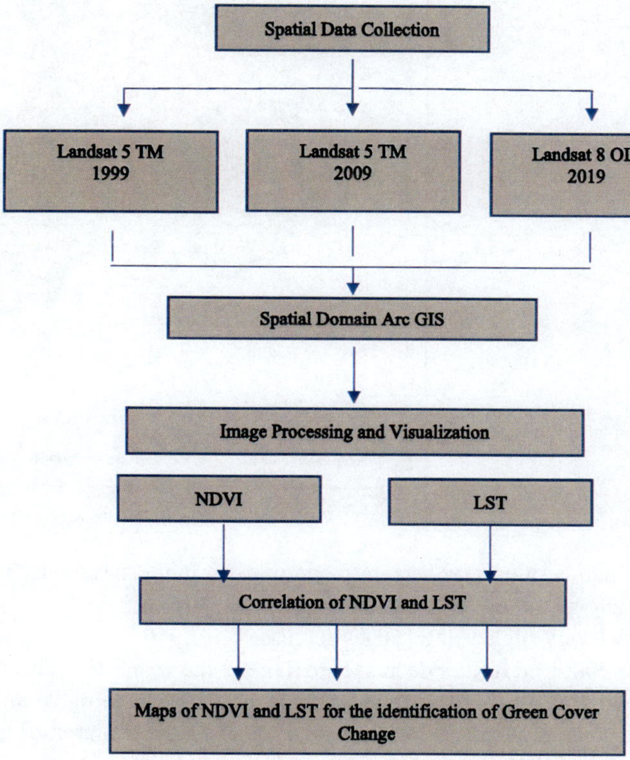

26.4 Result and Discussion

26.4.1 NDVI

NDVI Map for three different years has been shown below in Figs. 26.2, 26.3, and 26.4. It is clearly visible in Fig. 26.4 that the very high (>0.4) value of NDVI decreases with time as compared with Figs. 26.2 and 26.3. Whereas, the very low (<0) values of NDVI increase with time. The result implies that the loss of green cover in the last

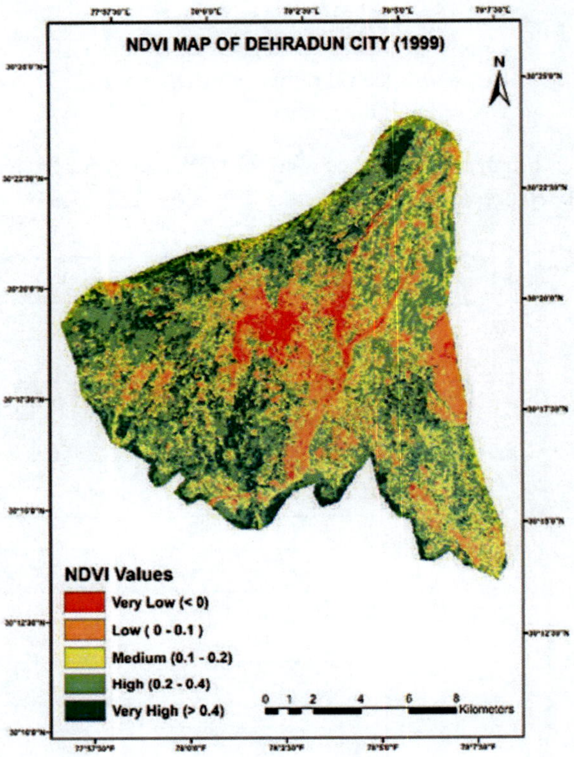

Fig. 26.2 NDVI map of 1999

two decades is mainly due to urbanization and tourism in those regions. Table 26.1 will show the damage of vegetation with respect to the area.

Area of NDVI very high value (>0.4) decreases from 10% in the year 1999 to 3.7% in the year 2009 and further decreases to 0.6% in the year 2019. This shows the decrease of green cover. Whereas, NDVI very low value (<0) in the year 1999 has 2.5% area which increases to 3.9% in the year 2009 and further increases to 6.1% in the year 2019. This shows the increase in the urban area.

26.4.2 LST

LST Map for three different years has been shown below in Figs. 26.5, 26.6, and 26.7. It is clearly visible in Fig. 26.7 that land surface temperature of range (>35 °C) increases with time as compared with Figs. 26.5 and 26.6. The result states that the surface temperature increases over a time period due to deforestation, urbanization, A.C of automobiles, etc. Table 26.2 will show the Surface temperature region with respect to area.

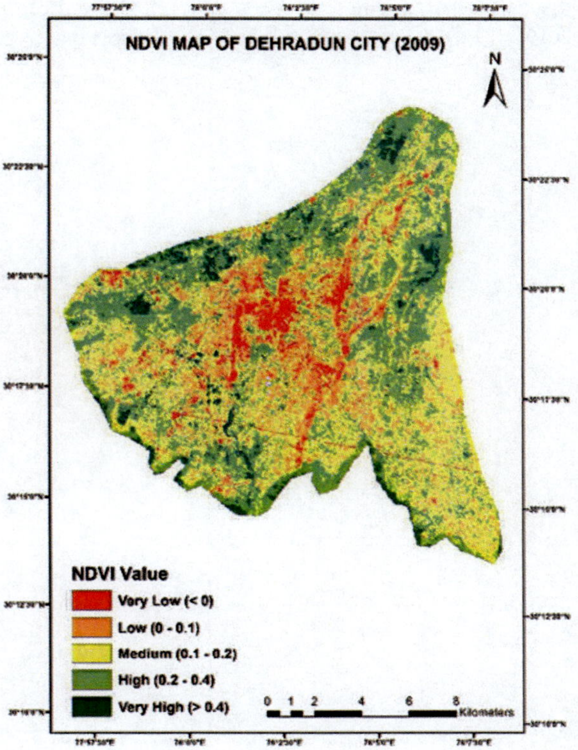

Fig. 26.3 NDVI map of 2009

From this table, we come to know that area of surface temperature ranges (very high >35 °C) has been increased from 1.68% in the year 1999 to 16.55% in the year 2009 and further increased from 16.55% in the year 2009 to 39.75% in the year 2019. On the contrary, the area of surface temperature ranges very low (<29 °C) has been decreased from 3.9% in the year 1999 to 0.94% in the year 2009 and further decreased 0.09% in the year 2019. This shows that surface temperature increases continuously with the time period. Many anthropogenic activities like construction of roads, institution, and higher automobiles production lead to increase in surface temperature.

26.4.3 Correlation Between LST and NDVI

As we know that the LST is inversely proportional to NDVI it implies that surface temperature increases with a decrease in NDVI value. The correlation graph of LST and NDVI of the years 1999, 2009, and 2019 is shown below in Figs. 26.8, 26.9, and 26.10.

Fig. 26.4 NDVI map of 2019

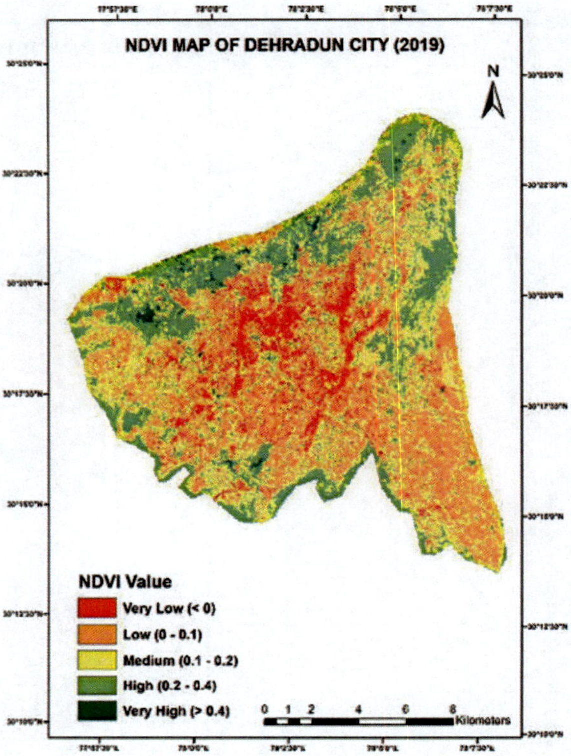

Table 26.1 NDVI area in percentage of the years 1999, 2009, and 2019

NDVI values	1999		2009		2019	
	Area (km^2)	Area (%)	Area (km^2)	Area (%)	Area (km^2)	Area (%)
Very low (<0)	4.9	2.5	7.76	3.9	11.9	6.1
Low (0–0.1)	39.7	20.3	33.2	17	73.55	37.67
Medium (0.1–0.2)	48.3	25.2	75.19	38.6	63.67	32.63
High (0.2–0.4)	82.1	42	71.8	36.8	44.89	23
Very high (>0.4)	20.23	10	7.28	3.7	1.22	0.6
Total	195.23	100	195.23	100	195.23	100

From these three figures (Figs. 26.8, 26.9, and 26.10), we came to know that temperature ranges between 35 and 45 °C have low (−0.1 to +0.1) values of NDVI and temperature ranges between 20 and 30 °C have very high (>0.4) values of NDVI. Through this correlation, it is proved that land surface temperature increases with a decrease in NDVI and vice versa.

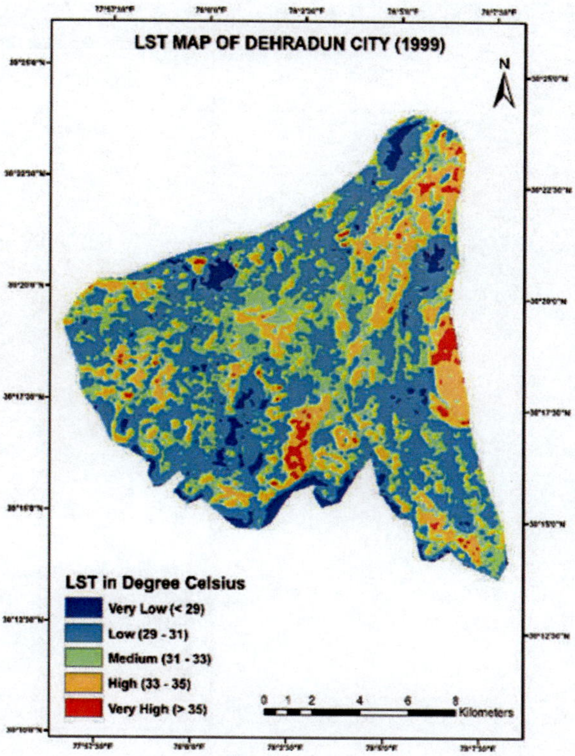

Fig. 26.5 LST map of 1999

26.5 Conclusion

The present study examined the relationship between NDVI and LST parameters for the Dehradun city. The results indicated that the destruction of green cover during the last two decades directly influenced surface temperature around the Dehradun city. Decadal map of NDVI for three different years exhibited the decrease in the area with very high value (>0.4) from 10% in year1999 to 3.7% for the year 2009 and further 0.6% in the year 2019. On the other hand, the decadal map of LST for the three different years exhibited the upsurge in surface temperature (range >35 °C) from 1.68% in the year 1999 to 16.55% for the year 2009 and further 39.75% for the year 2019. The results of the study can be utilized for natural resource management, climate change organization, or any other government/private organizations involved in the urban planning and climate change analysis for sustainable development of the Dehradun city.

Fig. 26.6 LST map of 2009

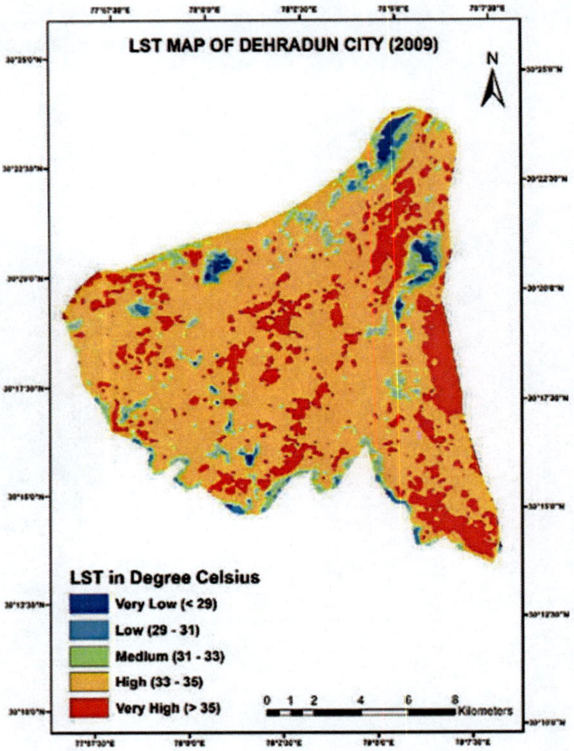

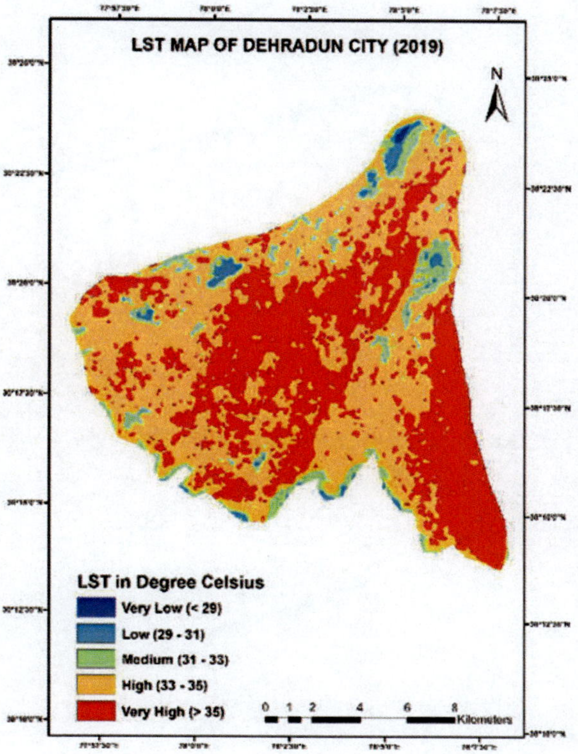

Fig. 26.7 LST map of 2019

Table 26.2 LST area in percentage of the years 1999, 2009, and 2019

LST (°C)	1999		2009		2019	
	Area (km^2)	Area (%)	Area (km^2)	Area (%)	Area (km^2)	Area (%)
Very low (<29)	7.6	3.9	1.84	0.94	0.19	0.09
Low (29–31)	83.6	42.82	3.38	1.73	2.86	1.46
Medium (31–33)	62	31.75	18.16	9.3	15.19	7.78
High (33–35)	38.7	19.85	139.56	71.48	99.42	50.92
Very high (>35)	3.28	1.68	32.29	16.55	77.57	39.75
Total	195.23	100	195.23	100	195.23	100

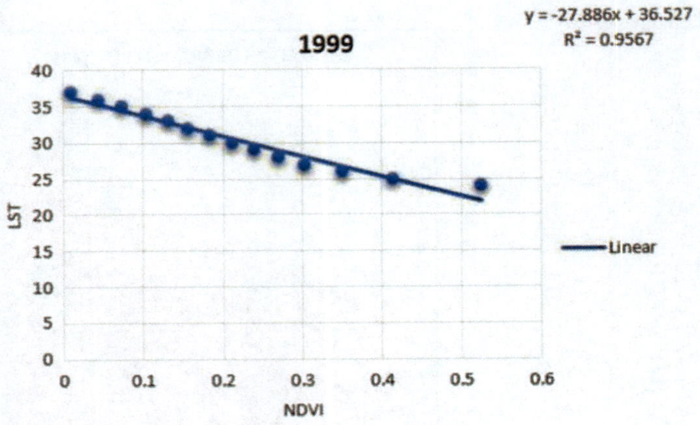

Fig. 26.8 Correlation graph of NDVI and LST in 1999

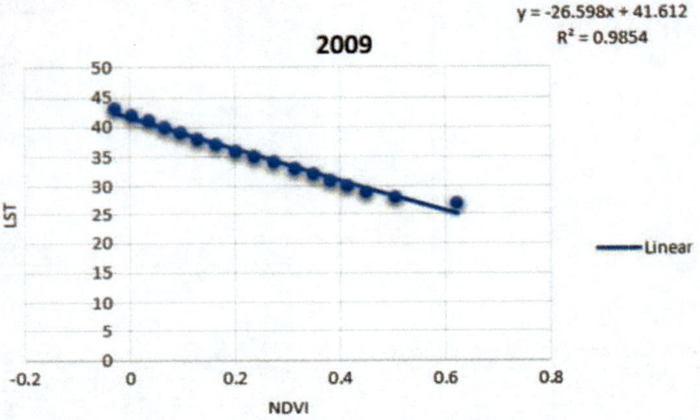

Fig. 26.9 Correlation graph of NDVI and LST in 2009

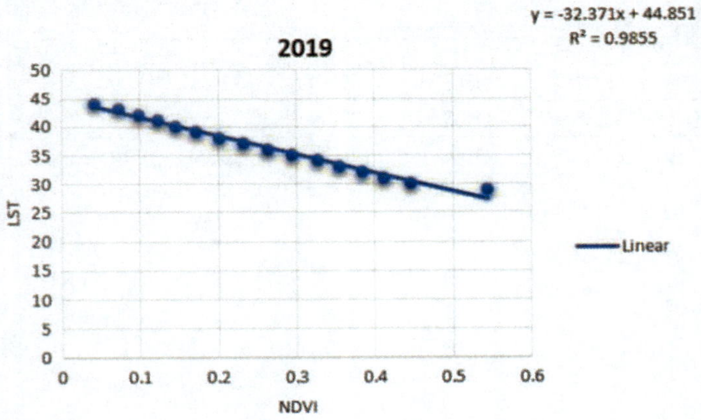

Fig. 26.10 Correlation graph of NDVI and LST in 2019

References

A-Du G et al (2006) Spatial distribution patterns of the urban heat island based on remote sensing images: a case study in Beijing, China. In: 2006 IEEE International symposium on geoscience and remote sensing, vol 1, pp 2321–2323. https://ieeexplore.ieee.org/lpdocs/epic03/wrapper.htm?arnumber=4241748

Alemu H, Senay GB, Kaptue AT, Kovalskyy V (2014) Evapotranspiration variability and its association with vegetation dynamics in the Nile Basin, 2002–2011. Remote Sens 6:5885–5908

Baniya B, Techato K, Ghimire SK, Chhipi-shrestha G (2018) A review of green roofs to mitigate urban heat island and Kathmandu Valley in Nepal. Appl Ecol Environ Sci 6(4):137–52. https://pubs.sciepub.com/aees/6/4/5

Blaschke T et al (2019) Framework for fusion of ascending and descending. Remote Sens 11(5):1–14. https://doi.org/10.1007/s12524-018-0873-0

Chen X, Zhao H, Li P, Yin Z (2006) Remote sensing image-based analysis of the relationship between urban heat island and land use/cover changes. 104:133–146

Gaohong Y, Gregoire, M, McCabe MF (2015) A multiple-point geostatistics method for filling gaps in landsat E TM + SLC-off images. In: 21st International congress on modelling and simulation, Gold Coast, Australia, pp 180–186

Irfan M et al (2020) Assessing the energy dynamics of pakistan: prospects of biomass energy. Energy Rep 6:80–93. https://doi.org/10.1016/j.egyr.2019.11.161

Jensen RR, Gonser RA, Joyner C (2014) Landscape factors that contribute to animal–vehicle collisions in two Northern Utah Canyons. Appl Geogr 50:74–79. https://www.sciencedirect.com/science/article/pii/S0143622814000290. Accessed 9 April 2014

Jiang Z, Huete AR, Didan K, Miura T (2008) Development of a two-band enhanced vegetation index without a blue band. Remote Sens Environ 112(10):3833–3845

Khandelwal S et al (2017) Detecting urban growth using remote sensing and GIS techniques in Al Gharbiya Governorate, Egypt. Egypt J Remote Sens Space Sci 20(1): 571–575. https://www.sciencedirect.com/science/article/pii/S2212609015000060.

Li ZL et al (2013) Satellite-derived land surface temperature: current status and perspectives. Remote Sens Environ 131:14–37. https://doi.org/10.1016/j.rse.2012.12.008

Narayanan P (2013) Analysing the urban sprawl through entropy of Gulbarga City and its spatial promoters of growth through geoinformatics cartosat imagery of Gulbarga IRS 1D pan imagery of Gulbarga 1998 extract built up through ENVI extract prepare sector grids for covering. XXXIII

Nkwunonwo UC (2013) Land use/land cover mapping of the Lagos metropolis of Nigeria using 2012 SLC-off Landsata ETM+ satellite images. Int J Sci Eng Res 4(11):1217–1223

Rashed T, Jürgens C (2010) Remote sensing of urban and suburban areas. Remote Sens Digit Image Process 10(42):181–192. http://www.springerlink.com/index/https://doi.org/10.1007/978-1-4020-4385-7.

Weng Q, Lu D, Schubring J (2004) Estimation of land surface temperature–vegetation abundance relationship for urban heat island studies. Remote Sens Environ 89(4):467–483. https://linkinghub.elsevier.com/retrieve/pii/S0034425703003390. Accessed 25 March 2014

World Bank, Washington DC (2005) Multifunctional agroforestry systems in India for livelihoods: current knowledge and future challenges. Carbon (2)

Yuan F, Bauer ME (2007) Comparison of impervious surface area and normalized difference vegetation index as indicators of surface urban heat island effects in landsat imagery. Remote Sens Environ 106(3):375–386. https://linkinghub.elsevier.com/retrieve/pii/S0034425706003191. Accessed 23 March 2014

Zhang Y, Yiyun C, Qing D, Jiang P (2012) Study on Urban heat island effect based on normalized difference vegetated index: a case study of Wuhan City. Procedia Environ Sci 13:574–581. https://linkinghub.elsevier.com/retrieve/pii/S1878029612000497. Accessed 26 Aug 2014

Zhuo L et al (2018) An improved temporal mixture analysis unmixing method for estimating impervious surface area based on MODIS and DMSP-OLS data. ISPRS J Photogramm Remote Sens 142(Oct 2017): 64–77. https://doi.org/https://doi.org/10.1016/j.isprsjprs.2018.05.016

Chapter 27
Summary and Way Forward

Poonam Sharma

Abstract The chapter summarizes the entire discussion on smart city, geo-informatics, and digitalization of spatial data. Authors have raised and analyzed, through conceptual, review, case study, and research technique based papers on varied aspects. The geospatial extent of the book encompasses studies from different parts of the globe including US, Poland, Norway, Malaysia, Iraq; case studies from various cities of India viz. Chennai, Bengaluru, and Kochi from southern parts; Mumbai and Pune from the western region; Dehradun, Allahabad, Lucknow, Bhopal, Bhilwara, and Delhi from north and central; Itanagar and Kamrup from northeastern India. An attempt has been made to incorporate studies from a variety of perspectives to explore and understand the urban complex systems and the journey toward making these urban complexities into smart functioning.

Keywords Smart city · Geospatial technology · Land transformation · Urban planning · Sustainable development

An urban system is comprised of massive and diverse demography. Socio-cultural and economic intricacies are interwoven through mobility, physical and virtual networks, connectivity among people, places, and resources. There is a need to have an insight into cities in a completely new perspective through the understanding of the processes that the city has evolved with and currently ongoing. To decipher the impact of slowly imprinting development and fast conquering technological innovations, it is important to know the adaptability indices of the city. There are cities that are capable to keep pace with the new inventive development and also the cities that are just not prepared to move ahead. The practitioners and policymakers have to understand cities from trans-disciplinary approaches and integrated methodologies, such as demographic, infrastructure, utilities, socio-economic-cultural aspects, and ecological dimensions. In all these and many more areas, it is not only managing the existing but to create new; and generate the adaptability of people and structures to the ever-evolving technological developments. It is very substantial to understand city functioning relating to its growth, congestion, accessibility, and connectivity in

P. Sharma (✉)
Department of Geography, Shaheed Bhagat Singh College, University of Delhi, New Delhi, India

P. Sharma (ed.), *Geospatial Technology and Smart Cities*, The Urban Book Series,
https://doi.org/10.1007/978-3-030-71945-6_27

terms of local and external factors. The cities are struggling with rapidly growing population on one hand and scarcity of infrastructure on the other hand. The major problems of urbanization that the world is facing include scarcity of resources, poor infrastructure, energy crisis, environmental problems, and human health. The smart city concept offers to manage these challenges, adding efficiency and quality to citizen's life.

Urban spatial complexity and aggregate patterns range from travel design of individual people to mass transit for a domestic and global movement. Similarly for the economy from basic neighborhood retail to specialized mass production, consumption, and trade. From the physical quality of life to the well-being of citizens becomes important which further comprises the environmental issues to be taken care of. Since cities are centers with high natural resource consumption and waste generation; resource efficiency and sustainability at the local and global levels are essential. The global climate change also pressurize the world to look for re-strategizing the production and consumption patterns. The Smart City concept is seen as key to achieving the efficiency of citizens' quality of life index and ecology; economy and inclusiveness; and infrastructure and governance. Intelligent and responsive cities will play a vital role in future urban transformations. The technological advancement in data generation and management like big data, artificial intelligence, geographical information system, ground-penetrating radar, and many others will create an entirely new urban experiment.

The scholar discourses and narratives across the disciplines of knowledge are ever-evolving. This edited book comprises of a total of twenty-seven erudite papers which characterize the evolving studies on different aspects in the field of smart city and application of geospatial technology. The analysis of urban forms and the spatial distribution of macroeconomic quantities that characterize a city such as population, built environment, and energy use are covered. The broad areas of research that are encompassed here include land-use changes and land transformation; disaster management and preparedness; energy use and solutions; transportations studies; healthcare management; and environmental and ecological research.

The urban sprawl, land-use change, land transformation, land consumption, the built-up area of the city, and fast-growing population demand the need for sustainable urban design. The geographical information system provides effective monitoring and decision-support tool in urban planning. The management of fast urbanization pace is a massive task in different parts of the world as it leads to land-use changes in the peri-urban zone that transforms the dynamics of these areas. Changes in the land use of place are complex of many elements as socio-economic structure, culture, economy, environment, policy, and governance. Urban expansion is a dynamic process resulting from the increased demographic pressure that involves the multidimensional interface of people with their surroundings. Sometimes the process of urbanization happens in an unplanned manner which leads to degradations of natural habitats, congestion, pollution, lack of housing, and other infrastructure. This phenomenon often results in a significant absence of rural areas and hinterlands and thus leads to loss of agriculture and converting these areas into vulnerable and fragile ecosystems. The unplanned and haphazard urban expansion affects the ecology and biodiversity of the surrounding areas. Particularly if such development is happening in coastal areas, it may completely devastate the wetland ecology and resultant socio-economic impact along with the environmental repercussions. The urban growth process needs proper monitoring of sustainable development.

The cities are moving toward becoming smart and smarter about public transportation, mass transit system, and app-based transportation. Transportation is one of the important parts of everyday life for people, society, and the economy. It also has multiple modes and various scales of operations. Another aspect of quality of life and smart city benchmark connected with transportation is resultant environmental effects. Therefore, transportation includes multifaceted attributes associated with it, i.e., intelligent, digital, convenience to residents, solution oriented, and sustainable. Technology provides solutions to the reliability of massive data and supporting real-time data management regarding the public transport system, managing the large fleet, schedule, traffic congestion, traffic signals, emissions of various kinds, conditions of the roads, etc. The Traffic Management Systems (TMS) are useful to manage traffic congestions on the road, better management in emergencies, and functions toward enhancing the travel experience for people. The intelligent transport system and road infrastructure are the main constituents of urban planning and management. The use of ICT and real-time data processing has made it possible for things like global airways hubs, intelligent road, and railway networks; protected routes and paths for buses, cycles, and pedestrians; and efficient and integrated public transport. Unprecedented technological innovations in ICT in the last few years have contributed immensely to the intelligent transportation system. The provisions like Vehicular Ad hoc Networks (VANETs), car to infrastructure (C2I) social networks, sensor networks, and car to car system (C2C) have transformed the transportation efficiency of smart cities.

The cutting-edge knowledge for power electronics, new approaches to produce, and marmalade energy from accumulators and renewable sources are important areas for a city. The smart energy grids are integrated ICT-based system that serves to provide quality, efficiency, and cost management. It also minimizes the environmental problems associated with energy generation, storage, infrastructure, facilities, transportation, and consumption. Smart energy grids work on a system of two-way flow of energy and information communication right from the source to the user. In a smart grid system, every device and subsystem works in the format of auto-real-time information of energy generation or consumption with the programmed factors of schedule load, costing, and other related details. The requirement for energy-efficient environments demands monitoring and management, and thus buildings, transportation, and industries have to be technologically enabled. This is accepted as essential constituents for reaching the goal of smart cities wherein ICT will play a dominant role. ICT helps to establish the energy system and its implementation which has a multi-layer structure, i.e., the physical devices (hardware), control and management (intelligence), and communication. Infrastructure for integrating renewable resources is another important area of concern. Bringing diverse energy options into cities becomes important not only for the economic aspect but also for the environmental considerations. The integration of different renewable energy types into a smart energy grid and achieving efficiency are challenging for city planners and governance.

The growth and expansion of cities pose challenges to services and infrastructure. One of the essential services is health care. The access to the urban health care centers, identifying disease patterns, hotspot detection, vulnerability, spatial diffusion of an outbreak, and roadmap for the geographical epidemiological model, analysis, and visualization are important areas of concern for city managers. The health services are not single-point issues in it; multiple other subsystems of the area are associated with it. The systematic management and disposal of hazardous and non-hazardous bio-medical waste are important to be taken care of. The current COVID-19 pandemic has exposed the big cities, the smart cities, and the business capitals across the world. These urban centers not only were the hotspots but also became gravity points for further spread of the disease. There is an increasing requirement for intelligent, remote, real-time, and smart health care services in smart cities. To reach these levels, cities need to equip with devices, sensors, big data, skilled people, policy, and governance.

Disasters are natural as well as man-made and lead to irretrievable loss of human life, massive damage to infrastructure, and economic loss. Vulnerability to fire and floods has witnessed a lot of increase which has added uncertainties and risks to people. Building safe and disasters resilient smart cities is a big challenge to the world. The meteorological disasters as cyclone, hurricane, heavy rainfall/snow, etc. occurs often in many major cities and cause heavy devastations to city resources, infrastructure, and loss of life. Even frequent inundation has become a serious problem in many urban areas which directly indicates the failure of the rainwater infiltration system due to concretization and shrinking watershed. Many cities are prone to earthquakes and a resultant tsunami has been experienced in parts of South Asia in 2006. Emergency response system and resilience in case of disasters are vital areas for smart cities whether it is a man-made disaster like fire, terror, blasts, some hazardous emissions, and radiations or natural like earthquakes or tsunami. The high-rise buildings and skyscrapers demand for efficient and integrated disaster management planning with building intelligence and design. During a disaster situation, it becomes the paramount importance for governments, response and relief agencies, police, and local bodies that the efficient communication system helps to share and receive information for each affected location to manage relief work. Even for the disaster preparedness plan, a complete integrated approach needs to be seen for effective working.

Smart city initiatives significantly emphasize on environmental aspects of the development. Urban landscapes indeed represent the cultural character of the city. In the name of new developments, infrastructure additions, the urbanscape unremittingly evolves and poses challenges to the ecological as well cultural front. The technology driven smart city concept works on the core model to increase environmental sustainability and efficient use of natural resources. Green cities are seen as wholesome integration toward environmental issues for smart development. The technological solutions, innovative forms of green governance, and policy amalgamation are considered for a green urban ecosystem. The major concerns are CO_2 footprint reduction and hybrid grids for energy mix so that alternative energy is also integrated into the city consumption with smart transport and building intelligence.

The international bodies such as World Economic Forum, World Bank, Organization for Economic Co-operation and Development (OECD), and European Union (EU) have proposed digital infrastructure development and growth for the city environment with ecological sustainability. Future urban strategies focus more on the use of innovative technology for infrastructure that supports eco-friendly growth. Intelligent technologies are expected to provide economically, resource consumption wise, and energy-efficient infrastructure toward the target of a sustainable urban environment. The huge energy consumptions and the resultant emissions are importantly associated with urban areas. A sustainable economy has to work as a core to foster growth and good quality of life for citizens along with a healthy environmental condition. Whether it is reducing energy consumption, better waste management, and renewable energy resources anything may be achieved only with city management practices. Crowdsourcing has become a useful alternative for air monitoring stations where a smooth flow of data to the system is contributed easily. Technology like IoT gives an abundance of feasibility for monitoring and collecting the data related to environmental parameters. The real-time processing with integration to other social-economic and well-being of citizens seems possible.

The surge of Information and Communication Technologies (ICT) has become pivotal in making cities smarter. This model for urban development has led to the total revamping of existing critical infrastructure and facilitating new ways of city management and governance. ICT has completely reformed cities' governance with high-tech communication and information services for planners and policymakers. The processing of data is from various sources and mechanisms viz satellite imagery and sensors and other devices. The vast urban systems with spatially and vertically heterogeneous communication flow and physical network of machines will require the integration of ICT platforms as per the urban scale. The management of the enormous volume of data provided in real time is by a large number of IoT devices in a variety of smart systems, the integration, communication standards, and synchronization among different segments. City management has to handle the issues connected with privacy and security of information. Managing the life-cycle of huge data that a city system generates at ubiquitous and heterogeneous cityscapes; at different spatio-temporal scales; static, dynamic, and real-time information is the major challenge that defines the success of the smart city concept. Infrastructure networks, intelligent systems for traffic, smart building, energy resource management, and real-time services to enable citizens are all achieved in many smart cities. Big data has taken a giant leap, it is estimated that around 90% of the world's digitized data are created and captured in the last two years. Big data is seen as an expedient hope to the world to support the management of smart cities. Blockchain-based management and computing services are used massively to look for smart city solutions. The complete idea behind smart cities revolves around two things, i.e., data and connectivity. RFID (radio frequency identification) system functions based on electromagnetic fields to automatically identify and track tags attached to objects. The WSN (wireless sensor network) system is used to connect devices that have improved the possibility of using a sensor network with numerous intelligent sensors. This data handling has raised a variety of privacy and security issues for people, organizations, and governance. Many of the large corporate promoters like IBM, Siemens, Oracle, Atos, and Microsoft to name some have developed technologies that have brought the technologies for sustainable smart infrastructure development. This has brought the smart city concept from research into existence, workable, and evolving. In the current scenario, the world is working on the Internet of Things (IoT), Cloud of Things (CoT), and Web of things (WoT).

A smart city aims to provide citizens with a healthy living environment and an enhanced quality of life. The citizens are not only the end-users of the infrastructure and services but also an important part of the complete technological-driven processes. While a city becomes a smart city, its citizens are expected to become smart communities and stakeholders for planning and management. Smart communities involve all factors and use ICT technology and human capital who contribute to innovation and alteration in a progressive urban environment.

This phenomenon involves fundamental behavioral change for people to adopt the new innovative ways of life. Citizen interacts with the smart city in various forms viz. as a consumer to avail/purchase services, which are developed by companies. The citizen performs as a resident when it is to be assessed if it is affordable to live in a smart building or smart spaces. A citizen also becomes a data product when the data is created through their use of smart city technologies. The urban systems are functioning in an intelligent digital environment where citizens are considered as enabled participants and consumers for innovative technology. This situation has an innate prejudice imbibed in which part of the city and its people are left behind with the pace the city is moving. This is also termed as a digital divide that has emerged as the major disconnect in terms of socio-political and economic perspective in the era of the fourth industrial revolution. The data-driven smart city system where the enabled citizens are consumers of technology as well as they generate immense data as users might be as a well-informed citizen or might not be. Social media, smartphone apps, various search engines, merchandise sites, utility sites, credit card transactions, CCTVs, and many more such digital actions create huge data. The concern of citizens or consumers regarding their privacy and data security are important debates if information connected to them is used or misused for any kind of commercial and political purpose. The networked technology and information revolution which is a fundamental base of the smart city provides an abundance of prospects for development, convenience, and efficiency to people's lives but also debated against individual rights, privacy, and surveillance.

Geospatial technology, Information Communication Technology (ICT), Geomatics, Geospatial Intelligence (GEOINT), Wireless Server Network (WSN) Internet of Things (IoT), Cloud of Things (CoT), Web of Things (WoT,) and many more innovative technological applications are functioning to achieve the objectives of the smart city development. In different part of the world, there are success stories regarding the role of these technologies in managing smart, inclusive, and sustainable cities. But at the same time, this is completely tech-know-how based at both the levels, i.e. the developers and end-users. The higher financial stakes of the new direction of urban development that appears to give solutions to many problems raise the challenges as well. It is very difficult for developing countries to follow the bandwagon for this type of development. Since these technology-based solutions expect that it is catering to innovative and enabled communities, this assumption appears a bit nonrealistic and cities are facing the intense challenge of the social exclusion and digital divide.

Index

P. Sharma (ed.), *Geospatial Technology and Smart Cities*, The Urban Book Series,
https://doi.org/10.1007/978-3-030-71945-6